Accessing the E-book edition

Using the VitalSource® ebook

Access to the VitalBook™ ebook accompanying this book is via VitalSource® Bookshelf — an ebook reader which allows you to make and share notes and highlights on your ebooks and search across all of the ebooks that you hold on your VitalSource Bookshelf. You can access the ebook online or offline on your smartphone, tablet or PC/Mac and your notes and highlights will automatically stay in sync no matter where you make them.

1. **Create a VitalSource Bookshelf account at** *https://online.vitalsource.com/user/new* or log into your existing account if you already have one.

2. **Redeem the code provided in the panel below to get online access to the ebook.**
   Log in to Bookshelf and select **Redeem** at the top right of the screen. Enter the redemption code shown on the scratch-off panel below in the **Redeem Code** pop-up and press **Redeem**. Once the code has been redeemed your ebook will download and appear in your library.

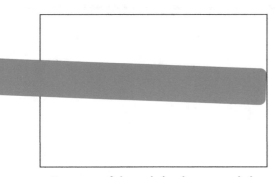

No returns if this code has been revealed.

DOWNLOAD AND READ OFFLINE

To use your ebook offline, download BookShelf to your PC, Mac, iOS device, Android device or Kindle Fire, and log in to your Bookshelf account to access your ebook:

On your PC/Mac

Go to *https://support.vitalsource.com/hc/en-us* and follow the instructions to download the free **VitalSource Bookshelf** app to your PC or Mac and log into your Bookshelf account.

On your iPhone/iPod Touch/iPad

Download the free **VitalSource Bookshelf** App available via the iTunes App Store and log into your Bookshelf account. You can find more information at *https://support.vitalsource.com/hc/en-us/categories/200134217-Bookshelf-for-iOS*

On your Android™ smartphone or tablet

Download the free **VitalSource Bookshelf** App available via Google Play and log into your Bookshelf account. You can find more information at *https://support.vitalsource.com/hc/en-us/categories/200139976-Bookshelf-for-Android-and-Kindle-Fire*

On your Kindle Fire

Download the free **VitalSource Bookshelf** App available from Amazon and log into your Bookshelf account. You can find more information at *https://support.vitalsource.com/hc/en-us/categories/200139976-Bookshelf-for-Android-and-Kindle-Fire*

N.B. The code in the scratch-off panel can only be used once. When you have created a Bookshelf account and redeemed the code you will be able to access the ebook online or offline on your smartphone, tablet or PC/Mac.

SUPPORT

If you have any questions about downloading Bookshelf, creating your account, or accessing and using your ebook edition, please visit *http://support.vitalsource.com/*

Modern Data Science with R

CHAPMAN & HALL/CRC
Texts in Statistical Science Series

Series Editors
Francesca Dominici, *Harvard School of Public Health, USA*
Julian J. Faraway, *University of Bath, UK*
Martin Tanner, *Northwestern University, USA*
Jim Zidek, *University of British Columbia, Canada*

Texts in Statistical Science

Modern Data Science with R

Benjamin S. Baumer

Daniel T. Kaplan

Nicholas J. Horton

CRC Press
Taylor & Francis Group
Boca Raton   London   New York

CRC Press is an imprint of the
Taylor & Francis Group an **informa** business

A CHAPMAN & HALL BOOK

CRC Press
Taylor & Francis Group
6000 Broken Sound Parkway NW, Suite 300
Boca Raton, FL 33487-2742

© 2017 by Taylor & Francis Group, LLC
CRC Press is an imprint of Taylor & Francis Group, an Informa business

No claim to original U.S. Government works

Printed on acid-free paper
Version Date: 20161215

International Standard Book Number-13: 978-1-4987-2448-7 (Pack - Book and Ebook)

Visit the Taylor & Francis Web site at
http://www.taylorandfrancis.com

and the CRC Press Web site at
http://www.crcpress.com

Contents

II   Statistics and Modeling   147

List of Tables

List of Figures

Preface

Background and motivation

The increasing volume and sophistication of data poses new challenges for analysts, who need to be able to transform complex data sets to answer important statistical questions. The widely-cited McKinsey & Company report stated that "by 2018, the United States alone could face a shortage of 140,000 to 190,000 people with deep analytical skills as well as 1.5 million managers and analysts with the know-how to use the analysis of big data to make effective decisions." There is a pressing need for additional resources to train existing analysts as well as the next generation to be able to pose questions, suggest hypotheses, collect, transform, and analyze data, then communicate results. According to the online company ratings site *Glassdoor*, "data scientist" was the best job in America in 2016 [142].

Statistics can be defined as the science of learning from data [203]. Michael Jordan has described data science as the marriage of computational thinking and inferential thinking. Without the skills to be able to "wrangle" the increasingly rich and complex data that surround us, analysts will not be able to use these data to make better decisions.

New data technologies and database systems facilitate scraping and merging data from different sources and formats and restructuring it into a form suitable for analysis. State-of-the-art workflow tools foster well-documented and reproducible analysis. Modern statistical methods allow the analyst to fit and assess models as well as to undertake supervised or unsupervised learning to extract information. Contemporary data science requires tight integration of these statistical, computing, data-related, and communication skills.

The book is intended for readers to develop and reinforce the appropriate skills to tackle complex data science projects and "think with data" (as coined by Diane Lambert). The ability to solve problems using data is at the heart of our approach.

We feature a series of complex, real-world extended case studies and examples from a broad range of application areas, including politics, transportation, sports, environmental science, public health, social media, and entertainment. These rich data sets require the use of sophisticated data extraction techniques, modern data visualization approaches, and refined computational approaches.

It is impossible to cover all these topics in any level of detail within a single book: Many of the chapters could productively form the basis for a course or series of courses. Our goal is to lay a foundation for analysis of real-world data and to ensure that analysts see the power of statistics and data analysis. After reading this book, readers will have greatly expanded their skill set for working with these data, and should have a newfound confidence about their ability to learn new technologies on-the-fly.

Key role of technology

While many tools can be used effectively to undertake data science, and the technologies to undertake analyses are quickly changing, R and Python have emerged as two powerful and extensible environments. While it is important for data scientists to be able to use multiple technologies for their analyses, we have chosen to focus on the use of R and RStudio to avoid cognitive overload. By use of a "Less Volume, More Creativity" approach [162], we intend to develop a small set of tools that can be mastered within the confines of a single semester and that facilitate sophisticated data management and exploration.

We take full advantage of the RStudio environment. This powerful and easy-to-use front end adds innumerable features to R including package support, code-completion, integrated help, a debugger, and other coding tools. In our experience, the use of RStudio dramatically increases the productivity of R users, and by tightly integrating reproducible analysis tools, helps avoid error-prone "cut-and-paste" workflows. Our students and colleagues find RStudio an extremely comfortable interface. No prior knowledge or experience with R or RStudio is required: we include an introduction within the Appendix.

We used a reproducible analysis system (`knitr`) to generate the example code and output in this book. Code extracted from these files is provided on the book's website. We provide a detailed discussion of the philosophy and use of these systems. In particular, we feel that the `knitr` and `markdown` packages for R, which are tightly integrated with RStudio, should become a part of every R user's toolbox. We can't imagine working on a project without them (and we've incorporated reproducibility into all of our courses).

Modern data science is a team sport. To be able to fully engage, analysts must be able to pose a question, seek out data to address it, ingest this into a computing environment, model and explore, then communicate results. This is an iterative process that requires a blend of statistics and computing skills.

Context is king for such questions, and we have structured the book to foster the parallel developments of statistical thinking, data-related skills, and communication. Each chapter focuses on a different extended example with diverse applications, while exercises allow for the development and refinement of the skills learned in that chapter.

Intended audiences

This book was originally conceived to support a one-semester, 13-week upper-level course in data science. We also intend that the book will be useful for more advanced students in related disciplines, or analysts who want to bolster their data science skills. The book is intended to be accessible to a general audience with some background in statistics (completion of an introductory statistics course).

In addition to many examples and extended case studies, the book incorporates exercises at the end of each chapter. Many of the exercises are quite open-ended, and are designed to allow students to explore their creativity in tackling data science questions.

The book has been structured with three main sections plus supplementary appendices. Part I provides an introduction to data science, an introduction to visualization, a foundation for data management (or 'wrangling'), and ethics. Part II extends key modeling notions including regression modeling, classification and prediction, statistical foundations, and simulation. Part III introduces more advanced topics, including interactive data visualization, SQL and relational databases, spatial data, text mining, and network science.

We conclude with appendices that introduce the book's R package, R and RStudio, key aspects of algorithmic thinking, reproducible analysis, a review of regression, and how to set up a local SQL database.

We have provided two indices: one organized by subject and the other organized by R function and package. In addition, the book features extensive cross-referencing (given the inherent connections between topics and approaches).

Website

The book website at `https://mdsr-book.github.io` includes the table of contents, subject and R indices, example datasets, code samples, exercises, additional activities, and a list of errata.

How to use this book

The material from this book has supported several courses to date at Amherst, Smith, and Macalester Colleges. This includes an intermediate course in data science (2013 and 2014 at Smith), an introductory course in data science (2016 at Smith), and a capstone course in advanced data analysis (2015 and 2016 at Amherst). The intermediate data science course required an introductory statistics course and some programming experience, and discussed much of the material in this book in one semester, culminating with an integrated final project [20]. The introductory data science course had no prerequisites and included the following subset of material:

- Data Visualization: three weeks, covering Chapters 2 and 3

- Data Wrangling: four weeks, covering Chapters 4 and 5

- Database Querying: two weeks, covering Chapter 12

- Spatial Data: two weeks, covering Chapter 14

- Text Mining: two weeks, covering Chapter 15

The capstone course covered the following material:

- Data Visualization: two weeks, covering Chapters 2, 3, and 11

- Data Wrangling: two weeks, covering Chapters 4 and 5

- Ethics: one week, covering Chapter 6

- Simulation: one week, covering Chapter 10

- Statistical Learning: two weeks, covering Chapters 8 and 9

- Databases: one week, covering Chapter 12 and Appendix F

- Text Mining: one week, covering Chapter 15

- Spatial Data: one week, covering Chapter 14

- Big Data: one week, covering Chapter 17

We anticipate that this book could serve as the primary text for a variety of other courses, with or without additional supplementary material.

The content in Part I—particularly the `ggplot2` visualization concepts presented in Chapter 3 and the `dplyr` data wrangling operations presented in Chapter 4—is fundamental and is assumed in Parts II and III. Each of the chapters in Part III are independent of each other and the material in Part II. Thus, while most instructors will want to cover most (if not all) of Part I in any course, the material in Parts II and III can be added with almost total freedom.

The material in Part II is designed to expose students with a beginner's understanding of statistics (i.e., basic inference and linear regression) to a richer world of statistical modeling and statistical inference.

Acknowledgments

We would like to thank John Kimmel at Informa CRC/Chapman and Hall for his support and guidance. We also thank Jim Albert, Nancy Boynton, Jon Caris, Mine Çetinkaya–Rundel, Jonathan Che, Patrick Frenett, Scott Gilman, Johanna Hardin, John Horton, Azka Javaid, Andrew Kim, Eunice Kim, Caroline Kusiak, Ken Kleinman, Priscilla (Wencong) Li, Amelia McNamara, Tasheena Narraido, Melody Owen, Randall Pruim, Tanya Riseman, Gabriel Sosa, Katie St. Clair, Amy Wagaman, Susan (Xiaofei) Wang, Hadley Wickham, J. J. Allaire and the RStudio developers, the anonymous reviewers, the Spring 2015 SDS192 class, the Fall 2016 STAT495 class, and many others for contributions to the R and RStudio environment, comments, guidance, and/or helpful suggestions on drafts of the manuscript.

Above all we greatly appreciate Cory, Maya, and Julia for their patience and support.

Northampton, MA and St. Paul, MN
December 2016

Part I

Introduction to Data Science

Chapter 1

Prologue: Why data science?

Information is what we want, but data are what we've got. The techniques for transforming data into information go back hundreds of years. A good starting point is 1592 with the publication of John Graunt's weekly "bills of mortality" in London. (See Figure 1.1.) These "bills" are tabulations—a condensation of data on individual events into a form more readily assimilated by the human reader. Constructing such tabulations was a manual operation.

Over the centuries, as data became larger, machines were introduced to speed up the tabulations. A major step was Herman Hollerith's development of punched cards and an electrical tabulating system for the United States Census of 1890. This was so successful that Hollerith started a company, International Business Machines Corporation (IBM), that came to play an important role in the development of today's electronic computers.

Also in the late 19th century, statistical methods began to develop rapidly. These methods have been tremendously important in interpreting data, but they were not intrinsically tied to mechanical data processing. Generations of students have learned to carry out statistical operations by hand on small sets of data.

Nowadays, it is common to have data sets that are so large they can be processed only by machine. In this era of "big data," data are amassed by networks of instruments and computers. The settings where such data arise are diverse: the genome, satellite observations of Earth, entries by web users, sales transactions, etc. There are new opportunities for finding and characterizing patterns using techniques described as data mining, machine learning, data visualization, and so on. Such techniques require computer processing. Among the tasks that need performing are data cleaning, combining data from multiple sources, and reshaping data into a form suitable as input to data-summarization operations for visualization and modeling.

In writing this book we hope to help people gain the understanding and skills for *data wrangling* (a process of preparing data for visualization and other modern techniques of statistical interpretation) and using those data to answer statistical questions via modeling and visualization. Doing so inevitably involves, at the center, the ability to reason statistically and utilize computational and algorithmic capacities.

Is an extended study of computer programming necessary to engage in sophisticated computing? Our view is that it is not. First, over the last half century, a coherent set of simple data operations have been developed that can be used as the building blocks of sophisticated data wrangling processes. The trick is not mastering programming but rather learning to think in terms of these operations. Much of this book is intended to help you master such thinking.

Second, it is possible to use recent developments in software to vastly reduce the amount of programming needed to use these data operations. We have drawn on such software—

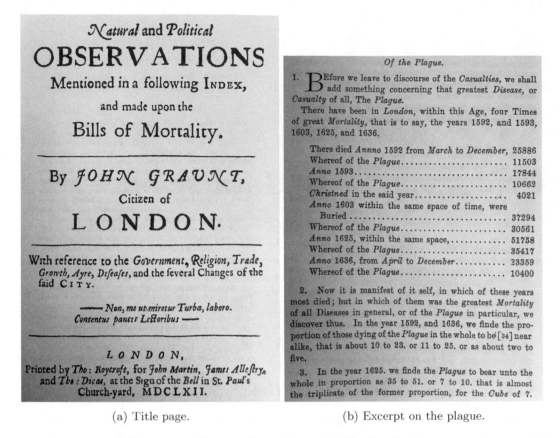

(a) Title page.   (b) Excerpt on the plague.

Figure 1.1: Excerpt from Graunt's bills of mortality.

particularly R and the packages `dplyr` and `ggplot2`—to focus on a small subset of functions that accomplish data wrangling tasks in a concise and expressive way. The programming syntax is consistent enough that, with a little practice, you should be able to adapt the code contained in this book to solve your own problems. (Experienced R programmers will note the distinctive style of R statements in this book, including a consistent focus on a small set of functions and extensive use of the "pipe" operator.) Part I of this book focuses on data wrangling and data visualization as key building blocks for data science.

1.1 What is data science?

We hold a broad view of *data science*—we see it as the science of extracting meaningful information from data. There are several key ideas embedded in that simple definition. First, data science is a *science*, a rigorous discipline combining elements of statistics and computer science, with roots in mathematics. Michael Jordan from the University of California, Berkeley has described data science as a fine-grained blend of intellectual traditions from statistics and computer science:

> Computer science is more than just programming; it is the creation of appropriate abstractions to express computational structures and the development of algorithms that operate on those abstractions. Similarly, statistics is more than just collections of estimators and tests; it is the interplay of general notions of

> sampling, models, distributions and decision-making. [Data science] is based on
> the idea that these styles of thinking support each other [159].

Second, data science is best applied in the context of expert knowledge about the domain from which the data originate. This domain might be anything from astronomy to zoology; business and health care are two particularly important domain areas. Third, the distinction between *data* and *information* is the *raison d'etre* of data science. Data scientists are people who are interested in converting the data that is now abundant into actionable information that always seems to be scarce.

Many statisticians will say: "But we already have a field for that: it's called *statistics!*" The goals of data scientists and statisticians are the same: They both want to extract meaningful information from data. Much of statistical technique was originally developed in an environment where data were scarce and difficult or expensive to collect, so statisticians focused on creating methods that would maximize the strength of inference one is able to make, given the least amount of data. These techniques were often ingenious, involved sophisticated mathematics, and have proven invaluable to the empirical sciences for going on a century. While several of the most influential early statisticians saw computing as an integral part of statistics, it is also true that much of the development of statistical theory was to find mathematical approximations for things that we couldn't yet compute [56].

Today, the manner in which we extract meaning from data is different in two ways—both due primarily to advances in computing:

1. we are able to compute many more things than we could before, and;

2. we have a *lot* more data than we had before.

The first change means that some of the techniques that were ubiquitous in statistics education in the 20th century (e.g., t-tests, ANOVA) are being replaced by computational techniques that are conceptually simpler, but were simply infeasible until the microcomputer revolution (e.g., the bootstrap, permutation tests). The second change means that many of the data we now collect are *observational*—they don't come from a designed experiment and they aren't really sampled at random. This makes developing realistic probability models for these data much more challenging, which in turn makes formal statistical inference a more challenging (and perhaps less relevant) problem. In some settings (e.g., clinical trials and A/B testing) the careful estimation of a model parameter is still the goal, and inferential statistics are still the primary tools of the trade. But in an array of academic, government, and industrial settings, the end result may instead be a *predictive model*, an interactive visualization of the data, or a web application that allows the user to slice-and-dice the data to make simple comparisons. We explore issues related to statistical inference and modeling in greater depth in Part II of this book.

The increasing complexity and heterogeneity of modern data means that each data analysis project needs to be custom-built. Simply put, the modern data analyst needs to be able to read and write computer instructions, the so-called "code" from which data analysis projects are built. Part I of this book develops foundational abilities in data visualization and data wrangling—two essential skills for the modern data scientist. These chapters focus on the traditional two-dimensional representation of data: rows and columns in a data table, and horizontal and vertical in a data graphic. In Part III, we explore a variety of non-traditional data types (e.g., spatial, text, network, "big") and interactive data graphics.

As you work through this book, you will develop computational skills that we describe as "precursors" to big data [107]. In Chapter 17, we point to some tools for working with truly big data. One has to learn to crawl before one can walk, and we argue that for most people the skills developed herein are more germane to the kinds of problems that you are likely to encounter.

1.2   Case study: The evolution of sabermetrics

The evolution of baseball analytics (often called *sabermetrics*) in many ways recapitulates the evolution of analytics in other domains. Although *domain knowledge* is always useful in data science, no background in baseball is required for this section[1].

The *use* of statistics in baseball has a long and storied history—in part because the game itself is naturally discrete, and in part because Henry Chadwick began publishing boxscores in the early 1900s [184]. For these reasons, a rich catalog of baseball data began to accumulate.

However, while more and more baseball data were piling up, *analysis* of that data was not so prevalent. That is, the extant data provided a means to keep records, and as a result some numerical elements of the game's history took on a life of their own (e.g., Babe Ruth's 714 home runs). But it is not as clear how much people were learning about the game of baseball from the data. Knowing that Babe Ruth hit more home runs than Mel Ott tells us something about two players, but doesn't provide any insight into the nature of the game itself.

In 1947—Jackie Robinson's rookie season—Brooklyn Dodgers' general manager Branch Rickey made another significant innovation: He hired Allan Roth to be baseball's first statistical analyst. Roth's *analysis* of baseball data led to insights that the Dodgers used to win more games. In particular, Roth convinced Rickey that a measurement of how often a batter reaches first base via any means (e.g., hit, walk, or being hit by the pitch) was a better indicator of that batter's value than how often he reaches first base via a hit (which was—and probably still is—the most commonly cited batting statistic). The logic supporting this insight was based on both Roth's understanding of the game of baseball (what we call *domain knowledge*) and his statistical analysis of baseball data.

During the next 50 years, many important contributions to baseball analytics were made by a variety of people (most notably "The Godfather of Sabermetrics" Bill James [119]), most of whom had little formal training in statistics, whose weapon of choice was a spreadsheet. They were able to use their creativity, domain knowledge, and a keen sense of what the interesting questions were to make interesting discoveries.

The 2003 publication of *Moneyball* [131]—which showcased how Billy Beane and Paul DePodesta used statistical analysis to run the Oakland A's—triggered a revolution in how front offices in baseball were managed [27]. Over the next decade, the size of the data expanded so rapidly that a spreadsheet was no longer a viable mechanism for storing—let alone analyzing—all of the available data. Today, many professional sports teams have research and development groups headed by people with Ph.D.'s in statistics or computer science along with graduate training in machine learning [16]. This is not surprising given that revenue estimates for major league baseball top $8 billion per year.

The contributions made by the next generation of baseball analysts will *require* coding ability. The creativity and domain knowledge that fueled the work of Allan Roth and Bill James remain necessary traits for success, but they are no longer sufficient. There is nothing special about baseball in this respect—a similar profusion of data are now available in many other areas, including astronomy, health services research, genomics, and climate change,

---

[1]The main rules of baseball are these: Two teams of nine players alternate trying to score runs on a field with four bases (first base, second base, third base, or home). The defensive team pitches while one member of the offensive team bats while standing by home base). A run is scored when an offensive player crosses home plate after advancing in order through the other bases.

among others. For data scientists of all application domains, creativity, domain knowledge, and technical ability are absolutely essential.

1.3 Datasets

There are many data sets used in this book. The smaller ones are available through either the mdsr (see Appendix A) or mosaic packages for R. Some other data used in this book are pulled directly from the Internet—URLs for these data are embedded in the text. There a few larger, more complicated data sets that we use repeatedly and that warrant some explication here.

Airline Delays The U.S. Bureau of Transportation Statistics has collected data on more than 169 million domestic flights dating back to October 1987. We have developed the airlines package to allow R users to download and process these data with minimal hassle. (Instructions as to how to set up a database can be found in Appendix F.) These data were originally used for the 2009 ASA Data Expo [213]. The nycflights13 package contains a subset of these data (only flights leaving the three most prominent New York City airports in 2013).

Baseball The Lahman database is maintained by Sean Lahman, a self-described database journalist. Compiled by a team of volunteers, it contains complete seasonal records going back to 1871 and is usually updated yearly. It is available for download both as a pre-packaged SQL file and as an R package [80].

Baby Names The babynames package for R provides data about the popularity of individual baby names from the U.S. Social Security Administration [221]. These data can be used, for example, to track the popularity of certain names over time.

Federal Election Commission The fec package provides access to campaign spending data for recent federal elections maintained by the Federal Election Commission. These data include contributions by individuals to committees, spending by those committees on behalf, or against individual candidates for president, the Senate, and the House of Representatives, as well information about those committees and candidates.

MacLeish The Ada and Archibald MacLeish Field Station is a 260-acre plot of land owned and operated by Smith College. It is used by faculty, students, and members of the local community for environmental research, outdoor activities, and recreation. The macleish R package allows you to download and process weather data (as a time series) from the MacLeish Field Station using the etl framework. It also contains shapefiles for contextualizing spatial information.

Movies The Internet Movie Database is a massive repository of information about movies [117]. The easiest way to get the IMDb data into SQL is by using the open-source IMDbPY Python package [1].

Restaurant Violations The mdsr package contains data on restaurant health inspections made by the New York City Health Department.

Twitter The micro-blogging social networking service Twitter has an application programming interface (API) accessed using the twitteR package that can be used to access data of short 140-character messages (called *tweets*) along with retweets and responses. Approximately 500 million tweets are shared daily on the service.

1.4   Further resources

Each chapter features a list of additional resources that provide further depth or serve as a definitive reference for a given topic. Other definitions of data science and analytics can be found in [158, 64, 57, 109, 95, 77, 160, 54].

Chapter 2

Data visualization

Data graphics provide one of the most accessible, compelling, and expressive modes to investigate and depict patterns in data. This chapter will motivate the importance of well-designed data graphics and describe a taxonomy for understanding their composition. If you are seeing this material for the first time, you will never look at data graphics the same way again—yours will soon be a more critical lens.

2.1 The 2012 federal election cycle

Every four years, the presidential election draws an enormous amount of interest in the United States. The most prominent candidates announce their candidacy nearly two years before the November elections, beginning the process of raising the hundreds of millions of dollars necessary to orchestrate a national campaign. In many ways, the experience of running a successful presidential campaign is in itself evidence of the leadership and organizational skills necessary to be commander-in-chief.

Voices from all parts of the political spectrum are critical of the influence of money upon political campaigns. While the contributions from individual citizens to individual candidates are limited in various ways, the Supreme Court's decision in Citizens United v. Federal Election Commission allows unlimited political spending by corporations (non-profit or otherwise). This has resulted in a system of committees (most notably, political action committees (PACs)) that can accept unlimited contributions and spend them on behalf of (or against) a particular candidate or set of candidates. Unraveling the complicated network of campaign spending is a subject of great interest.

To perform that unraveling is an exercise in data science. The Federal Election Commission (FEC) maintains a website with logs of not only all of the ($200 or more) contributions made by individuals to candidates and committees, but also of spending by committees on behalf of (and against) candidates. Of course, the FEC also maintains data on which candidates win elections, and by how much. These data sources are separate and it requires some ingenuity to piece them together. We will develop these skills in Chapters 4 and 5, but for now, we will focus on graphical displays of the information that can be gleaned from these data. Our emphasis at this stage is on making intelligent decisions about how to display certain data, so that a clear (and correct) message is delivered.

Among the most basic questions is: How much money did each candidate raise? However, the convoluted campaign finance network makes even this simple question difficult to answer, and—perhaps more importantly—less meaningful than we might think. A better question is: On whose candidacy was the most money spent? In Figure 2.1, we show a bar

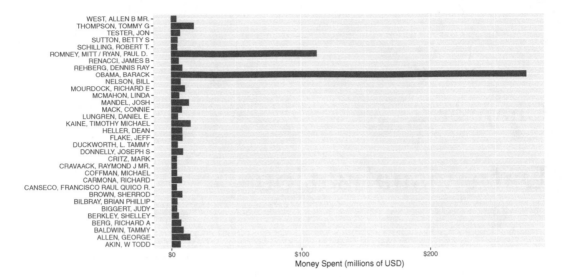

Figure 2.1: Amount of money spent on individual candidates in the general election phase of the 2012 federal election cycle, in millions of dollars. Candidacies with at least four million dollars in spending are depicted.

graph of the amount of money (in millions of dollars) that were spent by committees on particular candidates during the general election phase of the 2012 federal election cycle. This includes candidates for president, the Senate, and the House of Representatives. Only candidates on whose campaign at least $4 million was spent are included in Figure 2.1.

It seems clear from Figure 2.1 that President Barack Obama's re-election campaign spent far more money than any other candidate, in particular more than doubling the amount of money spent by his Republican challenger, Mitt Romney. However, committees are not limited to spending money in support of a candidate—they can also spend money *against* a particular candidate (i.e., on attack ads). In Figure 2.2 we separate the same spending shown in Figure 2.1 by whether the money was spent for or against the candidate.

In these elections, most of the money was spent against each candidate, and in particular, $251 million of the $274 million spent on President Obama's campaign was spent against his candidacy. Similarly, most of the money spent on Mitt Romney's campaign was against him, but the percentage of negative spending on Romney's campaign (70%) was lower than that of Obama (92%).

The difference between Figure 2.1 and Figure 2.2 is that in the latter we have used color to bring a third variable (type of spending) into the plot. This allows us to make a clear comparison that importantly changes the conclusions we might draw from the former plot. In particular, Figure 2.1 makes it appear as though President Obama's war chest dwarfed that of Romney, when in fact the opposite was true.

2.1.1   Are these two groups different?

Since so much more money was spent attacking Obama's campaign than Romney's, you might conclude from Figure 2.2 that Republicans were more successful in fundraising during this election cycle. In Figure 2.3 we can confirm that this was indeed the case, since more money was spent supporting Republican candidates than Democrats, and more money was spent attacking Democratic candidates than Republican. It also seems clear from Figure 2.3 that nearly all of the money was spent on either Democrats or Republicans.

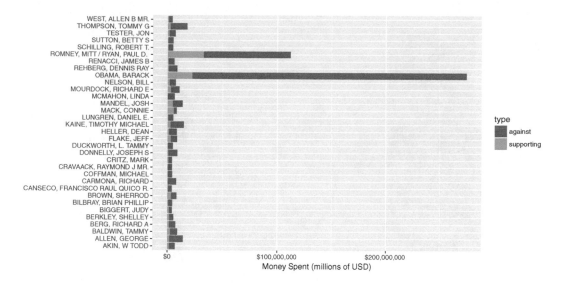

Figure 2.2: Amount of money spent on individual candidates in the general election phase of the 2012 federal election cycle, in millions of dollars, broken down by type of spending. Candidacies with at least four million dollars in spending are depicted.

However, the question of whether the money spent on candidates really differed by party affiliation is a bit thornier. As we saw above, the presidential election dominated the political donations in this election cycle. Romney faced a serious disadvantage in trying to unseat an incumbent president. In this case, the office being sought is a confounding variable. By further subdividing the contributions in Figure 2.3 by the office being sought, we can see in Figure 2.4 that while more money was spent supporting Republican candidates for all three houses of government, it was only in the presidential election that more money was spent attacking Democratic candidates. In fact, slightly more money was spent attacking Republican House and Senate candidates.

Note that Figures 2.3 and 2.4 display the same data. In Figure 2.4 we have an additional variable that provides and important clue into the mystery of campaign finance. Our choice to include that variable results in Figure 2.4 conveying substantially more meaning than Figure 2.3, even though both figures are "correct." In this chapter, we will begin to develop a framework for creating principled data graphics.

2.1.2 Graphing variation

One theme that arose during the presidential election was the allegation that Romney's campaign was supported by a few rich donors, whereas Obama's support came from people across the economic spectrum. If this were true, then we would expect to see a difference in the distribution of donation amounts between the two candidates. In particular, we would expect to see this in the histograms shown in Figure 2.5, which summarize the more than one million donations made by individuals to the two major committees that supported each candidate (for Obama, Obama for America, and the Obama Victory Fund 2012; for Romney, Romney for President, and Romney Victory 2012). We do see some evidence for this claim in Figure 2.5, Obama did appear to receive more smaller donations, but the evidence is far from conclusive. One problem is that both candidates received many small donations but just a few larger donations; the scale on the horizontal axis makes it difficult to actually see what is going on. Secondly, the histograms are hard to compare in a side-

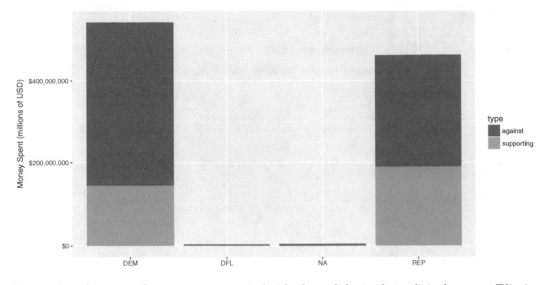

Figure 2.3: Amount of money spent on individual candidacies by political party affiliation during the general election phase of the 2012 federal election cycle.

by-side placement. Finally, we have lumped all of the donations from both phases of the presidential election (i.e., primary vs. general) in together.

In Figure 2.6, we remedy these issues by (1) using density curves instead of histograms, so that we can compare the distributions directly, (2) plotting the logarithm of the donation amount on the horizontal scale to focus on the data that are important, and (3) separating the donations by the phase of the election. Figure 2.6 allows us to make more nuanced conclusions. The right panel supports the allegation that Obama's donations came from a broader base during the primary election phase. It does appear that more of Obama's donations came in smaller amounts during this phase of the election. However, in the general phase, there is virtually no difference in the distribution of donations made to either campaign.

2.1.3   Examining relationships among variables

Naturally, the biggest questions raised by the *Citizens United* decision are about the influence of money in elections. If campaign spending is unlimited, does this mean that the candidate who generates the most spending on their behalf will earn the most votes? One way that we might address this question is to compare the amount of money spent on each candidate in each election with the number of votes that candidate earned. Statisticians will want to know the *correlation* between these two quantities—when one is high, is the other one likely to be high as well?

Since all 435 members of the United States House of Representatives are elected every two years, and the districts contain roughly the same number of people, House elections provide a nice data set to make this type of comparison. In Figure 2.7, we show a simple scatterplot relating the number of dollars spent on behalf of the Democratic candidate against the number of votes that candidate earned for each of the House elections.

The relationship between the two quantities depicted in Figure 2.7 is very weak. It does not appear that candidates who benefited more from campaign spending earned more votes. However, the comparison in Figure 2.7 is misleading. On both axes, it is not the *amount* that is important, but the *percentage*. Although the population of each congressional district is

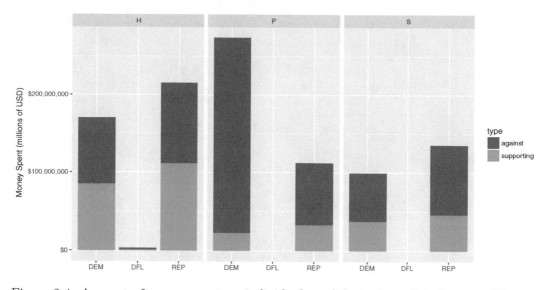

Figure 2.4: Amount of money spent on individual candidacies by political party affiliation during the general election phase of the 2012 federal election cycle, broken down by office being sought.

similar, they are not the same, and voter turnout will vary based on a variety of factors. By comparing the percentage of the vote, we can control for the size of the voting population in each district. Similarly, it makes less sense to focus on the total amount of money spent, as opposed to the percentage of money spent. In Figure 2.8 we present the same comparison, but with both axes scaled to percentages.

Figure 2.8 captures many nuances that were impossible to see in Figure 2.7. First, there *does* appear to be a positive association between the percentage of money supporting a candidate and the percentage of votes that they earn. However, that relationship is of greatest interest towards the center of the plot, where elections are actually contested. Outside of this region, one candidate wins more than 55% of the vote. In this case, there is usually very little money spent. These are considered "safe" House elections—you can see these points on the plot because most of them are close to $x = 0$ or $x = 1$, and the dots are very small. For example, in the lower right corner is the 8th district in Ohio, which was won by the then-current Speaker of the House John Boehner, who ran unopposed. The election in which the most money was spent (over $11 million) was also in Ohio. In the 16th district, Republican incumbent Jim Renacci narrowly defeated Democratic challenger Betty Sutton, who was herself an incumbent from the 13th district. This battle was made possible through decennial redistricting (see Chapter 14). Of the money spent in this election, 51.2% was in support of Sutton but she earned only 48.0% of the votes.

In the center of the plot, the dots are bigger, indicating that more money is being spent on these contested elections. Of course this makes sense, since candidates who are fighting for their political lives are more likely to fundraise aggressively. Nevertheless, the evidence that more financial support correlates with more votes in contested elections is relatively weak.

2.1.4 Networks

Not all relationships among variables are sensibly expressed by a scatterplot. Another way in which variables can be related is in the form of a network (we will discuss these in more

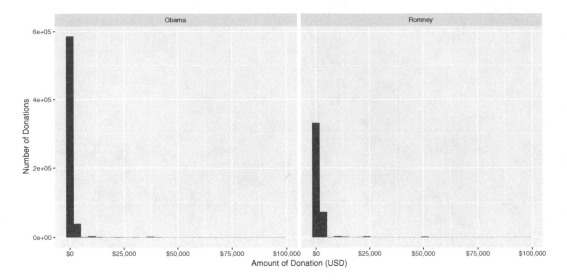

Figure 2.5: Donations made by individuals to the PACs supporting the two major presidential candidates in the 2012 election.

detail in Chapter 16). In this case, campaign funding has a network structure in which individuals donate money to committees, and committees then spend money on behalf of candidates. While the national campaign funding network is far too complex to show here, in Figure 2.9 we display the funding network for candidates from Massachusetts.

In Figure 2.9, we see that the two campaigns that benefited the most from committee spending were Republicans Mitt Romney and Scott Brown. This is not surprising, since Romney was running for president, and received massive donations from the Republican National Committee, while Brown was running to keep his Senate seat in a heavily Democratic state against a strong challenger, Elizabeth Warren. Both men lost their elections. The constellation of blue dots are the congressional delegation from Massachusetts, all of whom are Democrats.

2.2   Composing data graphics

Former *New York Times* intern and FlowingData.com creator Nathan Yau makes the analogy that creating data graphics is like cooking: Anyone can learn to type graphical commands and generate plots on the computer. Similarly, anyone can heat up food in a microwave. What separates a high-quality visualization from a plain one are the same elements that separate great chefs from novices: mastery of their tools, knowledge of their ingredients, insight, and creativity [243]. In this section, we present a framework—rooted in scientific research—for understanding data graphics. Our hope is that by internalizing these ideas you will refine your data graphics palette.

2.2.1   A taxonomy for data graphics

The taxonomy presented in [243] provides a systematic way of thinking about how data graphics convey specific pieces of information, and how they could be improved. A complementary *grammar* of graphics [238] is implemented by Hadley Wickham in the `ggplot2` graphics package [212], albeit using slightly different terminology. For clarity, we will post-

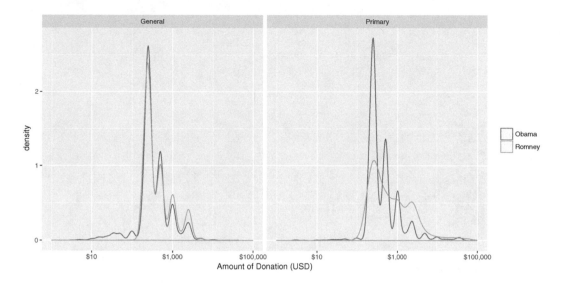

Figure 2.6: Donations made by individuals to the PACs supporting the two major presidential candidates in the 2012 election, separated by election phase.

pone discussion of `ggplot2` until Chapter 3. (To extend our cooking analogy, you must learn to taste before you can learn to cook well.)

In this framework, data graphics can be understood in terms of four basic elements: visual cues, coordinate system, scale, and context. In what follows we explicate this vision and append a few additional items (facets and layers). This section should equip the careful reader with the ability to systematically break down data graphics, enabling a more critical analysis of their content.

Visual Cues

Visual cues are graphical elements that draw the eye to what you want your audience to focus upon. They are the fundamental building blocks of data graphics, and the choice of which visual cues to use to represent which quantities is the central question for the data graphic composer. Yau identifies nine distinct visual cues, for which we also list whether that cue is used to encode a numerical or categorical quantity:

Position (numerical) where in relation to other things?

Length (numerical) how big (in one dimension)?

Angle (numerical) how wide? parallel to something else?

Direction (numerical) at what slope? In a time series, going up or down?

Shape (categorical) belonging to which group?

Area (numerical) how big (in two dimensions)?

Volume (numerical) how big (in three dimensions)?

Shade (either) to what extent? how severely?

Color (either) to what extent? how severely? Beware of red/green color blindness (see Section 2.2.2)

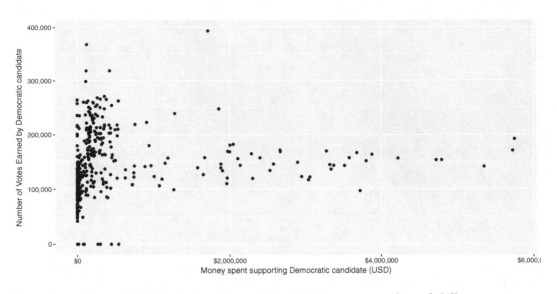

Figure 2.7: Scatterplot illustrating the relationship between number of dollars spent supporting and number of votes earned by Democrats in 2012 elections for the House of Representatives.

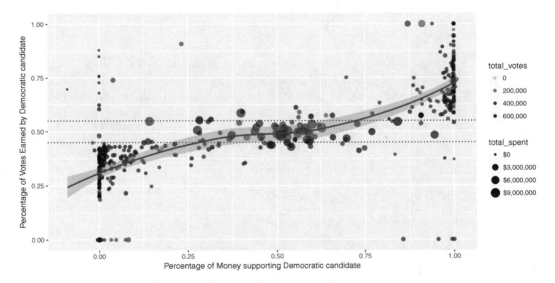

Figure 2.8: Scatterplot illustrating the relationship between percentage of dollars spent supporting and percentage of votes earned by Democrats in the 2012 House of Representatives elections. Each dot represents one district. The size of each dot is proportional to the total spending in that election, and the alpha transparency of each dot is proportional to the total number of votes in that district.

Figure 2.9: Campaign funding network for candidates from Massachusetts, 2012 federal elections. Each edge represents a contribution from a PAC to a candidate.

Research into graphical perception (dating back to the mid-1980s) has shown that human beings' ability to perceive differences in magnitude accurately descends in this order [55]. That is, humans are quite good at accurately perceiving differences in position (e.g., how much taller one bar is than another), but not as good at perceiving differences in angles. This is one reason why many people prefer bar charts to pie charts. Our relatively poor ability to perceive differences in color is a major factor in the relatively low opinion of heat maps that many data scientists have.

Coordinate systems

How are the data points organized? While any number of coordinate systems are possible, three are most common:

Cartesian This is the familiar (x, y)-rectangular coordinate system with two perpendicular axes.

Polar The radial analog of the Cartesian system with points identified by their radius ρ and angle θ.

Geographic This is the increasingly important system in which we have locations on the curved surface of the Earth, but we are trying to represent these locations in a flat two-dimensional plane. We will discuss such spatial analyses in Chapter 14.

An appropriate choice for a coordinate system is critical in representing one's data accurately, since, for example, displaying spatial data like airline routes on a flat Cartesian plane can lead to gross distortions of reality (see Section 14.3.2).

Scale

Scales translate values into visual cues. The choice of scale is often crucial. The central question is *how* does distance in the data graphic translate into meaningful differences in quantity? Each coordinate axis can have its own scale, for which we have three different choices:

Numeric A numeric quantity is most commonly set on a *linear, logarithmic,* or *percentage* scale. Note that a logarithmic scale does not have the property that, say, a one-centimeter difference in position corresponds to an equal difference in quantity anywhere on the scale.

Categorical A categorical variable may have no ordering (e.g., Democrat, Republican, or Independent), or it may be *ordinal* (e.g., never, former, or current smoker).

Time Time is a numeric quantity that has some special properties. First, because of the calendar, it can be demarcated by a series of different units (e.g., year, month, day, etc.). Second, it can be considered periodically (or cyclically) as a "wrap-around" scale. Time is also so commonly used and misused that it warrants careful consideration.

Misleading with scale is easy, since it has the potential to completely distort the relative positions of data points in any graphic.

Context

The purpose of data graphics is to help the viewer make *meaningful* comparisons, but a bad data graphic can do just the opposite: It can instead focus the viewer's attention on meaningless artifacts, or ignore crucial pieces of relevant but external knowledge. Context can be added to data graphics in the form of titles or subtitles that explain what is being shown, axis labels that make it clear how units and scale are depicted, or reference points or lines that contribute relevant external information. While one should avoid cluttering up a data graphic with excessive annotations, it is necessary to provide proper context.

Small multiples and layers

One of the fundamental challenges of creating data graphics is condensing multivariate information into a two-dimensional image. While three-dimensional images are occasionally useful, they are often more confusing than anything else. Instead, here are three common ways of incorporating more variables into a two-dimensional data graphic:

Small multiples Also known as *facets*, a single data graphic can be composed of several small multiples of the same basic plot, with one (discrete) variable changing in each of the small sub-images.

Layers It is sometimes appropriate to draw a new layer on top of an existing data graphic. This new layer can provide context or comparison, but there is a limit to how many layers humans can reliably parse.

Animation If time is the additional variable, then an animation can sometimes effectively convey changes in that variable. Of course, this doesn't work on the printed page, and makes it impossible for the user to see all the data at once.

2.2.2 Color

Color is one of the flashiest, but most misperceived and misused visual cues. In making color choices, there are a few key ideas that are important for any data scientist to understand.

First, as we saw above, color and its monochromatic cousin *shade* are two of the most poorly perceived visual cues. Thus, while potentially useful for a small number of levels of a categorical variable, color and shade are not particularly faithful ways to represent numerical variables—especially if small differences in those quantities are important to distinguish. This means that while color can be visually appealing to humans, it often isn't as informative as we might hope. For two numeric variables, it is hard to think of examples where color and shade would be more useful than position. Where color can be most effective is to represent a *third* or *fourth* numeric quantity on a scatterplot—once the two position cues have been exhausted.

Second, approximately 8 percent of the population—most of whom are men—have some form of color blindness. Most commonly, this renders them incapable of seeing colors accurately, most notably of distinguishing between red and green. Compounding the problem, many of these people do not know that they are color-blind. Thus, for professional graphics it is worth thinking carefully about which colors to use. The NFL famously failed to account for this in a 2015 game in which the Buffalo Bills wore all-red jerseys and the New York Jets wore all-green, leaving colorblind fans unable to distinguish one team from the other!

Pro Tip: Avoid contrasting red with green in data graphics (Bonus: your plots won't seem Christmas-y).

RdBu (divergent)

Figure 2.10: Diverging red-blue color palette.

Thankfully, we have been freed from the burden of having to create such intelligent palettes by the research of Cynthia Brewer, creator of the ColorBrewer website (and R package). Brewer has created colorblind-safe palettes in a variety of hues for three different types of numeric data in a single variable:

Sequential The ordering of the data has only one direction. Positive integers are sequential because they can only go up: they can't go past 0. (Thus, if 0 is encoded as white, then any darker shade of gray indicates a larger number.)

Diverging The ordering of the data has two directions. In an election forecast, we commonly see states colored based on how they are expected to vote for the president. Since red is associated with Republicans and blue with Democrats, states that are solidly red or blue are on opposite ends of the scale. But "swing states" that could go either way may appear purple, white, or some other neutral color that is "between" red and blue (see Figure 2.10).

Qualitative There is no ordering of the data, and we simply need color to differentiate different categories.

The `RColorBrewer` package provides functionality to use these palettes directly in R. Figure 2.11 illustrates the sequential, qualitative, and diverging palettes built into `RColorBrewer`.

Pro Tip: Take the extra time to use a well-designed color palette. Accept that those who work with color for a living will probably choose better colors than you.

2.2.3   Dissecting data graphics

With a little practice, one can learn to dissect data graphics in terms of the taxonomy outlined above. For example, your basic scatterplot uses *position* in the *Cartesian* plane with *linear* scales to show the relationship between two variables. In what follows, we identify the visual cues, coordinate system, and scale in a series of simple data graphics.

1. The bar graph in Figure 2.12 displays the average score on the math portion of the 1994–1995 SAT (with possible scores ranging from 200 to 800) among states for whom at least two-thirds of the students took the SAT.

   This plot uses the visual cue of *position* to represent the math SAT score on the vertical axis with a *linear* scale. The *categorical* variable of `state` is arrayed on the horizontal axis. Although the states are ordered alphabetically, it would not be appropriate to consider the `state` variable to be ordinal, since the ordering is not meaningful in the

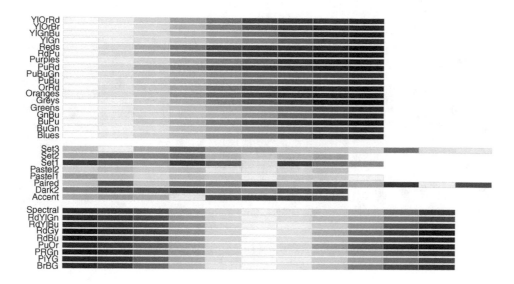

Figure 2.11: Palettes available through the RColorBrewer package.

context of math SAT scores. The coordinate system is *Cartesian*, although as noted previously, the horizontal coordinate is meaningless. Context is provided by the axis labels and title. Note also that since 200 is the minimum score possible on each section of the SAT, the vertical axis has been constrained to start at 200.

2. Next, we consider a time series that shows the progression of the world record times in the 100-meter freestyle swimming event for men and women. Figure 2.13 displays the times as a function of the year in which the new record was set.

   At some level this is simply a scatterplot that uses *position* on both the vertical and horizontal axes to indicate swimming time and chronological time, respectively, in a *Cartesian* plane. The numeric scale on the vertical axis is linear, in units of seconds, while the scale on the horizontal axis is also linear, measured in years. But there is more going on here. Color is being used as a visual cue to distinguish the categorical variable sex. Furthermore, since the points are connected by lines, *direction* is being used to indicate the progression of the record times. (In this case, the records can only get faster, so the direction is always down.) One might even argue that *angle* is being used to compare the descent of the world records across time and/or gender. In fact, in this case *shape* is also being used to distinguish sex.

3. Next, we present two pie charts in Figure 2.14 indicating the different substance of abuse for subjects in the Health Evaluation and Linkage to Primary Care (HELP) clinical trial. Each subject was identified with involvement with one primary substance (alcohol, cocaine, or heroin). On the right, we see the distribution of substance for housed (no nights in shelter or on the street) participants is fairly evenly distributed,

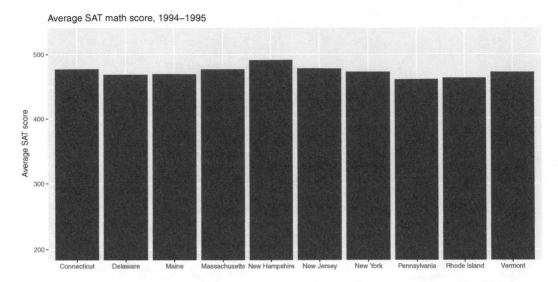

Figure 2.12: Bar graph of average SAT scores among states with at least two-thirds of students taking the test.

while on the left, we see the same distribution for those who were homeless one or more nights (more likely to have alcohol as their primary substance of abuse).

This graphic uses a *radial* coordinate system and the visual cue of *color* to distinguish the three levels of the *categorical* variable substance. The visual cue of *angle* is being used to quantify the differences in the proportion of patients using each substance. Are you able to accurately identify these percentages from the figure? The actual percentages are shown below.

Pro Tip: Don't use pie charts, except perhaps in small multiples.

```
          homeless
substance homeless housed
   alcohol    0.4928 0.3033
   cocaine    0.2823 0.3811
   heroin     0.2249 0.3156
```

This is a case where a simple table of these proportions is more effective at communicating the true differences than this—and probably any—data graphic. Note that there are only six data points presented, so any graphic is probably gratuitous.

4. Finally, in Figure 2.15 we present a *choropleth* map showing the population of Massachusetts by the 2010 Census tracts.

   Clearly, we are using a *geographic* coordinate system here, with latitude and longitude on the vertical and horizontal axes, respectively. (This plot is not projected: More information about projection systems is provided in Chapter 14.) *Shade* is once again being used to represent the quantity population, but here the scale is more complicated. The ten shades of blue have been mapped to the deciles of the census tract populations, and since the distribution of population across these tracts is right-skewed, each shade does not correspond to a range of people of the same width, but

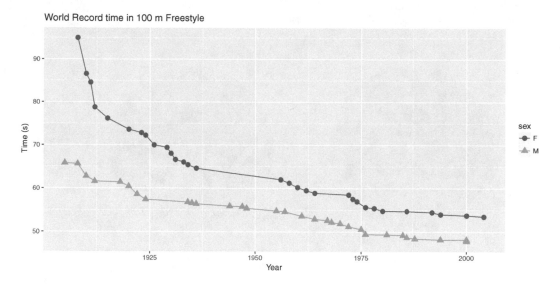

Figure 2.13: Scatterplot of world record time in 100-meter freestyle swimming.

rather to the same number of tracts that have a population in that range. Helpful context is provided by the title, subtitles, and legend.

2.3 Importance of data graphics: *Challenger*

On January 27th, 1986, engineers at Morton Thiokol, who supplied solid rocket motors (SRMs) to NASA for the space shuttle, recommended that NASA delay the launch of the space shuttle *Challenger* due to concerns that the cold weather forecast for the next day's launch would jeopardize the stability of the rubber O-rings that held the rockets together. These engineers provided 13 charts that were reviewed over a two-hour conference call involving the engineers, their managers, and NASA. The engineers' recommendation was overruled due to a lack of persuasive evidence, and the launch proceeded on schedule. The O-rings failed in exactly the manner the engineers had feared 73 seconds after launch, *Challenger* exploded, and all seven astronauts on board died [195].

In addition to the tragic loss of life, the incident was a devastating blow to NASA and the United States space program. The hand-wringing that followed included a two-and-a-half year hiatus for NASA and the formation of the Rogers Commission to study the disaster. What became clear is that the Morton Thiokol engineers had correctly identified the key causal link between *temperature* and *O-ring damage*. They did this using statistical data analysis combined with a plausible physical explanation: in short, that the rubber O-rings became brittle in low temperatures. (This link was famously demonstrated by legendary physicist and Rogers Commission member Richard Feynman during the hearings, using a glass of water and some ice cubes [195].) Thus, the engineers were able to identify the critical weakness using their *domain knowledge*—in this case, rocket science—and their data analysis. Their failure—and its horrific consequences—was one of persuasion: They simply did not present their evidence in a convincing manner to the NASA officials who ultimately made the decision to proceed with the launch. More than 30 years later this tragedy remains critically important. The evidence brought to the discussions about whether to launch was in the form of hand-written data tables (or "charts") but none were graphical. In his sweeping critique of the incident, Edward Tufte creates a powerful scatterplot similar to

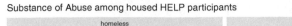

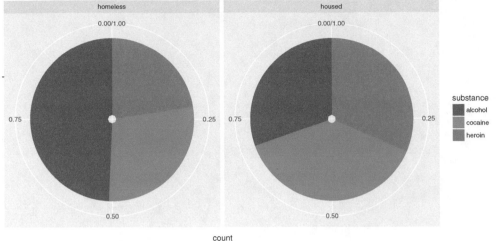

Figure 2.14: Pie charts showing the breakdown of substance of abuse among HELP study participants, faceted by homeless status.

the one shown in Figure 2.17, which can be derived from data that the engineers had at the time, but in a far more effective presentation [195].

Figure 2.16 indicates a clear relationship between the ambient temperature and O-ring damage on the solid rocket motors. To demonstrate the dramatic extrapolation made to the predicted temperature on January 27th, 1986, Tufte extended the horizontal axis in his scatterplot (Figure 2.17) to include the forecasted temperature. The huge gap makes plain the problem with extrapolation.

Tufte provided a full critique of the engineers' failures [195], many of which are instructive for data scientists.

Lack of authorship There were no names on any of the charts. This creates a lack of accountability. No single person was willing to take responsibility for the data contained in any of the charts. It is much easier to refute an argument made by a group of nameless people, than to a single or group of named people.

Univariate analysis The engineers provided several data tables, but all were essentially univariate. That is, they presented data on a single variable, but did not illustrate the relationship between two variables. Note that while Figure 2.18a does show data for two different variables, it is very hard to see the connection between the two in tabular form. Since the crucial connection here was between temperature and O-ring damage, this lack of bivariate analysis was probably the single most damaging omission in the engineers' presentation.

Anecdotal evidence With such a small sample size, anecdotal evidence can be particularly challenging to refute. In this case, a bogus comparison was made based on two observations. While the engineers argued that SRM-15 had the most damage on the coldest previous launch date (see Figure 2.17), NASA officials were able to counter that SRM-22 had the second-most damage on one of the warmer launch dates. These anecdotal pieces of evidence fall apart when all of the data are considered in context—in Figure 2.17 it is clear that SRM-22 is an outlier that deviates from the general pattern—but the engineers never presented all of the data in context.

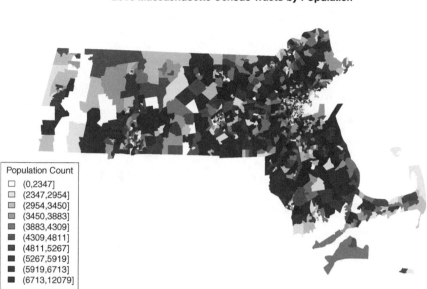

Figure 2.15: Choropleth map of population among Massachusetts Census tracts, based on 2010 U.S. Census.

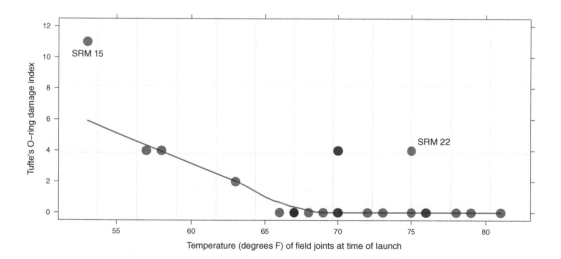

Figure 2.16: A scatterplot with smoother demonstrating the relationship between temperature and O-ring damage on solid rocket motors. The dots are semi-transparent, so that darker dots indicate multiple observations with the same values.

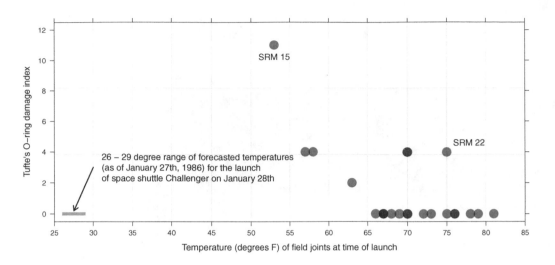

Figure 2.17: A recreation of Tufte's scatterplot demonstrating the relationship between temperature and O-ring damage on solid rocket motors.

Omitted data For some reason, the engineers chose not to present data from 22 other flights, which collectively represented 92% of launches. This may have been due to time constraints. This dramatic reduction in the accumulated evidence played a role in enabling the anecdotal evidence outlined above.

Confusion No doubt working against the clock, and most likely working in tandem, the engineers were not always clear about two different types of damage: *erosion* and *blow-by*. A failure to clearly define these terms may have hindered understanding on the part of NASA officials.

Extrapolation Most forcefully, the failure to include a simple scatterplot of the full data obscured the "stupendous extrapolation" [195] necessary to justify the launch. The bottom line was that the forecasted launch temperatures (between 26 and 29 degrees Fahrenheit) were so much colder than anything that had occurred previously, any model for O-ring damage as a function of temperature would be untested.

Pro Tip: When more than a handful of observations are present, data graphics are often more revealing than tables. Always consider alternative representations to improve communication.

Pro Tip: Always ensure that graphical displays are clearly described with appropriate axis labels, additional text descriptions, and a caption.

Tufte notes that the cardinal sin of the engineers was a failure to frame the data *in relation to what?* The notion that certain data may be understood in relation to something, is perhaps the fundamental and defining characteristic of statistical reasoning. We will follow this thread throughout the book.

We present this tragic episode in this chapter as motivation for a careful study of data visualization. It illustrates a critical truism for practicing data scientists: Being right isn't

HISTORY OF O-RING TEMPERATURES
(DEGREES - F)

MOTOR	MBT	AMB	O-RING	WIND
QM-4	68	36	47	10 MPH
QM-2	76	45	52	10 MPH
QM-3	72.5	40	48	10 MPH
QM-4	76	48	51	10 MPH
SRM-15	52	64	53	10 MPH
SRM-22	77	78	75	10 MPH
SRM-25	55	26	29 27	10 MPH 25 MPH

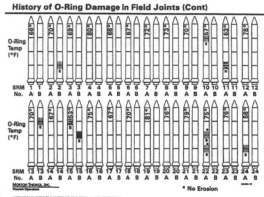

(a) One of the original 13 charts presented by Morton Thiokol engineers to NASA on the conference call the night before the Challenger launch. This is one of the more data-intensive charts.

(b) Evidence presented during the congressional hearings after the Challenger explosion. This is a classic example of "chartjunk."

Figure 2.18: Reprints of two Morton Thiokol data graphics. [195]

enough—you have to be *convincing*. Note that Figure 2.18b contains the same data that are present in Figure 2.17, but in a far less suggestive format. It just so happens that for most human beings, graphical explanations are particularly persuasive. Thus, to be a successful data analyst, one must master at least the basics of data visualization.

2.4 Creating effective presentations

Giving effective presentations is an important skill for a data scientist. Whether these presentations are in academic conferences, in a classroom, in a boardroom, or even on stage, the ability to communicate to an audience is of immeasurable value. While some people may be naturally more comfortable in the limelight, everyone can improve the quality of their presentations.

A few pieces of general advice are warranted [136]:

Budget your time You only have x minutes to talk, and usually 1 or 2 minutes to answer questions. If your talk runs too short or too long, it makes you seem unprepared. Rehearse your talk several times in order to get a better feel for your timing. Note also that you may have a tendency to talk faster during your actual talk than you will during your rehearsal. Talking faster in order to speed up is not a good strategy—you are much better off simply cutting material ahead of time. You will probably have a hard time getting through x slides in x minutes.

Pro Tip: Talking faster in order to speed up is not a good strategy—you are much better off simply cutting material ahead of time or moving to a key slide or conclusion.

Don't write too much on each slide You don't want people to have to read your slides, because if the audience is reading your slides, then they aren't listening to you. You want your slides to provide visual cues to the points that you are making—not substitute for your spoken words. Concentrate on graphical displays and bullet-pointed lists of ideas.

Put your problem in context Remember that (in most cases) most of your audience
will have little or no knowledge of your subject matter. The easiest way to lose people
is to dive right into technical details that require prior domain knowledge. Spend a
few minutes at the beginning of your talk introducing your audience to the most basic
aspects of your topic and presenting some motivation for what you are studying.

Speak loudly and clearly Remember that (in most cases) you know more about your
topic that anyone else in the room, so speak and act with confidence!

Tell a story, but not necessarily the whole story It is unrealistic to expect that you
can tell your audience everything that you know about your topic in x minutes. You
should strive to convey the big ideas in a clear fashion, but not dwell on the details.
Your talk will be successful if your audience is able to walk away with an understanding
of what your research question was, how you addressed it, and what the implications
of your findings are.

2.5   The wider world of data visualization

Thus far our discussion of data visualization has been limited to static, two-dimensional
data graphics. However, there are many additional ways to visualize data. While Chapter 3
focuses on static data graphics, Chapter 11 presents several cutting-edge tools for making
interactive data visualizations. Even more broadly, the field of visual analytics is concerned
with the science behind building interactive visual interfaces that enhance one's ability to
reason about data. Finally, we have data art.

You can do many things with data. On one end of the spectrum, you might be focused on
predicting the outcome of a specific response variable. In such cases, your goal is very well-
defined and your success can be quantified. On the other end of the spectrum are projects
called *data art*, wherein the meaning of what you are doing with the data is elusive, but
the experience of viewing the data in a new way is in itself meaningful.

Consider Memo Akten and Quayola's *Forms*, which was inspired by the physical movement of athletes in the Commonwealth Games. Through video analysis, these movements
were translated into 3D digital objects shown in Figure 2.19. Note how the image in the
upper-left is evocative of a swimmer surfacing after a dive. When viewed as a movie, *Forms*
is an arresting example of data art.

Successful data art projects require both artistic talent and technical ability. *Before Us is
the Salesman's House* is a live, continuously-updating exploration of the online marketplace
eBay. This installation was created by statistician Mark Hansen and digital artist Jer
Thorpe and is projected on a big screen as you enter eBay's campus. The display begins
by pulling up Arthur Miller's classic play *Death of a Salesman*, and "reading" the text
of the first chapter. Along the way, several nouns are plucked from the text (e.g., flute,
refrigerator, chair, bed, trophy, etc.). For each in succession, the display then shifts to a
geographic display of where things with that noun in the description are *currently* being
sold on eBay, replete with price and auction information. (Note that these descriptions are
not always perfect. In the video, a search for "refrigerator" turns up a T-shirt of former
Chicago Bears defensive end William "Refrigerator" Perry). Next, one city where such an
item is being sold is chosen, and any classic books of American literature being sold nearby
are collected. One is chosen, and the cycle returns to the beginning by "reading" the first
page of that book. This process continues indefinitely. When describing the exhibit, Hansen
spoke of "one data set reading another." It is this interplay of data and literature that makes
such data art projects so powerful.

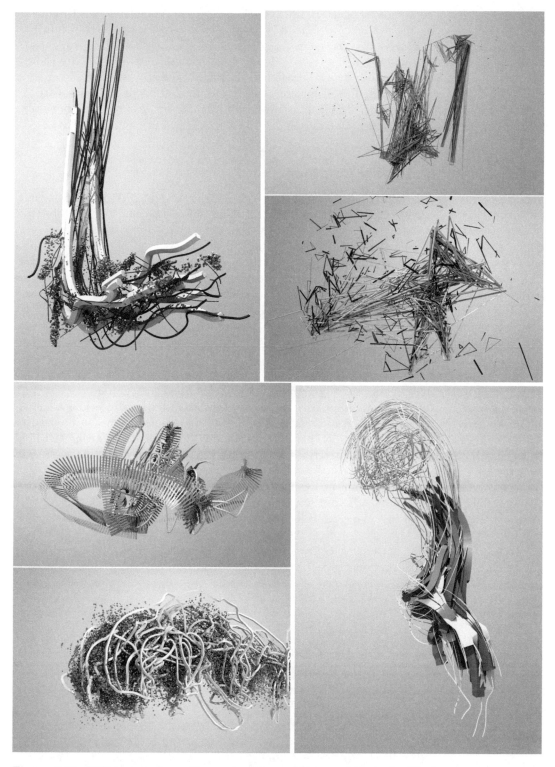

Figure 2.19: Still images from *Forms*, by Memo Akten and Quayola. Each image represents an athletic movement made by a competitor at the Commonwealth Games, but reimagined as a collection of moving 3D digital objects. Reprinted with permission.

Finally, we consider another Mark Hansen collaboration, this time with Ben Rubin and Michele Gorman. In *Shakespeare Machine*, 37 digital LCD blades—each corresponding to one of Shakespeare's plays—are arrayed in a circle. The display on each blade is a pattern of words culled from the text of these plays. First, pairs of hyphenated words are shown. Next, Boolean pairs (e.g., "good or bad") are found. Third, articles and adjectives modifying nouns (e.g., "the holy father"). In this manner, the artistic masterpieces of Shakespeare are shattered into formulaic chunks. In Chapter 15 we will learn how to use *regular expressions* to find the data for *Shakespeare Machine*.

2.6 Further resources

While issues related to data visualization pervade this entire text, they will be the particular focus of Chapters 3 (Data visualization II), 11 (Data visualization III), and 14 (Spatial data).

No education in data graphics is complete without reading Tufte's *Visual Display of Quantitative Information* [196], which also contains a description of John Snow's cholera map (see Chapter 14). For a full description of the *Challenger* incident, see *Visual Explanations* [195]. Tufte has also published two other landmark books [194, 198], as well as reasoned polemics about the shortcomings of PowerPoint [197]. Bill Cleveland's work on visual perception [55] provides the foundation for Yau's taxonomy [243]. Yau's text [242] provides many examples of thought-provoking data visualizations, particularly data art. The grammar of graphics was first described by Wilkinson [238]. Hadley Wickham implemented `ggplot2` based on this formulation [212].

Many important data graphics were developed by John Tukey [199]. Andrew Gelman [87] has also written persuasively about data graphics in statistical journals. Gelman discusses a set of canonical data graphics as well as Tufte's suggested modifications to them. Nolan and Perrett discuss data visualization assignments and rubrics that can be used to grade them [147]. Steven J. Murdoch has created some R functions for drawing the kind of modified diagrams that Tufte describes in [196]. These also appear in the `ggthemes` package [9].

Cynthia Brewer's color palettes are available at `http://colorbrewer2.org` and through the `RColorBrewer` package for R. Her work is described in more detail in [38, 39]. Wickham and others created the whimsical color palette that evokes Wes Anderson's distinctive movies [173].

Technically Speaking (Denison University) is an NSF-funded project for presentation advice that contains instructional videos for students [136].

2.7 Exercises

Exercise 2.1

What would a Cartesian plot that used colors to convey categorical values look like?

Exercise 2.2

Consider the two graphics related to *The New York Times* "Taxmageddon" article at `http://www.nytimes.com/2012/04/15/sunday-review/coming-soon-taxmageddon.html`. The first is "Whose Tax Rates Rose or Fell" and the second is "Who Gains Most From Tax Breaks."

1. Examine the two graphics carefully. Discuss what you think they convey. What story do the graphics tell?

2. Evaluate both graphics in terms of the taxonomy described in this chapter. Are the scales appropriate? Consistent? Clearly labelled? Do variable dimensions exceed data dimensions?

3. What, if anything, is misleading about these graphics?

Exercise 2.3

Choose *one* of the data graphics listed at `http://mdsr-book.github.io/exercises.html#exercise_23` and answer the following questions. Be sure to indicate which graphical display you picked.

1. Identify the visual cues, coordinate system, and scale(s).

2. How many variables are depicted in the graphic? Explicitly link each variable to a visual cue that you listed above.

3. Critique this data graphic using the taxonomy described in this chapter.

Exercise 2.4

Answer the following questions for each of the following collections of data graphics listed at `http://mdsr-book.github.io/exercises.html#exercise_24`.

Briefly (one paragraph) critique the designer's choices. Would you have made different choices? Why or why not?

Note: Each link contains a collection of many data graphics, and we don't expect (or want) you to write a dissertation on each individual graphic. But each collection shares some common stylistic elements. You should comment on a few things that you notice about the design of the collection.

Exercise 2.5

Consider one of the more complicated data graphics listed at `http://mdsr-book.github.io/exercises.html#exercise_25`.

1. What story does the data graphic tell? What is the main message that you take away from it?

2. Can the data graphic be described in terms of the taxonomy presented in this chapter? If so, list the visual cues, coordinate system, and scales(s) as you did in Problem 2(a). If not, describe the feature of this data graphic that lies outside of that taxonomy.

3. Critique and/or praise the visualization choices made by the designer. Do they work? Are they misleading? Thought-provoking? Brilliant? Are there things that you would have done differently? Justify your response.

Exercise 2.6

Consider the data graphic (`http://tinyurl.com/nytimes-unplanned`) about birth control methods.

1. What quantity is being shown on the y-axis of each plot?

2. List the variables displayed in the data graphic, along with the units and a few typical values for each.

3. List the visual cues used in the data graphic and explain how each visual cue is linked to each variable.

4. Examine the graphic carefully. Describe, in words, what *information* you think the data graphic conveys. Do not just summarize the *data*—interpret the data in the context of the problem and tell us what it means.

Chapter 3

A grammar for graphics

In Chapter 2, we presented a taxonomy for understanding data graphics. In this chapter, we illustrate how the ggplot2 package can be used to create data graphics. Other packages for creating static, two-dimensional data graphics in R include base graphics and the lattice system. We employ the ggplot2 system because it provides a unifying framework—a grammar—for describing and specifying graphics. The grammar for specifying graphics will allow the creation of custom data graphics that support visual display in a purposeful way. We note that while the terminology used in ggplot2 is not the same as the taxonomy we outlined in Chapter 2, there are many close parallels, which we will make explicit.

3.1   A grammar for data graphics

The ggplot2 package is one of the many creations of prolific R programmer Hadley Wickham. It has become one of the most widely-used R packages, in no small part because of the way it builds data graphics incrementally from small pieces of code.

In the grammar of ggplot2, an *aesthetic* is an explicit mapping between a variable and the visual cues that represent its values. A *glyph* is the basic graphical element that represents one case (other terms used include "mark" and "symbol"). In a scatterplot, the *positions* of a glyph on the plot—in both the horizontal and vertical senses—are the *visual cues* that help the viewer understand how big the corresponding quantities are. The *aesthetic* is the mapping that defines these correspondences. When more than two variables are present, additional aesthetics can marshal additional visual cues. Note also that some visual cues (like *direction* in a time series) are implicit and do not have a corresponding aesthetic.

For many of the chapters in this book, the first step in following these examples will be to load the mdsr package for R, which contains all of the data sets referenced in this book. In particular, loading mdsr also loads the mosaic package, which in turn loads dplyr and ggplot2. (For more information about the mdsr package see Appendix A. If you are using R for the first time, please see Appendix B for an introduction.)

```
library(mdsr)
```

Pro Tip: If you want to learn how to use a particular command, we highly recommend running the example code on your own.

We begin with a data set that includes measures that are relevant to answer questions about economic productivity. The `CIACountries` data table contains seven variables collected for each of 236 countries: population (pop), area (`area`), gross domestic product (gdp), percentage of GDP spent on education (educ), length of roadways per unit area (`roadways`), Internet use as a fraction of the population (`net_users`), and the number of barrels of oil produced per day (`oil_prod`). Table 3.1 displays a selection of variables for the first six countries.

country	oil_prod	gdp	educ	roadways	net_users
Afghanistan	0.00	1900.00		0.06	>5%
Albania	20510.00	11900.00	3.30	0.63	>35%
Algeria	1420000.00	14500.00	4.30	0.05	>15%
American Samoa	0.00	13000.00		1.21	
Andorra		37200.00		0.68	>60%
Angola	1742000.00	7300.00	3.50	0.04	>15%

Table 3.1: A selection of variables from the first six rows of the `CIACountries` data table.

3.1.1   Aesthetics

In the simple scatterplot shown in Figure 3.1, we employ the grammar of graphics to build a multivariate data graphic. In ggplot2, a plot is created with the `ggplot()` command, and any arguments to that function are applied across any subsequent plotting directives. In this case, this means that any variables mentioned anywhere in the plot are understood to be within the `CIACountries` data frame, since we have specified that in the data argument. Graphics in `ggplot2` are built incrementally by elements. In this case, the only elements are points, which are plotted using the `geom_point()` function. The arguments to `geom_point()` specify *where* and *how* the points are drawn. Here, the two *aesthetics* (`aes()`) map the vertical (y) coordinate to the gdp variable, and the horizontal (x) coordinate to the educ variable. The size argument to `geom_point()` changes the size of all of the glyphs. Note that here, every dot is the same size. Thus, size is *not* an aesthetic, since it does not map a variable to a visual cue. Since each case (i.e., row in the data frame) is a country, each dot represents one country.

In Figure 3.1 the glyphs are simple. Only position in the frame distinguishes one glyph from another. The shape, size, etc. of all of the glyphs are identical—there is nothing about the glyph itself that identifies the country.

However, it is possible to use a glyph with several attributes. We can define additional aesthetics to create new visual cues. In Figure 3.2, we have extended the previous example by mapping the color of each dot to the categorical `net_users` variable.

Changing the glyph is as simple as changing the function that draws that glyph—the aesthetic can often be kept exactly the same. In Figure 3.3, we plot text instead of a dot.

Of course, we can employ multiple aesthetics. There are four aesthetics in Figure 3.4. Each of the four aesthetics is set in correspondence with a variable—we say the variable is *mapped* to the aesthetic. Educational attainment is being mapped to horizontal position, GDP to vertical position, Internet connectivity to color, and length of roadways to size. Thus, we encode four variables (gdp, educ, net_users, and roadways) using the visual cues of position, position, color, and area, respectively.

A data table provides the basis for drawing a data graphic. The relationship between a data table and a graphic is simple: Each case in the data table becomes a mark in the graph (we will return to the notion of *glyph-ready data* in Chapter 5). As the designer of

```
g <- ggplot(data = CIACountries, aes(y = gdp, x = educ))
g + geom_point(size = 3)
```

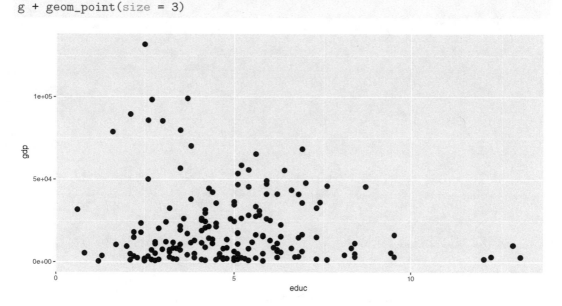

Figure 3.1: Scatterplot using only the position aesthetic for glyphs.

```
g + geom_point(aes(color = net_users), size = 3)
```

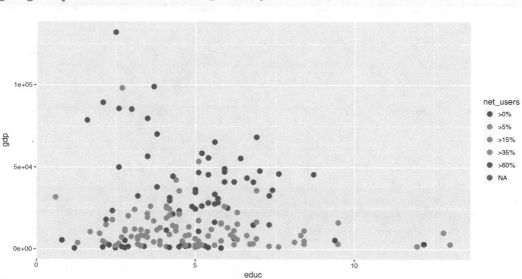

Figure 3.2: Scatterplot in which net_users is mapped to color.

```
g + geom_text(aes(label = country, color = net_users), size = 3)
```

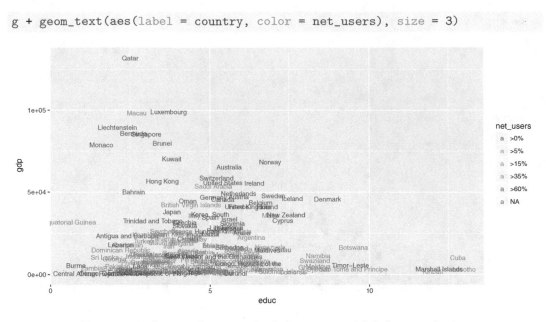

Figure 3.3: Scatterplot using both location and label as aesthetics.

```
g + geom_point(aes(color = net_users, size = roadways))
```

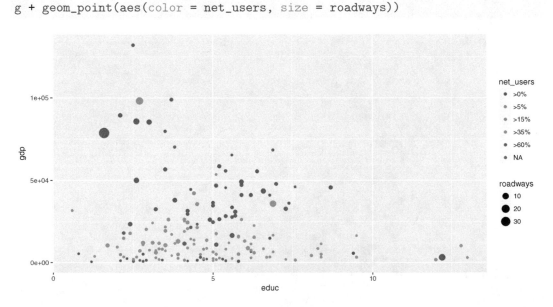

Figure 3.4: Scatterplot in which net_users is mapped to color and roadways mapped to size. Compare this graphic to Figure 3.6, which displays the same data using facets.

the graphic, you choose which variables the graphic will display and how each variable is to be represented graphically: position, size, color, and so on.

3.1.2 Scale

Compare Figure 3.4 to Figure 3.5. In the former, it is hard to discern differences in GDP due to its right-skewed distribution and the choice of a *linear* scale. In the latter, the *logarithmic* scale on the vertical axis makes the scatterplot more readable. Of course, this makes interpreting the plot more complex, so we must be very careful when doing so. Note that the only difference in the code is the addition of the coord_trans() directive.

```
g + geom_point(aes(color = net_users, size = roadways)) +
  coord_trans(y = "log10")
```

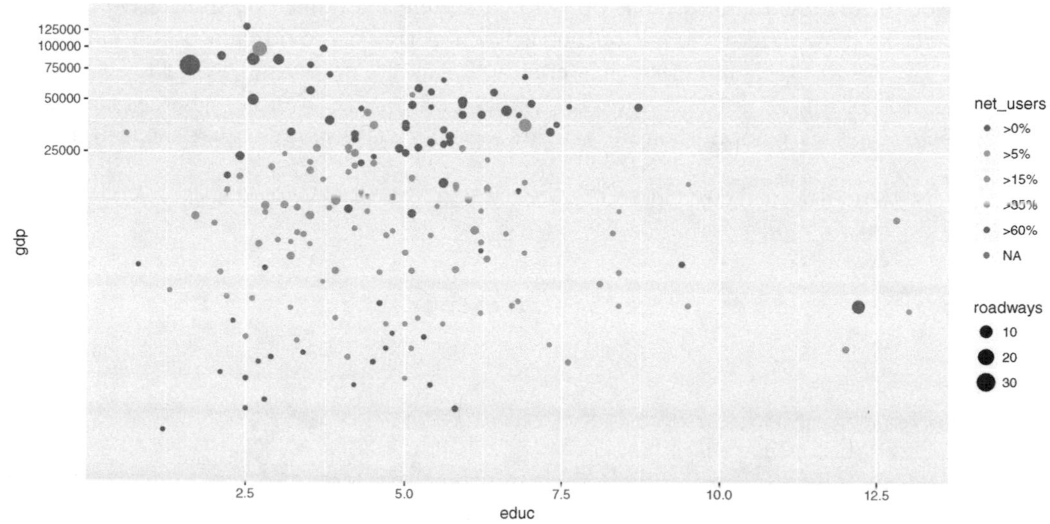

Figure 3.5: Scatterplot using a logarithmic transformation of GDP that helps to mitigate visual clustering caused by the right-skewed distribution of GDP among countries.

Scales can also be manipulated in ggplot2 using any of the scale() functions. For example, instead of using the coord_trans() function as we did above, we could have achieved a similar plot through the use of the scale_y_continuous() function, as illustrated below. In either case, the points will be drawn in the same location—the difference in the two plots is how and where the major tick marks and axis labels are drawn. We prefer to use coord_trans() in Figure 3.5 because it draws attention to the use of the log scale. Similarly named functions (e.g., scale_x_continuous(), scale_x_discrete(), scale_color(), etc.) perform analogous operations on different aesthetics.

```
g + geom_point(aes(color = net_users, size = roadways)) +
  scale_y_continuous(name = "Gross Domestic Product", trans = "log10")
```

Not all scales are about position. For instance, in Figure 3.4, net_users is translated to color. Similarly, roadways is translated to size: the largest dot corresponds to a value of five roadways per unit area.

3.1.3   Guides

Context is provided by *guides* (more commonly called legends). A guide helps a human reader to understand the meaning of the visual cues by providing context.

For position visual cues, the most common sort of guide is the familiar axis with its tick marks and labels. But other guides exist. In Figures 3.4 and 3.5, legends relate how dot color corresponds to Internet connectivity, and how dot size corresponds to length of roadways (note the use of a log scale). The geom_text() and geom_annotate() functions can also be used to provide specific textual annotations on the plot. Examples of how to use these functions for annotations are provide in Section 3.3.

3.1.4   Facets

Using multiple aesthetics such as shape, color, and size to display multiple variables can produce a confusing, hard-to-read graph. *Facets*—multiple side-by-side graphs used to display levels of a categorical variable—provide a simple and effective alternative. Figure 3.6 uses facets to show different levels of Internet connectivity, providing a better view than Figure 3.4. There are two functions that create facets: facet_wrap() and facet_grid(). The former creates a facet for each level of a single categorical variable, whereas the latter creates a facet for each combination of two categorical variables, arranging them in a grid.

```
g + geom_point(alpha = 0.9, aes(size = roadways)) + coord_trans(y="log10") +
   facet_wrap(~net_users, nrow = 1) + theme(legend.position = "top")
```

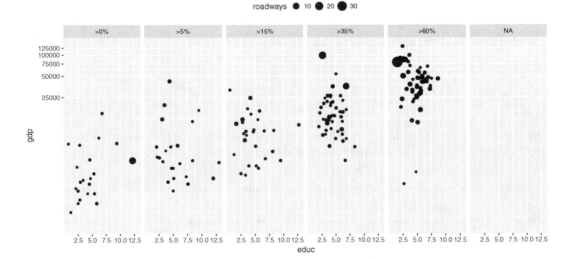

Figure 3.6: Scatterplot using facets for different ranges of Internet connectivity.

3.1.5   Layers

On occasion, data from more than one data table are graphed together. For example, the MedicareCharges and MedicareProviders data tables provide information about the average cost of each medical procedure in each state. If you live in New Jersey, you might wonder how providers in your state charge for different medical procedures. However, you will certainly want to understand those averages in the context of the averages across all states. In the `MedicareCharges` table, each row represents a different medical procedure (`drg`) with its associated average cost in each state. We also create a second data table called `ChargesNJ`, which contains only those rows corresponding to providers in the state of New Jersey. Do not worry if these commands aren't familiar—we will learn these in Chapter 4.

```
data(MedicareCharges)
ChargesNJ <- MedicareCharges %>% filter(stateProvider == "NJ")
```

The first few rows from the data table for New Jersey are shown in Table 3.2. This glyph-ready table (see Chapter 5) can be translated to a chart (Figure 3.7) using bars to represent the average charges for different medical procedures in New Jersey. The `geom_bar()` function creates a separate bar for each of the 100 different medical procedures.

drg	stateProvider	num_charges	mean_charge
039	NJ	31	35103.81
057	NJ	55	45692.07
064	NJ	55	87041.64
065	NJ	59	59575.74
066	NJ	56	45819.13
069	NJ	61	41916.70
074	NJ	41	42992.81
101	NJ	58	42314.18
149	NJ	50	34915.54
176	NJ	36	58940.98

Table 3.2: Glyph-ready data for the barplot layer in Figure 3.7.

How do the charges in New Jersey compare to those in other states? The two data tables, one for New Jersey and one for the whole country, can be plotted with different glyph types: bars for New Jersey and dots for the states across the whole country as in Figure 3.8. With the context provided by the individual states, it is easy to see that the charges in New Jersey are among the highest in the country for each medical procedure.

3.2 Canonical data graphics in R

Over time, statisticians have developed standard data graphics for specific use cases [199]. While these data graphics are not always mesmerizing, they are hard to beat for simple effectiveness. Every data scientist should know how to make and interpret these canonical data graphics—they are ignored at your peril.

3.2.1 Univariate displays

It is generally useful to understand how a single variable is distributed. If that variable is numeric, then its distribution is commonly summarized graphically using a *histogram* or *density plot*. Using the `ggplot2` package, we can display either plot for the `Math` variable in the `SAT_2010` data frame by binding the `Math` variable to the x aesthetic.

```
g <- ggplot(data = SAT_2010, aes(x = math))
```

```
p <- ggplot(data = ChargesNJ,
            aes(x = reorder(drg, mean_charge), y = mean_charge)) +
   geom_bar(fill = "gray", stat = "identity") +
   ylab("Statewide Average Charges ($)") + xlab("Medical Procedure (DRG)") +
   theme(axis.text.x = element_text(angle = 90, hjust = 1))
p
```

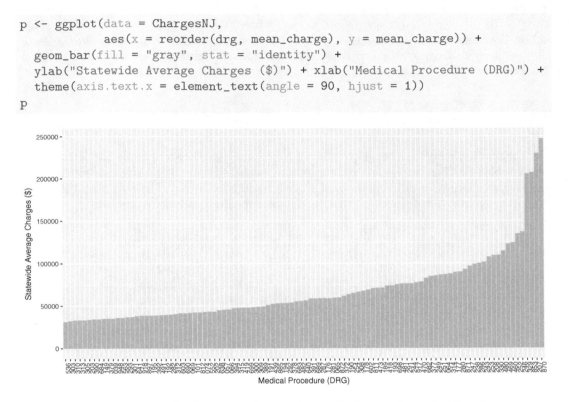

Figure 3.7: Bar graph of average charges for medical procedures in New Jersey.

```
p + geom_point(data = MedicareCharges, size = 1, alpha = 0.3)
```

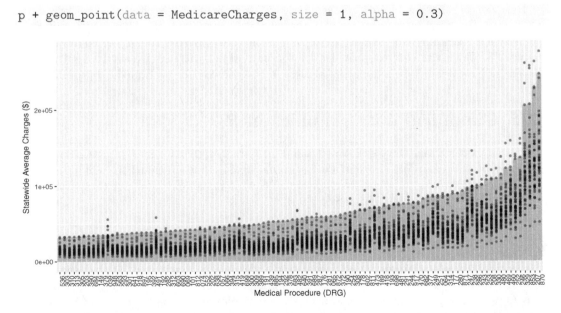

Figure 3.8: Bar graph adding a second layer to provide a comparison of New Jersey to other states. Each dot represents one state, while the bars represent New Jersey.

```
g + geom_histogram(binwidth = 10)
```

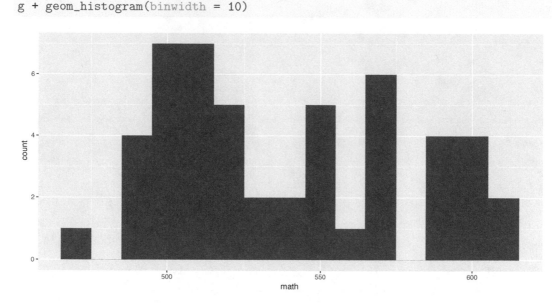

Figure 3.9: Histogram showing the distribution of Math SAT scores by state.

Then we only need to choose either geom_histogram() or geom_density(). Both Figures 3.9 and 3.10 convey the same information, but whereas the histogram uses pre-defined bins to create a discrete distribution, a density plot uses a kernel smoother to make a continuous curve.

Note that the binwidth argument is being used to specify the width of bins in the histogram. Here, each bin contains a ten–point range of SAT scores. In general, the appearance of a histogram can vary considerably based on the choice of bins, and there is no one "best" choice. You will have to decide what bin width is most appropriate for your data.

Similarly, in the density plot shown in Figure 3.10 we use the adjust argument to modify the *bandwidth* being used by the kernel smoother. In the taxonomy defined above, a density plot uses position and direction in a Cartesian plane with a horizontal scale defined by the units in the data.

If your variable is categorical, it doesn't make sense to think about the values as having a continuous density. Instead, we can use *bar graphs* to display the distribution of a categorical variable. To make a simple bar graph for math, identifying each bar by the label state, we use the geom_bar() command, as displayed in Figure 3.11. Note that we add a few wrinkles to this plot. First, we use the head() function to display only the first 10 states (in alphabetical order). Second, we use the reorder() function to sort the state names in order of their average math SAT score. Third, we set the stat argument to identity to force ggplot2 to use the y aesthetic, which is mapped to math.

As noted earlier, we recommend against the use of pie charts to display the distribution of a categorical variable since, in most cases, a table of frequencies is more informative. An informative graphical display can be achieved using a *stacked bar plot*, such as the one shown in Figure 3.12. Note that we have used the coord_flip() function to display the bars horizontally instead of vertically.

This method of graphical display enables a more direct comparison of proportions than would be possible using two pie charts. In this case, it is clear that homeless participants were more likely to identify as being involved with alcohol as their primary substance of

```
g + geom_density(adjust = 0.3)
```

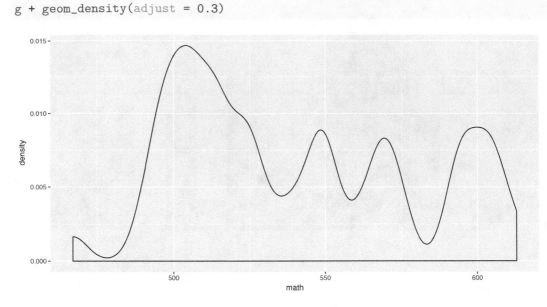

Figure 3.10: Density plot showing the distribution of Math SAT scores by state.

```
ggplot(data = head(SAT_2010, 10), aes(x = reorder(state, math), y = math)) +
  geom_bar(stat = "identity")
```

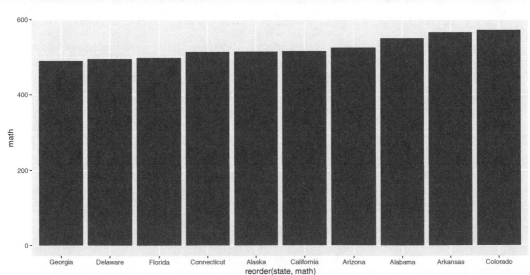

Figure 3.11: A bar plot showing the distribution of Math SAT scores for a selection of states.

```
ggplot(data = HELPrct, aes(x = homeless)) +
  geom_bar(aes(fill = substance), position = "fill") +
  coord_flip()
```

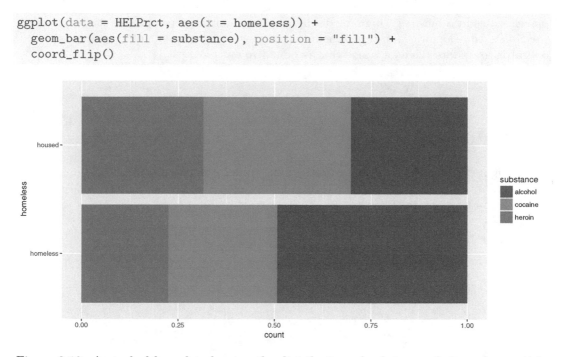

Figure 3.12: A stacked bar plot showing the distribution of substance of abuse for participants in the HELP study. Compare this to Figure 2.14.

abuse. However, like pie charts, bar charts are sometimes criticized for having a low data-to-ink ratio. That is, they use a comparatively large amount of ink to depict relatively few data points.

3.2.2 Multivariate displays

Multivariate displays are the most effective way to convey the relationship between more than one variable. The venerable *scatterplot* remains an excellent way to display observations of two quantitative (or numerical) variables. The scatterplot is provided in `ggplot2` by the `geom_point()` command. The main purpose of a scatterplot is to show the relationship between two variables across many cases. Most often, there is a Cartesian coordinate system in which the x-axis represents one variable and the y-axis the value of a second variable.

```
g <- ggplot(data = SAT_2010, aes(x = expenditure, y = math)) + geom_point()
```

We will also add a smooth trend line and some more specific axis labels.

```
g <- g + geom_smooth(method = "lm", se = 0) +
  xlab("Average expenditure per student ($1000)") +
  ylab("Average score on math SAT")
```

In Figures 3.13 and 3.14 we plot the relationship between the average SAT math score and the expenditure per pupil (in thousands of United States dollars) among states in 2010. A third (categorical) variable can be added through *faceting* and/or *layering*. In this case, we use the `mutate()` function (see Chapter 4) to create a new variable called `SAT_rate` that places states into bins (e.g., high, medium, low) based on the percentage of students taking the SAT. Additionally, in order to include that new variable in our plots, we use the %+% operator to update the data frame that is bound to our plot.

```
SAT_2010 <- SAT_2010 %>%
  mutate(SAT_rate = cut(sat_pct, breaks = c(0,30,60,100),
    labels = c("low", "medium", "high")))
g <- g %+% SAT_2010
```

In Figure 3.13, we use the color aesthetic to separate the data by SAT_rate on a single plot (i.e., layering). Compare this with Figure 3.14 where we add a facet_wrap() mapped to SAT_rate to separate by facet.

```
g + aes(color = SAT_rate)
```

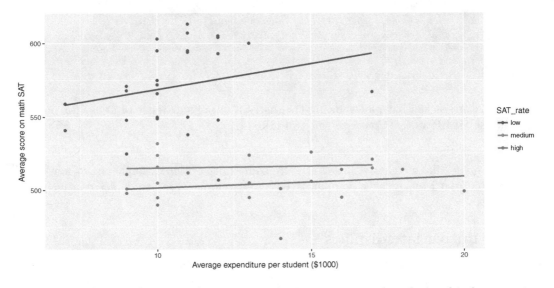

Figure 3.13: Scatterplot using the color aesthetic to separate the relationship between two numeric variables by a third categorical variable.

Note for these two plots we have used the geom_smooth() function in order to plot the simple linear regression line (method = "lm") through those points (see Section 7.6 and Appendix E).

The NHANES data table provides medical, behavioral, and morphometric measurements of individuals. The scatterplot in Figure 3.15 shows the relationship between two of the variables, height and age. Each dot represents one person and the position of that dot signifies the value of the two variables for that person. Scatterplots are useful for visualizing a simple relationship between two variables. For instance, you can see in Figure 3.15 the familiar pattern of growth in height from birth to the late teens.

Some scatterplots have special meanings. A *time series*—such as the one shown in Figure 3.16—is just a scatterplot with time on the horizontal axis and points connected by lines to indicate temporal continuity. In Figure 3.16, the temperature at a weather station in western Massachusetts is plotted over the course of the year. The familiar fluctuations based on the seasons are evident. Be especially aware of dubious causality in these plots: Is time really a good explanatory variable?

```
g + facet_wrap(~ SAT_rate)
```

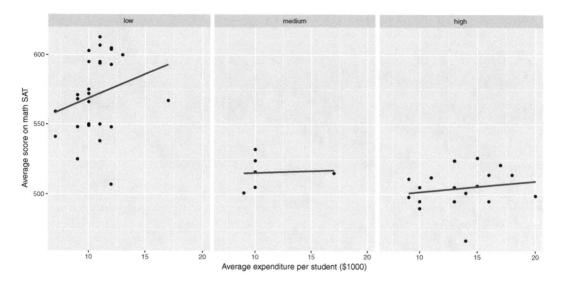

Figure 3.14: Scatterplot using a `facet_wrap()` to separate the relationship between two numeric variables by a third categorical variable.

```
library(NHANES)
ggplot(data = sample_n(NHANES, size = 1000),
  aes(x = Age, y = Height, color = Gender)) +
  geom_point() + geom_smooth() + xlab("Age (years)") + ylab("Height (cm)")
```

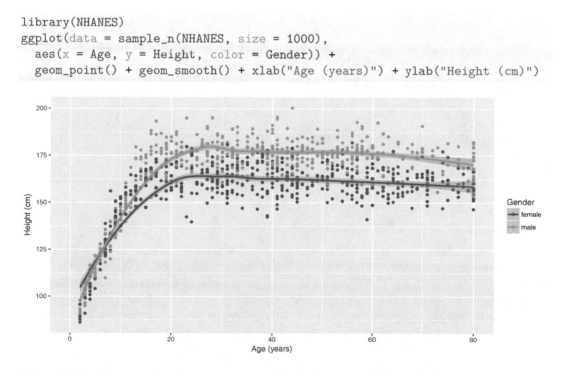

Figure 3.15: A scatterplot for 1,000 random individuals from the NHANES study. Note how mapping gender to color illuminates the differences in height between men and women.

```
library(macleish)
ggplot(data = whately_2015, aes(x = when, y = temperature)) +
  geom_line(color = "darkgray") + geom_smooth() +
  xlab(NULL) + ylab("Temperature (degrees Fahrenheit)")
```

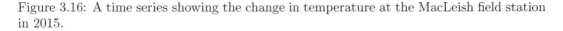

Figure 3.16: A time series showing the change in temperature at the MacLeish field station in 2015.

For displaying a numerical response variable against a categorical explanatory variable, a common choice is a *box-and-whisker* (or box) plot, as shown in Figure 3.17. It may be easiest to think about this as simply a graphical depiction of the five-number summary (minimum, Q1, median, Q3, and maximum).

```
favstats(length ~ sex, data = KidsFeet)

  sex  min    Q1 median    Q3  max  mean    sd  n missing
1   B 22.9 24.35  24.95  25.8 27.5 25.11 1.217 20       0
2   G 21.6 23.65  24.20  25.1 26.7 24.32 1.330 19       0
```

When both the explanatory and response variables are categorical (or binned), points and lines don't work as well. How likely is a person to have diabetes, based on their age and BMI (body mass index)? In the *mosaicplot* (or eikosogram) shown in Figure 3.18 the number of observations in each cell is proportional to the area of the box. Thus, you can see that diabetes tends to be more common for older people as well as for those who are obese, since the blue shaded regions are larger than expected under an independence model while the pink are less than expected. These provide a more accurate depiction of the intuitive notions of probability familiar from Venn diagrams [152].

In Table 3.3 we summarize the use of `ggplot2` plotting commands and their relationship to canonical data graphics. Note that the `mosaicplot()` function is not part of `ggplot2`, but rather is available through the built-in graphics system.

```
ggplot(data = KidsFeet, aes(x = sex, y = length)) + geom_boxplot()
```

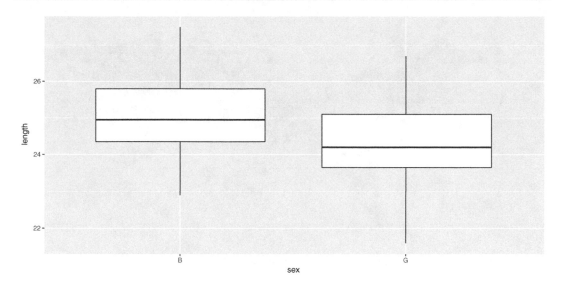

Figure 3.17: A box-and-whisker plot showing the distribution of foot length by gender for 39 children.

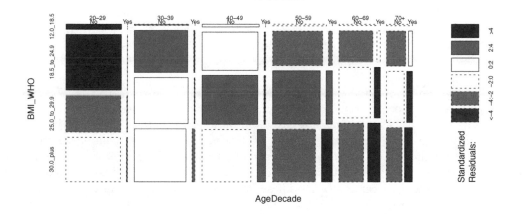

Figure 3.18: Mosaic plot (eikosogram) of diabetes by age and weight status (BMI).

response (y)	explanatory (x)	plot type	ggplot2 geom()
	numeric	histogram, density	geom_histogram, geom_density()
	categorical	stacked bar	geom_bar()
numeric	numeric	scatter	geom_point()
numeric	categorical	box	geom_boxplot()
categorical	categorical	mosaic	graphics::mosaicplot()

Table 3.3: Table of canonical data graphics and their corresponding ggplot2 commands. Note that mosaicplot() is not part of the ggplot2 package.

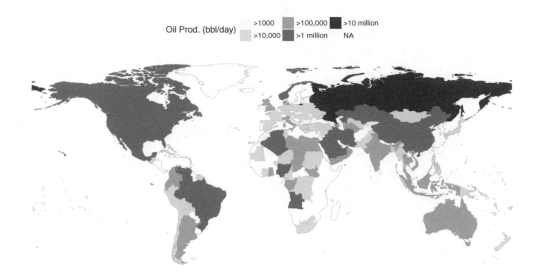

Figure 3.19: A choropleth map displaying oil production by countries around the world in barrels per day.

3.2.3   Maps

Using a map to display data geographically helps both to identify particular cases and to show spatial patterns and discrepancies. In Figure 3.19, the shading of each country represents its oil production. This sort of map, where the fill color of each region reflects the value of a variable, is sometimes called a *choropleth* map. We will learn more about mapping and how to work with spatial data in Chapter 14.

3.2.4   Networks

A *network* is a set of connections, called *edges*, between nodes, called *vertices*. A vertex represents an entity. The edges indicate pairwise relationships between those entities.

The NCI60 data set is about the genetics of cancer. The data set contains more than 40,000 probes for the expression of genes, in each of 60 cancers. In the network displayed in Figure 3.20, a vertex is a given cell line, and each is depicted as a dot. The dot's color and label gives the type of cancer involved. These are ovarian, colon, central nervous system, melanoma, renal, breast, and lung cancers. The edges between vertices show pairs of cell lines that had a strong correlation in gene expression.

The network shows that the melanoma cell lines (ME) are closely related to each other but not so much to other cell lines. The same is true for colon cancer cell lines (CO) and for central nervous system (CN) cell lines. Lung cancers, on the other hand, tend to have associations with multiple other types of cancers. We will explore the topic of network science in greater depth in Chapter 16.

3.3   Extended example: Historical baby names

For many of us, there are few things that are more personal than your name. It is impossible to remember a time when you didn't have your name, and you carry it with you wherever you go. You instinctively react when you hear it. And yet, you didn't choose your name—your parents did (unless you've legally changed your name).

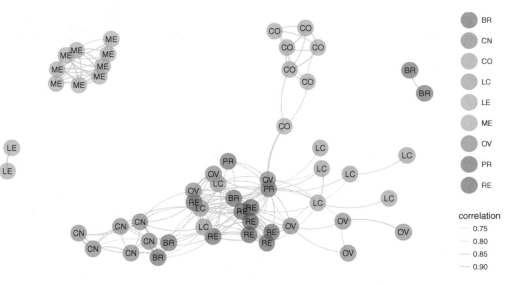

Figure 3.20: A network diagram displaying the relationship between types of cancer cell lines.

How do parents go about choosing names? Clearly, there seem to be both short and long-term trends in baby names. The popularity of the name "Bella" spiked after the lead character in *Twilight* became a cultural phenomenon. Other once-popular names seem to have fallen out of favor—writers at FiveThirtyEight asked, "where have all the Elmer's gone?"

Using data from the `babynames` package, which uses public data from the Social Security Administration (SSA), we can re-create many of the plots presented in the FiveThirtyEight blog post, and in the process learn how to use `ggplot2` to make production-quality data graphics.

In Figure 3.21, we have reprinted an informative, annotated FiveThirtyEight data graphic that shows the relative ages of American males named "Joseph." Drawing on what you have learned in Chapter 2, take a minute to jot down the visual cues, coordinate system, scales, and context present in this plot. This diagnosis will facilitate our use of `ggplot2` to re-construct it.

The key insight of the FiveThirtyEight work is the estimation of the number of people with each name who are currently alive. The `lifetables` table from the `babynames` package contains actuarial estimates of the number of people per 100,000 who are alive at age x, for every $0 \leq x \leq 114$. The `make_babynames_dist()` function in the `mdsr` package adds some more convenient variables and filters for only the data that is relevant to people alive in 2014.[1]

```
library(babynames)
BabynamesDist <- make_babynames_dist()
head(BabynamesDist, 2)

# A tibble: 2 9
   year   sex  name     n    prop alive_prob count_thousands age_today
  <dbl> <chr> <chr> <int>   <dbl>      <dbl>           <dbl>     <dbl>
```

---

[1]See the SSA documentation `https://www.ssa.gov/oact/NOTES/as120/LifeTables_Body.html` for more information.

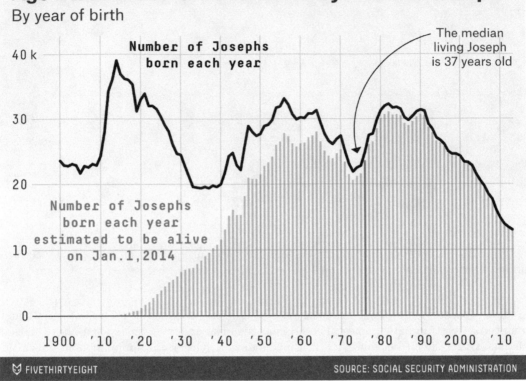

Figure 3.21: Popularity of the name "Joseph" as constructed by FiveThirtyEight.

```
1   1900      F  Mary 16707 0.05257          0            16.707       114
2   1900      F Helen  6343 0.01996          0             6.343       114
# ... with 1 more variables: est_alive_today <dbl>
```

To find information about a specific name, we can just use the `filter()` function.

```
BabynamesDist %>% filter(name == "Benjamin")
```

3.3.1   Percentage of people alive today

What was your diagnosis of Figure 3.21? There are two main data elements in that plot: a thick black line indicating the number of Josephs born each year, and the thin light blue bars indicating the number of Josephs born in each year that are expected to still be alive today. In both cases, the vertical axis corresponds to the number of people (in thousands), and the horizontal axis corresponds to the year of birth.

We can compose a similar plot in `ggplot2`. First we take the relevant subset of the data and set up the initial `ggplot2` object. The data frame `joseph` is bound to the plot, since this contains all of the data that we need for this plot, but we will be using it with multiple geoms. Moreover, the `year` variable is mapped to the x-axis as an aesthetic. This will ensure that everything will line up properly.

```
joseph <- BabynamesDist %>%
  filter(name == "Joseph" & sex == "M")
name_plot <- ggplot(data = joseph, aes(x = year))
```

Next, we will add the bars.

```
name_plot <- name_plot +
  geom_bar(stat = "identity", aes(y = count_thousands * alive_prob),
            fill = "#b2d7e9", colour = "white")
```

The geom_bar() function adds bars, which are filled with a light blue color and a white border. The height of the bars is an aesthetic that is mapped to the estimated number of people alive today who were born in each year. The stat argument is set to identity, since we want the actual y values to be used—not the number of each (which is the default). The black line is easily added using the geom_line() function.

```
name_plot <- name_plot + geom_line(aes(y = count_thousands), size = 2)
```

Adding an informative label for the vertical axis and removing an uninformative label for the horizontal axis will improve the readability of our plot.

```
name_plot <- name_plot +
  ylab("Number of People (thousands)") + xlab(NULL)
```

Inspecting the summary() of our plot at this point can help us keep things straight. Does this accord with what you jotted down previously?

```
summary(name_plot)

data: year, sex, name, n, prop, alive_prob, count_thousands,
  age_today, est_alive_today [111x9]
mapping:  x = year
faceting: <ggproto object: Class FacetNull, Facet>
    compute_layout: function
    draw_back: function
    draw_front: function
    draw_labels: function
    draw_panels: function
    finish_data: function
    init_scales: function
    map: function
    map_data: function
    params: list
    render_back: function
    render_front: function
    render_panels: function
    setup_data: function
    setup_params: function
    shrink: TRUE
    train: function
    train_positions: function
```

```
    train_scales: function
    vars: function
    super:  <ggproto object: Class FacetNull, Facet>
------------------------------------
mapping: y = count_thousands * alive_prob
geom_bar: width = NULL, na.rm = FALSE
stat_identity: na.rm = FALSE
position_stack

mapping: y = count_thousands
geom_line: na.rm = FALSE
stat_identity: na.rm = FALSE
position_identity
```

The final data-driven element of Figure 3.21 is a darker blue bar indicating the median year of birth. We can compute this with the `wtd.quantile()` function in the `Hmisc` package. Setting the `probs` argument to 0.5 will give us the median year of birth, weighted by the number of people estimated to be alive today (`est_alive_today`).

```
wtd.quantile <- Hmisc::wtd.quantile
median_yob <-
  with(joseph, wtd.quantile(year, est_alive_today, probs = 0.5))
median_yob
```

```
 50%
1975
```

We can then overplot a single bar in a darker shade of blue. Here, we are using the `ifelse()` function cleverly. If the year is equal to the median year of birth, then the height of the bar is the estimated number of Josephs alive today. Otherwise, the height of the bar is zero (so you can't see it at all). In this manner we plot only the one darker blue bar that we want to highlight.

```
name_plot <- name_plot +
  geom_bar(stat = "identity", colour = "white", fill = "#008fd5",
           aes(y = ifelse(year == median_yob, est_alive_today / 1000, 0)))
```

Lastly, Figure 3.21 contains many contextual elements specific to the name Joseph. We can add a title, annotated text, and an arrow providing focus to a specific element of the plot. Figure 3.22 displays our reproduction of Figure 3.21. There are a few differences in the presentation of fonts, title, etc. These can be altered using ggplot2's theming framework, but we won't explore these subtleties here (see Section 11.4).[2]

```
name_plot +
  ggtitle("Age Distribution of American Boys Named Joseph") +
  geom_text(x = 1935, y = 40, label = "Number of Josephs\nborn each year") +
```

---

[2]You may note that our number of births per year are lower than FiveThirtyEight's beginning in about 1940. It is explained in a footnote in their piece that some of the SSA records are incomplete for privacy reasons, and thus they pro-rated their data based on United States Census estimates for the early years of the century. We have omitted this step, but the `births` table in the babynames package will allow you to perform it.

```
geom_text(x = 1915, y = 13, label =
  "Number of Josephs\nborn each year\nestimated to be alive\non 1/1/2014",
  colour = "#b2d7e9") +
geom_text(x = 2003, y = 40,
  label = "The median\nliving Joseph\nis 37 years old",
        colour = "darkgray") +
geom_curve(x = 1995, xend = 1974, y = 40, yend = 24,
  arrow = arrow(length = unit(0.3,"cm")), curvature = 0.5) + ylim(0, 42)
```

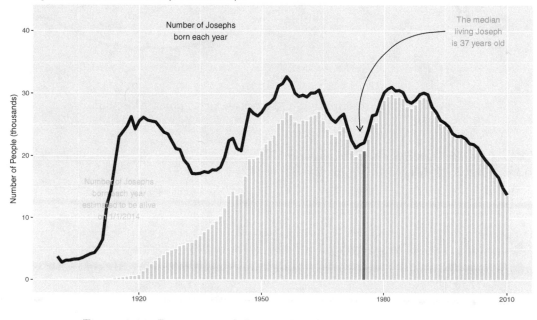

Figure 3.22: Recreation of the age distribution of "Joseph" plot.

Notice that we did not update the `name_plot` object with this contextual information. This was intentional, since we can update the `data` argument of `name_plot` and obtain an analogous plot for another name. This functionality makes use of the special `%+%` operator. As shown in Figure 3.23, the name "Josephine" enjoyed a spike in popularity around 1920 that later subsided.

```
name_plot %+% filter(BabynamesDist, name == "Josephine" & sex == "F")
```

While some names are almost always associated with a particular gender, many are not. More interestingly, the proportion of people assigned male or female with a given name often varies over time. These data were presented nicely by Nathan Yau at FlowingData.

We can compare how our `name_plot` differs by gender for a given name using a *facet*. To do this, we will simply add a call to the `facet_wrap()` function, which will create small multiples based on a single categorical variable, and then feed a new data frame to the plot that contains data for both sexes. In Figure 3.24, we show the prevalence of "Jessie" changed for the two sexes.

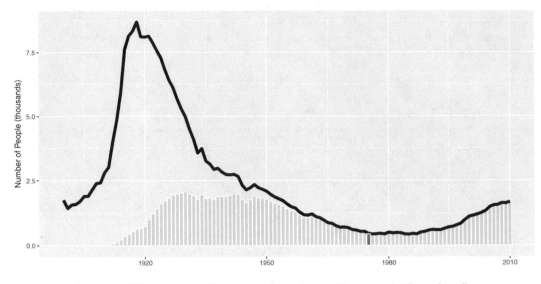

Figure 3.23: Age distribution of American girls named "Josephine".

```
names_plot <- name_plot + facet_wrap(~sex)
names_plot %+% filter(BabynamesDist, name == "Jessie")
```

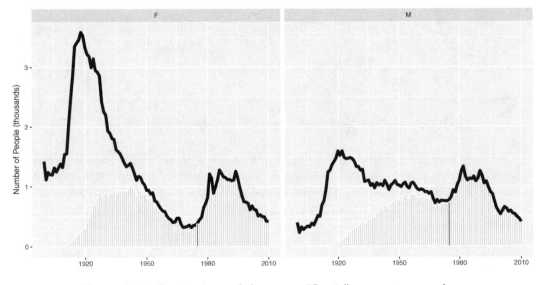

Figure 3.24: Comparison of the name "Jessie" across two genders.

The plot at FlowingData shows the 35 most common "unisex" names—that is, the names that have historically had the greatest balance between males and females. We can use a facet_grid() to compare the gender breakdown for a few of the most common of these, as shown in Figures 3.25 and 3.26.

```
many_names_plot <- name_plot + facet_grid(name ~ sex)
mnp <- many_names_plot %+% filter(BabynamesDist, name %in%
```

```
  c("Jessie", "Marion", "Jackie"))
mnp
```

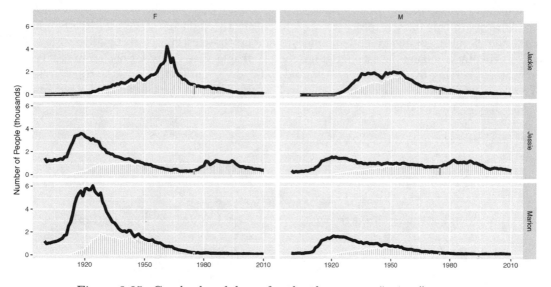

Figure 3.25: Gender breakdown for the three most "unisex" names.

Reversing the order of the variables in the call to facet_grid() flips the orientation of the facets.

```
mnp + facet_grid(sex ~ name)
```

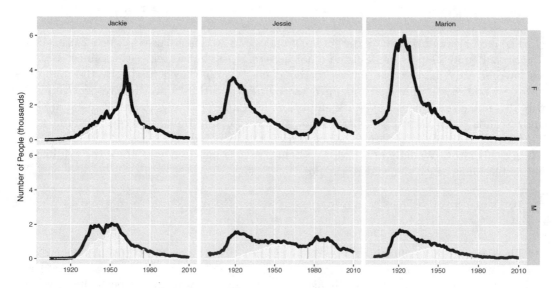

Figure 3.26: Gender breakdown for the three most "unisex" names, oriented vertically.

3.3.2   Most common women's names

A second interesting data graphic from the same FiveThirtyEight articles is shown in Figure 3.27. Take a moment to analyze this data graphic. What are visual cues? What are the variables? How are the variables being mapped to the visual cues? What geom()s are present?

To recreate this data graphic, we need to collect the right data. We need to figure out what the 25 most common female names are among those estimated to be alive today. We can do this by counting the estimated number of people alive today for each name, filtering for women, sorting by the number estimated to be alive, and then taking the top 25 results. We also need to know the median age, as well as the first and third quartiles for age among people having each name.

```
com_fem <- BabynamesDist %>%
  filter(sex == "F") %>%
  group_by(name) %>%
  summarise(
    N = n(), est_num_alive = sum(est_alive_today),
    q1_age = wtd.quantile(age_today, est_alive_today, probs = 0.25),
    median_age = wtd.quantile(age_today, est_alive_today, probs = 0.5),
    q3_age = wtd.quantile(age_today, est_alive_today, probs = 0.75)) %>%
  arrange(desc(est_num_alive)) %>%
  head(25)
```

This data graphic is a bit trickier than the previous one. We'll start by binding the data, and defining the x and y aesthetics. Contrary to Figure 3.27, we put the names on the x-axis and the median_age on the y—the reasons for doing so will be made clearer later. We will also define the title of the plot, and remove the x-axis label, since it is self-evident.

```
w_plot <- ggplot(data = com_fem, aes(x = reorder(name, -median_age),
  y = median_age)) + xlab(NULL) + ylab("Age (in years)") +
  ggtitle("Median ages for females with the 25 most common names")
```

The next element to add are the gold rectangles. To do this, we use the geom_linerange() function. It may help to think of these not as rectangles, but as really thick lines. Because we have already mapped the names to the x-axis, we only need to specify the mappings for ymin and ymax. These are mapped to the first and third quartiles, respectively. We will also make these lines very thick and color them appropriately. geom_linerange() only understands ymin and ymax—there is not a corresponding function with xmin and xmax. This is the reason that we are drawing our plot transposed to Figure 3.27. However, we will fix this later. We have also added a slight alpha transparency to allow the gridlines to be visible underneath the gold rectangles.

```
w_plot <- w_plot + geom_linerange(aes(ymin = q1_age, ymax = q3_age),
  color = "#f3d478", size = 10, alpha = 0.8)
```

There is a red dot indicating the median age for each of these names. If you look carefully, you can see a white border around each red dot. The default glyph for geom_point() is a solid dot, which is shape 19. By changing it to shape 21, we can use both the fill and colour arguments.

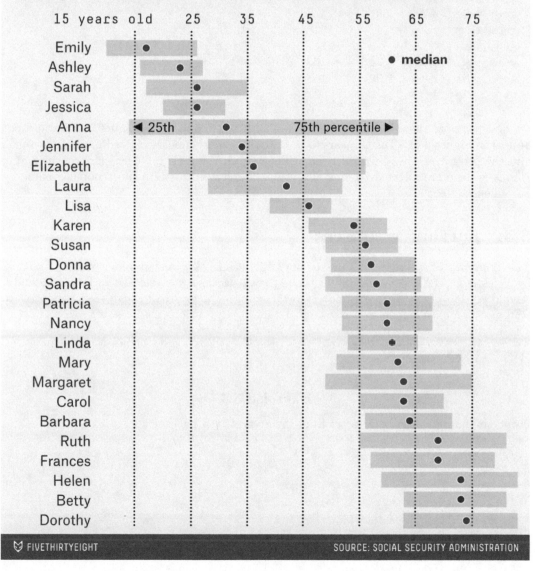

Figure 3.27: FiveThirtyEight's depiction of the age ranges for the 25 most common female names.

```
w_plot <- w_plot +
  geom_point(fill = "#ed3324", colour = "white", size = 4, shape = 21)
```

It remains only to add the context and flip our plot around so the orientation matches that of Figure 3.27. The coord_flip() function does exactly that.

```
w_plot +
  geom_point(aes(y = 55, x = 24), fill = "#ed3324", colour = "white",
    size = 4, shape = 21) +
  geom_text(aes(y = 58, x = 24, label = "median")) +
  geom_text(aes(y = 26, x = 16, label = "25th")) +
  geom_text(aes(y = 51, x = 16, label = "75th percentile")) +
  geom_point(aes(y = 24, x = 16), shape = 17) +
  geom_point(aes(y = 56, x = 16), shape = 17) +
  coord_flip()
```

You will note that the name "Anna" was fifth most common in Figure 3.27 but did not appear in Figure 3.28. This appears to be a result of that name's extraordinarily large range and the pro-rating that FiveThirtyEight did to their data. The "older" names—including Anna—were more affected by this alteration. Anna was the 47th most popular name by our calculations.

3.4   Further resources

The grammar of graphics was created by Wilkinson [238], and implemented in ggplot2 by Wickham [212]. Version 2.0.0 of the ggplot2 package was released in late 2015 and a second edition of the ggplot2 book is forthcoming. The ggplot2 cheat sheet produced by RStudio is an excellent reference for understanding the various features of ggplot2.

3.5   Exercises

Exercise 3.1

Using the famous Galton data set from the mosaicData package:

```
library(mosaic)
head(Galton)
```

```
  family father mother sex height nkids
1      1   78.5   67.0   M   73.2     4
2      1   78.5   67.0   F   69.2     4
3      1   78.5   67.0   F   69.0     4
4      1   78.5   67.0   F   69.0     4
5      2   75.5   66.5   M   73.5     4
6      2   75.5   66.5   M   72.5     4
```

1. Create a scatterplot of each person's height against their father's height

2. Separate your plot into facets by sex

3. Add regression lines to all of your facets

Recall that you can find out more about the data set by running the command ?Galton.

Exercise 3.2

Using the RailTrail data set from the mosaicData package:

```
library(mosaic)
head(RailTrail)
```

```
  hightemp lowtemp avgtemp spring summer fall cloudcover precip volume
1       83      50    66.5      0      1    0        7.6   0.00    501
2       73      49    61.0      0      1    0        6.3   0.29    419
3       74      52    63.0      1      0    0        7.5   0.32    397
4       95      61    78.0      0      1    0        2.6   0.00    385
5       44      52    48.0      1      0    0       10.0   0.14    200
6       69      54    61.5      1      0    0        6.6   0.02    375
  weekday
1       1
2       1
3       1
4       0
5       1
6       1
```

1. Create a scatterplot of the number of crossings per day volume against the high temperature that day

2. Separate your plot into facets by weekday

3. Add regression lines to the two facets

Exercise 3.3

Angelica Schuyler Church (1756–1814) was the daughter of New York Governer Philip Schuyler and sister of Elizabeth Schuyler Hamilton. Angelica, New York was named after her. Generate a plot of the reported proportion of babies born with the name Angelica over time and interpret the figure.

Exercise 3.4

The following questions use the Marriage data set from the mosaicData package.

```
library(mosaic)
head(Marriage, 2)
```

```
  bookpageID  appdate ceremonydate delay     officialTitle person      dob
1  B230p539 10/29/96      11/9/96    11     CIRCUIT JUDGE   Groom 4/11/64
2  B230p677 11/12/96     11/12/96     0 MARRIAGE OFFICIAL   Groom  8/6/64
    age  race prevcount prevconc hs college dayOfBirth  sign
1 32.60 White         0     <NA> 12       7        102 Aries
2 32.29 White         1  Divorce 12       0        219   Leo
```

1. Create an informative and meaningful data graphic.

2. Identify each of the visual cues that you are using, and describe how they are related to each variable.

3. Create a data graphic with at least *five* variables (either quantitative or categorical). For the purposes of this exercise, do not worry about making your visualization meaningful—just try to encode five variables into one plot.

Exercise 3.5

The `MLB_teams` data set in the `mdsr` package contains information about Major League Baseball teams in the past four seasons. There are several quantitative and a few categorical variables present. See how many variables you can illustrate on a single plot in R. The current record is 7. (Note: This is *not* good graphical practice—it is merely an exercise to help you understand how to use visual cues and aesthetics!)

```
library(mdsr)
head(MLB_teams, 4)

# A tibble: 4   11
  yearID teamID  lgID     W     L   WPct attendance normAttend    payroll
   <int>  <chr> <fctr> <int> <int>  <dbl>      <int>      <dbl>      <int>
1   2008    ARI     NL    82    80 0.5062    2509924     0.5839   66202712
2   2008    ATL     NL    72    90 0.4444    2532834     0.5892  102365683
3   2008    BAL     AL    68    93 0.4224    1950075     0.4536   67196246
4   2008    BOS     AL    95    67 0.5864    3048250     0.7091  133390035
# ... with 2 more variables: metroPop <dbl>, name <chr>
```

Exercise 3.6

Use the `MLB_teams` data in the `mdsr` package to create an informative data graphic that illustrates the relationship between winning percentage and payroll in context.

Exercise 3.7

Use the `make_babynames_dist()` function in the `mdsr` package to recreate the "Deadest Names" graphic from FiveThirtyEight (http://tinyurl.com/zcbcl9o).

```
library(mdsr)
babynames_dist <- make_babynames_dist()
```

```
babynames_dist

# A tibble: 1,639,368   9
   year sex         name     n    prop alive_prob count_thousands
   <dbl> <chr>      <chr> <int>   <dbl>      <dbl>           <dbl>
1  1900   F          Mary 16707 0.05257          0          16.707
2  1900   F         Helen  6343 0.01996          0           6.343
3  1900   F          Anna  6114 0.01924          0           6.114
4  1900   F      Margaret  5306 0.01670          0           5.306
5  1900   F          Ruth  4765 0.01499          0           4.765
6  1900   F     Elizabeth  4096 0.01289          0           4.096
7  1900   F      Florence  3920 0.01234          0           3.920
8  1900   F         Ethel  3896 0.01226          0           3.896
9  1900   F         Marie  3856 0.01213          0           3.856
10 1900   F       Lillian  3414 0.01074          0           3.414
# ... with 1,639,358 more rows, and 2 more variables: age_today <dbl>,
#   est_alive_today <dbl>
```

Exercise 3.8

The macleish package contains weather data collected every ten minutes in 2015 from two weather stations in Whately, MA.

```
library(macleish)
head(whately_2015)

# A tibble: 6   8
                 when temperature wind_speed wind_dir rel_humidity
               <dttm>       <dbl>      <dbl>    <dbl>        <dbl>
1 2015-01-01 00:00:00       -9.32      1.399    225.4        54.55
2 2015-01-01 00:10:00       -9.46      1.506    248.2        55.38
3 2015-01-01 00:20:00       -9.44      1.620    258.3        56.18
4 2015-01-01 00:30:00       -9.30      1.141    243.8        56.41
5 2015-01-01 00:40:00       -9.32      1.223    238.4        56.87
6 2015-01-01 00:50:00       -9.34      1.090    241.7        57.25
# ... with 3 more variables: pressure <int>, solar_radiation <dbl>,
#   rainfall <int>
```

Using ggplot2, create a data graphic that displays the average temperature over each 10-minute interal (temperature) as a function of time (when).

Exercise 3.9

Using data from the nasaweather package, create a scatterplot between wind and pressure, with color being used to distinguish the type of storm.

Exercise 3.10

Using data from the nasaweather package, use the geom_path() function to plot the path of each tropical storm in the storms data table. Use color to distinguish the storms from one another, and use facetting to plot each year in its own panel.

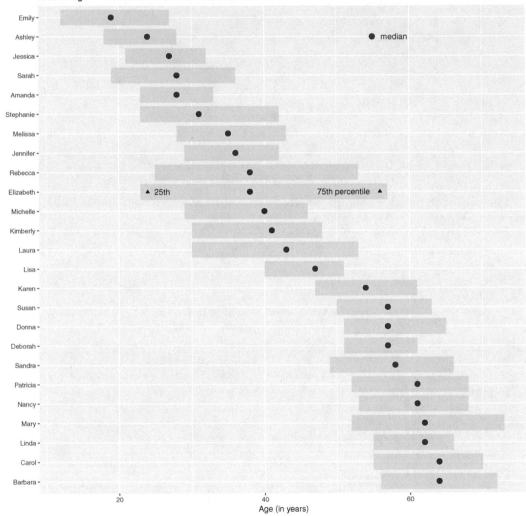

Figure 3.28: Recreation of FiveThirtyEight's plot of the age distributions for the 25 most common women's names.

Chapter 4

Data wrangling

This chapter introduces basics of how to wrangle data in R. Wrangling skills will provide an intellectual and practical foundation for working with modern data.

4.1 A grammar for data wrangling

In much the same way that `ggplot2` presents a grammar for data graphics, the `dplyr` package presents a grammar for data wrangling [234]. Hadley Wickham, one of the authors of `dplyr`, has identified five *verbs* for working with data in a data frame:

`select()` take a subset of the columns (i.e., features, variables)

`filter()` take a subset of the rows (i.e., observations)

`mutate()` add or modify existing columns

`arrange()` sort the rows

`summarize()` aggregate the data across rows (e.g., group it according to some criteria)

Each of these functions takes a data frame as its first argument, and returns a data frame. Thus, these five verbs can be used in conjunction with each other to provide a powerful means to slice-and-dice a single table of data. As with any grammar, what these verbs mean on their own is one thing, but being able to combine these verbs with nouns (i.e., data frames) creates an infinite space for data wrangling. Mastery of these five verbs can make the computation of most any descriptive statistic a breeze and facilitate further analysis. Wickham's approach is inspired by his desire to blur the boundaries between R and the ubiquitous relational database querying syntax SQL. When we revisit SQL in Chapter 12, we will see the close relationship between these two computing paradigms. A related concept more popular in business settings is the OLAP (online analytical processing) hypercube, which refers to the process by which multidimensional data is "sliced-and-diced."

4.1.1 `select()` and `filter()`

The two simplest of the five verbs are `filter()` and `select()`, which allow you to return only a subset of the rows or columns of a data frame, respectively. Generally, if we have a data frame that consists of n rows and p columns, Figures 4.1 and 4.2 illustrate the effect of filtering this data frame based on a condition on one of the columns, and selecting a subset of the columns, respectively.

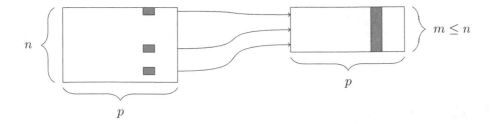

Figure 4.1: The `filter()` function. At left, a data frame that contains matching entries in a certain column for only a subset of the rows. At right, the resulting data frame after filtering.

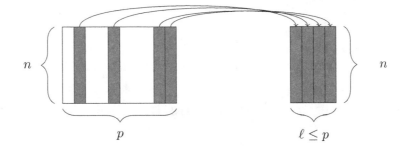

Figure 4.2: The `select()` function. At left, a data frame, from which we retrieve only a few of the columns. At right, the resulting data frame after selecting those columns.

Specifically, we will demonstrate the use of these functions on the `presidential` data frame (from the `ggplot2` package), which contains $p = 4$ variables about the terms of $n = 11$ recent U.S. Presidents.

```
library(mdsr)
presidential
```

```
# A tibble: 11   4
          name        start         end       party
         <chr>       <date>      <date>       <chr>
1   Eisenhower  1953-01-20  1961-01-20  Republican
2      Kennedy  1961-01-20  1963-11-22  Democratic
3      Johnson  1963-11-22  1969-01-20  Democratic
4        Nixon  1969-01-20  1974-08-09  Republican
5         Ford  1974-08-09  1977-01-20  Republican
6       Carter  1977-01-20  1981-01-20  Democratic
7       Reagan  1981-01-20  1989-01-20  Republican
8         Bush  1989-01-20  1993-01-20  Republican
9      Clinton  1993-01-20  2001-01-20  Democratic
10        Bush  2001-01-20  2009-01-20  Republican
11       Obama  2009-01-20  2017-01-20  Democratic
```

To retrieve only the names and party affiliations of these presidents, we would use `select()`. The first *argument* to the `select()` function is the data frame, followed by an arbitrarily long list of column names, separated by commas. Note that it is not necessary to wrap the column names in quotation marks.

```
select(presidential, name, party)
```

```
# A tibble: 11  2
          name       party
         <chr>       <chr>
1   Eisenhower  Republican
2      Kennedy  Democratic
3      Johnson  Democratic
4        Nixon  Republican
5         Ford  Republican
6       Carter  Democratic
7       Reagan  Republican
8         Bush  Republican
9      Clinton  Democratic
10        Bush  Republican
11       Obama  Democratic
```

Similarly, the first argument to `filter()` is a data frame, and subsequent arguments are logical conditions that are evaluated on any involved columns. Thus, if we want to retrieve only those rows that pertain to Republican presidents, we need to specify that the value of the party variable is equal to `Republican`.

```
filter(presidential, party == "Republican")
```

```
# A tibble: 6  4
          name       start         end       party
         <chr>      <date>      <date>       <chr>
1 Eisenhower  1953-01-20  1961-01-20  Republican
2      Nixon  1969-01-20  1974-08-09  Republican
3       Ford  1974-08-09  1977-01-20  Republican
4     Reagan  1981-01-20  1989-01-20  Republican
5       Bush  1989-01-20  1993-01-20  Republican
6       Bush  2001-01-20  2009-01-20  Republican
```

Note that the `==` is a *test for equality*. If we were to use only a single equal sign here, we would be asserting that the value of party was `Republican`. This would cause all of the rows of `presidential` to be returned, since we would have overwritten the actual values of the party variable. Note also the quotation marks around `Republican` are necessary here, since `Republican` is a literal value, and not a variable name.

Naturally, combining the `filter()` and `select()` commands enables one to drill down to very specific pieces of information. For example, we can find which Democratic presidents served since Watergate.

```
select(filter(presidential, start > "1973-01-01" & party == "Democratic"), name)
```

```
# A tibble: 3  1
      name
     <chr>
1   Carter
2  Clinton
3    Obama
```

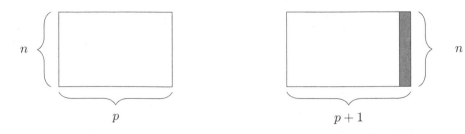

Figure 4.3: The mutate() function. At left, a data frame. At right, the resulting data frame after adding a new column.

In the syntax demonstrated above, the filter() operation is *nested* inside the select() operation. As noted above, each of the five verbs takes and returns a data frame, which makes this type of nesting possible. Shortly, we will see how these verbs can be chained together to make rather long expressions that can become very difficult to read. Instead, we recommend the use of the %>% (pipe) operator. Pipe-forwarding is an alternative to nesting that yields code that can be easily read from top to bottom. With the pipe, we can write the same expression as above in this more readable syntax.

```
presidential %>%
   filter(start > 1973 & party == "Democratic") %>%
   select(name)

# A tibble: 3   1
      name
      <chr>
1   Carter
2 Clinton
3    Obama
```

This expression is called a *pipeline*. Notice how the expression

```
dataframe %>% filter(condition)
```

is equivalent to filter(dataframe, condition). In later examples we will see how this operator can make our code more readable and efficient, particularly for complex operations on large data sets.

4.1.2 mutate() and rename()

Frequently, in the process of conducting our analysis, we will create, re-define, and rename some of our variables. The functions mutate() and rename() provide these capabilities. A graphical illustration of the mutate() operation is shown in Figure 4.3.

While we have the raw data on when each of these presidents took and relinquished office, we don't actually have a numeric variable giving the length of each president's term. Of course, we can derive this information from the dates given, and add the result as a new column to our data frame. This date arithmetic is made easier through the use of the lubridate package, which we use to compute the number of exact years (eyears(1)()) that elapsed since during the interval() from the start until the end of each president's term.

In this situation, it is generally considered good style to create a new object rather than clobbering the one that comes from an external source. To preserve the existing presidential data frame, we save the result of mutate() as a new object called mypresidents.

```
library(lubridate)
mypresidents <- presidential %>%
  mutate(term.length = interval(start, end) / eyears(1))
mypresidents
```

```
# A tibble: 11  5
        name        start          end      party term.length
        <chr>       <date>       <date>      <chr>        <dbl>
1  Eisenhower  1953-01-20  1961-01-20 Republican        8.01
2     Kennedy  1961-01-20  1963-11-22 Democratic        2.84
3     Johnson  1963-11-22  1969-01-20 Democratic        5.17
4       Nixon  1969-01-20  1974-08-09 Republican        5.55
5        Ford  1974-08-09  1977-01-20 Republican        2.45
6      Carter  1977-01-20  1981-01-20 Democratic        4.00
7      Reagan  1981-01-20  1989-01-20 Republican        8.01
8        Bush  1989-01-20  1993-01-20 Republican        4.00
9     Clinton  1993-01-20  2001-01-20 Democratic        8.01
10       Bush  2001-01-20  2009-01-20 Republican        8.01
11      Obama  2009-01-20  2017-01-20 Democratic        8.01
```

The mutate() function can also be used to modify the data in an existing column. Suppose that we wanted to add to our data frame a variable containing the year in which each president was elected. Our first naïve attempt is to assume that every president was elected in the year before he took office. Note that mutate() returns a data frame, so if we want to modify our existing data frame, we need to overwrite it with the results.

```
mypresidents <- mypresidents %>% mutate(elected = year(start) - 1)
mypresidents
```

```
# A tibble: 11  6
        name        start          end      party term.length elected
        <chr>       <date>       <date>      <chr>        <dbl>   <dbl>
1  Eisenhower  1953-01-20  1961-01-20 Republican        8.01    1952
2     Kennedy  1961-01-20  1963-11-22 Democratic        2.84    1960
3     Johnson  1963-11-22  1969-01-20 Democratic        5.17    1962
4       Nixon  1969-01-20  1974-08-09 Republican        5.55    1968
5        Ford  1974-08-09  1977-01-20 Republican        2.45    1973
6      Carter  1977-01-20  1981-01-20 Democratic        4.00    1976
7      Reagan  1981-01-20  1989-01-20 Republican        8.01    1980
8        Bush  1989-01-20  1993-01-20 Republican        4.00    1988
9     Clinton  1993-01-20  2001-01-20 Democratic        8.01    1992
10       Bush  2001-01-20  2009-01-20 Republican        8.01    2000
11      Obama  2009-01-20  2017-01-20 Democratic        8.01    2008
```

Some aspects of this data set are wrong, because presidential elections are only held every four years. Lyndon Johnson assumed the office after President Kennedy was assassinated in 1963, and Gerald Ford took over after President Nixon resigned in 1974. Thus, there were no presidential elections in 1962 or 1973, as suggested in our data frame. We should overwrite these values with NA's—which is how R denotes missing values. We can use the `ifelse()` function to do this. Here, if the value of `elected` is either 1962 or 1973, we overwrite that value with NA.[1] Otherwise, we overwrite it with the same value that it currently has. In this case, instead of checking to see whether the value of `elected` equals 1962 or 1973, for brevity we can use the `%in%` operator to check to see whether the value of `elected` belongs to the vector consisting of 1962 and 1973.

```
mypresidents <- mypresidents %>%
  mutate(elected = ifelse((elected %in% c(1962, 1973)), NA, elected))
mypresidents
```

```
# A tibble: 11  6
         name      start        end       party term.length elected
         <chr>     <date>     <date>       <chr>       <dbl>   <dbl>
1  Eisenhower 1953-01-20 1961-01-20  Republican        8.01    1952
2     Kennedy 1961-01-20 1963-11-22  Democratic        2.84    1960
3     Johnson 1963-11-22 1969-01-20  Democratic        5.17      NA
4       Nixon 1969-01-20 1974-08-09  Republican        5.55    1968
5        Ford 1974-08-09 1977-01-20  Republican        2.45      NA
6      Carter 1977-01-20 1981-01-20  Democratic        4.00    1976
7      Reagan 1981-01-20 1989-01-20  Republican        8.01    1980
8        Bush 1989-01-20 1993-01-20  Republican        4.00    1988
9     Clinton 1993-01-20 2001-01-20  Democratic        8.01    1992
10       Bush 2001-01-20 2009-01-20  Republican        8.01    2000
11      Obama 2009-01-20 2017-01-20  Democratic        8.01    2008
```

Finally, it is considered bad practice to use periods in the name of functions, data frames, and variables in R. Ill-advised periods could conflict with R's use of *generic* functions (i.e., R's mechanism for *method overloading*). Thus, we should change the name of the `term.length` column that we created earlier. In this book, we will use snake_case for function and variable names. We can achieve this using the `rename()` function.

Pro Tip: Don't use periods in the names of functions, data frames, or variables, as this can conflict with R's programming model.

```
mypresidents <- mypresidents %>% rename(term_length = term.length)
mypresidents
```

```
# A tibble: 11  6
         name      start        end       party term_length elected
         <chr>     <date>     <date>       <chr>       <dbl>   <dbl>
1  Eisenhower 1953-01-20 1961-01-20  Republican        8.01    1952
2     Kennedy 1961-01-20 1963-11-22  Democratic        2.84    1960
3     Johnson 1963-11-22 1969-01-20  Democratic        5.17      NA
4       Nixon 1969-01-20 1974-08-09  Republican        5.55    1968
5        Ford 1974-08-09 1977-01-20  Republican        2.45      NA
6      Carter 1977-01-20 1981-01-20  Democratic        4.00    1976
7      Reagan 1981-01-20 1989-01-20  Republican        8.01    1980
8        Bush 1989-01-20 1993-01-20  Republican        4.00    1988
```

---

[1] Incidentally, Johnson was elected in 1964 as an incumbent.

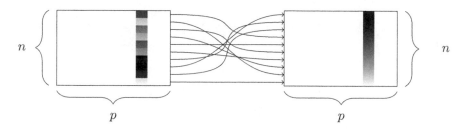

Figure 4.4: The arrange() function. At left, a data frame with an ordinal variable. At right, the resulting data frame after sorting the rows in descending order of that variable.

```
9       Clinton 1993-01-20 2001-01-20 Democratic       8.01    1992
10         Bush 2001-01-20 2009-01-20 Republican       8.01    2000
11        Obama 2009-01-20 2017-01-20 Democratic       8.01    2008
```

4.1.3 arrange()

The function sort() will sort a vector, but not a data frame. The function that will sort a data frame is called arrange(), and its behavior is illustrated in Figure 4.4.

In order to use arrange() on a data frame, you have to specify the data frame, and the column by which you want it to be sorted. You also have to specify the direction in which you want it to be sorted. Specifying multiple sort conditions will result in any ties being broken. Thus, to sort our presidential data frame by the length of each president's term, we specify that we want the column term_length in descending order.

```
mypresidents %>% arrange(desc(term_length))

# A tibble: 11  6
          name        start         end        party term_length elected
          <chr>      <date>      <date>       <chr>        <dbl>    <dbl>
1   Eisenhower 1953-01-20 1961-01-20 Republican         8.01     1952
2       Reagan 1981-01-20 1989-01-20 Republican         8.01     1980
3      Clinton 1993-01-20 2001-01-20 Democratic         8.01     1992
4         Bush 2001-01-20 2009-01-20 Republican         8.01     2000
5        Obama 2009-01-20 2017-01-20 Democratic         8.01     2008
6        Nixon 1969-01-20 1974-08-09 Republican         5.55     1968
7      Johnson 1963-11-22 1969-01-20 Democratic         5.17       NA
8       Carter 1977-01-20 1981-01-20 Democratic         4.00     1976
9         Bush 1989-01-20 1993-01-20 Republican         4.00     1988
10     Kennedy 1961-01-20 1963-11-22 Democratic         2.84     1960
11        Ford 1974-08-09 1977-01-20 Republican         2.45       NA
```

A number of presidents completed either one or two full terms, and thus have the exact same term length (4 or 8 years, respectively). To break these ties, we can further sort by party and elected.

```
mypresidents %>% arrange(desc(term_length), party, elected)

# A tibble: 11  6
```

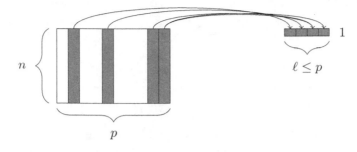

Figure 4.5: The summarize() function. At left, a data frame. At right, the resulting data frame after aggregating three of the columns.

	name	start	end	party	term_length	elected
	<chr>	<date>	<date>	<chr>	<dbl>	<dbl>
1	Clinton	1993-01-20	2001-01-20	Democratic	8.01	1992
2	Obama	2009-01-20	2017-01-20	Democratic	8.01	2008
3	Eisenhower	1953-01-20	1961-01-20	Republican	8.01	1952
4	Reagan	1981-01-20	1989-01-20	Republican	8.01	1980
5	Bush	2001-01-20	2009-01-20	Republican	8.01	2000
6	Nixon	1969-01-20	1974-08-09	Republican	5.55	1968
7	Johnson	1963-11-22	1969-01-20	Democratic	5.17	NA
8	Carter	1977-01-20	1981-01-20	Democratic	4.00	1976
9	Bush	1989-01-20	1993-01-20	Republican	4.00	1988
10	Kennedy	1961-01-20	1963-11-22	Democratic	2.84	1960
11	Ford	1974-08-09	1977-01-20	Republican	2.45	NA

Note that the default sort order is ascending order, so we do not need to specify an order if that is what we want.

4.1.4 summarize() with group_by()

Our last of the five verbs for single-table analysis is summarize(), which is nearly always used in conjunction with group_by(). The previous four verbs provided us with means to manipulate a data frame in powerful and flexible ways. But the extent of the analysis we can perform with these four verbs alone is limited. On the other hand, summarize() with group_by() enables us to make comparisons.

When used alone, summarize() collapses a data frame into a single row. This is illustrated in Figure 4.5. Critically, we have to specify *how* we want to reduce an entire column of data into a single value. The method of aggregation that we specify controls what will appear in the output.

```
mypresidents %>%
  summarize(
    N = n(), first_year = min(year(start)), last_year = max(year(end)),
    num_dems = sum(party == "Democratic"),
    years = sum(term_length),
    avg_term_length = mean(term_length))
```

```
# A tibble: 1  6
```

```
    N first_year last_year num_dems years avg_term_length
  <int>      <dbl>     <dbl>    <int> <dbl>          <dbl>
1    11       1953      2017        5    64           5.82
```

The first argument to summarize() is a data frame, followed by a list of variables that will appear in the output. Note that every variable in the output is defined by operations performed on *vectors*—not on individual values. This is essential, since if the specification of an output variable is not an operation on a vector, there is no way for R to know how to collapse each column.

In this example, the function n() simply counts the number of rows. This is almost always useful information.

Pro Tip: To help ensure that data aggregation is being done correctly, use n() every time you use summarize().

The next two variables determine the first year that one of these presidents assumed office. This is the smallest year in the start column. Similarly, the most recent year is the largest year in the end column. The variable num_dems simply counts the number of rows in which the value of the party variable was Democratic. Finally, the last two variables compute the sum and average of the term_length variable. Thus, we can quickly see that 5 of the 11 presidents who served from 1953 to 2017 were Democrats, and the average term length over these 64 years was about 5.8 years.

This begs the question of whether Democratic or Republican presidents served a longer average term during this time period. To figure this out, we can just execute summarize() again, but this time, instead of the first argument being the data frame mypresidents, we will specify that the rows of the mypresidents data frame should be grouped by the values of the party variable. In this manner, the same computations as above will be carried out for each party separately.

```
mypresidents %>%
  group_by(party) %>%
  summarize(
    N = n(), first_year = min(year(start)), last_year = max(year(end)),
    num_dems = sum(party == "Democratic"),
    years = sum(term_length),
    avg_term_length = mean(term_length))
```

```
# A tibble: 2  7
       party   N first_year last_year num_dems years avg_term_length
       <chr> <int>      <dbl>     <dbl>    <int> <dbl>          <dbl>
1 Democratic     5       1961      2017        5    28            5.6
2 Republican     6       1953      2009        0    36            6.0
```

This provides us with the valuable information that the six Republican presidents served an average of 6 years in office, while the five Democratic presidents served an average of only 5.6. As with all of the dplyr verbs, the final output is a data frame.

Pro Tip: In this chapter we are using the dplyr package. The most common way to extract data from data tables is with SQL (structured query language). We'll introduce SQL in Chapter 12. The dplyr package provides a new interface that fits more smoothly into an overall data analysis workflow and is, in our opinion, easier to learn. Once you

understand data wrangling with `dplyr`, it's straightforward to learn SQL if needed. And
`dplyr` can work as an interface to many systems that use SQL internally.

4.2   Extended example: Ben's time with the Mets

In this extended example, we will continue to explore Sean Lahman's historical baseball
database, which contains complete seasonal records for all players on all Major League
Baseball teams going back to 1871. These data are made available in R via the `Lahman`
package [80]. Here again, while domain knowledge may be helpful, it is not necessary to
follow the example. To flesh out your understanding, try reading the Wikipedia entry on
Major League Baseball.

```
library(Lahman)
dim(Teams)
```

```
[1] 2805    48
```

The `Teams` table contains the seasonal results of every major league team in every season
since 1871. There are 2805 rows and 48 columns in this table, which is far too much to show
here, and would make for a quite unwieldy spreadsheet. Of course, we can take a peek at
what this table looks like by printing the first few rows of the table to the screen with the
`head()` command, but we won't print that on the page of this book.

Ben worked for the New York Mets from 2004 to 2012. How did the team do during
those years? We can use `filter()` and `select()` to quickly identify only those pieces of
information that we care about.

```
mets <- Teams %>% filter(teamID == "NYN")
myMets <- mets %>% filter(yearID %in% 2004:2012)
myMets %>% select(yearID, teamID, W, L)
```

```
  yearID teamID  W  L
1   2004    NYN 71 91
2   2005    NYN 83 79
3   2006    NYN 97 65
4   2007    NYN 88 74
5   2008    NYN 89 73
6   2009    NYN 70 92
7   2010    NYN 79 83
8   2011    NYN 77 85
9   2012    NYN 74 88
```

Notice that we have broken this down into three steps. First, we filter the rows of the
`Teams` data frame into only those teams that correspond to the New York Mets.[2] There are
54 of those, since the Mets joined the National League in 1962.

```
nrow(mets)
```

```
[1] 54
```

---

[2]The `teamID` value of `NYN` stands for the **N**ew **Y**ork **N**ational League club.

Next, we filtered these data so as to include only those seasons in which Ben worked for the team—those with `yearID` between 2004 and 2012. Finally, we printed to the screen only those columns that were relevant to our question: the year, the team's ID, and the number of wins and losses that the team had.

While this process is logical, the code can get unruly, since two ancillary data frames (`mets` and `myMets`) were created during the process. It may be the case that we'd like to use data frames later in the analysis. But if not, they are just cluttering our workspace, and eating up memory. A more streamlined way to achieve the same result would be to *nest* these commands together.

```
select(filter(mets, teamID == "NYN" & yearID %in% 2004:2012),
  yearID, teamID, W, L)

  yearID teamID  W  L
1  2004   NYN   71 91
2  2005   NYN   83 79
3  2006   NYN   97 65
4  2007   NYN   88 74
5  2008   NYN   89 73
6  2009   NYN   70 92
7  2010   NYN   79 83
8  2011   NYN   77 85
9  2012   NYN   74 88
```

This way, no additional data frames were created. However, it is easy to see that as we nest more and more of these operations together, this code could become difficult to read. To maintain readability, we instead *chain* these operations, rather than nest them (and get the same exact results).

```
Teams %>%
  select(yearID, teamID, W, L) %>%
  filter(teamID == "NYN" & yearID %in% 2004:2012)
```

This *piping* syntax (introduced in Section 4.1.1) is provided by the `dplyr` package. It retains the step-by-step logic of our original code, while being easily readable, and efficient with respect to memory and the creation of temporary data frames. In fact, there are also performance enhancements under the hood that make this the most efficient way to do these kinds of computations. For these reasons we will use this syntax whenever possible throughout the book. Note that we only have to type `Teams` once—it is implied by the pipe operator (`%>%`) that the subsequent command takes the previous data frame as its first argument. Thus, `df %>% f(y)` is equivalent to `f(df, y)`.

We've answered the simple question of how the Mets performed during the time that Ben was there, but since we are data scientists, we are interested in deeper questions. For example, some of these seasons were subpar—the Mets had more losses than wins. Did the team just get unlucky in those seasons? Or did they actually play as badly as their record indicates?

In order to answer this question, we need a model for *expected winning percentage*. It turns out that one of the most widely used contributions to the field of baseball analytics (courtesy of Bill James) is exactly that. This model translates the number of runs [3] that

---

[3]In baseball, a team scores a run when a player traverses the bases and return to home plate. The team with the most runs in each game wins, and no ties are allowed.

a team scores and allows *over the course of an entire season* into an expectation for how many games they should have won. The simplest version of this model is this:

$$\widehat{WPct} = \frac{1}{1 + \left(\frac{RA}{RS}\right)^2} \, ,$$

where RA is the number of runs the team allows, RS is the number of runs that the team scores, and $\widehat{WPct}$ is the team's expected winning percentage. Luckily for us, the runs scored and allowed are present in the Teams table, so let's grab them and save them in a new data frame.

```
metsBen <- Teams %>% select(yearID, teamID, W, L, R, RA) %>%
  filter(teamID == "NYN" & yearID %in% 2004:2012)
metsBen
```

```
  yearID teamID  W  L   R  RA
1   2004    NYN 71 91 684 731
2   2005    NYN 83 79 722 648
3   2006    NYN 97 65 834 731
4   2007    NYN 88 74 804 750
5   2008    NYN 89 73 799 715
6   2009    NYN 70 92 671 757
7   2010    NYN 79 83 656 652
8   2011    NYN 77 85 718 742
9   2012    NYN 74 88 650 709
```

First, note that the runs-scored variable is called R in the Teams table, but to stick with our notation we want to rename it RS.

```
metsBen <- metsBen %>% rename(RS = R)     # new name = old name
metsBen
```

```
  yearID teamID  W  L  RS  RA
1   2004    NYN 71 91 684 731
2   2005    NYN 83 79 722 648
3   2006    NYN 97 65 834 731
4   2007    NYN 88 74 804 750
5   2008    NYN 89 73 799 715
6   2009    NYN 70 92 671 757
7   2010    NYN 79 83 656 652
8   2011    NYN 77 85 718 742
9   2012    NYN 74 88 650 709
```

Next, we need to compute the team's actual winning percentage in each of these seasons. Thus, we need to add a new column to our data frame, and we do this with the mutate() command.

```
metsBen <- metsBen %>% mutate(WPct = W / (W + L))
metsBen
```

```
  yearID teamID  W  L  RS  RA  WPct
1   2004    NYN 71 91 684 731 0.438
```

```
2    2005    NYN 83 79 722 648 0.512
3    2006    NYN 97 65 834 731 0.599
4    2007    NYN 88 74 804 750 0.543
5    2008    NYN 89 73 799 715 0.549
6    2009    NYN 70 92 671 757 0.432
7    2010    NYN 79 83 656 652 0.488
8    2011    NYN 77 85 718 742 0.475
9    2012    NYN 74 88 650 709 0.457
```

We also need to compute the model estimates for winning percentage.

```
metsBen <- metsBen %>% mutate(WPct_hat = 1 / (1 + (RA/RS)^2))
metsBen
```

```
  yearID teamID  W  L  RS  RA  WPct WPct_hat
1  2004    NYN 71 91 684 731 0.438    0.467
2  2005    NYN 83 79 722 648 0.512    0.554
3  2006    NYN 97 65 834 731 0.599    0.566
4  2007    NYN 88 74 804 750 0.543    0.535
5  2008    NYN 89 73 799 715 0.549    0.555
6  2009    NYN 70 92 671 757 0.432    0.440
7  2010    NYN 79 83 656 652 0.488    0.503
8  2011    NYN 77 85 718 742 0.475    0.484
9  2012    NYN 74 88 650 709 0.457    0.457
```

The expected number of wins is then equal to the product of the expected winning percentage times the number of games.

```
metsBen <- metsBen %>% mutate(W_hat = WPct_hat * (W + L))
metsBen
```

```
  yearID teamID  W  L  RS  RA  WPct WPct_hat W_hat
1  2004    NYN 71 91 684 731 0.438    0.467  75.6
2  2005    NYN 83 79 722 648 0.512    0.554  89.7
3  2006    NYN 97 65 834 731 0.599    0.566  91.6
4  2007    NYN 88 74 804 750 0.543    0.535  86.6
5  2008    NYN 89 73 799 715 0.549    0.555  90.0
6  2009    NYN 70 92 671 757 0.432    0.440  71.3
7  2010    NYN 79 83 656 652 0.488    0.503  81.5
8  2011    NYN 77 85 718 742 0.475    0.484  78.3
9  2012    NYN 74 88 650 709 0.457    0.457  74.0
```

In this case, the Mets' fortunes were better than expected in three of these seasons, and worse than expected in the other six.

```
filter(metsBen, W >= W_hat)
```

```
  yearID teamID  W  L  RS  RA  WPct WPct_hat W_hat
1  2006    NYN 97 65 834 731 0.599    0.566  91.6
2  2007    NYN 88 74 804 750 0.543    0.535  86.6
3  2012    NYN 74 88 650 709 0.457    0.457  74.0
```

```
filter(metsBen, W < W_hat)
```

```
  yearID teamID  W  L  RS  RA  WPct WPct_hat W_hat
1   2004    NYN 71 91 684 731 0.438    0.467  75.6
2   2005    NYN 83 79 722 648 0.512    0.554  89.7
3   2008    NYN 89 73 799 715 0.549    0.555  90.0
4   2009    NYN 70 92 671 757 0.432    0.440  71.3
5   2010    NYN 79 83 656 652 0.488    0.503  81.5
6   2011    NYN 77 85 718 742 0.475    0.484  78.3
```

Naturally, the Mets experienced ups and downs during Ben's time with the team. Which seasons were best? To figure this out, we can simply sort the rows of the data frame.

```
arrange(metsBen, desc(WPct))
```

```
  yearID teamID  W  L  RS  RA  WPct WPct_hat W_hat
1   2006    NYN 97 65 834 731 0.599    0.566  91.6
2   2008    NYN 89 73 799 715 0.549    0.555  90.0
3   2007    NYN 88 74 804 750 0.543    0.535  86.6
4   2005    NYN 83 79 722 648 0.512    0.554  89.7
5   2010    NYN 79 83 656 652 0.488    0.503  81.5
6   2011    NYN 77 85 718 742 0.475    0.484  78.3
7   2012    NYN 74 88 650 709 0.457    0.457  74.0
8   2004    NYN 71 91 684 731 0.438    0.467  75.6
9   2009    NYN 70 92 671 757 0.432    0.440  71.3
```

In 2006, the Mets had the best record in baseball during the regular season and nearly made the World Series. But how do these seasons rank in terms of the team's performance relative to our model?

```
metsBen %>%
  mutate(Diff = W - W_hat) %>%
  arrange(desc(Diff))
```

```
  yearID teamID  W  L  RS  RA  WPct WPct_hat W_hat    Diff
1   2006    NYN 97 65 834 731 0.599    0.566  91.6  5.3840
2   2007    NYN 88 74 804 750 0.543    0.535  86.6  1.3774
3   2012    NYN 74 88 650 709 0.457    0.457  74.0  0.0199
4   2008    NYN 89 73 799 715 0.549    0.555  90.0 -0.9605
5   2009    NYN 70 92 671 757 0.432    0.440  71.3 -1.2790
6   2011    NYN 77 85 718 742 0.475    0.484  78.3 -1.3377
7   2010    NYN 79 83 656 652 0.488    0.503  81.5 -2.4954
8   2004    NYN 71 91 684 731 0.438    0.467  75.6 -4.6250
9   2005    NYN 83 79 722 648 0.512    0.554  89.7 -6.7249
```

So 2006 was the Mets' most fortunate year—since they won five more games than our model predicts—but 2005 was the least fortunate—since they won almost seven games fewer than our model predicts. This type of analysis helps us understand how the Mets performed in individual seasons, but we know that any randomness that occurs in individual years is likely to average out over time. So while it is clear that the Mets performed well in some seasons and poorly in others, what can we say about their overall performance?

We can easily summarize a single variable with the `favstats()` command from the `mosaic` package.

```
favstats(~ W, data = metsBen)

 min Q1 median Q3 max mean  sd n missing
  70 74     79 88  97 80.9 9.1 9       0
```

This tells us that the Mets won nearly 81 games on average during Ben's tenure, which corresponds almost exactly to a 0.500 winning percentage, since there are 162 games in a regular season. But we may be interested in aggregating more than one variable at a time. To do this, we use `summarize()`.

```
metsBen %>%
  summarize(
    num_years = n(), total_W = sum(W), total_L = sum(L),
    total_WPct = sum(W) / sum(W + L), sum_resid = sum(W - W_hat))

  num_years total_W total_L total_WPct sum_resid
1         9     728     730      0.499     -10.6
```

In these nine years, the Mets had a combined record of 728 wins and 730 losses, for an overall winning percentage of .499. Just one extra win would have made them exactly 0.500! (If we could pick which game, we would definitely pick the final game of the 2007 season. A win there would have resulted in a playoff berth.) However, we've also learned that the team under-performed relative to our model by a total of 10.6 games over those nine seasons.

Usually, when we are summarizing a data frame like we did above, it is interesting to consider different groups. In this case, we can discretize these years into three chunks: one for each of the three general managers under whom Ben worked. Jim Duquette was the Mets' general manager in 2004, Omar Minaya from 2005 to 2010, and Sandy Alderson from 2011 to 2012. We can define these eras using two nested `ifelse()` functions (the `case_when()` function in the `dplyr` package is helpful in such a setting).

```
metsBen <- metsBen %>%
  mutate(
    gm = ifelse(yearID == 2004, "Duquette",
        ifelse(yearID >= 2011, "Alderson", "Minaya")))
```

Next, we use the gm variable to define these groups with the `group_by()` operator. The combination of summarizing data by groups can be very powerful. Note that while the Mets were far more successful during Minaya's regime (i.e., many more wins than losses), they did not meet expectations in any of the three periods.

```
metsBen %>%
  group_by(gm) %>%
  summarize(
    num_years = n(), total_W = sum(W), total_L = sum(L),
    total_WPct = sum(W) / sum(W + L), sum_resid = sum(W - W_hat)) %>%
  arrange(desc(sum_resid))

# A tibble: 3  6
```

```
       gm num_years total_W total_L total_WPct sum_resid
    <chr>      <int>   <int>   <int>      <dbl>     <dbl>
1 Alderson         2     151     173      0.466     -1.32
2 Duquette         1      71      91      0.438     -4.63
3   Minaya         6     506     466      0.521     -4.70
```

The full power of the chaining operator is revealed below, where we do all the analysis at once, but retain the step-by-step logic.

```
Teams %>%
  select(yearID, teamID, W, L, R, RA) %>%
  filter(teamID == "NYN" & yearID %in% 2004:2012) %>%
  rename(RS = R) %>%
  mutate(
    WPct = W / (W + L), WPct_hat = 1 / (1 + (RA/RS)^2),
    W_hat = WPct_hat * (W + L),
    gm = ifelse(yearID == 2004, "Duquette",
         ifelse(yearID >= 2011, "Alderson", "Minaya"))) %>%
  group_by(gm) %>%
  summarize(
    num_years = n(), total_W = sum(W), total_L = sum(L),
    total_WPct = sum(W) / sum(W + L), sum_resid = sum(W - W_hat)) %>%
  arrange(desc(sum_resid))
```

```
# A tibble: 3   6
       gm num_years total_W total_L total_WPct sum_resid
    <chr>      <int>   <int>   <int>      <dbl>     <dbl>
1 Alderson         2     151     173      0.466     -1.32
2 Duquette         1      71      91      0.438     -4.63
3   Minaya         6     506     466      0.521     -4.70
```

Even more generally, we might be more interested in how the Mets performed relative to our model, in the context of all teams during that nine year period. All we need to do is remove the `teamID` filter and group by franchise (`franchID`) instead.

```
Teams %>% select(yearID, teamID, franchID, W, L, R, RA) %>%
  filter(yearID %in% 2004:2012) %>%
  rename(RS = R) %>%
  mutate(
    WPct = W / (W + L), WPctHat = 1 / (1 + (RA/RS)^2),
    WHat = WPctHat * (W + L)) %>%
  group_by(franchID) %>%
  summarize(
    numYears = n(), totalW = sum(W), totalL = sum(L),
    totalWPct = sum(W) / sum(W + L), sumResid = sum(W - WHat)) %>%
  arrange(sumResid) %>%
  print(n = 6)
```

```
# A tibble: 30   6
  franchID numYears totalW totalL totalWPct sumResid
    <fctr>    <int>  <int>  <int>     <dbl>    <dbl>
```

1	TOR	9	717	740	0.492	-29.2
2	ATL	9	781	677	0.536	-24.0
3	COL	9	687	772	0.471	-22.7
4	CHC	9	706	750	0.485	-14.5
5	CLE	9	710	748	0.487	-13.9
6	NYM	9	728	730	0.499	-10.6

```
# ... with 24 more rows
```

We can see now that only five other teams fared worse than the Mets,[4] relative to our model, during this time period. Perhaps they are cursed!

4.3 Combining multiple tables

In the previous section, we illustrated how the five verbs can be chained to perform operations on a single table. This single table is reminiscent of a single well-organized spreadsheet. But in the same way that a workbook can contain multiple spreadsheets, we will often work with multiple tables. In Chapter 12, we will describe how multiple tables related by unique identifiers called *keys* can be organized into a *relational database management system*.

It is more efficient for the computer to store and search tables in which "like is stored with like." Thus, a database maintained by the Bureau of Transportation Statistics on the arrival times of U.S. commercial flights will consist of multiple tables, each of which contains data about different things. For example, the `nycflights13` package contains one table about `flights`—each row in this table is a single flight. As there are many flights, you can imagine that this table will get very long—hundreds of thousands of rows per year. But there are other related kinds of information that we will want to know about these flights. We would certainly be interested in the particular airline to which each flight belonged. It would be inefficient to store the complete name of the airline (e.g., `American Airlines Inc.`) in every row of the flights table. A simple code (e.g., `AA`) would take up less space on disk. For small tables, the savings of storing two characters instead of 25 is insignificant, but for large tables, it can add up to noticeable savings both in terms of the size of data on disk, and the speed with which we can search it. However, we still want to have the full names of the airlines available if we need them. The solution is to store the data *about airlines* in a separate table called `airlines`, and to provide a *key* that links the data in the two tables together.

4.3.1 inner_join()

If we examine the first few rows of the `flights` table, we observe that the `carrier` column contains a two-character string corresponding to the airline.

```
library(nycflights13)
head(flights, 3)
```

```
# A tibble: 3  19
  year month   day dep_time sched_dep_time dep_delay arr_time
  <int> <int> <int>    <int>          <int>     <dbl>    <int>
1 2013     1     1      517            515         2      830
2 2013     1     1      533            529         4      850
```

---

[4]Note that whereas the `teamID` that corresponds to the Mets is `NYN`, the value of the `franchID` variable is `NYM`.

```
3   2013       1      1      542          540           2       923
# ... with 12 more variables: sched_arr_time <int>, arr_delay <dbl>,
#   carrier <chr>, flight <int>, tailnum <chr>, origin <chr>, dest <chr>,
#   air_time <dbl>, distance <dbl>, hour <dbl>, minute <dbl>,
#   time_hour <dttm>
```

In the airlines table, we have those same two-character strings, but also the full names of the airline.

```
head(airlines, 3)
```

```
# A tibble: 3   2
  carrier                      name
    <chr>                     <chr>
1      9E         Endeavor Air Inc.
2      AA     American Airlines Inc.
3      AS       Alaska Airlines Inc.
```

In order to retrieve a list of flights and the full names of the airlines that managed each flight, we need to match up the rows in the flights table with those rows in the airlines table that have the corresponding values for the carrier column in *both* tables. This is achieved with the function inner_join().

```
flightsJoined <- flights %>%
  inner_join(airlines, by = c("carrier" = "carrier"))
glimpse(flightsJoined)
```

```
Observations: 336,776
Variables: 20
$ year          <int> 2013, 2013, 2013, 2013, 2013, 2013, 2013, 2013,...
$ month         <int> 1, 1, 1, 1, 1, 1, 1, 1, 1, 1, 1, 1, 1, 1, 1, 1,...
$ day           <int> 1, 1, 1, 1, 1, 1, 1, 1, 1, 1, 1, 1, 1, 1, 1, 1,...
$ dep_time      <int> 517, 533, 542, 544, 554, 554, 555, 557, 557, 55...
$ sched_dep_time <int> 515, 529, 540, 545, 600, 558, 600, 600, 600, 60...
$ dep_delay     <dbl> 2, 4, 2, -1, -6, -4, -5, -3, -3, -2, -2, -2, -2...
$ arr_time      <int> 830, 850, 923, 1004, 812, 740, 913, 709, 838, 7...
$ sched_arr_time <int> 819, 830, 850, 1022, 837, 728, 854, 723, 846, 7...
$ arr_delay     <dbl> 11, 20, 33, -18, -25, 12, 19, -14, -8, 8, -2, -...
$ carrier       <chr> "UA", "UA", "AA", "B6", "DL", "UA", "B6", "EV",...
$ flight        <int> 1545, 1714, 1141, 725, 461, 1696, 507, 5708, 79...
$ tailnum       <chr> "N14228", "N24211", "N619AA", "N804JB", "N668DN...
$ origin        <chr> "EWR", "LGA", "JFK", "JFK", "LGA", "EWR", "EWR"...
$ dest          <chr> "IAH", "IAH", "MIA", "BQN", "ATL", "ORD", "FLL"...
$ air_time      <dbl> 227, 227, 160, 183, 116, 150, 158, 53, 140, 138...
$ distance      <dbl> 1400, 1416, 1089, 1576, 762, 719, 1065, 229, 94...
$ hour          <dbl> 5, 5, 5, 5, 6, 5, 6, 6, 6, 6, 6, 6, 6, 6, 6, 5,...
$ minute        <dbl> 15, 29, 40, 45, 0, 58, 0, 0, 0, 0, 0, 0, 0, 0, ...
$ time_hour     <dttm> 2013-01-01 05:00:00, 2013-01-01 05:00:00, 2013...
$ name          <chr> "United Air Lines Inc.", "United Air Lines Inc....
```

Notice that the flightsJoined data frame now has an additional variable called name.

This is the column from `airlines` that is now attached to our combined data frame. Now we can view the full names of the airlines instead of the cryptic two-character codes.

```
flightsJoined %>%
  select(carrier, name, flight, origin, dest) %>%
  head(3)
```

```
# A tibble: 3  5
  carrier                      name flight origin  dest
    <chr>                     <chr>  <int>  <chr> <chr>
1      UA  United Air Lines Inc.    1545    EWR   IAH
2      UA  United Air Lines Inc.    1714    LGA   IAH
3      AA  American Airlines Inc.   1141    JFK   MIA
```

In an `inner_join()`, the result set contains only those rows that have matches in both tables. In this case, all of the rows in `flights` have exactly one corresponding entry in `airlines`, so the number of rows in `flightsJoined` is the same as the number of rows in `flights` (this will not always be the case).

```
nrow(flights)
```

```
[1] 336776
```

```
nrow(flightsJoined)
```

```
[1] 336776
```

Pro Tip: It is always a good idea to carefully check that the number of rows returned by a join operation is what you expected. In particular, you often want to check for rows in one table that matched to more than one row in the other table.

4.3.2 `left_join()`

Another commonly used type of join is a `left_join()`. Here the rows of the first table are *always* returned, regardless of whether there is a match in the second table.

Suppose that we are only interested in flights from the NYC airports to the West Coast. Specifically, we're only interested in airports in the Pacific Time Zone. Thus, we filter the airports data frame to only include those 152 airports.

```
airportsPT <- filter(airports, tz == -8)
nrow(airportsPT)
```

```
[1] 152
```

Now, if we perform an `inner_join()` on `flights` and `airportsPT`, matching the destinations in `flights` to the FAA codes in `airports`, we retrieve only those flights that flew to our airports in the Pacific Time Zone.

```
nycDestsPT <- flights %>% inner_join(airportsPT, by = c("dest" = "faa"))
nrow(nycDestsPT)
```

```
[1] 46324
```

However, if we use a `left_join()` with the same conditions, we retrieve all of the rows of `flights`. NA's are inserted into the columns where no matched data was found.

```
nycDests <- flights %>% left_join(airportsPT, by = c("dest" = "faa"))
nrow(nycDests)
```

```
[1] 336776
```

```
sum(is.na(nycDests$name))
```

```
[1] 290452
```

Left joins are particularly useful in databases in which *referential integrity* is broken (not all of the *keys* are present—see Chapter 12).

4.4   Extended example: Manny Ramirez

In the context of baseball and the `Lahman` package, multiple tables are used to store information. The batting statistics of players are stored in one table (`Batting`), while information about people (most of whom are players) is in a different table (`Master`).

Every row in the `Batting` table contains the statistics accumulated by a single player during a single stint for a single team in a single year. Thus, a player like Manny Ramirez has many rows in the `Batting` table (21, in fact).

```
manny <- filter(Batting, playerID == "ramirma02")
nrow(manny)
```

```
[1] 21
```

Using what we've learned, we can quickly tabulate Ramirez's most common career offensive statistics. For those new to baseball, some additional background may be helpful. A hit (`H`) occurs when a batter reaches base safely. A home run (`HR`) occurs when the ball is hit out of the park or the runner advances through all of the bases during that play. Barry Bonds has the record for most home runs (762) hit in a career. A player's batting average (`BA`) is the ratio of the number of hits to the number of eligible at-bats. The highest career batting average in major league baseball history of 0.366 was achieved by Ty Cobb—season averages above 0.300 are impressive. Finally, runs batted in (`RBI`) is the number of runners (including the batter in the case of a home run) that score during that batter's at-bat. Hank Aaron has the record for most career RBIs with 2,297.

```
manny %>% summarize(
  span = paste(min(yearID), max(yearID), sep = "-"),
  numYears = n_distinct(yearID), numTeams = n_distinct(teamID),
  BA = sum(H)/sum(AB), tH = sum(H), tHR = sum(HR), tRBI = sum(RBI))
```

```
        span numYears numTeams    BA   tH tHR tRBI
1 1993-2011       19        5 0.312 2574 555 1831
```

Notice how we have used the `paste()` function to combine results from multiple variables into a new variable, and how we have used the `n_distinct()` function to count the number of distinct rows. In his 19-year career, Ramirez hit 555 home runs, which puts him in the top 20 among all Major League players.

However, we also see that Ramirez played for five teams during his career. Did he perform equally well for each of them? Breaking his statistics down by team, or by league, is as easy as adding an appropriate group_by() command.

```
manny %>%
  group_by(teamID) %>%
  summarize(
    span = paste(min(yearID), max(yearID), sep = "-"),
    numYears = n_distinct(yearID), numTeams = n_distinct(teamID),
    BA = sum(H)/sum(AB), tH = sum(H), tHR = sum(HR), tRBI = sum(RBI)) %>%
  arrange(span)
```

```
# A tibble: 5  8
  teamID        span numYears numTeams      BA    tH   tHR  tRBI
  <fctr>       <chr>    <int>    <int>   <dbl> <int> <int> <int>
1    CLE 1993-2000        8        1  0.3130  1086   236   804
2    BOS 2001-2008        8        1  0.3117  1232   274   868
3    LAN 2008-2010        3        1  0.3224   237    44   156
4    CHA 2010-2010        1        1  0.2609    18     1     2
5    TBA 2011-2011        1        1  0.0588     1     0     1
```

While Ramirez was very productive for Cleveland, Boston, and the Los Angeles Dodgers, his brief tours with the Chicago White Sox and Tampa Bay Rays were less than stellar. In the pipeline below, we can see that Ramirez spent the bulk of his career in the American League.

```
manny %>%
  group_by(lgID) %>%
  summarize(
    span = paste(min(yearID), max(yearID), sep = "-"),
    numYears = n_distinct(yearID), numTeams = n_distinct(teamID),
    BA = sum(H)/sum(AB), tH = sum(H), tHR = sum(HR), tRBI = sum(RBI)) %>%
  arrange(span)
```

```
# A tibble: 2  8
  lgID        span numYears numTeams     BA    tH   tHR  tRBI
  <fctr>     <chr>    <int>    <int>  <dbl> <int> <int> <int>
1   AL 1993-2011       18        4  0.311  2337   511  1675
2   NL 2008-2010        3        1  0.322   237    44   156
```

If Ramirez played in only 19 different seasons, why were there 21 rows attributed to him? Notice that in 2008, he was traded from the Boston Red Sox to the Los Angeles Dodgers, and thus played for both teams. Similarly, in 2010 he played for both the Dodgers and the Chicago White Sox. When summarizing data, it is critically important to understand exactly how the rows of your data frame are organized. To see what can go wrong here, suppose we were interested in tabulating the number of seasons in which Ramirez hit at least 30 home runs. The simplest solution is:

```
manny %>%
  filter(HR >= 30) %>%
  nrow()
```

```
[1] 11
```

But this answer is wrong, because in 2008, Ramirez hit 20 home runs for Boston before being traded and then 17 more for the Dodgers afterwards. Neither of those rows were counted, since they were *both* filtered out. Thus, the year 2008 does not appear among the 11 that we counted in the previous pipeline. Recall that each row in the manny data frame corresponds to one stint with one team in one year. On the other hand, the question asks us to consider each year, *regardless of team*. In order to get the right answer, we have to aggregate the rows by team. Thus, the correct solution is:

```
manny %>%
  group_by(yearID) %>%
  summarize(tHR = sum(HR)) %>%
  filter(tHR >= 30) %>%
  nrow()
```

```
[1] 12
```

Note that the filter() operation is applied to tHR, the total number of home runs in a season, and not HR, the number of home runs in a single stint for a single team in a single season. (This distinction between filtering the rows of the original data versus the rows of the aggregated results will appear again in Chapter 12.)

We began this exercise by filtering the Batting table for the player with playerID equal to ramirma02. How did we know to use this identifier? This player ID is known as a *key*, and in fact, playerID is the *primary key* defined in the Master table. That is, every row in the Master table is uniquely identified by the value of playerID. Thus there is exactly one row in that table for which playerID is equal to ramirma02.

But how did we know that this ID corresponds to Manny Ramirez? We can search the Master table. The data in this table include characteristics about Manny Ramirez that do not change across multiple seasons (with the possible exception of his weight).

```
Master %>% filter(nameLast == "Ramirez" & nameFirst == "Manny")
```

```
    playerID birthYear birthMonth birthDay birthCountry        birthState
1 ramirma02      1972          5       30        D.R. Distrito Nacional
     birthCity deathYear deathMonth deathDay deathCountry deathState
1 Santo Domingo        NA         NA       NA         <NA>       <NA>
  deathCity nameFirst nameLast       nameGiven weight height bats throws
1      <NA>     Manny  Ramirez Manuel Aristides    225     72    R      R
       debut  finalGame  retroID   bbrefID deathDate   birthDate
1 1993-09-02 2011-04-06 ramim002 ramirma02      <NA> 1972-05-30
```

The playerID column forms a primary key in the Master table, but it does not in the Batting table, since as we saw previously, there were 21 rows with that playerID. In the Batting table, the playerID column is known as a *foreign key*, in that it references a primary key in another table. For our purposes, the presence of this column in both tables allows us to link them together. This way, we can combine data from the Batting table with data in the Master table. We do this with inner_join() by specifying the two tables that we want to join, and the corresponding columns in each table that provide the link. Thus, if we want to display Ramirez's name in our previous result, as well as his age, we must join the Batting and Master tables together.

```
Batting %>%
  filter(playerID == "ramirma02") %>%
  inner_join(Master, by = c("playerID" = "playerID")) %>%
  group_by(yearID) %>%
  summarize(
    Age = max(yearID - birthYear), numTeams = n_distinct(teamID),
    BA = sum(H)/sum(AB), tH = sum(H), tHR = sum(HR), tRBI = sum(RBI)) %>%
  arrange(yearID)
```

```
# A tibble: 19  7
   yearID   Age numTeams     BA    tH   tHR  tRBI
    <int> <int>    <int>  <dbl> <int> <int> <int>
1    1993    21        1 0.1698     9     2     5
2    1994    22        1 0.2690    78    17    60
3    1995    23        1 0.3079   149    31   107
4    1996    24        1 0.3091   170    33   112
5    1997    25        1 0.3280   184    26    88
6    1998    26        1 0.2942   168    45   145
7    1999    27        1 0.3333   174    44   165
8    2000    28        1 0.3508   154    38   122
9    2001    29        1 0.3062   162    41   125
10   2002    30        1 0.3486   152    33   107
11   2003    31        1 0.3251   185    37   104
12   2004    32        1 0.3081   175    43   130
13   2005    33        1 0.2924   162    45   144
14   2006    34        1 0.3207   144    35   102
15   2007    35        1 0.2961   143    20    88
16   2008    36        2 0.3315   183    37   121
17   2009    37        1 0.2898   102    19    63
18   2010    38        2 0.2981    79     9    42
19   2011    39        1 0.0588     1     0     1
```

Pro Tip: Always specify the `by` argument that defines the join condition. Don't rely on the defaults.

Notice that even though Ramirez's age is a constant for each season, we have to use a vector operation (i.e., `max()` in order to reduce any potential vector to a single number.

Which season was Ramirez's best as a hitter? One relatively simple measurement of batting prowess is OPS, or On-Base Plus Slugging Percentage, which is the simple sum of two other statistics: On-Base Percentage (OBP) and Slugging Percentage (SLG). The former basically measures the percentage of time that a batter reaches base safely, whether it comes via a hit (`H`), a base on balls (`BB`), or from being hit by the pitch (`HBP`). The latter measures the average number of bases advanced per at-bat (`AB`), where a single is worth one base, a double (`X2B`) is worth two, a triple (`X3B`) is worth three, and a home run (`HR`) is worth four. (Note that every hit is exactly one of a single, double, triple, or home run.) Let's add this statistic to our results and use it to rank the seasons.

```
mannyBySeason <- Batting %>%
  filter(playerID == "ramirma02") %>%
  inner_join(Master, by = c("playerID" = "playerID")) %>%
```

```
  group_by(yearID) %>%
  summarize(
    Age = max(yearID - birthYear), numTeams = n_distinct(teamID),
    BA = sum(H)/sum(AB), tH = sum(H), tHR = sum(HR), tRBI = sum(RBI),
    OBP = sum(H + BB + HBP) / sum(AB + BB + SF + HBP),
    SLG = sum(H + X2B + 2*X3B + 3*HR) / sum(AB)) %>%
  mutate(OPS = OBP + SLG) %>%
  arrange(desc(OPS))
mannyBySeason
```

```
# A tibble: 19  10
   yearID   Age numTeams      BA    tH   tHR  tRBI    OBP    SLG    OPS
    <int> <int>    <int>   <dbl> <int> <int> <int>  <dbl>  <dbl>  <dbl>
1    2000    28        1  0.3508   154    38   122 0.4568 0.6970  1.154
2    1999    27        1  0.3333   174    44   165 0.4422 0.6628  1.105
3    2002    30        1  0.3486   152    33   107 0.4498 0.6468  1.097
4    2006    34        1  0.3207   144    35   102 0.4391 0.6192  1.058
5    2008    36        2  0.3315   183    37   121 0.4297 0.6014  1.031
6    2003    31        1  0.3251   185    37   104 0.4271 0.5870  1.014
7    2001    29        1  0.3062   162    41   125 0.4048 0.6087  1.014
8    2004    32        1  0.3081   175    43   130 0.3967 0.6127  1.009
9    2005    33        1  0.2924   162    45   144 0.3877 0.5939  0.982
10   1996    24        1  0.3091   170    33   112 0.3988 0.5818  0.981
11   1998    26        1  0.2942   168    45   145 0.3771 0.5989  0.976
12   1995    23        1  0.3079   149    31   107 0.4025 0.5579  0.960
13   1997    25        1  0.3280   184    26    88 0.4147 0.5383  0.953
14   2009    37        1  0.2898   102    19    63 0.4176 0.5312  0.949
15   2007    35        1  0.2961   143    20    88 0.3884 0.4928  0.881
16   1994    22        1  0.2690    78    17    60 0.3571 0.5207  0.878
17   2010    38        2  0.2981    79     9    42 0.4094 0.4604  0.870
18   1993    21        1  0.1698     9     2     5 0.2000 0.3019  0.502
19   2011    39        1  0.0588     1     0     1 0.0588 0.0588  0.118
```

We see that Ramirez's OPS was highest in 2000. But 2000 was the height of the steroid era, when many sluggers were putting up tremendous offensive numbers. As data scientists, we know that it would be more instructive to put Ramirez's OPS in context by comparing it to the league average OPS in each season—the resulting ratio is often called OPS+. To do this, we will need to compute those averages. Because there is missing data in some of these columns in some of these years, we need to invoke the na.rm argument to ignore that data.

```
mlb <- Batting %>%
  filter(yearID %in% 1993:2011) %>%
  group_by(yearID) %>%
  summarize(lgOPS =
    sum(H + BB + HBP, na.rm = TRUE) / sum(AB + BB + SF + HBP, na.rm = TRUE) +
    sum(H + X2B + 2*X3B + 3*HR, na.rm = TRUE) / sum(AB, na.rm = TRUE))
```

Next, we need to match these league average OPS values to the corresponding entries for Ramirez. We can do this by joining these tables together, and computing the ratio of Ramirez's OPS to that of the league average.

```
mannyRatio <- mannyBySeason %>%
  inner_join(mlb, by = c("yearID" = "yearID")) %>%
  mutate(OPSplus = OPS / lgOPS) %>%
  select(yearID, Age, OPS, lgOPS, OPSplus) %>%
  arrange(desc(OPSplus))
mannyRatio
```

```
# A tibble: 19   5
   yearID   Age   OPS lgOPS OPSplus
    <int> <int> <dbl> <dbl>   <dbl>
1    2000    28 1.154 0.782   1.475
2    2002    30 1.097 0.748   1.466
3    1999    27 1.105 0.778   1.420
4    2006    34 1.058 0.768   1.377
5    2008    36 1.031 0.749   1.376
6    2003    31 1.014 0.755   1.344
7    2001    29 1.014 0.759   1.336
8    2004    32 1.009 0.763   1.323
9    2005    33 0.982 0.749   1.310
10   1998    26 0.976 0.755   1.292
11   1996    24 0.981 0.767   1.278
12   1995    23 0.960 0.755   1.272
13   2009    37 0.949 0.751   1.264
14   1997    25 0.953 0.756   1.261
15   2010    38 0.870 0.728   1.194
16   2007    35 0.881 0.758   1.162
17   1994    22 0.878 0.763   1.150
18   1993    21 0.502 0.736   0.682
19   2011    39 0.118 0.720   0.163
```

In this case, 2000 still ranks as Ramirez's best season relative to his peers, but notice that his 1999 season has fallen from 2nd to 3rd. Since by definition a league batter has an OPS+ of 1, Ramirez posted 17 consecutive seasons with an OPS that was at least 15% better than the average across the major leagues—a truly impressive feat.

Finally, not all joins are the same. An inner_join() requires corresponding entries in *both* tables. Conversely, a left_join() returns at least as many rows as there are in the first table, regardless of whether there are matches in the second table. Thus, an inner_join() is bidirectional, whereas in a left_join(), the order in which you specify the tables matters.

Consider the career of Cal Ripken, who played in 21 seasons from 1981 to 2001. His career overlapped with Ramirez's in the nine seasons from 1993 to 2001, so for those, the league averages we computed before are useful.

```
ripken <- Batting %>% filter(playerID == "ripkeca01")
nrow(inner_join(ripken, mlb, by = c("yearID" = "yearID")))
```

```
[1] 9
```

```
nrow(inner_join(mlb, ripken, by = c("yearID" = "yearID"))) #same
```

```
[1] 9
```

For seasons when Ramirez did not play, NA's will be returned.

```
ripken %>%
  left_join(mlb, by = c("yearID" = "yearID")) %>%
  select(yearID, playerID, lgOPS) %>%
  head(3)

  yearID  playerID lgOPS
1   1981 ripkeca01    NA
2   1982 ripkeca01    NA
3   1983 ripkeca01    NA
```

Conversely, by reversing the order of the tables in the join, we return the 19 seasons for which we have already computed the league averages, regardless of whether there is a match for Ripken (results not displayed).

```
mlb %>%
  left_join(ripken, by = c("yearID" = "yearID")) %>%
  select(yearID, playerID, lgOPS)
```

4.5   Further resources

Hadley Wickham is an enormously influential innovator in the field of statistical computing. Along with his colleagues at RStudio and other organizations, he has made significant contributions to improve data wrangling in R. These packages are sometimes called the "Hadleyverse" or the "*tidyverse*," and are now manageable through a single tidyverse [231] package. His papers and vignettes describing widely used packages such as dplyr [234] and tidyr [230] are highly recommended reading. In particular, his paper on tidy data [218] builds upon notions of normal forms—common to database designers from computer science—to describe a process of thinking about how data should be stored and formatted. Finzer [77] writes of a "data habit of mind" that needs to be inculcated among data scientists. The RStudio data wrangling cheat sheet is a useful reference.

Sean Lahman, a self-described "database journalist," has long curated his baseball data set, which feeds the popular website baseball-reference.com. Michael Friendly maintains the Lahman R package [80]. For the baseball enthusiast, Cleveland Indians analyst Max Marchi and Jim Albert have written an excellent book on analyzing baseball data in R [140]. Albert has also written a book describing how baseball can be used as a motivating example for teaching statistics [2].

4.6   Exercises

Exercise 4.1

Each of these tasks can be performed using a single data verb. For each task, say which verb it is:

1. Find the average of one of the variables.

2. Add a new column that is the ratio between two variables.

3. Sort the cases in descending order of a variable.

4. Create a new data table that includes only those cases that meet a criterion.

5. From a data table with three categorical variables A, B, and C, and a quantitative variable X, produce a data frame that has the same cases but only the variables A and X.

Exercise 4.2

Use the nycflights13 package and the flights data frame to answer the following questions: What month had the highest proportion of cancelled flights? What month had the lowest? Interpret any seasonal patterns.

Exercise 4.3

Use the nycflights13 package and the flights data frame to answer the following question: What plane (specified by the tailnum variable) traveled the most times from New York City airports in 2013? Plot the number of trips per week over the year.

Exercise 4.4

Use the nycflights13 package and the flights and planes tables to answer the following questions: What is the oldest plane (specified by the tailnum variable) that flew from New York City airports in 2013? How many airplanes that flew from New York City are included in the planes table?

Exercise 4.5

Use the nycflights13 package and the flights and planes tables to answer the following questions: How many planes have a missing date of manufacture? What are the five most common manufacturers? Has the distribution of manufacturer changed over time as reflected by the airplanes flying from NYC in 2013? (Hint: you may need to recode the manufacturer name and collapse rare vendors into a category called Other.)

Exercise 4.6

Use the nycflights13 package and the weather table to answer the following questions: What is the distribution of temperature in July, 2013? Identify any important outliers in terms of the wind_speed variable. What is the relationship between dewp and humid? What is the relationship between precip and visib?

Exercise 4.7

Use the nycflights13 package and the weather table to answer the following questions: On how many days was there precipitation in the New York area in 2013? Were there differences in the mean visibility (visib) based on the day of the week and/or month of the year?

Exercise 4.8

Define two new variables in the Teams data frame from the Lahman package: batting average (BA) and slugging percentage (SLG). Batting average is the ratio of hits (H) to at-bats (AB), and slugging percentage is total bases divided by at-bats. To compute total bases, you get 1 for a single, 2 for a double, 3 for a triple, and 4 for a home run.

Exercise 4.9

Plot a time series of SLG since 1954 conditioned by lgID. Is slugging percentage typically higher in the American League (AL) or the National League (NL)? Can you think of why this might be the case?

Exercise 4.10

Display the top 15 teams ranked in terms of slugging percentage in MLB history. Repeat this using teams since 1969.

Exercise 4.11

The Angels have at times been called the California Angels (CAL), the Anaheim Angels (ANA), and the Los Angeles Angels of Anaheim (LAA). Find the 10 most successful seasons in Angels history. Have they ever won the World Series?

Exercise 4.12

Create a factor called election that divides the yearID into four-year blocks that correspond to U.S. presidential terms. During which term have the most home runs been hit?

Exercise 4.13

Name every player in baseball history who has accumulated at least 300 home runs (HR) and at least 300 stolen bases (SB).

Exercise 4.14

Name every pitcher in baseball history who has accumulated at least 300 wins (W) and at least 3,000 strikeouts (SO).

Exercise 4.15

Identify the name and year of every player who has hit at least 50 home runs in a single season. Which player had the lowest batting average in that season?

Exercise 4.16

The Relative Age Effect is an attempt to explain anomalies in the distribution of birth month among athletes. Briefly, the idea is that children born just after the age cut-off for participation will be as much as 11 months older than their fellow athletes, which is enough of a disparity to give them an advantage. That advantage will then be compounded over the years, resulting in notably more professional athletes born in these months. Display the distribution of birth months of baseball players who batted during the decade of the 2000s. How are they distributed over the calendar year? Does this support the notion of a relative age effect? Use the Births78 data set from the mosaicData package as a reference.

Exercise 4.17

The Violations data set in the mdsr package contains information regarding the outcome of health inspections of restaurants in New York City. Use these data to calculate the median violation score by zip code for zip codes in Manhattan with 50 or more inspections. What pattern do you see between the number of inspections and the median score?

Exercise 4.18

Download data on the number of deaths by firearm from the Florida Department of Law Enforcement. Wrangle these data and use ggplot2 to re-create Figure 6.1.

Chapter 5

Tidy data and iteration

In this chapter, we will continue to develop data wrangling skills. In particular, we will discuss tidy data, how to automate iterative processes, common file formats, and techniques for scraping and cleaning data, especially dates. Together with the material from Chapter 4, these skills will provide facility with wrangling data that is foundational for data science.

5.1 Tidy data

5.1.1 Motivation

One popular source of data is Gapminder [180], the brainchild of Swedish physician and public health researcher Hans Rosling. Gapminder contains data about countries over time for a variety of different variables such as the prevalence of HIV (human immunodeficiency virus) among adults aged 15 to 49 and other health and economic indicators. These data are stored in Google Spreadsheets, or one can download them as Microsoft Excel workbooks. The typical presentation of a small subset of such data is shown below, where we have used the googlesheets package to pull these data directly into R.

```
library(mdsr)
library(googlesheets)
hiv_key <- "pyj6tScZqmEfbZylOqjbiRQ"
hiv <- gs_key(hiv_key, lookup = FALSE) %>%
  gs_read(ws = "Data", range = cell_limits(c(1, 1), c(276, 34)))
names(hiv)[1] <- "Country"
hiv %>%
  filter(Country %in% c("United States", "France", "South Africa")) %>%
  select(Country, `1979`, `1989`, `1999`, `2009`)
```

```
# A tibble: 3  5
        Country `1979` `1989` `1999` `2009`
          <chr>  <dbl>  <lgl>  <dbl>  <dbl>
1          France     NA     NA    0.3    0.4
2   South Africa     NA     NA   14.8   17.2
3  United States 0.0318     NA    0.5    0.6
```

The data set has the form of a two-dimensional array where each of the $n = 3$ rows represents a country and each of the $p = 4$ columns is a year. Each entry represents the percentage of adults aged 15 to 49 living with HIV in the ith country in the jth year. This presentation of the data has some advantages. First, it is possible (with a big enough monitor) to *see* all of the data. One can quickly follow the trend over time for a particular country, and one can also estimate quite easily the percentage of data that is missing (e.g., NA). Thus, if visual inspection is the primary analytical technique, this *spreadsheet*-style presentation can be convenient.

Alternatively, consider this presentation of those same data.

```
library(tidyr)
hiv_long <- hiv %>% gather(key = Year, value = hiv_rate, -Country)
hiv_long %>%
  filter(Country %in% c("United States", "France", "South Africa")) %>%
  filter(Year %in% c(1979, 1989, 1999, 2009))
```

```
# A tibble: 12  3
           Country  Year hiv_rate
             <chr> <chr>    <dbl>
1           France  1979       NA
2     South Africa  1979       NA
3    United States  1979   0.0318
4           France  1989       NA
5     South Africa  1989       NA
6    United States  1989       NA
7           France  1999   0.3000
8     South Africa  1999  14.8000
9    United States  1999   0.5000
10          France  2009   0.4000
11    South Africa  2009  17.2000
12   United States  2009   0.6000
```

While our data can still be represented by a two-dimensional array, it now has $np = 12$ rows and just three columns. Visual inspection of the data is now more difficult, since our data are long and very narrow—the aspect ratio is not similar to that of our screen.

It turns out that there are substantive reasons to prefer the long (or tall), narrow version of these data. With multiple tables (see Chapter 12), it is a more efficient way for the computer to store and retrieve the data. It is more convenient for the purpose of data analysis. And it is more scalable, in that the addition of a second variable simply contributes another column, whereas to add another variable to the spreadsheet presentation would require a confusing three-dimensional view, multiple tabs in the spreadsheet, or worse, merged cells.

These gains come at a cost: we have relinquished our ability to *see all the data at once*. When data sets are small, being able to see them all at once can be useful, and even comforting. But in this era of big data, a quest to see all the data at once in a spreadsheet layout is a fool's errand. Learning to manage data via programming frees us from the click-and-drag paradigm popularized by spreadsheet applications, allows us to work with data of arbitrary size, and reduces errors. Recording our data management operations in code also makes them reproducible (see Appendix D)—an increasingly necessary trait in this era of collaboration. It enables us to fully separate the raw data from our analysis, which is difficult to achieve using a spreadsheet.

Pro Tip: Always keep your raw data and your analysis in separate files. Store the uncorrected data file (with errors and problems) and make corrections with a script (see Appendix D) file that transforms the raw data into the data that will actually be analyzed. This process will maintain the provenance of your data and allow analyses to be updated with new data without having to start data wrangling from scratch.

The long, narrow format for the Gapminder data that we have outlined above is called *tidy* [218]. In what follows we will further expand upon this notion, and develop more sophisticated techniques for wrangling data.

5.1.2   What are tidy data?

Data can be as simple as a column of numbers in a spreadsheet file or as complex as the electronic medical records collected by a hospital. A newcomer to working with data may expect each source of data to be organized in a unique way and to require unique techniques. The expert, however, has learned to operate with a small set of standard tools. As you'll see, each of the standard tools performs a comparatively simple task. Combining those simple tasks in appropriate ways is the key to dealing with complex data.

One reason the individual tools can be simple is that each tool gets applied to data arranged in a simple but precisely defined pattern called *tidy data*. Tidy data exists in systematically defined *data tables* (e.g., the rectangular arrays of data seen previously), but not all data tables are tidy.

To illustrate, Table 5.1 shows a handful of entries from a large United States Social Security Administration tabulation of names given to babies. In particular, the table shows how many babies of each sex were given each name in each year.

year	sex	name	n
1955	F	Judine	5
2002	M	Kadir	6
1935	F	Jerre	11
1935	F	Elynor	12
1910	M	Bertram	33
1985	F	Kati	212
1942	M	Grafton	22

Table 5.1: A data table showing how many babies were given each name in each year in the U.S., for a few names.

Table 5.1 shows that there were 6 boys named Kadir born in the U.S. in 2002 and 12 girls named Elynor born in 1935. As a whole, the babynames data table covers the years 1880 through 2014 and includes a total of 337,135,426 individuals, somewhat larger than the current population of the U.S.

The data in Table 5.1 are *tidy* because they are organized according to two simple rules.

1. The rows, called *cases* or observations, each refer to a specific, unique, and similar sort of thing, e.g., girls named Elynor in 1935.

2. The columns, called *variables*, each have the same sort of value recorded for each row. For instance, n gives the number of babies for each case; sex tells which gender was assigned at birth.

When data are in tidy form, it is relatively straightforward to transform the data into arrangements that are more useful for answering interesting questions. For instance, you might wish to know which were the most popular baby names over all the years. Even though Table 5.1 contains the popularity information implicitly, we need to re-arrange these data by adding up the counts for a name across all the years before the popularity becomes obvious, as in Table 5.2.

```
popular_names <- babynames %>%
    group_by(sex, name) %>%
    summarize(total_births = sum(n)) %>%
    arrange(desc(total_births))
```

	sex	name	total_births
1	M	James	5105919
2	M	John	5084943
3	M	Robert	4796695
4	M	Michael	4309198
5	F	Mary	4115282
6	M	William	4055473
7	M	David	3577704
8	M	Joseph	2570095
9	M	Richard	2555330
10	M	Charles	2364332

Table 5.2: The most popular baby names across all years.

The process of transforming information that is implicit in a data table into another data table that gives the information explicitly is called *data wrangling*. The wrangling itself is accomplished by using *data verbs* that take a tidy data table and transform it into another tidy data table in a different form. In Chapter 4, you were introduced to several *data verbs*.

Table 5.3 displays results from the Minneapolis mayoral election. Unlike `babynames`, it is not in tidy form, though the display is attractive and neatly laid out. There are helpful labels and summaries that make it easy for a person to read and draw conclusions. (For instance, Ward 1 had a higher voter turnout than Ward 2, and both wards were lower than the city total.)

However, being neat is not what makes data *tidy*. Table 5.3 violates the first rule for tidy data.

1. Rule 1: The rows, called *cases*, each must represent the same underlying attribute, that is, the same kind of thing.

   That's not true in Table 5.3. For most of the table, the rows represent a single precinct. But other rows give ward or city-wide totals. The first two rows are captions describing the data, not cases.

2. Rule 2: Each column is a variable containing the same type of value for each case.

   That's mostly true in Table 5.3, but the tidy pattern is interrupted by labels that are not variables. For instance, the first two cells in row 15 are the label "Ward 1 Subtotal," which is different from the ward/precinct identifiers that are the values in most of the first column.

Conforming to the rules for tidy data simplifies summarizing and analyzing data. For instance, in the tidy babynames table, it is easy (for a computer) to find the total number

	A	B	E	F	G	H	I	J
1			**City of Minneapolis Statistics**					
2			**General Election November 5, 2013**					
3	Ward	Precinct	Voters Registering by Absentee	Total Registrations	Voters at Polls	Absentee Voters	Total Ballots Cast	Total Turnout
4	**City-Wide Total**		**708**	**6,634**	**75,145**	**4,954**	**80,099**	**33.38%**
5								
6	1	1	3	28	492	27	519	27.23%
7	1	2	1	44	836	56	892	31.71%
8	1	3	0	40	905	19	924	38.87%
9	1	4	5	29	768	26	794	36.62%
10	1	5	0	31	683	31	714	37.46%
11	1	6	0	69	739	20	759	32.62%
12	1	7	0	47	291	8	299	15.79%
13	1	8	0	43	415	5	420	30.55%
14	1	9	0	42	596	25	621	25.42%
15	**Ward 1 Subtotal**		**9**	**373**	**5,725**	**217**	**5,942**	**30.93%**
16								
17	2	1	1	63	1,011	39	1,050	36.42%
18	2	2	5	44	679	37	716	50.39%
19	2	3	4	48	324	18	342	18.88%
20	2	4	0	53	117	3	120	7.34%
21	2	5	2	50	495	26	521	25.49%
22	2	6	1	36	433	19	452	39.10%
23	2	7	0	39	138	7	145	13.78%
24	2	8	1	50	1,206	36	1,242	47.90%
25	2	9	2	39	351	16	367	30.56%
26	2	10	0	87	196	5	201	6.91%
27	**Ward 2 Subtotal**		**16**	**509**	**4,950**	**206**	**5,156**	**27.56%**
28								
29	3	1	0	52	165	1	166	7.04%

Table 5.3: Ward and precinct votes cast in the 2013 Minneapolis mayoral election.

of babies: just add up all the numbers in the n variable. It is similarly easy to find the number of cases: just count the rows. And if you want to know the total number of Ahmeds or Sherinas across the years, there is an easy way to do that.

In contrast, it would be more difficult in the Minneapolis election data to find, say, the total number of ballots cast. If you take the seemingly obvious approach and add up the numbers in column I of Table 5.3 (labelled "Total Ballots Cast"), the result will be *three times* the true number of ballots, because some of the rows contain summaries, not cases.

Indeed, if you wanted to do calculations based on the Minneapolis election data, you would be far better off to put it in a tidy form.

The tidy form in Table 5.4 is, admittedly, not as attractive as the form published by the Minneapolis government. But it is much easier to use for the purpose of generating summaries and analyses.

Once data are in a tidy form, you can present them in ways that can be more effective than a formatted spreadsheet. For example, the data graphic in Figure 5.1 presents the turnout in each ward in a way that makes it easy to see how much variation there is within and among precincts.

The tidy format also makes it easier to bring together data from different sources. For instance, to explain the variation in voter turnout, you might want to consider variables such as party affiliation, age, income, etc. Such data might be available on a ward-by-ward basis from other records, such as public voter registration logs and census records. Tidy data can be wrangled into forms that can be connected to one another (i.e., using the inner_join() function from Chapter 4). This task would be difficult if you had to deal with an idiosyncratic format for each different source of data.

ward	precinct	registered	voters	absentee	total_turnout
1	1	28	492	27	0.27
1	4	29	768	26	0.37
1	7	47	291	8	0.16
2	1	63	1011	39	0.36
2	4	53	117	3	0.07
2	7	39	138	7	0.14
2	10	87	196	5	0.07
3	3	71	893	101	0.37
3	6	102	927	71	0.35

Table 5.4: A selection from the Minneapolis election data in tidy form.

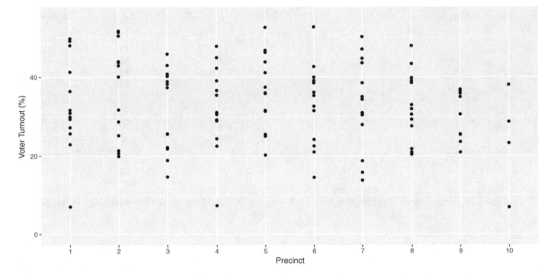

Figure 5.1: A graphical depiction of voter turnout in the different wards.

Variables

In data science, the word *variable* has a different meaning than in mathematics. In algebra, a variable is an unknown quantity. In data, a variable is known—it has been measured. Rather, the word *variable* refers to a specific quantity or quality that can vary from case to case. There are two major types of variables:

- Categorical variables record type or category and often take the form of a word.

- Quantitative variables record a numerical attribute. A quantitative variable is just what it sounds like: a number.

A *categorical variable* tells into which category or group a case falls. For instance, in the baby names data table, sex is a categorical variable with two levels F and M, standing for female and male. Similarly, the name variable is categorical. It happens that there are 93,889 different levels for name, ranging from Aaron, Ab, and Abbie to Zyhaire, Zylis, and Zymya.

	Precinct	First	Second	Third	Ward
6	P-04	undervote	undervote	undervote	W-6
2	P-06	BOB FINE	MARK ANDREW	undervote	W-10
10	P-02D	NEAL BAXTER	BETSY HODGES	DON SAMUELS	W-7
5	P-01	DON SAMUELS	undervote	undervote	W-5
27	P-03	CAM WINTON	DON SAMUELS	OLE SAVIOR	W-1

Table 5.5: Individual ballots in the Minneapolis election. Each voter votes in one ward in one precinct. The ballot marks the voter's first three choices for mayor.

Cases and what they represent

As noted previously, a row of a tidy data table refers to a case. To this point, you may have little reason to prefer the word *case* to *row*. When working with a data table, it is important to keep in mind what a case stands for in the real world. Sometimes the meaning is obvious. For instance, Table 5.5 is a tidy data table showing the ballots in the Minneapolis mayoral election in 2013. Each case is an individual voter's ballot. (The voters were directed to mark their ballot with their first choice, second choice, and third choice among the candidates. This is part of a procedure called rank choice voting.)

The case in Table 5.5 is a different sort of thing than the case in Table 5.4. In Table 5.4, a case is a ward in a precinct. But in Table 5.5, the case is an individual ballot. Similarly, in the baby names data (Table 5.1), a case is a name and sex and year while in Table 5.2 the case is a name and sex.

When thinking about cases, ask this question: What description would make every case unique? In the vote summary data, a precinct does not uniquely identify a case. Each individual precinct appears in several rows. But each precinct and ward combination appears once and only once. Similarly, in Table 5.1, name and sex do not specify a unique case. Rather, you need the combination of name-sex-year to identify a unique row.

Runners and races

Table 5.6 displays some of the results from a 10-mile running race held each year in Washington, D.C.

What is the meaning of a case here? It is tempting to think that a case is a person. After all, it is people who run road races. But notice that individuals appear more than once: Jane Poole ran each year from 2003 to 2007. (Her times improved consistently as she got older!) Jane Smith ran in the races from 1999 to 2006, missing only the year 2000 race. This suggests that the case is a runner in one year's race.

Codebooks

Data tables do not necessarily display all the variables needed to figure out what makes each row unique. For such information, you sometimes need to look at the documentation of how the data were collected and what the variables mean.

The codebook is a document—separate from the data table—that describes various aspects of how the data were collected, what the variables mean and what the different levels of categorical variables refer to. The word *codebook* comes from the days when data was encoded for the computer in ways that make it hard for a human to read. A codebook should include information about how the data were collected and what constitutes a case. Figure 5.2 shows the codebook for the babynames data in Table 5.1. In R, codebooks for data tables are available from the help() function.

	name.yob	sex	age	year	gun
1	jane polanek 1974	F	32	2006	114.50
2	jane poole 1948	F	55	2003	92.72
3	jane poole 1948	F	56	2004	87.28
4	jane poole 1948	F	57	2005	85.05
5	jane poole 1948	F	58	2006	80.75
6	jane poole 1948	F	59	2007	78.53
7	jane schultz 1964	F	35	1999	91.37
8	jane schultz 1964	F	37	2001	79.13
9	jane schultz 1964	F	38	2002	76.83
10	jane schultz 1964	F	39	2003	82.70
11	jane schultz 1964	F	40	2004	87.92
12	jane schultz 1964	F	41	2005	91.47
13	jane schultz 1964	F	42	2006	88.43
14	jane smith 1952	F	47	1999	90.60
15	jane smith 1952	F	49	2001	97.87

Table 5.6: An excerpt of runners' performance over time in a 10-mile race.

```
help(HELPrct)
```

For the runners data in Table 5.6, a codebook should tell you that the meaning of the gun variable is the time from when the start gun went off to when the runner crosses the finish line and that the unit of measurement is *minutes*. It should also state what might be obvious: that `age` is the person's age in years and `sex` has two levels, male and female, represented by `M` and `F`.

Multiple tables

It is often the case that creating a meaningful display of data involves combining data from different sources and about different kinds of things. For instance, you might want your analysis of the runners' performance data in Table 5.6 to include temperature and precipitation data for each year's race. Such weather data is likely contained in a table of daily weather measurements.

In many circumstances, there will be multiple tidy tables, each of which contains information relative to your analysis but has a different kind of thing as a case. We saw in Chapter 4 how the `inner_join()` and `left_join()` functions can be used to combine multiple tables, and in Chapter 12 we will further develop skills for working with relational databases. For now, keep in mind that being tidy is not about shoving everything into one table.

5.2   Reshaping data

Each row of a tidy data table is an individual case. It is often useful to re-organize the same data in a such a way that a case has a different meaning. This can make it easier to perform wrangling tasks such as comparisons, joins, and the inclusion of new data.

Consider the format of `BP_wide` shown in Table 5.7, in which each case is a research study subject and there are separate variables for the measurement of systolic blood pressure (SBP) before and after exposure to a stressful environment. Exactly the same data can

Description: The HELP study was a clinical trial for adult inpatients recruited from a detoxification unit. Patients with no primary care physician were randomized to receive a multidisciplinary assessment and a brief motivational intervention or usual care, with the goal of linking them to primary medical care.

Usage: data(HELPrct)

Format: Data frame with 453 observations on the following variables.

   age: subject age at baseline (in years)

   anysub: use of any substance post-detox: a factor with levels no yes

   cesd: Center for Epidemiologic Studies Depression measure at baseline (possible range 0-60: high scores indicate more depressive symptoms)

   d1: lifetime number of hospitalizations for medical problems (measured at baseline)

   daysanysub: time (in days) to first use of any substance post-detox

   ...

Details: Eligible subjects were adults, who spoke Spanish or English, reported alcohol, heroin or cocaine as their first or second drug of choice, resided in proximity to the primary care clinic to which they would be referred or were homeless. Patients with established primary care relationships they planned to continue, significant dementia, specific plans to leave the Boston area that would prevent research participation, failure to provide contact information for tracking purposes, or pregnancy were excluded.

Source: http://nhorton.people.amherst.edu/help

Figure 5.2: Part of the codebook for the HELPrct data table from the mosaicData package.

be presented in the format of the BP_narrow data table (Table 5.8), where the case is an individual occasion for blood-pressure measurement.

subject	before	after
BHO	160	115
GWB	120	135
WJC	105	145

Table 5.7: BP_wide: a data table in a wide format

Each of the formats BP_wide and BP_narrow has its advantages and its disadvantages. For example, it is easy to find the before-and-after change in blood pressure using BP_wide.

```
BP_wide %>% mutate(change = after - before)
```

On the other hand, a narrow format is more flexible for including additional variables, for example the date of the measurement or the diastolic blood pressure as in Table 5.9. The narrow format also makes it feasible to add in additional measurement occasions. For instance, Table 5.9 shows several "after" measurements for subject WJC. (Such *repeated measures* are a common feature of scientific studies.) A simple strategy allows you to get the benefits of either format: convert from wide to narrow or from narrow to wide as suits your purpose.

subject	when	sbp
BHO	before	160
GWB	before	120
WJC	before	105
BHO	after	115
GWB	after	135
WJC	after	145

Table 5.8: `BP_narrow`: a tidy data table in a narrow format.

subject	when	sbp	dbp	date
BHO	before	160	69	13683.00
GWB	before	120	54	10337.00
BHO	before	155	65	13095.00
WJC	after	145	75	12006.00
WJC	after	NA	65	14694.00
WJC	after	130	60	15963.00
GWB	after	135	NA	14372.00
WJC	before	105	60	7533.00
BHO	after	115	78	17321.00

Table 5.9: A data table extending the information in Tables 5.8 and 5.7 to include additional variables and repeated measurements. The narrow format facilitates including new cases or variables.

5.2.1   Data verbs for converting wide to narrow and *vice versa*

Transforming a data table from wide to narrow is the action of the gather() data verb: A wide data table is the input and a narrow data table is the output. The reverse task, transforming from narrow to wide, involves the data verb spread(). Both functions are implemented in the tidyr package.

5.2.2   Spreading

The spread() function converts a data table from narrow to wide. Carrying out this operation involves specifying some information in the arguments to the function. The value is the variable in the narrow format that is to be divided up into multiple variables in the resulting wide format. The key is the name of the variable in the narrow format that identifies for each case individually which column in the wide format will receive the value.

For instance, in the narrow form of BP_narrow (Table 5.8) the value variable is sbp. In the corresponding wide form, BP_wide (Table 5.7), the information in sbp will be spread between two variables: before and after. The key variable in BP_narrow is when. Note that the different categorical levels in when specify which variable in BP_wide will be the destination for the sbp value of each case. Only the key and value variables are involved in the transformation from narrow to wide. Other variables in the narrow table, such as subject in BP_narrow, are used to define the cases. Thus, to translate from BP_narrow to BP_wide we would write this code:

```
BP_narrow %>% spread(key = when, value = sbp)
```

5.2.3 Gathering

Now consider how to transform BP_wide into BP_narrow. The names of the variables to be gathered together, before and after, will become the categorical levels in the narrow form. That is, they will make up the key variable in the narrow form. The data analyst has to invent a name for this variable. There are all sorts of sensible possibilities, for instance before_or_after. In gathering BP_wide into BP_narrow, the concise variable name when was chosen.

Similarly, a name must be specified for the variable that is to hold the values in the variables being gathered. Again, there are many reasonable possibilities. It is sensible to choose a name that reflects the kind of thing those values are, in this case systolic blood pressure. So, sbp is a good choice.

Finally, the analyst needs to specify which variables are to be gathered. For instance, it hardly makes sense to gather subject with the other variables; it will remain as a separate variable in the narrow result. Values in subject will be repeated as necessary to give each case in the narrow format its own correct value of subject. In summary, to convert BP_wide into BP_narrow, we run the following command.

```
BP_wide %>% gather(key = when, value = sbp, before, after)
```

The names of the key and value arguments are given as arguments. These are the names invented by the data analyst; those names are not part of the wide input to gather(). The arguments after the key and value are the names of the variables to be gathered.

5.2.4 Example: Gender-neutral names

In "A Boy Named Sue" country singer Johnny Cash famously told the story of a boy toughened in life—eventually reaching gratitude—by being given a girl's name. The conceit is of course the rarity of being a boy with the name Sue, and indeed, Sue is given to about 300 times as many girls as boys (at least being recorded in this manner: Data entry errors may account for some of these names).

```
babynames %>%
   filter(name == "Sue") %>%
   group_by(name, sex) %>%
   summarise(total = sum(n))

Source: local data frame [2 x 3]
Groups: name [?]

   name   sex   total
   <chr> <chr> <int>
1   Sue     F  144424
2   Sue     M     519
```

On the other hand, some names that are predominantly given to girls are also commonly given to boys. Although only 15% of people named Robin are male, it is easy to think of a few famous men with that name: the actor Robin Williams, the singer Robin Gibb, and the basketball player Robin Lopez (not to mention Batman's sidekick) come to mind.

```
babynames %>%
  filter(name == "Robin") %>%
  group_by(name, sex) %>%
  summarise(total = sum(n))

Source: local data frame [2 x 3]
Groups: name [?]

    name   sex   total
   <chr> <chr>   <int>
1 Robin     F  288636
2 Robin     M   44026
```

This computational paradigm (e.g., filtering) works well if you want to look at gender balance in one name at a time, but suppose you want to find the most gender-neutral names from all 93,889 names in babynames? For this, it would be useful to have the results in a wide format, like the one shown below.

```
babynames %>%
  filter(name %in% c("Sue", "Robin", "Leslie")) %>%
  group_by(name, sex) %>%
  summarise(total = sum(n)) %>%
  spread(key = sex, value = total, fill=0)

Source: local data frame [3 x 3]
Groups: name [3]

      name      F      M
*    <chr>  <dbl>  <dbl>
1 Leslie  264054 112533
2  Robin  288636  44026
3    Sue  144424    519
```

The spread() function can help us generate the wide format. Note that the sex variable is the key used in the conversion. A fill of zero is appropriate here: For a name like Aaban or Aadam, where there are no females, the entry for F should be zero.

```
BabyWide <- babynames %>%
  group_by(sex, name) %>%
  summarize(total = sum(n)) %>%
  spread(key = sex, value = total, fill = 0)
head(BabyWide, 3)

# A tibble: 3  3
    name      F     M
   <chr> <dbl> <dbl>
1 Aaban      0    72
2 Aabha     21     0
3 Aabid      0     5
```

One way to define "approximately the same" is to take the smaller of the ratios M/F and F/M. If females greatly outnumber males, then F/M will be large, but M/F will be small. If the sexes are about equal, then both ratios will be near one. The smaller will never be greater than one, so the most balanced names are those with the smaller of the ratios near one.

The code to identify the most balanced gender-neutral names out of the names with more than 50,000 babies of each sex are shown below. Remember, a ratio of one means exactly balanced; a ratio of 0.5 means two to one in favor of one sex; 0.33 means three to one. (The pmin() transformation function returns the smaller of the two arguments for each individual case.)

```
BabyWide %>%
  filter(M > 50000, F > 50000) %>%
  mutate(ratio = pmin(M / F, F / M) ) %>%
  arrange(desc(ratio)) %>%
  head(3)
```

```
# A tibble: 3  4
    name      F       M ratio
   <chr> <dbl>   <dbl> <dbl>
1  Riley 81605   87494 0.933
2 Jackie 90337   78148 0.865
3  Casey 75060  108595 0.691
```

Riley has been the most gender-balanced name, followed by Jackie. Where does your name fall on this list?

5.3 Naming conventions

Like any language, R has some rules that you cannot break, but also many conventions that you can—but should not—break. There are a few simple rules that apply when creating a *name* for an object:

- The name cannot start with a digit. So you cannot assign the name 100NCHS to a data frame, but NCHS100 is fine. This rule is to make it easy for R to distinguish between object names and numbers. It also helps you avoid mistakes such as writing 2pi when you mean 2*pi.

- The name cannot contain any punctuation symbols other than . and _. So ?NCHS or N*Hanes are not legitimate names. However, you can use . and _ in a name. For reasons that will be explained later, the use of . in function names has a specific meaning, but should otherwise be avoided. The use of _ is preferred.

- The case of the letters in the name matters. So NCHS, nchs, Nchs, and nChs, etc., are all different names that only look similar to a human reader, not to R.

Pro Tip: Do not use . in function names, to avoid conflicting with internal functions.

One of R's strengths is its modularity—many people have contributed many packages that do many different things. However, this decentralized paradigm has resulted in many *different* people writing code using many *different* conventions. The resulting lack of uniformity can make code harder to read. We suggest adopting a style guide and sticking to it—we have attempted to do that in this book. However, the inescapable use of other people's code results in inevitable deviations from that style.

Two public style guides for R are widely adopted and influential: Google's R Style Guide and the Style Guide in Hadley Wickham's *Advanced R* book [220]. Needless to say, they don't always agree. In this book, we follow the latter as closely as possible. This means:

- We use underscores (_) in variable and function names. The use of periods (.) in function names is restricted to S3 methods.

- We use spaces liberally and prefer multiline, narrow blocks of code to single lines of wide code (although we have relaxed this in many of our examples to save space).

- We use CamelCase for the names of data tables. This means that each "word" in a name starts with a capital letter, but there are no spaces (e.g., `Teams`, `MedicareCharges`, `WorldCities`, etc.).

5.4   Automation and iteration

Calculators free human beings from having to perform arithmetic computations *by hand*. Similarly, programming languages free humans from having to perform iterative computations by re-running chunks of code, or worse, copying-and-pasting a chunk of code many times, while changing just one or two things in each chunk.

For example, in Major League Baseball there are 30 teams, and the game has been played for over 100 years. There are a number of natural questions that we might want to ask about *each team* (e.g., which player has accrued the most hits for that team?) or about each season (e.g., which seasons had the highest levels of scoring?). If we can write a chunk of code that will answer these questions for a single team or a single season, then we should be able to generalize that chunk of code to work for *all* teams or seasons. Furthermore, we should be able to do this without having to re-type that chunk of code. In this section, we present a variety of techniques for automating these types of iterative operations.

5.4.1   Vectorized operations

In every programming language that we can think of, there is a way to write a *loop*. For example, you can write a `for()` loop in R the same way you can with most programming languages. Recall that the `Teams` data frame contains one row for each team in each MLB season.

```
library(Lahman)
names(Teams)
```

```
 [1] "yearID"    "lgID"      "teamID"    "franchID"
 [5] "divID"     "Rank"      "G"         "Ghome"
 [9] "W"         "L"         "DivWin"    "WCWin"
[13] "LgWin"     "WSWin"     "R"         "AB"
[17] "H"         "X2B"       "X3B"       "HR"
[21] "BB"        "SO"        "SB"        "CS"
[25] "HBP"       "SF"        "RA"        "ER"
[29] "ERA"       "CG"        "SHO"       "SV"
[33] "IPouts"    "HA"        "HRA"       "BBA"
[37] "SOA"       "E"         "DP"        "FP"
```

```
[41] "name"           "park"          "attendance"    "BPF"
[45] "PPF"            "teamIDBR"      "teamIDlahman45" "teamIDretro"
```

What might not be immediately obvious is that columns 15 through 40 of this data frame contain numerical data about how each team performed in that season. To see this, you can execute the str() command to see the **structure** of the data frame, but we suppress that output here. For data frames, a similar alternative that is a little cleaner is glimpse().

```
str(Teams)
glimpse(Teams)
```

Regardless of your prior knowledge of baseball, you might be interested in computing the averages of these 26 numeric columns. However, you don't want to have to type the names of each of them, or re-type the mean() command 26 times. Thus, most programmers will immediately identify this as a situation in which a *loop* is a natural and efficient solution.

```
averages <- NULL
for (i in 15:40) {
  averages[i - 14] <- mean(Teams[, i], na.rm = TRUE)
}
names(averages) <- names(Teams)[15:40]
averages
```

R	AB	H	X2B	X3B	HR	BB	SO
681.946	5142.492	1346.273	227.625	47.104	101.137	473.649	737.949
SB	CS	HBP	SF	RA	ER	ERA	CG
112.272	48.766	56.096	44.677	681.946	570.895	3.815	50.481
SHO	SV	IPouts	HA	HRA	BBA	SOA	E
9.664	23.668	4022.383	1346.084	101.137	474.011	731.229	186.337
DP	FP						
140.186	0.962						

This certainly works. However, it is almost always possible (and usually preferable) to perform such operations in R without explicitly defining a loop. R programmers prefer to use the concept of applying an operation to each element in a vector. This often requires only one line of code, with no appeal to indices.

It is important to understand that the fundamental architecture of R is based on *vectors*. That is, in contrast to general-purpose programming languages like C++ or Python that distinguish between single items—like strings and integers—and arrays of those items, in R a "string" is just a character vector of length 1. There is no special kind of atomic object. Thus, if you assign a single "string" to an object, R still stores it as a vector.

```
a <- "a string"
class(a)
```

```
[1] "character"
```

```
length(a)
```

```
[1] 1
```

As a consequence of this construction, R is highly optimized for vectorized operations (see Appendix B for more detailed information about R internals). Loops, by their nature, do not take advantage of this optimization. Thus, R provides several tools for performing loop-like operations without actually writing a loop. This can be a challenging conceptual hurdle for those who are used to more general-purpose programming languages.

Pro Tip: Try to avoid writing `for()` loops in R, even when it seems like the easiest solution.

5.4.2   The `apply()` family of functions

To apply a function to the rows or columns of a matrix or data frame, use `apply()`. In this example, we calculate the mean of each of the statistics defined above, all at once. Compare this to the `for()` loop written above.

```
Teams %>%
  select(15:40) %>%
  apply(MARGIN = 2, FUN = mean, na.rm = TRUE)

        R         AB          H        X2B        X3B         HR         BB         SO
  681.946   5142.492   1346.273    227.625     47.104    101.137    473.649    737.949
       SB         CS        HBP         SF         RA         ER        ERA         CG
  112.272     48.766     56.096     44.677    681.946    570.895      3.815     50.481
      SHO         SV     IPouts         HA        HRA        BBA        SOA          E
    9.664     23.668   4022.383   1346.084    101.137    474.011    731.229    186.337
       DP         FP
  140.186      0.962
```

The first argument to `apply()` is the matrix or data frame that you want to do something to. The second argument specifies whether you want to apply the function FUN to the rows or the columns of the matrix. Any further arguments are passed as options to FUN. Thus, this command applies the `mean()` function to the 15th through the 40th columns of the Teams data frame, while removing any NAs that might be present in any of those columns.

Note that the row-wise averages have no meaning in this case, but you could calculate them by setting the MARGIN argument to 1 instead of 2:

```
Teams %>%
  select(15:40) %>%
  apply(MARGIN = 1, FUN = mean, na.rm = TRUE)
```

Of course, we began by taking the subset of the columns that were all numeric values. If you tried to take the `mean()` of a non-numeric vector, you would get a *warning* (and a value of NA).

```
Teams %>%
  select(teamID) %>%
  apply(MARGIN = 2, FUN = mean, na.rm = TRUE)

Warning in mean.default(x, ..., na.rm = na.rm):
argument is not numeric or logical:  returning NA
```

```
teamID
    NA
```

sapply() and lapply()

Often you will want to apply a function to each element of a vector or list. For example, the franchise now known as the Los Angeles Angels of Anaheim has gone by several names during its time in MLB.

```
angels <- Teams %>%
  filter(franchID == "ANA") %>%
  group_by(teamID, name) %>%
  summarise(began = first(yearID), ended = last(yearID)) %>%
  arrange(began)
angels
```

```
Source: local data frame [4 x 4]
Groups: teamID [3]

  teamID                           name began ended
  <fctr>                          <chr> <int> <int>
1    LAA            Los Angeles Angels  1961  1964
2    CAL            California Angels   1965  1996
3    ANA              Anaheim Angels    1997  2004
4    LAA Los Angeles Angels of Anaheim  2005  2015
```

The franchise began as the Los Angeles Angels (LAA) in 1961, then became the California Angels (CAL) in 1965, the Anaheim Angels (ANA) in 1997, before taking their current name (LAA again) in 2005. This situation is complicated by the fact that the teamID LAA was re-used. This sort of schizophrenic behavior is unfortunately common in many data sets.

Now, suppose we want to find the length, in number of characters, of each of those team names. We could check each one manually using the function nchar():

```
angels_names <- angels$name
nchar(angels_names[1])
```

```
[1] 18
```

```
nchar(angels_names[2])
```

```
[1] 17
```

```
nchar(angels_names[3])
```

```
[1] 14
```

```
nchar(angels_names[4])
```

```
[1] 29
```

But this would grow tiresome if we had many names. It would be simpler, more efficient, more elegant, and scalable to apply the function nchar() to each element of the vector angel_names. We can accomplish this using either sapply() or lapply().

```
sapply(angels_names, FUN = nchar)
```

```
       Los Angeles Angels               California Angels
                     18                               17
          Anaheim Angels Los Angeles Angels of Anaheim
                     14                               29
```

```
lapply(angels_names, FUN = nchar)
```

```
[[1]]
[1] 18

[[2]]
[1] 17

[[3]]
[1] 14

[[4]]
[1] 29
```

The key difference between sapply() and lapply() is that the former will try to return a vector or matrix, whereas the latter will always return a list. Recall that the main difference between lists and data.frames is that the elements (columns) of a data.frame have to have the same length, whereas the elements of a list are arbitrary. So while lapply() is more versatile, we usually find sapply() to be more convenient when it is appropriate.

Pro Tip: Use sapply() whenever you want to do something to each element of a vector, and get a vector in return.

One of the most powerful uses of these iterative functions is that you can apply *any* function, including a function that you have defined (see Appendix C for a discussion of how to write user-defined functions). For example, suppose we want to display the top 5 seasons in terms of wins for each of the Angels teams.

```
top5 <- function(x, teamname) {
  x %>%
    filter(name == teamname) %>%
    select(teamID, yearID, W, L, name) %>%
    arrange(desc(W)) %>%
    head(n = 5)
}
```

We can now do this for each element of our vector with a single call to lapply().

```
angels_list <- lapply(angels_names, FUN = top5, x = Teams)
angels_list
```

```
[[1]]
  teamID yearID  W  L                name
1    LAA   1962 86 76 Los Angeles Angels
```

```
2    LAA   1964 82 80 Los Angeles Angels
3    LAA   1961 70 91 Los Angeles Angels
4    LAA   1963 70 91 Los Angeles Angels

[[2]]
  teamID yearID W  L            name
1    CAL   1982 93 69 California Angels
2    CAL   1986 92 70 California Angels
3    CAL   1989 91 71 California Angels
4    CAL   1985 90 72 California Angels
5    CAL   1979 88 74 California Angels

[[3]]
  teamID yearID W  L           name
1    ANA   2002 99 63 Anaheim Angels
2    ANA   2004 92 70 Anaheim Angels
3    ANA   1998 85 77 Anaheim Angels
4    ANA   1997 84 78 Anaheim Angels
5    ANA   2000 82 80 Anaheim Angels

[[4]]
  teamID yearID  W  L                        name
1    LAA   2008 100 62 Los Angeles Angels of Anaheim
2    LAA   2014  98 64 Los Angeles Angels of Anaheim
3    LAA   2009  97 65 Los Angeles Angels of Anaheim
4    LAA   2005  95 67 Los Angeles Angels of Anaheim
5    LAA   2007  94 68 Los Angeles Angels of Anaheim
```

Finally, we can collect the results into a data frame by passing the resulting `list` to the `bind_rows()` function. Below, we do this and then compute the average number of wins in a top 5 seasons for each Angels team name. Based on these data, the Los Angeles Angels of Anaheim has been the most successful incarnation of the franchise, when judged by average performance in the best five seasons.

```
angels_list %>% bind_rows() %>%
  group_by(teamID, name) %>%
  summarize(N = n(), mean_wins = mean(W)) %>%
  arrange(desc(mean_wins))

Source: local data frame [4 x 4]
Groups: teamID [3]

  teamID                          name  N mean_wins
  <fctr>                         <chr> <int>    <dbl>
1    LAA Los Angeles Angels of Anaheim     5     96.8
2    CAL               California Angels   5     90.8
3    ANA                  Anaheim Angels   5     88.4
4    LAA              Los Angeles Angels   4     77.0
```

Once you've read Chapter 12, think about how you might do this operation in SQL. It is not that easy!

5.4.3   Iteration over subgroups with dplyr::do()

In Chapter 4 we introduced data *verbs* that could be chained to perform very powerful data wrangling operations. These functions—which come from the dplyr package—operate on data frames and return data frames. The do() function in dplyr allows you to apply an arbitrary function to the *groups* of a data frame. That is, you will first define a grouping using the group_by() function, and then apply a function to all of those groups. Note that this is similar to sapply(), in that you are mapping a function over a collection of values, but whereas the values used in sapply() are individual elements of a vector, in dplyr::do() they are groups defined on a data frame.

One of the more enduring models in sabermetrics is Bill James's formula for estimating a team's expected winning percentage, given knowledge only of the team's runs scored and runs allowed to date (recall that the team that scores the most runs wins a given game). This statistic is known—unfortunately—as Pythagorean Winning Percentage, even though it has nothing to do with Pythagoras. The formula is simple, but non-linear:

$$\widehat{WPct} = \frac{RS^2}{RS^2 + RA^2} = \frac{1}{1 + (RA/RS)^2},$$

where RS and RA are the number of runs the team has scored and allowed, respectively. If we define $x = RS/RA$ to be the team's *run ratio*, then this is a function of one variable having the form $f(x) = \frac{1}{1+(1/x)^2}$.

This model seems to fit quite well upon visual inspection—in Figure 5.3 we show the data since 1954, along with a line representing the model. Indeed, this model has also been successful in other sports, albeit with wholly different exponents.

```
exp_wpct <- function (x) {
  return(1/(1 + (1/x)^2))
}
TeamRuns <- Teams %>%
  filter(yearID >= 1954) %>%
  rename(RS = R) %>%
  mutate(WPct = W / (W + L), run_ratio = RS/RA) %>%
  select(yearID, teamID, lgID, WPct, run_ratio)
ggplot(data = TeamRuns, aes(x = run_ratio, y = WPct)) +
  geom_vline(xintercept = 1, color= "darkgray", linetype = 2) +
  geom_hline(yintercept = 0.5, color= "darkgray", linetype = 2) +
  geom_point(alpha = 0.3) +
  stat_function(fun = exp_wpct, size = 2, color = "blue") +
  xlab("Ratio of Runs Scored to Runs Allowed") + ylab("Winning Percentage")
```

However, the exponent of 2 was posited by James. One can imagine having the exponent become a parameter k, and trying to find the optimal fit. Indeed, researchers have found that in baseball, the optimal value of k is not 2, but something closer to 1.85 [208]. It is easy enough for us to find the optimal value using the fitModel() function from the mosaic package.

```
exWpct <- fitModel(WPct ~ 1/(1 + (1/run_ratio)^k), data = TeamRuns)
coef(exWpct)
```

```
   k
1.84
```

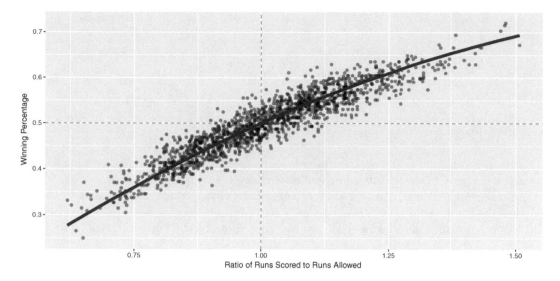

Figure 5.3: Fit for the Pythagorean Winning Percentage model for all teams since 1954.

Furthermore, researchers investigating this model have found that the optimal value of the exponent differs based on the era during which the model is fit. We can use the `dplyr::do()` function to do this for all decades in baseball history. First, we must write a short function that will return a data frame containing the optimal exponent.

```
fit_k <- function(x) {
  mod <- fitModel(formula = WPct ~ 1/(1 + (1/run_ratio)^k), data = x)
  return(data.frame(k = coef(mod)))
}
```

Note that this function will return the optimal value of the exponent over any time period.

```
fit_k(TeamRuns)

    k
k 1.84
```

Finally, we compute the decade for each year, and apply `fit_k()` to those decades. In the code below, the `.` refers to the result of the previous command, which in this case is the data frame containing the information for a single decade.

```
TeamRuns %>%
  mutate(decade = yearID %/% 10 * 10) %>%
  group_by(decade) %>%
  do(fit_k(x = .))

Source: local data frame [7 x 2]
Groups: decade [7]

  decade     k
```

```
     <dbl> <dbl>
1     1950  1.69
2     1960  1.90
3     1970  1.74
4     1980  1.93
5     1990  1.88
6     2000  1.94
7     2010  1.78
```

Note the variation in the optimal value of k. Even though the exponent is not the same in each decade, it varies within a fairly narrow range between 1.70 and 1.95.

As a second example, consider the problem of identifying the team in each season that led their league in home runs. We can easily write a function that will, for a specific year and league, return a data frame with one row that contains the team with the most home runs.

```
hr_leader <- function (x) {
# x is a subset of Teams for a single year and league
  x %>%
    select(yearID, lgID, teamID, HR) %>%
    arrange(desc(HR)) %>%
    head(n = 1)
}
```

We can verify that in 1961, the New York Yankees led the American League in home runs.

```
Teams %>%
  filter(yearID == 1961 & lgID == "AL") %>%
  hr_leader()

  yearID lgID teamID  HR
1   1961   AL    NYA 240
```

We can use dplyr::do() to quickly find all the teams that led their league in home runs.

```
hr_leaders <- Teams %>%
  group_by(yearID, lgID) %>%
  do(hr_leader(.))
head(hr_leaders, 4)

Source: local data frame [4 x 4]
Groups: yearID, lgID [4]

  yearID   lgID teamID     HR
   <int> <fctr> <fctr>  <int>
1   1871     NA    CH1     10
2   1872     NA    BL1     14
3   1873     NA    BS1     13
4   1874     NA    BS1     18
```

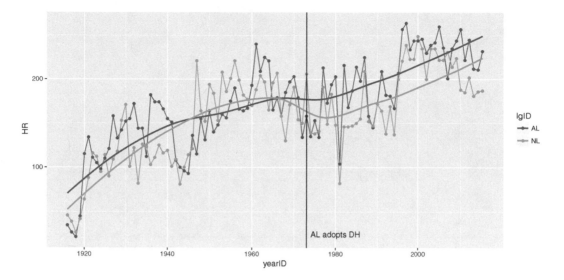

Figure 5.4: Number of home runs hit by the team with the most home runs, 1916–2014. Note how the AL has consistently bested the NL since the introduction of the designated hitter (DH) in 1973.

In this manner, we can compute the average number of home runs hit in a season by the team that hit the most.

```
mean(HR ~ lgID, data = hr_leaders)

   AA    AL    FL    NA    NL    PL    UA
 40.6 153.3  51.0  13.8 126.1  66.0  32.0

mean(HR ~ lgID, data = filter(hr_leaders, yearID >= 1916))

 AA  AL  FL  NA  NL  PL  UA
NaN 171 NaN NaN 158 NaN NaN
```

In Figure 5.4 we show how this number has changed over time. We restrict our attention to the years since 1916, during which only the AL and NL leagues have existed. We note that while the top HR hitting teams were comparable across the two leagues until the mid 1970s, the AL teams have dominated since their league adopted the designated hitter rule in 1973.

```
hr_leaders %>%
  filter(yearID >= 1916) %>%
  ggplot(aes(x = yearID, y = HR, color = lgID)) + geom_line() +
    geom_point() + geom_smooth(se = 0) + geom_vline(xintercept = 1973) +
    annotate("text", x=1974, y=25, label="AL adopts DH", hjust="left")
```

5.4.4 Iteration with `mosaic::do`

In the previous section we learned how to repeat operations while iterating over the elements of a vector. It can also be useful to simply repeat an operation many times and

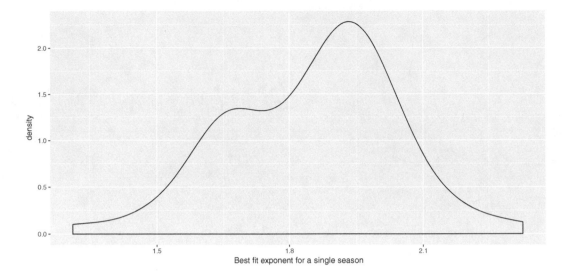

Figure 5.5: Distribution of best-fitting exponent across single seasons from 1961–2014.

collect the results. Obviously, if the result of the operation is deterministic (i.e., you get the same answer every time) then this is pointless. On the other hand, if this operation involves randomness, then you won't get the same answer every time, and understanding the distribution of values that your random operation produces can be useful. We will flesh out these ideas further in Chapter 10.

For example, in our investigation into the expected winning percentage in baseball, we determined that the optimal exponent fit to the 61 seasons worth of data from 1954 to 2014 was 1.85. However, we also found that if we fit this same model separately for each decade, that optimal exponent varies from 1.69 to 1.94. This gives us a rough sense of the variability in this exponent—we observed values between 1.6 and 2, which may give some insights as to plausible values for the exponent.

Nevertheless, our choice to stratify by decade was somewhat arbitrary. A more natural question might be: What is the distribution of optimal exponents fit to a *single-season*'s worth of data? How confident should we be in that estimate of 1.85?

We can use dplyr::do() and the function we wrote previously to compute the 61 actual values. The resulting distribution is summarized in Figure 5.5.

```
k_actual <- TeamRuns %>%
  group_by(yearID) %>%
  do(fit_k(.))
favstats(~ k, data = k_actual)
```

```
 min   Q1 median   Q3  max mean   sd  n missing
1.31 1.69   1.89 1.97 2.33 1.85 0.19 62       0
```

```
ggplot(data = k_actual, aes(x = k)) + geom_density() +
  xlab("Best fit exponent for a single season")
```

Since we only have 61 samples, we might obtain a better understanding of the sampling distribution of the mean k by *resampling*—sampling with replacement—from these 61 val-

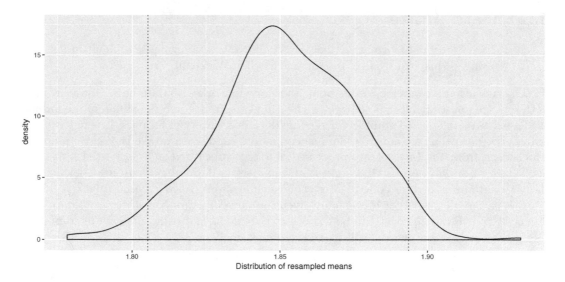

Figure 5.6: Bootstrap distribution of mean optimal exponent.

ues. (This is a statistical technique known as the *bootstrap*, which we describe further in Chapter 7.) A simple way to do this is with the do() function in the mosaic package.

```
bstrap <- do(1000) * mean(~ k, data = resample(k_actual))
head(bstrap, 3)

   mean
1 1.85
2 1.84
3 1.85

civals <- qdata(~ mean, c(0.025, .975), data = bstrap)
civals

       quantile     p
2.5%       1.81 0.025
97.5%      1.89 0.975
```

After repeating the resampling 1,000 times, we found that 95% of the resampled exponents were between 1.805 and 1.893, with our original estimates of 1.85 lying somewhere near the center of that distribution. This distribution, along the boundaries of the middle 95%, is depicted in Figure 5.6.

```
ggplot(data = bstrap, aes(x = mean)) + geom_density() +
  xlab("Distribution of resampled means") +
  geom_vline(data = civals, aes(xintercept = quantile), color = "red",
    linetype = 3)
```

5.5   Data intake

> Every easy data format is alike. Every difficult data format is difficult in its own way. —inspired by Leo Tolstoy and Hadley Wickham

The tools that we develop in this book allow one to work with data in R. However, most data sets are not available in R to begin with—they are often stored in a different file format. While R has sophisticated abilities for reading data in a variety of formats, it is not without limits. For data that are not in a file, one common form of data intake is *Web scraping*, in which data from the Internet are processed as (structured) text and converted into data. Such data often have errors that stem from blunders in data entry or from deficiencies in the way data are stored or coded. Correcting such errors is called *data cleaning.*

The native file format for R is usually given the suffix `.Rda` (or sometimes, `.RData`). Any object in your R environment can be written to this file format using the `save()` command. Using the `compress` argument will make these files smaller.

```
save(hr_leaders, file = "hr_leaders.rda", compress = "xz")
```

This file format is usually an efficient means for storing data, but it is not the most portable. To load a stored object into your R environment, use the `load()` command.

```
load(file = "hr_leaders.rda")
```

Pro Tip: Maintaining the provenance of data from beginning to the end of an analysis is an important part of a reproducible workflow. This can be facilitated by creating one R Markdown file or notebook that undertakes the data wrangling and generates an analytic data set (using `save()`) that can be read (using `load()`) into a second R Markdown file.

5.5.1   Data-table friendly formats

Many formats for data are essentially equivalent to data tables. When you come across data in a format that you don't recognize, it is worth checking whether it is one of the data-table friendly formats. Sometimes the filename extension provides an indication. Here are several, each with a brief description:

CSV: a non-proprietary comma separated text format that is widely used for data exchange between different software packages. CSVs are easy to understand, but are not compressed, and therefore can take up more space on disk than other formats.

Pro Tip: Be careful with date and time variables in CSV format: these can sometimes be formatted in inconsistent ways that make it more challenging to ingest.

Software-package specific format some common examples include:

    Octave (and through that, MATLAB): widely used in engineering and physics

    Stata: commonly used for economic research

    SPSS: commonly used for social science research

    Minitab: often used in business applications

SAS: often used for large data sets

Epi: used by the Centers for Disease Control (CDC) for health and epidemiology data

Relational databases: the form that much of institutional, actively-updated data are stored in. This includes business transaction records, government records, Web logs, and so on. (See Chapter 12 for a discussion of relational database management systems.)

Excel: a set of proprietary spreadsheet formats heavily used in business. Watch out, though. Just because something is stored in an Excel format doesn't mean it is a data table. Excel is sometimes used as a kind of tablecloth for writing down data with no particular scheme in mind.

Web-related: For example:

- HTML (hypertext markup language): `<table>` format
- XML (extensible markup language) format, a tree-based document structure
- JSON (JavaScript Object Notation) is an increasingly common data format that breaks the "rows-and-columns" paradigm (see Section 17.2.4)
- Google spreadsheets: published as HTML
- Application programming interfaces (API)

The procedure for reading data in one of these formats varies depending on the format. For Excel or Google spreadsheet data, it is sometimes easiest to use the application software to export the data as a CSV file. There are also R packages for reading directly from either (`readxl` and `googlesheets`, respectively), which are useful if the spreadsheet is being updated frequently. For the technical software package formats, the `foreign` R package provides useful reading and writing functions. For relational databases, even if they are on a remote server, there are several useful R packages that allow you to connect to these databases directly, most notably `dplyr` and `DBI`. CSV and HTML `<table>` formats are frequently encountered sources for data scraping. The next subsections give a bit more detail about how to read them into R.

CSV (comma separated value) files

This text format can be read with a huge variety of software. It has a data table format, with the values of variables in each case separated by commas. Here is an example of the first several lines of a CSV file:

```
"year","sex","name","n","prop"
1880,"F","Mary",7065,0.0723835869064085
1880,"F","Anna",2604,0.0266789611187951
1880,"F","Emma",2003,0.0205214896777829
1880,"F","Elizabeth",1939,0.0198657855642641
1880,"F","Minnie",1746,0.0178884278469341
1880,"F","Margaret",1578,0.0161672045489473
```

The top row usually (but not always) contains the variable names. Quotation marks are often used at the start and end of character strings—these quotation marks are not part of the content of the string, but are useful if, say, you want to include a comma in the text of

a field. CSV files are often named with the .csv suffix; it is also common for them to be
named with .txt, .dat, or other things. You will also see characters other than commas
being used to delimit the fields: Tabs and vertical bars are particularly common.

Since reading from a CSV file is so common, several implementations are available. The
read.csv() function in the base package is perhaps the most widely used, but the more
recent read_csv() function in the readr package is noticeably faster for large CSVs. CSV
files need not exist on your local hard drive. For example, here is a way to access a .csv
file over the Internet using a URL (universal resource locator).

```
myURL <- "http://tiny.cc/dcf/houses-for-sale.csv"
Houses <- readr::read_csv(myURL)
head(Houses, 3)
```

```
# A tibble: 3  16
    price lot_size waterfront   age land_value construction air_cond  fuel
    <int>    <dbl>      <int> <int>      <int>        <int>    <int> <int>
1 132500     0.09          0    42      50000            0        0     3
2 181115     0.92          0     0      22300            0        0     2
3 109000     0.19          0   133       7300            0        0     2
# ... with 8 more variables: heat <int>, sewer <int>, living_area <int>,
#   pct_college <int>, bedrooms <int>, fireplaces <int>, bathrooms <dbl>,
#   rooms <int>
```

Just as reading a data file from the Internet uses a URL, reading a file on your computer
uses a complete name, called a *path* to the file. Although many people are used to using a
mouse-based selector to access their files, being specific about the full path to your files is
important to ensure the reproducibility of your code (see Appendix D).

HTML tables

Web pages are HTML documents, which are then translated by a browser to the formatted
content that users see. HTML includes facilities for presenting tabular content. The HTML
<table> markup is often the way human-readable data is arranged.

When you have the URL of a page containing one or more tables, it is sometimes easy
to read them into R as data tables. Since they are not CSVs, we can't use read_csv().
Instead, we use functionality in the rvest package to ingest the HTML as a data structure
in R. Once you have the content of the Web page, you can translate any tables in the page
from HTML to data table format.

In this brief example, we will investigate the progression of the world record time in the
mile run, as detailed on the Wikipedia. This page (see Figure 5.7) contains several tables,
each of which contains a list of new world records for a different class of athlete (e.g., men,
women, amateur, professional, etc.).

```
library(rvest)
library(methods)
url <- "http://en.wikipedia.org/wiki/Mile_run_world_record_progression"
tables <- url %>%
  read_html() %>%
  html_nodes("table")
```

The result, tables, is not a data table. Instead, it is a *list* (see Appendix B) of the
tables found in the Web page. Use length() to find how many items there are in the list
of tables.

progression before that year. One version starts with Richard Webster (GBR) who ran 4:36.5 in 1865, surpassed by Chinnery in 1868.[3]

Another variation of the amateur record progression pre-1862 is as follows:[4]

Time	Athlete	Nationality	Date	Venue
4:52	Cadet Marshall	United Kingdom	2 September 1852	Addiscome
4:45	Thomas Finch	United Kingdom	3 November 1858	Oxford
4:45	St. Vincent Hammick	United Kingdom	15 November 1858	Oxford
4:40	Gerald Surman	United Kingdom	24 November 1859	Oxford
4:33	George Farran	United Kingdom	23 May 1862	Dublin

IAAF era [edit]

The first **world record** in the **mile for men** (athletics) was recognized by the International Amateur Athletics Federation, now known as the International Association of Athletics Federations, in 1913.

To June 21, 2009, the IAAF has ratified 32 world records in the event.[5]

Time	Auto	Athlete	Nationality	Date	Venue
4:14.4		John Paul Jones	United States	31 May 1913[5]	Allston, Mass.
4:12.6		Norman Taber	United States	16 July 1915[5]	Allston, Mass.
4:10.4		Paavo Nurmi	Finland	23 August 1923[5]	Stockholm
4:09.2		Jules Ladoumègue	France	4 October 1931[5]	Paris
4:07.6		Jack Lovelock	New Zealand	15 July 1933[5]	Princeton, N.J.
4:06.8		Glenn Cunningham	United States	16 June 1934[5]	Princeton, N.J.
4:06.4		Sydney Wooderson	United Kingdom	28 August 1937[5]	Motspur Park
4:06.2		Gunder Hägg	Sweden	1 July 1942[5]	Gothenburg

Figure 5.7: Part of a page on mile-run world records from Wikipedia. Two separate data tables are visible. You can't tell from this small part of the page, but there are seven tables altogether on the page. These two tables are the third and fourth in the page.

```
length(tables)
```

```
[1] 7
```

You can access any of those tables using the [[() operator. The first table is `tables[[1]]`, the second table is `tables[[2]]`, and so on. The third table—which corresponds to amateur men up until 1862—is shown in Table 5.10.

```
Table3 <- html_table(tables[[3]])
```

Time	Athlete	Nationality	Date	Venue
4:52	Cadet Marshall	United Kingdom	2 September 1852	Addiscome
4:45	Thomas Finch	United Kingdom	3 November 1858	Oxford
4:45	St. Vincent Hammick	United Kingdom	15 November 1858	Oxford
4:40	Gerald Surman	United Kingdom	24 November 1859	Oxford
4:33	George Farran	United Kingdom	23 May 1862	Dublin

Table 5.10: The third table embedded in the Wikipedia page on running records.

Likely of greater interest is the information in the fourth table, which corresponds to the current era of International Amateur Athletics Federation world records. The first few rows of that table are shown in Table 5.11. The last row of that table (now shown) contains the current world record of 3:43.13, which was set by Hicham El Guerrouj of Morocco in Rome on July 7th, 1999.

```
Table4 <- html_table(tables[[4]])
Table4 <- select(Table4, -Auto)  # remove unwanted column
```

Time	Athlete	Nationality	Date	Venue
4:14.4	John Paul Jones	United States	31 May 1913[5]	Allston, Mass.
4:12.6	Norman Taber	United States	16 July 1915[5]	Allston, Mass.
4:10.4	Paavo Nurmi	Finland	23 August 1923[5]	Stockholm
4:09.2	Jules Ladoumgue	France	4 October 1931[5]	Paris
4:07.6	Jack Lovelock	New Zealand	15 July 1933[5]	Princeton, N.J.
4:06.8	Glenn Cunningham	United States	16 June 1934[5]	Princeton, N.J.

Table 5.11: The fourth table embedded in the Wikipedia page on running records.

5.5.2   APIs

An *application programming interface* (API) is a protocol for interacting with a computer program that you can't control. It is a set of agreed-upon instructions for using a "black-box"—not unlike the manual for a television's remote control.   APIs provide access to massive troves of public data on the Web, from a vast array of different sources. Not all APIs are the same, but by learning how to use them, you can dramatically increase your ability to pull data into R without having to "scrape" it.

If you want to obtain data from a public source, it is a good idea to check to see whether: a) the company has a public API; b) someone has already written an R package to said interface. These packages don't provide the actual data—they simply provide a series of R functions that allow you to access the actual data. The documentation for each package will explain how to use it to collect data from the original source.

5.5.3   Cleaning data

A person somewhat knowledgeable about running would have little trouble interpreting Tables 5.10 and 5.11 correctly. The Time is in minutes and seconds. The Date gives the day on which the record was set. When the data table is read into R, both Time and Date are stored as character strings. Before they can be used, they have to be converted into a format that the computer can process like a date and time. Among other things, this requires dealing with the footnote (listed as [5]) at the end of the date information.

Data cleaning refers to taking the information contained in a variable and transforming it to a form in which that information can be used.

Recoding

Table 5.12 displays a few variables from the Houses data table we downloaded earlier. It describes 1,728 houses for sale in Saratoga, NY.[1] The full table includes additional variables such as living_area, price, bedrooms, and bathrooms. The data on house systems such as sewer_type and heat_type have been stored as numbers, even though they are really categorical.

There is nothing fundamentally wrong with using integers to encode, say, fuel type, though it may be confusing to interpret results. What is worse is that the numbers imply a meaningful order to the categories when there is none.

---

[1]The example comes from Richard De Veaux at Williams College.

fuel	heat	sewer	construction
3	4	2	0
2	3	2	0
2	3	3	0
2	2	2	0
2	2	3	1

Table 5.12: Four of the variables from the `houses-for-sale.csv` file giving features of the Saratoga houses stored as integer codes. Each case is a different house.

To translate the integers to a more informative coding, you first have to find out what the various codes mean. Often, this information comes from the codebook, but sometimes you will need to contact the person who collected the data. Once you know the translation, you can use spreadsheet software to enter them into a data table, like this one for the houses:

```
Translations <- readr::read_csv("http://tiny.cc/dcf/house_codes.csv")
Translations %>% head(5)

# A tibble: 5  3
  code system_type meaning
  <int>       <chr>   <chr>
1     0   new_const      no
2     1   new_const     yes
3     1 sewer_type    none
4     2 sewer_type private
5     3 sewer_type  public
```

`Translations` describes the codes in a format that makes it easy to add new code values as the need arises. The same information can also be presented a wide format as in Table 5.13.

```
CodeVals <- Translations %>%
  spread(key = system_type, value = meaning, fill = "invalid")
```

code	central_air	fuel_type	heat_type	new_const	sewer_type
0	no	invalid	invalid	no	invalid
1	yes	invalid	invalid	yes	none
2	invalid	gas	hot air	invalid	private
3	invalid	electric	hot water	invalid	public
4	invalid	oil	electric	invalid	invalid

Table 5.13: The `Translations` data table rendered in a wide format.

In `CodeVals`, there is a column for each system type that translates the integer code to a meaningful term. In cases where the integer has no corresponding term, `invalid` has been entered. This provides a quick way to distinguish between incorrect entries and missing entries. To carry out the translation, we join each variable, one at a time, to the data table of interest. Note how the `by` value changes for each variable:

```
Houses <- Houses %>%
  left_join(CodeVals %>%
  select(code, fuel_type), by = c(fuel="code")) %>%
  left_join(CodeVals %>% select(code, heat_type), by = c(heat="code")) %>%
  left_join(CodeVals %>% select(code, sewer_type), by = c(sewer="code"))
```

Table 5.14 shows the re-coded data. We can compare this to the previous display in Table 5.12.

	fuel_type	heat_type	sewer_type
1	electric	electric	private
2	gas	hot water	private
3	gas	hot water	public
4	gas	hot air	private
5	gas	hot air	public
6	gas	hot air	private

Table 5.14: The Houses data with re-coded categorical variables.

From strings to numbers

You have seen two major types of variables: quantitative and categorical. You are used to using quoted character strings as the levels of categorical variables, and numbers for quantitative variables.

Often, you will encounter data tables that have variables whose meaning is numeric but whose representation is a character string. This can occur when one or more cases is given a non-numeric value, e.g., *not available*.

The as.numeric() function will translate character strings with numerical content into numbers. But as.character() goes the other way. For example, in the OrdwayBirds data, the Month, Day, and Year variables are all being stored as character vectors, even though their evident meaning is numeric.

```
OrdwayBirds %>%
  select(Timestamp, Year, Month, Day) %>%
  glimpse()
```

```
Observations: 15,829
Variables: 4
$ Timestamp <chr> "4/14/2010 13:20:56", "", "5/13/2010 16:00:30", "5/1...
$ Year      <chr> "1972", "", "1972", "1972", "1972", "1972", "1972", ...
$ Month     <chr> "7", "", "7", "7", "7", "7", "7", "7", "7", "7", "7"...
$ Day       <chr> "16", "", "16", "16", "16", "16", "16", "16", "16", ...
```

We can convert the strings to numbers using mutate() and parse_number(). Note how the empty strings (i.e., "") in those fields are automatically converted into NA's, since they cannot be converted into valid numbers.

```
library(readr)
OrdwayBirds <- OrdwayBirds %>%
  mutate(Month = parse_number(Month), Year = parse_number(Year),
```

```
         Day = parse_number(Day))
OrdwayBirds %>%
  select(Timestamp, Year, Month, Day) %>%
  glimpse()


Observations: 15,829
Variables: 4
$ Timestamp <chr> "4/14/2010 13:20:56", "", "5/13/2010 16:00:30", "5/1...
$ Year      <dbl> 1972, NA, 1972, 1972, 1972, 1972, 1972, 1972, 1972, ...
$ Month     <dbl> 7, NA, 7, 7, 7, 7, 7, 7, 7, 7, 7, 7, 7, 7, 7, 7, 7, ...
$ Day       <dbl> 16, NA, 16, 16, 16, 16, 16, 16, 16, 16, 17, 18, 18, ...
```

Dates

Unfortunately, dates are often recorded as character strings (e.g., 29 October 2014). Among other important properties, dates have a natural order. When you plot values such as 16 December 2015 and 29 October 2016, you expect the December date to come after the October date, even though this is not true alphabetically of the string itself.

When plotting a value that is numeric, you expect the axis to be marked with a few round numbers. A plot from 0 to 100 might have ticks at 0, 20, 40, 60, 100. It is similar for dates. When you are plotting dates within one month, you expect the day of the month to be shown on the axis. If you are plotting a range of several years, it would be appropriate to show only the years on the axis.

When you are given dates stored as a character vector, it is usually necessary to convert them to a data type designed specifically for dates. For instance, in the OrdwayBirds data, the Timestamp variable refers to the time the data were transcribed from the original lab notebook to the computer file. This variable is currently stored as a character string, but we can translate it into a genuine date using functions from the lubridate package.

These dates are written in a format showing month/day/year hour:minute:second. The mdy_hms() function from the lubridate package converts strings in this format to a date. Note that the data type of the When variable is now time.

```
library(lubridate)
WhenAndWho <- OrdwayBirds %>%
  mutate(When = mdy_hms(Timestamp)) %>%
  select(Timestamp, Year, Month, Day, When, DataEntryPerson) %>%
  glimpse()


Observations: 15,829
Variables: 6
$ Timestamp       <chr> "4/14/2010 13:20:56", "", "5/13/2010 16:00:30"...
$ Year            <dbl> 1972, NA, 1972, 1972, 1972, 1972, 1972, 1972, ...
$ Month           <dbl> 7, NA, 7, 7, 7, 7, 7, 7, 7, 7, 7, 7, 7, 7, 7, ...
$ Day             <dbl> 16, NA, 16, 16, 16, 16, 16, 16, 16, 16, 17, 18...
$ When            <dttm> 2010-04-14 13:20:56, NA, 2010-05-13 16:00:30,...
$ DataEntryPerson <chr> "Jerald Dosch", "Caitlin Baker", "Caitlin Bake...
```

With the When variable now recorded as a timestamp, we can create a sensible plot showing when each of the transcribers completed their work, as in Figure 5.8.

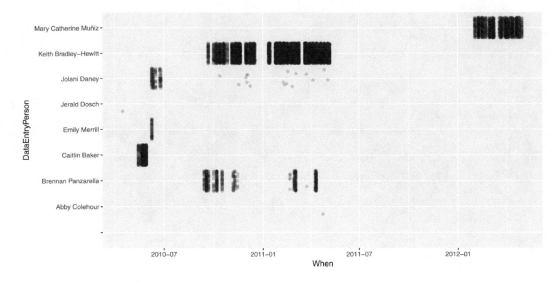

Figure 5.8: The transcribers of `OrdwayBirds` from lab notebooks worked during different time intervals.

```
WhenAndWho %>% ggplot(aes(x = When, y = DataEntryPerson)) +
  geom_point(alpha = 0.1, position = "jitter")
```

Many of the same operations that apply to numbers can be used on dates. For example, the range of dates that each transcriber worked can be calculated as a difference in times (i.e., an `interval()`), and shown in Table 5.15. This makes it clear that Jolani worked on the project for nearly a year (329 days), while Abby's first transcription was also her last.

```
WhenAndWho %>%
  group_by(DataEntryPerson) %>%
  summarize(start = first(When), finish = last(When)) %>%
  mutate(duration = interval(start, finish) / ddays(1))
```

DataEntryPerson	start	finish	duration
Abby Colehour	2011-04-23 15:50:24	2011-04-23 15:50:24	0.00
Brennan Panzarella	2010-09-13 10:48:12	2011-04-10 21:58:56	209.47
Emily Merrill	2010-06-08 09:10:01	2010-06-08 14:47:21	0.23
Jerald Dosch	2010-04-14 13:20:56	2010-04-14 13:20:56	0.00
Jolani Daney	2010-06-08 09:03:00	2011-05-03 10:12:59	329.05
Keith Bradley-Hewitt	2010-09-21 11:31:02	2011-05-06 17:36:38	227.25
Mary Catherine Muiz	2012-02-02 08:57:37	2012-04-30 14:06:27	88.21

Table 5.15: Starting and ending dates for each transcriber involved in the `OrdwayBirds` project.

There are many similar `lubridate` functions for converting strings in different formats into dates, e.g., `ymd()`, `dmy()`, and so on. There are also functions like `hour()`, `yday()`, etc. for extracting certain pieces of variables encoded as dates.

Internally, R uses several different classes to represent dates and times. For timestamps (also referred to as datetimes), these classes are POSIXct and POSIXlt. For most purposes, you can treat these as being the same, but internally, they are stored differently. A POSIXct object is stored as the number of seconds since the UNIX epoch (1970-01-01), whereas POSIXlt objects are stored as a list of year, month, day, etc. character strings.

```
now()
```

```
[1] "2016-11-23 11:19:59 EST"
```

```
class(now())
```

```
[1] "POSIXct" "POSIXt"
```

```
class(as.POSIXlt(now()))
```

```
[1] "POSIXlt" "POSIXt"
```

For dates that do not include times, the Date class is most commonly used.

```
as.Date(now())
```

```
[1] "2016-11-23"
```

Factors or strings?

R was designed with a special data type for holding categorical data: factor. Factors store categorical data efficiently and provide a means to put the categorical levels in whatever order is desired. Unfortunately, factors also make cleaning data more confusing. The problem is that it is easy to mistake a factor for a character string, but they have different properties when it comes to converting a numeric or date form. This is especially problematic when using the character processing techniques in Chapter 15.

By default, readr::read_csv() will interpret character strings as strings and not as factors. Other functions such as read.csv() convert character strings into factors by default. Cleaning such data often requires converting them back to a character format using as.character(). Failing to do this when needed can result in completely erroneous results without any warning.

For this reason, the data tables used in this book have been stored with categorical or text data in character format. Be aware that data provided by other packages do not necessarily follow this convention. If you get mysterious results when working with such data, consider the possibility that you are working with factors rather than character vectors. Recall that summary(), glimpse(), and str() will all reveal the data types of each variable in a data frame.

Pro Tip: It's always a good idea to carefully check all variables and data wrangling operations to ensure that reasonable values are generated.

CSV files in this book are typically read with read_csv() provided by the readr package. If, for some reason, you prefer to use the read.csv() function, we recommend setting the argument stringsAsFactors argument to FALSE to ensure that text data be stored as character strings.

Japan [edit]

See also: Nuclear power in Japan

Power station reactors   [edit]

Name	Reactor No.	Reactor		Status	Capacity in MW		Construction Start Date	Commercial Operation Date	Closure
		Type	Model		Net	Gross			
Fukushima Daiichi	1	BWR	BWR-3	Inoperable	439	460	25 July 1967	26 March 1971	19 May 2011
Fukushima Daiichi	2	BWR	BWR-4	Inoperable	760	784	9 June 1969	18 July 1974	19 May 2011
Fukushima Daiichi	3	BWR	BWR-4	Inoperable	760	784	28 December 1970	27 March 1976	19 May 2011
Fukushima Daiichi	4	BWR	BWR-4	Shut down/ Inoperable	760	784	12 February 1973	12 October 1978	19 May 2011
Fukushima Daiichi	5	BWR	BWR-4	Shut down	760	784	22 May 1972	18 April 1978	17 December 2013
Fukushima Daiichi	6	BWR	BWR-5	Shut down	1067	1100	26 October 1973	24 October 1979	17 December 2013
Fukushima Daini	1	BWR	BWR-5	Operation suspended	1067	1100	16 March 1976	20 April 1982	

Figure 5.9: Screenshot of Wikipedia's list of Japanese nuclear reactors.

5.5.4   Example: Japanese nuclear reactors

Dates and times are an important aspect of many analyses. In the example below, the vector
example contains human-readable datetimes stored as character by R. The ymd_hms()
function from lubridate will convert this into POSIXct—a datetime format. This makes
it possible for R to do date arithmetic.

```
library(lubridate)
example <- c("2017-04-29 06:00:00", "2017-12-31 12:00:00")
str(example)

 chr [1:2] "2017-04-29 06:00:00" "2017-12-31 12:00:00"
```

```
converted <- ymd_hms(example)
str(converted)

 POSIXct[1:2], format: "2017-04-29 06:00:00" "2017-12-31 12:00:00"
```

```
converted

[1] "2017-04-29 06:00:00 UTC" "2017-12-31 12:00:00 UTC"

converted[2] - converted[1]

Time difference of 246 days
```

We will use this functionality to analyze data on nuclear reactors in Japan. Figure 5.9
displays the first part of this table as of the summer of 2016.

```
my_html <-
  read_html("http://en.wikipedia.org/wiki/List_of_nuclear_reactors")
tables <- my_html %>% html_nodes(css = "table")
relevant_tables <- tables[grep("Fukushima Daiichi", tables)]
reactors <- html_table(relevant_tables[[1]], fill = TRUE)
names(reactors)[c(3,4,6,7)] <- c("Reactor Type",
  "Reactor Model", "Capacity Net", "Capacity Gross")
reactors <- reactors[-1,]
```

We see that the first entries are the ill-fated Fukushima Daiichi reactors. The `mutate()` function can be used in conjunction with the `dmy()` function from the `lubridate` package to wrangle these data into a better form. (Note the back ticks used to specify variable names that include space or special characters.)

```
library(readr)
reactors <- reactors %>%
  rename(capacity_net=`Capacity Net`, capacity_gross=`Capacity Gross`) %>%
  mutate(plantstatus = ifelse(grepl("Shut down", reactors$Status),
      "Shut down", "Not formally shut down"),
    capacity_net = parse_number(capacity_net),
    construct_date = dmy(`Construction Start Date`),
    operation_date = dmy(`Commercial Operation Date`),
    closure_date = dmy(Closure))
```

How have these plants evolved over time? It seems likely that as nuclear technology has progressed, plants should see an increase in capacity. A number of these reactors have been shut down in recent years. Are there changes in capacity related to the age of the plant? Figure 5.10 displays the data.

```
ggplot(data = reactors,
  aes(x = construct_date, y = capacity_net, color = plantstatus)) +
  geom_point() + geom_smooth() +
  xlab("Date of Plant Construction") + ylab("Net Plant Capacity (MW)")
```

Indeed, reactor capacity has tended to increase over time, while the older reactors were more likely to have been formally shut down. While it would have been straightforward to code these data by hand, automating data ingestion for larger and more complex tables is more efficient and less error-prone.

5.6 Further resources

The `tidyr` package, and in particular, the Tidy Data [230] paper provide principles for tidy data. We provide further statistical justification for resampling-based techniques in Chapter 7. The `feather` package provides an efficient mechanism for storing data frames that can be read and written by both R and Python.

There are many R packages that do nothing other than provide access to a public API from within R. There are far too many API packages to list here, but a fair number of them are maintained by the rOpenSci group. In fact, several of the packages referenced in this book, including the `twitteR` and `aRxiv` packages in Chapter 15, and the `plotly` package in Chapter 11, are APIs. The CRAN task view on Web Technologies lists hundreds more

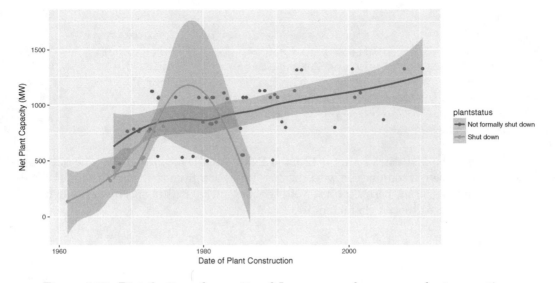

Figure 5.10: Distribution of capacity of Japanese nuclear power plants over time.

packages, including Rfacebook, instaR, Rflickr, tumblR, and Rlinkedin. The RSocrata package facilitates the use of Socrata, which is itself an API for querying—among other things—the NYC Open Data platform.

5.7   Exercises

Exercise 5.1

Consider the number of home runs hit (HR) and home runs allowed (HRA) for the Chicago Cubs (*CHN*) baseball team. Reshape the Teams data from the Lahman package into *long* format and plot a time series conditioned on whether the HRs that involved the Cubs were hit by them or allowed by them.

Exercise 5.2

Write a function called count_seasons() that, when given a teamID, will count the number of seasons the team played in the Teams data frame from the Lahman package.

Exercise 5.3

The team IDs corresponding to Brooklyn baseball teams from the Teams data frame from the Lahman package are listed below. Use sapply() to find the number of seasons in which each of those teams played.

```
bk_teams <- c("BR1", "BR2", "BR3", "BR4", "BRO", "BRP", "BRF")
```

Exercise 5.4

In the Marriage data set included in mosaicData, the appdate, ceremonydate, and dob variables are encoded as factors, even though they are dates. Use lubridate to convert those three columns into a date format.

```
library(mosaic)
Marriage %>%
  select(appdate, ceremonydate, dob) %>%
  glimpse()

Observations: 98
Variables: 3
$ appdate     <fctr> 10/29/96, 11/12/96, 11/19/96, 12/2/96, 12/9/96, ...
$ ceremonydate <fctr> 11/9/96, 11/12/96, 11/27/96, 12/7/96, 12/14/96, ...
$ dob         <fctr> 4/11/64, 8/6/64, 2/20/62, 5/20/56, 12/14/66, 2/2...
```

Exercise 5.5

Consider the values returned by the as.numeric() and readr::parse_number() functions when applied to the following vectors. Describe the results and their implication.

```
x1 <- c("1900.45", "$1900.45", "1,900.45", "nearly $2000")
x2 <- as.factor(x1)
```

Exercise 5.6

An analyst wants to calculate the pairwise differences between the Treatment and Control values for a small data set from a crossover trial (all subjects received both treatments) that consists of the following observations.

```
tab <- xtable(ds1)
print(tab, floating=FALSE)
```

	id	group	vals
1	1	T	4.00
2	2	T	6.00
3	3	T	8.00
4	1	C	5.00
5	2	C	6.00
6	3	C	10.00

They use the following code to create the new diff variable.

```
Treat <- filter(ds1, group=="T")
Control <- filter(ds1, group=="C")
all <- mutate(Treat, diff = Treat$vals - Control$vals)
all
```

Verify that this code works for this example and generates the correct values of -1, 0, and -2. Describe two problems that might arise if the data set is not sorted in a particular order or if one of the observations is missing for one of the subjects. Provide an alternative approach to generate this variable that is more robust (hint: use tidyr::spread()).

Exercise 5.7

Generate the code to convert the following data frame to wide format.

	grp	sex	meanL	sdL	meanR	sdR
1	A	F	0.22	0.11	0.34	0.08
2	A	M	0.47	0.33	0.57	0.33
3	B	F	0.33	0.11	0.40	0.07
4	B	M	0.55	0.31	0.65	0.27

The result should look like the following display.

	grp	F.meanL	F.meanR	F.sdL	F.sdR	M.meanL	M.meanR	M.sdL	M.sdR
1	A	0.22	0.34	0.11	0.08	0.47	0.57	0.33	0.33
2	B	0.33	0.40	0.11	0.07	0.55	0.65	0.31	0.27

Hint: use `gather()` in conjunction with `spread()`.

Exercise 5.8

Use the `dplyr::do()` function and the `HELPrct` data frame from the `mosaicData` package to fit a regression model predicting `cesd` as a function of `age` separately for each of the levels of the `substance` variable. Generate a table of results (estimates and confidence intervals) for each level of the grouping variable.

Exercise 5.9

Use the `dplyr::do()` function and the `Lahman` data to replicate one of these baseball records plots (`http://tinyurl.com/nytimes-records`) from the *The New York Times*.

Exercise 5.10

Use the `fec` package to download the Federal Election Commission data for 2012. Re-create Figure 2.1 and Figure 2.2 using `ggplot2`.

Exercise 5.11

Using the same FEC data as the previous exercise, re-create Figure 2.8.

Exercise 5.12

Using the approach described in Section 5.5.4, find another table in Wikipedia that can be scraped and visualized. Be sure to interpret your graphical display.

Exercise 5.13

Replicate the wrangling to create the `house_elections` table in the `fec` package from the original Excel source file.

Exercise 5.14

Replicate the functionality of `make_babynames_dist()` from the `mdsr` package to wrangle the original tables from the `babynames` package.

Chapter 6

Professional Ethics

6.1 Introduction

Work in data analytics involves expert knowledge, understanding, and skill. In much of your work, you will be relying on the trust and confidence that your clients place in you. The term *professional ethics* describes the special responsibilities not to take unfair advantage of that trust. This involves more than being thoughtful and using common sense; there are specific professional standards that should guide your actions.

The best known professional standards are those in the Hippocratic Oath for physicians, which were originally written in the 5th century B.C. Three of the eight principles in the modern version of the oath [237] are presented here because of similarity to standards for data analytics.

- "I will not be ashamed to say 'I know not,' nor will I fail to call in my colleagues when the skills of another are needed for a patient's recovery."

- "I will respect the privacy of my patients, for their problems are not disclosed to me that the world may know."

- "I will remember that I remain a member of society, with special obligations to all my fellow human beings, those sound of mind and body as well as the infirm."

Depending on the jurisdiction, these principles are extended and qualified by law. For instance, notwithstanding the need to "respect the privacy of my patients," health-care providers in the United States are required by law to report to appropriate government authorities evidence of child abuse or infectious diseases such as botulism, chicken pox, and cholera.

This chapter introduces principles of professional ethics for data analytics and gives examples of legal obligations as well as guidelines issued by professional societies. There is no data analyst's oath—only guidelines. Reasonable people can disagree about what actions are best, but the existing guidelines provide a description of the ethical expectations on which your clients can reasonably rely. As a consensus statement of professional ethics, the guidelines also establish standards of accountability.

6.2 Truthful falsehoods

The single best-selling book with "statistics" in the title is *How to Lie with Statistics* by Darrell Huff [114]. Written in the 1950s, the book shows graphical ploys to fool people

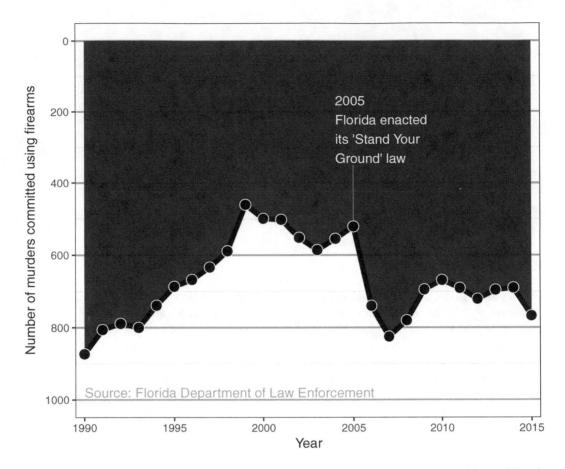

Figure 6.1: Reproduction of a data graphic reporting the number of gun deaths in Florida over time. The original image was published by Reuters.

even with accurate data. A general method is to violate conventions and tacit expectations that readers rely on when interpreting graphs. One way to think of *How to Lie* is a text to show the general public what these tacit expectations are and give tips for detecting when the trick is being played on them. The book's title, while compelling, has wrongly tarred the field of statistics. The "statistics" of the title are really just "numbers." The misleading graphical techniques are employed by politicians, journalists, and businessmen: not statisticians. More accurate titles would be "How to Lie with Numbers," or "Don't be misled by graphics."

Some of the graphical tricks in "How to Lie ..." are still in use. Consider these two recent examples.

In 2005, the Florida legislature passed the controversial "Stand Your Ground" law that broadened the situations in which citizens can use lethal force to protect themselves against perceived threats. Advocates believed that the new law would ultimately reduce crime; opponents feared an increase in the use of lethal force. What was the actual outcome?

The graphic in Figure 6.1 is a reproduction of one published by the news service Reuters showing the number of firearm murders in Florida over the years (see Exercise 4.18). Upon first glance, the graphic gives the visual impression that right after the passage of the 2005 law, the number of murders decreased substantially. However, the numbers tell a different story.

The convention in data graphics is that up corresponds to increasing values. This is not an obscure convention—rather, it's a standard part of the secondary school curriculum. Close inspection reveals that the y-axis in Figure 6.1 has been flipped upside down—the number of gun deaths increased sharply after 2005.

Figure 6.2 shows another example of misleading graphics: a tweet by the news magazine *National Review* on the subject of climate change. The dominant visual impression of the graphic is that global temperature has hardly changed at all.

National Review ✔
@NRO

⚲ **Follow**

The only #climatechange chart you need to see.
natl.re/wPKpro

(h/t @powerlineUS)

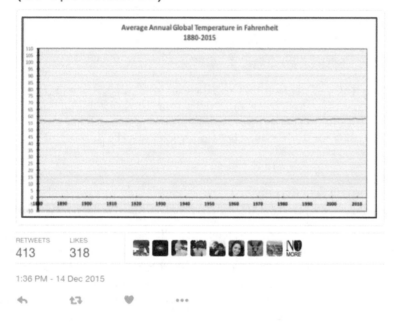

| RETWEETS | LIKES |
| 413 | 318 |

1:36 PM - 14 Dec 2015

Figure 6.2: A tweet by *National Review* on December 14, 2015 showing the change in global temperature over time.

There is a tacit graphical convention that the coordinate scales on which the data are plotted are relevant to an informed interpretation of the data. The x-axis follows the convention—1880 to 2015 is a reasonable choice when considering the relationship between human industrial activity and climate. The y-axis, however, is utterly misleading. The scale goes from -10 to 110 degrees Fahrenheit. While this is a relevant scale for showing *season-to-season* variation in temperature, that is not the salient issue with respect to climate change. The concern with climate change is about rising ocean levels, intensification of storms, ecological and agricultural disruption, etc. These are the anticipated results of a change in global *average* temperature on the order of 5 degrees Fahrenheit. The *National Review* graphic has obscured the data by showing them on an irrelevant scale where the actual changes in temperature are practically invisible. By graying out the numbers on the *y*-axis, the *National Review* makes it even harder to see the trick that's being played.

The examples in Figures 6.1 and 6.2 are not about lying with statistics. Statistical methodology doesn't enter into them. It's the professional ethics of journalism that the graphics violate, aided and abetted by an irresponsible ignorance of statistical methodology. Insofar as both graphics concern matters of political controversy, they can be seen as part of the blustering and bloviating of politics. While politics may be a profession, it's a profession without any comprehensive standard of professional ethics.

6.3   Some settings for professional ethics

Common sense is a good starting point for evaluating the ethics of a situation. Tell the truth. Don't steal. Don't harm innocent people. But professional ethics also require a neutral, unemotional, and informed assessment. A dramatic illustration of this comes from legal ethics: a situation where the lawyers for an accused murderer found the bodies of two victims whose deaths were unknown to authorities and to the victims' families. The responsibility to confidentiality for their client precluded the lawyers from following their hearts and reporting the discovery. The lawyers' careers were destroyed by the public and political recriminations that followed, yet courts and legal scholars have confirmed that the lawyers were right to do what they did, and have even held them up as heroes for their ethical behavior.

Such extreme drama is rare. This section describes in brief six situations that raise questions of the ethical course of action. Some are drawn from the authors' personal experience, others from court cases and other reports. The purpose of these short case reports is to raise questions. Principles for addressing those questions are the subject of the next section.

6.3.1   The chief executive officer

One of us once worked as a statistical consultant for a client who wanted a proprietary model to predict commercial outcomes. After reviewing the literature, an existing multiple linear regression model was found that matched the scenario well and available public data were used to fit the parameters of the model. The client's staff were pleased with the result, but the CEO wanted a model that would give a competitive advantage. After all, their competitors could easily follow the same process to the same model, so what advantage would the client's company have? The CEO asked the statistical consultant whether the coefficients in the model could be "tweaked" to reflect the specific values of his company. The consultant suggested that this would not be appropriate, that the fitted coefficients best match the data and to change them arbitrarily would be "playing God." In response, the CEO rose from his chair and asserted, "I want to play God."

How should the consultant respond?

6.3.2   Employment discrimination

One of us works with legal cases arising from audits of employers, conducted by the United States Office of Federal Contract Compliance Programs (OFCCP). In a typical case, the OFCCP asks for hiring and salary data from a company that has a contract with the United States government. The company usually complies, sometimes unaware that the OFCCP applies a method to identify "discrimination" through a two-standard-deviation test outlined in the Uniform Guidelines on Employee Selection Procedures (UGESP). A company that does not discriminate has some risk of being labeled as discriminating by the OFCCP method [41]. By using a questionable statistical method, is the OFCCP acting unethically?

6.3.3 Data scraping

In May 2016, the online OpenPsych Forum published a paper titled "The OkCupid data set: A very large public data set of dating site users". The resulting data set contained 2,620 variables—including usernames, gender, and dating preferences—from 68,371 people scraped from the OkCupid dating website. The ostensible purpose of the data dump was to provide an interesting open public data set to fellow researchers. These data might be used to answer questions such as this one suggested in the abstract of the paper: whether the Zodiac sign of each user was associated with any of the other variables (spoiler alert: it wasn't).

The data scraping did not involve any illicit technology such as breaking passwords. Nonetheless, the author received many comments on the OpenPsych Forum challenging the work as an ethical breach in *doxing* people by releasing personal data. Does the work raise actual ethical issues?

6.3.4 Reproducible spreadsheet analysis

In 2010, Harvard economists Carmen Reinhart and Kenneth Rogoff published a report entitled *Growth in a Time of Debt* [177], which argued that countries which pursued austerity measures did not necessarily suffer from slow economic growth. These ideas influenced the thinking of policymakers—notably United States Congressman Paul Ryan—during the time of the European debt crisis.

Graduate student Thomas Herndon requested access to the data and analysis contained in the paper. After receiving the original spreadsheet from Reinhart, Herndon found several errors.

> "I clicked on cell L51, and saw that they had only averaged rows 30 through 44, instead of rows 30 through 49." —Thomas Herndon [179]

In a critique [100] of the paper, Herndon, Ash, and Pollin point out coding errors, selective inclusion of data, and odd weighting of summary statistics that shaped the conclusions of the Reinhart/Rogoff paper.

Does publishing a flawed analysis raise ethical questions?

6.3.5 Drug dangers

In September 2004, drug company Merck withdrew from the market a popular product Vioxx because of evidence that the drug increases the risk of myocardial infarction (MI), a major type of heart attack. Approximately 20 million Americans had taken Vioxx up to that point. The leading medical journal *Lancet* later reported an estimate that Vioxx use resulted in 88,000 Americans having heart attacks, of whom 38,000 died.

Vioxx had been approved in May 1999 by the United States Food and Drug Administration based on tests involving 5,400 subjects. Slightly more than a year after the FDA approval, a study [36] of 8,076 patients published in another leading medical journal, *The New England Journal of Medicine*, established that Vioxx reduced the incidence of severe gastro-intestinal events substantially compared to the standard treatment, naproxen. That's good for Vioxx. In addition, the abstract reports these findings regarding heart attacks:

"The incidence of myocardial infarction was lower among patients in the naproxen group than among those in the [Vioxx] group (0.1 percent vs. 0.4 percent; relative risk, 0.2; 95% confidence interval, 0.1 to 0.7); the overall mortality rate and the rate of death from cardiovascular causes were similar in the two groups."

Read the abstract again carefully. The Vioxx group had a much *higher* rate of MI than the group taking the standard treatment. This influential report identified the high risk soon after the drug was approved for use. Yet Vioxx was not withdrawn for another three years. Something clearly went wrong here. Did it involve an ethical lapse?

6.3.6   Legal negotiations

Lawyers sometimes retain statistical experts to help plan negotiations. In a common scenario, the defense lawyer will be negotiating the amount of damages in a case with the plaintiff's attorney. Plaintiffs will ask the statistician to estimate the amount of damages, with a clear but implicit directive that the estimate should reflect the plaintiff's interests. Similarly, the defense will ask their own expert to construct a framework that produces an estimate at a lower level.

Is this a game statisticians should play?

6.4   Some principles to guide ethical action

As noted previously, lying, cheating, and stealing are common and longstanding unethical behaviors. To guide professional action, however, more nuance and understanding is needed. For instance, an essential aspect of the economy is that firms compete. As a natural part of such competition, firms hurt one another; they take away business that the competitor would otherwise have. We don't consider competition to be unethical, although there are certainly limits to ethical competition.

As a professional, you possess skills that are not widely available. A fundamental notion of professional ethics is to avoid using those skills in a way that is effectively lying—leading others to believe one thing when in fact something different is true. In every professional action you take, there is an implicit promise that you can be relied on—that you will use appropriate methods and draw appropriate conclusions. Non-professionals are not always in a position to make an informed judgment about whether your methods and conclusions are appropriate. Part of acting in a professionally ethical way is making sure that your methods and conclusions are indeed appropriate.

It is necessary to believe that your methods and conclusions are appropriate, but not sufficient. First, it's easy to mislead yourself, particularly in the heat and excitement of satisfying your client or your research team. Second, it's usually not a matter of absolutes: It's not always certain that a method is appropriate. Instead, there is almost always a risk that something is wrong.

An important way to deal with these issues is to draw on generally recognized professional standards. Some examples: Use software systems that have been vetted by the community. Check that your data are what you believe them to be. Don't use analytical methods that would not pass scrutiny by professional colleagues.

Note that the previous paragraph says "draw on" rather than "scrupulously follow." Inevitably there will be parts of your work that are not and cannot be vetted by the community. You write your own data wrangling statements: They aren't always vetted. In special circumstances you might reasonably choose to use software that is new or created just for the purpose at hand. You can look for internal consistency in your data, but it would be unreasonable in most circumstances to insist on tracking everything back to the original point at which it was measured.

Another important approach is to be open and honest. Don't overstate your confidence in results. Point out to clients substantial risks of error or unexpected outcome. If you would squirm if some aspect or another of your work came under expert scrutiny, it's likely that you should draw attention to that aspect yourself.

Still, there are limits. You generally can't usefully inform your clients of *every* possible risk and methodological limitation. The information would overwhelm them. And you usually will not have the resources—time, money, data—that you would need to make every aspect of your work perfect. You have to use good professional judgment to identify the most salient risks and to ensure that your work is good enough even if it's not perfect.

You have a professional responsibility to particular stakeholders. It's important that you consider and recognize all the various stakeholders to whom you have this responsibility. These vary depending on the circumstances. Sometimes, your main responsibility is simply to your employer or your client. In other circumstances, you will have a responsibility to the general public or to subjects in your study or individuals represented in your data. You may have a special responsibility to the research community or to your profession itself. The legal system can also impose responsibilities; there are laws that are relevant to your work. Expert witnesses in court cases have a particular responsibility to the court itself.

Another concern is the potential for a conflict of interest. A *conflict of interest* is not itself unethical. We all have such conflicts: We want to do work that will advance us professionally, which instills a temptation to satisfy the expectations of our employers or colleagues or the marketplace. The conflict refers to the *potential* that our personal goals may cloud or bias or otherwise shape our professional judgment.

Many professional fields have rules that govern actions in the face of a conflict of interest. Judges recuse themselves when they have a prior involvement in a case. Lawyers and law firms should not represent different clients whose interests are at odds with each other. Clear protocols and standards for analysis regulated by the FDA help ensure that potential conflicts of interest for researchers working for drug companies do not distort results. There's always a basic professional obligation to disclose potential conflicts of interest to your clients, to journals, etc.

For concreteness, here is a list of professional ethical precepts. It's simplistic; it's not feasible to capture every nuance in a brief exposition.

1. Do your work well by your own standards and by the standards of your profession.

2. Recognize the parties to whom you have a special professional obligation.

3. Report results and methods honestly and respect your responsibility to identify and report flaws and shortcomings in your work.

6.4.1 Applying the precepts

Let's explore how these precepts play out in the several scenarios outlined in the previous section.

The CEO

You've been asked by a company CEO to modify model coefficients from the correct values, that is, from the values found by a generally accepted method. The stakeholder in this setting is the company. If your work will involve a method that's not generally accepted by the professional community, you're obliged to point this out to the company.

Remember that your client also has substantial knowledge of how their business works. Statistical purity is not the issue. Your work is a tool for your client to use; they can use it as they want. Going a little further, it's important to realize that your client's needs may not map well onto a particular statistical methodology. The consultant should work genuinely to understand the client's whole set of interests. Often the problem that clients identify is not really the problem that needs to be solved when seen from an expert statistical perspective.

Employment discrimination

The procedures adopted by the OFCCP are stated using statistical terms like "standard deviation" that themselves suggest that they are part of a legitimate statistical method. Yet the methods raise significant questions, since by construction they will sometimes label a company that is not discriminating as a discriminator. OFCCP and others might argue that they are not a statistical organization. They are enforcing a law, not participating in research. The OFCCP has a responsibility to the courts. The courts themselves, including the United States Supreme Court, have not developed or even called for a coherent approach to the use of statistics (although in 1977 the Supreme Court labeled differences greater than two or three standard deviations as too large to attribute solely to chance).

Data scraping

OkCupid provides public access to data. A researcher uses legitimate means to acquire those data. What could be wrong?

There is the matter of the stakeholders. The collection of data was intended to support psychological research. The ethics of research involving humans requires that the human not be exposed to any risk for which consent has not been explicitly given. The OkCupid members did not provide such consent. Since the data contain information that makes it possible to identify individual humans, there is a realistic risk of the release of potentially embarrassing information, or worse, information that jeopardizes the physical safety of certain users.

Another stakeholder is OkCupid itself. Many information providers, like OkCupid, have *terms of use* that restrict how the data may be legitimately used. Such terms of use (see Section 6.5.3) form an explicit agreement between the service and the users of that service. They cannot ethically be disregarded.

Reproducible spreadsheet analysis

The scientific community as a whole is a stakeholder in public research. Insofar as the research is used to inform public policy, the public as a whole is a stakeholder. Researchers have an obligation to be truthful in their reporting of research. This is not just a matter of being honest, but also of participating in the process by which scientific work is challenged or confirmed. Reinhart and Rogoff honored this professional obligation by providing reasonable access to their software and data.

Note that it is not an ethical obligation to reach correct research results. The obligation is to do everything feasible to ensure that the conclusions faithfully reflect the data and the theoretical framework in which the data are analyzed. Scientific findings are often subject to dispute, reinterpretation, and refinement.

Since this book is specifically about data science, it can be helpful to examine the Reinhart and Rogoff findings with respect to the professional standards of data science. Note that these can be different from the professional standards of economics, which might reasonably be the ones that economists like Reinhart and Rogoff adopt. So the following is not a criticism of them, *per se*, but an opportunity to delineate standards relevant to data scientists.

Seen from the perspective of data science, Microsoft Excel, the tool used by Reinhart and Rogoff, is an unfortunate choice. It mixes the data with the analysis. It works at a low level of abstraction, so it's difficult to program in a concise and readable way. Commands are customized to a particular size and organization of data, so it's hard to apply to a new or modified data set. One of the major strategies in debugging is to work on a data set where the answer is known; this is impractical in Excel. Programming and revision in Excel generally involves lots of click-and-drag copying, which is itself an error-prone operation.

Data science professionals have an ethical obligation to use tools that are reliable, verifiable, and conducive to reproducible data analysis (see Appendix D). This is a good reason for professionals to eschew Excel.

Drug dangers

When something goes wrong on a large scale, it's tempting to look for a breach of ethics. This may indeed identify an offender, but we must also beware of creating scapegoats. With Vioxx, there were many claims, counterclaims, and lawsuits. The researchers failed to incorporate some data that were available and provided a misleading summary of results. The journal editors also failed to highlight the very substantial problem of the increased rate of myocardial infarction with Vioxx.

To be sure, it's unethical not to include data that undermines the conclusion presented in a paper. The Vioxx researchers were acting according to their original research protocol—a solid professional practice.

What seems to have happened with Vioxx is that the researchers had a theory that the higher rate of infarction was not due to Vioxx, *per se*, but to an aspect of the study protocol that excluded subjects who were being treated with aspirin to reduce the risk of heart attacks. The researchers believed with some justification that the drug to which Vioxx was being compared, naproxen, was acting as a substitute for aspirin. They were wrong, as subsequent research showed.

Professional ethics dictate that professional standards be applied in work. Incidents like Vioxx should remind us to work with appropriate humility and to be vigilant to the possibility that our own explanations are misleading us.

Legal negotiations

In legal cases such as the one described earlier in the chapter, the data scientist has ethical obligations to their client. Depending on the circumstances, they may also have obligations to the court.

As always, you should be forthright with your client. Usually you will be using methods that you deem appropriate, but on occasion you will be directed to use a method that you think is inappropriate. For instance, we've seen occasions when the client requested that the time period of data included in the analysis be limited in some way to produce a "better" result. We've had clients ask us to subdivide the data (in employment discrimination cases, say, by job title) in order to change p-values. Although such subdivision may be entirely legitimate, the decision about subdividing—seen from a purely statistical point of view—ought to be based on the situation, not the desired outcome (see the discussion of the "garden of forking paths" in Section 7.7).

Your client is entitled to make such requests. Whether or not you think the method being asked for is the right one doesn't enter into it. Your professional obligation is to inform the client what the flaws in the proposed method are and how and why you think another method would be better. (See the major exception that follows.)

The legal system in countries such as the U.S. is an *adversarial* system. Lawyers are allowed to frame legal arguments that may be dismissed: They are entitled to enter some facts and not others into evidence. Of course, the opposing legal team is entitled to create their own legal arguments and to cross-examine the evidence to show how it is incomplete and misleading. When you are working with a legal team as a data scientist, you are part of the team. The lawyers on the team are the experts about what negotiation strategies and legal theories to use, how to define the limits of the case (such as damages), and how to present their case or negotiate with the other party.

It is a different matter when you are presenting to the court. This might take the form of filing an expert report to the court, testifying as an expert witness, or being deposed. A deposition is when you are questioned, under oath, outside of the court room. You are obliged to answer all questions honestly. (Your lawyer may, however, direct you not to answer a question about privileged communications.)

If you are an expert witness or filing an expert report, the word "expert" is significant. A court will certify you as an expert in a case giving you permission to express your opinions. Now you have professional ethical obligations to apply your expertise honestly and openly in forming those opinions.

When working on a legal case, you should get advice from a legal authority, which might be your client. Remember that if you do shoddy work, or fail to reply honestly to the other side's criticisms of your work, your credibility as an expert will be imperiled.

6.5   Data and disclosure

6.5.1   Reidentification and disclosure avoidance

The ability to link multiple data sets and to use public information to identify individuals is a growing problem. A glaring example of this occurred in 1996 when then-Governor of Massachusetts William Weld collapsed while attending a graduation ceremony at Bentley College. An MIT graduate student used information from a public data release by the Massachusetts Group Insurance Commission to identify Weld's subsequent hospitalization records. The disclosure of this information was highly publicized and led to many changes in data releases. This was a situation where the right balance was not struck between disclosure (to help improve health care and control costs) and nondisclosure (to help ensure private information is not made public). There are many challenges to ensure disclosure avoidance [244, 151]: This remains an active and important area of research.

The Health Insurance Portability and Accountability Act (HIPAA) was passed by the United States Congress in 1996—the same year as Weld's illness. The law augmented and clarified the role that researchers and medical care providers had in maintaining protected health information (PHI). The HIPAA regulations developed since then specify procedures to ensure that individually identifiable PHI is protected when it is transferred, received, handled, analyzed, or shared. As an example, detailed geographic information (e.g., home or office location) is not allowed to be shared unless there is an overriding need. For research purposes, geographic information might be limited to state or territory, though for certain rare diseases or characteristics even this level of detail may lead to disclosure. Those whose PHI is not protected can file a complaint with the Office of Civil Rights.

The HIPAA structure, while limited to medical information, provides a useful model for disclosure avoidance that is relevant to other data scientists. Parties accessing PHI need to have privacy policies and procedures. They must identify a privacy official and undertake training of their employees. If there is a disclosure they must mitigate the effects to the extent practical. There must be reasonable data safeguards to prevent intentional or unintentional use. Covered entities may not retaliate against someone for assisting in investigations of disclosures. They must maintain records and documentation for six years after their last use of the data. Similar regulations protect information collected by the statistical agencies of the United States.

6.5.2 Safe data storage

Inadvertent disclosures of data can be even more damaging than planned disclosures. Stories abound of protected data being made available on the Internet with subsequent harm to those whose information is made accessible. Such releases may be due to misconfigured databases, malware, theft, or by posting on a public forum. Each individual and organization needs to practice safe computing, to regularly audit their systems, and to implement plans to address computer and data security. Such policies need to ensure that protections remain even when equipment is transferred or disposed of.

6.5.3 Data scraping and terms of use

A different issue arises relating to legal status of material on the Web. Consider Zillow.com, an online real-estate database company that combines data from a number of public and private sources to generate house price and rental information on more than 100 million homes across the United States. Zillow has made access to their database available through an API (see Section 5.5.2) under certain restrictions. The terms of use for Zillow are provided in a legal document. They require that users of the API consider the data on an "as is" basis, not replicate functionality of the Zillow website or mobile app, not retain any copies of the Zillow data, not separately extract data elements to enhance other data files, and not use the data for direct marketing.

Another common form for terms of use is a limit to the amount or frequency of access. Zillow's API is limited to 1,000 calls per day to the home valuations or property details. Another example: The Weather Underground maintains an API focused on weather information. They provide no-cost access limited to 500 calls per day and 10 calls per minute and with no access to historical information. They have a for-pay system with multiple tiers for accessing more extensive data.

Data points are not just content in tabular form. Text is also data. Many websites have restrictions on text mining. Slate.com, for example, states that users may not:

> "Engage in unauthorized spidering, scraping, or harvesting of content or information, or use any other unauthorized automated means to compile information."

Apparently, it violates the Slate.com terms of use to compile a compendium of Slate articles (even for personal use) without their authorization.

To get authorization, you need to ask for it. For instance, Albert Kim of Middlebury College published data with information for 59,946 San Francisco OkCupid users (a free online dating website) with the permission of the president of OkCupid [125]. To help minimize possible damage, he also removed certain variables (e.g., username) that would make it more straightforward to reidentify the profiles. Contrast the concern for privacy taken here to the careless doxing of OkCupid users mentioned above.

6.6　Reproducibility

Disappointingly often, even the original researchers are unable to reproduce their own results. This failure arises naturally enough when researchers use menu-driven software that does not keep an audit trail of each step in the process. For instance, in Excel, the process of sorting data is not recorded. You can't look at a spreadsheet and determine what range of data was sorted, so mistakes in selecting cases or variables for a sort are propagated untraceably through the subsequent analysis. Researchers commonly use tools like word processors that do not mandate an explicit tie between the result presented in a publication and the analysis that produced the result. These seemingly innocuous practices contribute to the loss of reproducibility: numbers may be copied by hand into a document and graphics are cut-and-pasted into the report. (Imagine that you have inserted a graphic into a report in this way. How could you, or anyone else, easily demonstrate that the correct graphic was selected for inclusion?)

Reproducible analysis is the practice of recording each and every step, no matter how trivial seeming, in a data analysis. The main elements of a reproducible analysis plan (as described by Project TIER (`https://www.haverford.edu/project-tier`) include:

Data: all original data files in the form in which they originated,

Metadata: codebooks and other information needed to understand the data,

Commands: the computer code needed to extract, transform, and load the data—then run analyses, fit models, generate graphical displays, and

Map: a file that maps between the output and the results in the report.

The American Statistical Association (ASA) notes the importance of reproducible analysis in its curricular guidelines. The development of new tools such as R Markdown and `knitr` have dramatically improved the usability of these methods in practice. See Appendix D for an introduction to these tools.

Individuals and organizations have been working to develop protocols to facilitate making the data analysis process more transparent and to integrate this into the workflow of practitioners and students. One of us has worked as part of a research project team at the Channing Laboratory at Harvard University. As part of the vetting process for all manuscripts, an analyst outside of the research team is required to review all programs used to generate results. In addition, another individual is responsible for checking each number in the paper to ensure that it was correctly transcribed from the results. Similar practice is underway at The Odum Institute for Research in Social Science at the University of North Carolina. This organization performs third-party code and data verification for several political science journals.

6.6.1　Example: Erroneous data merging

In Chapter 4, we discuss how the *join* operation can be used to merge two data tables together. Incorrect merges can be very difficult to unravel unless the exact details of the merge have been recorded. The `dplyr inner_join()` function simplifies this process.

In a 2013 paper published in the journal *Brain, Behavior, and Immunity*, Kern et al. reported a link between immune response and depression. To their credit, the authors later noticed that the results were the artifact of a faulty data merge between the lab results and other survey data. A retraction [124], as well as a corrected paper reporting negative results [123], were published in the same journal.

In some ways this is science done well—ultimately the correct negative result was published, and the authors acted ethically by alerting the journal editor to their mistake. However, the error likely would have been caught earlier had the authors adhered to stricter standards of reproducibility (see Appendix D) in the first place.

6.7 Professional guidelines for ethical conduct

This chapter has outlined basic principles of professional ethics. Usefully, several organizations have developed detailed statements on topics such as professionalism, integrity of data and methods, responsibilities to stakeholders, conflicts of interest, and the response to allegations of misconduct. One good source is the framework for professional ethics endorsed by the American Statistical Association (ASA) [58].

The Committee on Science, Engineering, and Public Policy of the National Academy of Sciences, National Academy of Engineering, and Institute of Medicine has published the third edition of *On Being a Scientist: A Guide to Responsible Conduct in Research*. The guide is structured into a number of chapters, many of which are highly relevant for data scientists (including "the Treatment of Data," "Mistakes and Negligence," "Sharing of Results," "Competing Interests, Commitment, and Values," and "The Researcher in Society").

The Association for Computing Machinery (ACM)—the world's largest computing society, with more than 100,000 members—adopted a code of ethics in 1992 (see `https://www.acm.org/about/code-of-ethics`). Other relevant statements and codes of conduct have been promulgated by the Data Science Association (`http://www.datascienceassn.org/code-of-conduct.html`), the International Statistical Institute (`http://www.isi-web.org/about-isi/professional-ethics`), and the United Nations Statistics Division (`http://unstats.un.org/unsd/dnss/gp/fundprinciples.aspx`). The Belmont Report outlines ethical principles and guidelines for the protection of human research subjects.

6.8 Ethics, collectively

Although science is carried out by individuals and teams, the scientific community as a whole is a stakeholder. Some of the ethical responsibilities faced by data scientists are created by the collective nature of the enterprise.

A team of Columbia University scientists discovered that a former post-doc in the group, unbeknownst to the others, had fabricated and falsified research reported in articles in the journals *Cell* and *Nature*. Needless to say, the post-doc had violated his ethical obligations both with respect to his colleagues and to the scientific enterprise as a whole. When the misconduct was discovered, the other members of the team incurred an ethical obligation to the scientific community. In fulfillment of this obligation, they notified the journals and retracted the papers, which had been highly cited. To be sure, such episodes can tarnish the reputation of even the innocent team members, but the ethical obligation outweighs the desire to protect one's reputation.

Perhaps surprisingly, there are situations where it is not ethical *not* to publish one's work. "Publication bias" (or the "file-drawer problem") refers to the situation where reports of statistically significant (i.e., $p < 0.05$) results are much more likely to be published than reports where the results are not statistically significant. In many settings, this bias is for the good; a lot of scientific work is in the pursuit of hypotheses that turn out to be wrong or ideas that turn out not to be productive.

But with many research teams investigating similar ideas, or even with a single research team that goes down many parallel paths, the meaning of "statistically significant" becomes clouded and corrupt. Imagine 100 parallel research efforts to investigate the effect of a drug that in reality has no effect at all. Roughly five of those efforts are expected to culminate in a misleadingly "statistically significant" ($p < 0.05$) result. Combine this with publication bias and the scientific literature might consist of reports on just the five projects that happened to be significant. In isolation, five such reports would be considered substantial evidence about the (non-null) effect of the drug. It might seem unlikely that there would be 100 parallel research efforts on the same drug, but at any given time there are tens of thousands of research efforts, any one of which has a 5% chance of producing a significant result even if there were no genuine effect.

The American Statistical Association's ethical guidelines state, "Selecting the one 'significant' result from a multiplicity of parallel tests poses a grave risk of an incorrect conclusion. Failure to disclose the full extent of tests and their results in such a case would be highly misleading." So, if you're examining the effect on five different measures of health by five different foods, and you find that broccoli consumption has a statistically significant relationship with the development of colon cancer, not only should you be skeptical but you should include in your report the null result for the other twenty-four tests or perform an appropriate statistical correction to account for the multiple tests. Often, there may be several different outcome measures, several different food types, and several potential covariates (age, sex, whether breastfed as an infant, smoking, the geographical area of residence or upbringing, etc.), so it's easy to be performing dozens or hundreds of different tests without realizing it.

For clinical health trials, there are efforts to address this problem through trial registries. In such registries (e.g., `https://clinicaltrials.gov`), researchers provide their study design and analysis protocol in advance and post results.

6.9   Further resources

For a book-length treatment of ethical issues in statistics, see [113]. A historical perspective on the ASA's Ethical Guidelines for Statistical Practice can be found in [70]. The University of Michigan provides an EdX course on "Data Science Ethics." Gelman has written a column on ethics in statistics in *CHANCE* for the past several years (see, for example [84, 86, 85]). *Weapons of Math Destruction: How Big Data Increases Inequality and Threatens Democracy* describes a number of frightening uses of big data and algorithms [153].

The Center for Open Science—which develops the Open Science Framework (OSF)—is an organization that promotes openness, integrity, and reproducibility in scientific research. The OSF provides an online platform for researchers to publish their scientific projects. Emil Kirkegaard used OSF to publish his OkCupid data set.

The Institute for Quantitative Social Science at Harvard and the Berkeley Initiative for Transparency in the Social Sciences are two other organizations working to promote reproducibility in social science research. The American Political Association has incorporated the Data Access and Research Transparency (DA-RT) principles into its ethics guide. The Consolidated Standards of Reporting Trials (CONSORT) statement at `http://www.consort-statement.org` provides detailed guidance on the analysis and reporting of clinical trials.

Many more examples of how irreproducibility has led to scientific errors are available at `http://retractionwatch.com/`. For example, a study linking severe illness and divorce rates was retracted due to a coding mistake.

6.10   Exercises

Exercise 6.1

A researcher is interested in the relationship of weather to sentiment on Twitter. They want to scrape data from `www.wunderground.com` and join that to Tweets in that geographic area at a particular time. One complication is that Weather Underground limits the number of data points that can be downloaded for free using their API (application program interface). The researcher sets up six free accounts to allow them to collect the data they want in a shorter time-frame. What ethical guidelines are violated by this approach to data scraping?

Exercise 6.2

A data analyst received permission to post a data set that was scraped from a social media site. The full data set included name, screen name, email address, geographic location, IP (Internet protocol) address, demographic profiles, and preferences for relationships. Why might it be problematic to post a deidentified form of this data set where name and email address were removed?

Exercise 6.3

A company uses a machine learning algorithm to determine which job advertisement to display for users searching for technology jobs. Based on past results, the algorithm tends to display lower paying jobs for women than for men (after controlling for other characteristics than gender). What ethical considerations might be considered when reviewing this algorithm?

Exercise 6.4

A reporter carried out a clinical trial of chocolate where a small number of overweight subjects who had received medical clearance were randomized to either eat dark chocolate or not to eat dark chocolate. They were followed for a period and their change in weight was recorded from baseline until the end of the study. More than a dozen outcomes were recorded and one proved to be significantly different in the treatment group than the outcome. This study was publicized and received coverage from a number of magazines and television programs. Outline the ethical considerations that arise in this situation.

Exercise 6.5

A data scientist compiled data from several public sources (voter registration, political contributions, tax records) that were used to predict sexual orientation of individuals in a community. What ethical considerations arise that should guide use of such data sets?

Exercise 6.6

A *Slate* article (`http://tinyurl.com/slate-ethics`) discussed whether race/ethnicity should be included in a predictive model for how long a homeless family would stay in homeless services. Discuss the ethical considerations involved in whether race/ethnicity should be included as a predictor in the model.

Exercise 6.7

In the United States, most students apply for grants or subsidized loans to finance their college education. Part of this process involves filling in a federal government form called the Free Application for Federal Student Aid (FAFSA). The form asks for information about family income and assets. The form also includes a place for listing the universities to which the information is to be sent. The data collected by FAFSA includes confidential financial information (listing the schools eligible to receive the information is effectively giving permission to share the data with them).

It turns out that the order in which the schools are listed carries important information. Students typically apply to several schools, but can attend only one of them. Until recently, admissions offices at some universities used the information as an important part of their models of whether an admitted student will accept admissions. The earlier in a list a school appears, the more likely the student is to attend that school.

Here's the catch from the student's point of view. Some institutions use statistical models to allocate grant aid (a scarce resource) where it is most likely to help ensure that a student enrolls. For these schools, the more likely a student is deemed to accept admissions, the lower the amount of grant aid they are likely to receive.

Is this ethical? Discuss.

Exercise 6.8

In 2006, AOL released a database of search terms that users had used in the prior month (see http://www.nytimes.com/2006/08/09/technology/09aol.html). Research this disclosure and the reaction that ensued. What ethical issues are involved? What potential impact has this disclosure had?

Exercise 6.9

In the United States, the Confidential Information Protection and Statistical Efficiency Act (CIPSEA) governs the confidentiality of data collected by agencies such as the Bureau of Labor Statistics and the Census Bureau. What are the penalties for willful and knowing disclosure of protected information to unauthorized persons?

Exercise 6.10

A statistical analyst carried out an investigation of the association of gender and teaching evaluations at a university. They undertook exploratory analysis of the data and carried out a number of bivariate comparisons. The multiple items on the teaching evaluation were consolidated to a single measure based on these exploratory analyses. They used this information to construct a multivariable regression model that found evidence for biases. What issues might arise based on such an analytic approach?

Exercise 6.11

An investigative team wants to winnow the set of variables to include in their final multiple regression model. They have 100 variables and one outcome measured for $n = 250$ observations). They use the following procedure:

1. Fit each of the 100 bivariate models for the outcome as a function of a single predictor, then

2. Include all of the significant predictors in the overall model.

What does the distribution of the p-value for the overall test look like, assuming that there are no associations between any of the predictors and the outcome (all are assumed to be multivariate normal and independent). Carry out a simulation to check your answer.

Part II

Statistics and Modeling

Chapter 7

Statistical foundations

The ultimate objective in data science is to extract meaning from data. Data wrangling and visualization are tools to this end. Wrangling re-organizes cases and variables to make data easier to interpret. Visualization is a primary tool for connecting our minds with the data, so that we humans can search for meaning.

Visualizations are powerful because human visual cognitive skills are strong. We are very good at seeing patterns even when partially obscured by random noise. On the other hand, we are also very good at seeing patterns even when they are not there. People can easily be misled by the accidental, evanescent patterns that appear in random noise. It's important therefore to be able to discern when the patterns we see are so strong and robust that we can be confident they are not mere accidents.

Statistical methods quantify patterns and their strength. They are essential tools for interpreting data. As we'll see later in this book, the methods are also crucial for finding patterns that are too complex or multi-faceted to be seen visually.

Some people think that *big data* has made statistics obsolete. The argument is that with lots of data, the data can speak clearly for themselves. This is wrong, as we shall see. The discipline for making efficient use of data that is a core of statistical methodology leads to deeper thinking about how to make use of data—that thinking applies to large data sets as well.

In this chapter we will introduce key ideas from statistics that permeate data science and that will be reinforced later in the book. At the same time, the extended example used in this chapter will illustrate a data science *workflow* that uses a cycle of wrangling, exploring, visualizing, and modeling.

7.1 Samples and populations

In previous chapters, we've considered data as being fixed. Indeed, the word "data" stems from the Latin word for "given"—any set of data is treated as given.

Statistical methodology is governed by a broader point of view. Yes, the data we have in hand are fixed, but the methodology assumes that the cases are drawn from a much larger set of potential cases. The given data are a *sample* of a larger *population* of potential cases. In statistical methodology, we view our sample of cases in the context of this population. We imagine other samples that might have been drawn from the population.

At the same time, we imagine that there might have been additional variables that could have been measured from the population. We permit ourselves to construct new variables that have a special feature: any patterns that appear involving the new variables are guaranteed to be random and accidental. The tools we will use to gain access to the imagined cases from the population and the contrived no-pattern variables involve the mathematics of probability or (more simply) random selection from a set.

In the next section, we'll elucidate some of the connections between the sample—the data we've got—and the population. To do this, we'll use an artifice: constructing a playground that contains the entire population. Then, we can work with data consisting of a smaller set of cases selected at random from this population. This lets us demonstrate and justify the statistical methods in a setting where we know the "correct" answer. That way, we can develop ideas about how much confidence statistical methods can give us about the patterns we see.

Example: Sampling from the population

Suppose you were asked to help develop a travel policy for business travelers based in New York City. Imagine that the traveler has a meeting in San Francisco (airport code SFO) at a specified time t. The policy to be formulated will say how much earlier than t an acceptable flight should arrive in order to avoid being late to the meeting due to a flight delay.

For the purpose of this example, recall from the previous section that we are going to pretend that we already have on hand the complete *population* of flights. For this purpose, we're going to use the set of 336,776 flights in 2013 in the `nycflights13` package, which gives airline delays from New York City airports in 2013. The policy we develop will be for 2013. Of course this is unrealistic in practice. If we had the complete population we could simply look up the best flight that arrived in time for the meeting!

More realistically, the problem would be to develop a policy for this year based on the sample of data that have already been collected. We're going to simulate this situation by drawing a sample from the population of flights into SFO. Playing the role of the population in our little drama, SF comprises the complete collection of such flights.

```
library(mdsr)
library(nycflights13)
SF <- flights %>%
  filter(dest == "SFO", !is.na(arr_delay))
```

We're going to work with just a sample from this population. For now, we'll set the sample size to be $n = 25$ cases.

```
set.seed(101)
Sample25 <- SF %>%
  sample_n(size = 25)
```

A simple (but naïve) way to set the policy is to look for the longest flight delay, and insist that travel be arranged to deal with this delay.

```
favstats( ~ arr_delay, data = Sample25)
```

```
 min  Q1 median Q3 max   mean   sd  n missing
 -50 -23     -7  4 124  -2.96 35.3 25       0
```

The maximum delay is 124 minutes, about 2 hours. So, should our travel policy be that the traveler should plan on arriving in SFO at least two hours ahead? In our example world, we can look at the complete set of flights to see what was the actual worst delay in 2013.

```
favstats( ~ arr_delay, data = SF)

 min   Q1 median Q3  max mean   sd      n missing
 -86 -23     -8 12 1007 2.67 47.7 13173       0
```

Notice that the results from the sample are different from the results for the population. In the population, the longest delay was 1,007 minutes—almost 17 hours. This suggests that to avoid missing a meeting, you should travel the day before the meeting. Safe enough, but then:

- an extra travel day is expensive in terms of lodging, meals, and the traveler's time;

- even at that, there's no guarantee that there will never be a delay of more than 1,007 minutes.

A sensible travel policy will trade off small probabilities of being late against the savings in cost and traveler's time. For instance, you might judge it acceptable to be late just 2% of the time—a 98% chance of being on time.

Here's the 98th percentile of the arrival delays in our data sample:

```
qdata( ~ arr_delay, p = 0.98, data = Sample25)

      p quantile
   0.98    87.52
```

A delay of 88 minutes is about an hour and a half. The calculation is easy, but how good is the answer? This is not a question about whether the 98th percentile was calculated properly—that will always be the case for any competent data scientist. The question is really along these lines: Suppose we used the 90-minute travel policy. How well would that have worked in achieving our intention to be late for meetings only 2% of the time?

With the population data in hand, it's easy to answer this question.

```
tally( ~ arr_delay < 90, data = SF, format = "proportion")

arr_delay < 90
  TRUE  FALSE
0.9514 0.0486
```

The 90-minute policy would miss its mark 5% of the time, much worse than we intended. To correctly hit the mark 2% of the time, we will want to increase the policy from 90 minutes to what value?

With the population, it's easy to calculate the 98th percentile of the arrival delays:

```
qdata( ~ arr_delay, p = 0.98, data = SF)

      p quantile
   0.98   153.00
```

It should have been about 150 minutes, not 90. But in many important real–world settings, we do not have access to the population data. We have only our sample. How can we use our sample to judge whether the result we get from the sample is going to be good enough to meet the 98% goal? And if it's not good enough, how large should a sample be to give a result that is likely to be good enough? This is where the concepts and methods from statistics come in.

We will continue exploring this example throughout the chapter. In addition to addressing our initial question, we'll examine the extent to which the policy should depend on the airline carrier, the time of year, hour of day, and day of the week.

The basic concepts we'll build on are sample statistics such as the mean and standard deviation. These topics are covered in introductory statistics books. Readers who have not yet encountered these should review an introductory statistics text such as the OpenIntro Statistics books (`http://openintro.org`), Appendix E, or the materials in Section 7.8 (Further resources).

7.2 Sample statistics

Statistics (plural) is a field that overlaps with and contributes to data science. A *statistic* (singular) is a number that summarizes data. Ideally, a statistic captures all of the useful information from the individual observations.

When we calculate the 98th percentile of a sample, we are calculating one of many possible sample statistics. Among the many sample statistics are the mean of a variable, the standard deviation, the median, the maximum, and the minimum. It turns out that sample statistics such as the maximum and minimum are not very useful. The reason is that there is not a reliable way to figure out how well the sample statistic reflects what is going on in the population. Similarly, the 98th percentile is not a reliable sample statistic for small samples (such as our 25 flights into SFO), in the sense that it will vary considerably in small samples.

On the other hand, a median is a more reliable sample statistic. Under certain conditions, the mean and standard deviation are reliable as well. In other words, there are established techniques for figuring out, from the sample itself, how well the sample statistic reflects the population.

The sampling distribution

Ultimately we need to figure out the reliability of a sample statistic from the sample itself. For now, though, we are going to use the population to develop some ideas about how to define reliability. So we will still be in the playground world where we have the population in hand.

If we were to collect a new sample from the population, how similar would the sample statistic on that new sample be to the same statistic calculated on the original sample? Or, stated somewhat differently, if we draw many different samples from the population, each of size n, and calculated the sample statistic on each of those samples, how similar would the sample statistic be across all the samples?

With the population in hand, it's easy to figure this out; use `sample_n()` many times and calculate the sample statistic on each trial. For instance, here are two trials in which we sample and calculate the mean arrival delay. (We'll explain the `replace = FALSE` in the next section. Briefly, it means to draw the sample as one would deal from a set of cards: None of the cards can appear twice in one hand.)

```
n <- 25
mean( ~ arr_delay, data = sample_n(SF, size = n, replace = FALSE))
```

```
[1] -7.4
```

```
mean( ~ arr_delay, data = sample_n(SF, size = n, replace = FALSE))
```

```
[1] 1.16
```

Perhaps it would be better to run many trials (though each one would require considerable effort in the real world). The do() function from the mosaic package lets us automate the process. Here are the results from 500 trials.

```
Trials <- do(500) *
  mean( ~ arr_delay, data = sample_n(SF, size = n, replace = FALSE))
head(Trials)
```

```
    mean
1 -14.64
2   7.40
3  19.24
4  10.96
5  20.16
6  -5.52
```

We now have 500 trials, for each of which we calculated the mean arrival delay. Let's examine how spread out the results are.

```
favstats( ~ mean, data = Trials)
```

```
  min    Q1 median   Q3  max mean   sd   n missing
-21.2 -3.86      1 8.74 51.9 3.35 10.3 500       0
```

To discuss reliability, it helps to have some standardized vocabulary.

- The *sample size* is the number of cases in the sample, usually denoted with n. In the above, the sample size is $n = 25$.

- The *sampling distribution* is the collection of the sample statistic from all of the trials. We carried out 500 trials here, but the exact number of trials is not important so long as it is large.

- The *shape* of the sampling distribution is worth noting. Here it is a little skewed to the right.

- The *standard error* is the standard deviation of the sampling distribution. It describes the width of the sampling distribution. For the trials calculating the sample mean in samples with $n = 25$, the standard error is 10.3 minutes. (You can see it in the output of favstats() above.)

- The *95% confidence interval* is another way of summarizing the sampling distribution. From Figure 7.1 (left panel) you can see it is about -10 to $+25$ minutes. As taught in introductory statistics courses, often the interval is calculated from the mean and standard error of the sampling distribution:

```
mean(~ mean, data = Trials) + 2 * sd(~ mean, data = Trials) * c(-1, 1)
```

```
[1] -17.3  24.0
```

Pro Tip: This vocabulary can be very confusing at first. Remember that "standard error" and "confidence interval" always refer to the sampling distribution, not to the population and not to a single sample. The standard error and confidence intervals are two different, but closely related, forms for describing the reliability of the calculated sample statistic.

An important question that statistical methods allow you to address is what size of sample n is needed to get a result with an acceptable reliability. What constitutes "acceptable" depends on the goal you are trying to accomplish. But measuring the reliability is a straightforward matter of finding the standard error and/or confidence interval.

Notice that the sample statistic varies considerably. For samples of size $n = 25$ they range from -21 to 52 minutes. This is important information. It illustrates the reliability of the sample mean for samples of arrival delays of size $n = 25$. Figure 7.1 (left) shows the distribution of the trials with a histogram.

In this example, we used a sample size of $n = 25$ and found a standard error of 10.3 minutes. What would happen if we used an even larger sample, say $n = 100$? The calculation is the same as before, but with a different n.

```
Trials_100 <- do(500) *
  mean( ~ arr_delay, data = SF %>% sample_n(size = 100, replace = FALSE))
```

```
rbind(Trials %>% mutate(n = 25), Trials_100 %>% mutate(n = 100)) %>%
  ggplot(aes(x = mean)) + geom_histogram(bins = 30) +
  facet_grid( ~ n) + xlab("Sample mean")
```

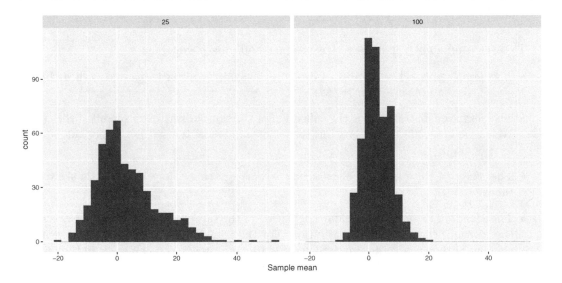

Figure 7.1: The sampling distribution of the mean arrival delay with a sample size of $n = 25$ (left) and also for a larger sample size of $n = 100$ (right).

Figure 7.1 (right panel) also displays the shape of the sampling distribution for samples of size $n = 25$ and $n = 100$. Comparing the two sampling distributions, one with $n = 25$ and the other with $n = 100$ shows some patterns that are generally true for statistics such as the mean:

- Both sampling distributions are centered at the same value.

- A larger sample size produces a standard error that is smaller. That is, a larger sample size is more reliable than a smaller sample size. You can see that the standard deviation for $n = 100$ is one-half that for $n = 25$. As a rule, the standard error of a sampling distribution scales as $1/\sqrt{n}$.

- For large sample sizes n, the shape of the sampling distribution tends to bell-shaped. In a bit of archaic terminology, this shape is often called the *normal distribution*. Indeed, the distribution arises very frequently in statistics, but there is nothing abnormal about any other distribution shape.

7.3 The bootstrap

In the previous examples, we had access to the population data and so we could find the sampling distribution by repeatedly sampling from the population. In practice, however, we have only one sample and not the entire population. The *bootstrap* is a statistical method that allows us to approximate the sampling distribution even without access to the population.

The logical leap involved in the bootstrap is to think of our sample itself as if it were the population. Just as in the previous examples we drew many samples from the population, now we will draw many new samples from our original sample. This process is called *resampling*: drawing a new sample from an existing sample.

When sampling from a population, we would of course make sure not to duplicate any of the cases, just as we would never deal the same playing card twice in one hand. When resampling, however, we do allow such duplication. That is, we *sample with replacement*.

To illustrate, consider Small, a very small sample ($n = 3$) from the flights data. Notice that each of the cases in Small is unique. There are no duplicates.

```
Small <- sample_n(SF, size = 3, replace = FALSE)
```

```
# A tibble: 3 7
  year month   day dep_time sched_dep_time dep_delay arr_time
  <int> <int> <int>    <int>          <int>     <dbl>    <int>
1  2013     4    27     1653           1700        -7     1952
2  2013     5    14     1810           1800        10     2104
3  2013     5    16     1729           1732        -3     2133
```

Resampling from Small is done by setting the `replace` argument to TRUE, which allows the sample to include duplicates.

```
Small %>% sample_n(size = 3, replace = TRUE)
```

```
# A tibble: 3 7
  year month   day dep_time sched_dep_time dep_delay arr_time
  <int> <int> <int>    <int>          <int>     <dbl>    <int>
1  2013     5    16     1729           1732        -3     2133
2  2013     5    16     1729           1732        -3     2133
3  2013     5    16     1729           1732        -3     2133
```

In this particular resample the same single case is repeated 3 times. That's a matter of luck. Let's try again.

```
Small %>% sample_n(size = 3, replace = TRUE)
```

```
# A tibble: 3   7
   year month   day dep_time sched_dep_time dep_delay arr_time
  <int> <int> <int>   <int>          <int>     <dbl>    <int>
1  2013     5    14    1810           1800        10     2104
2  2013     4    27    1653           1700        -7     1952
3  2013     5    14    1810           1800        10     2104
```

This resample has two instances of one case and a single instance of another.

Bootstrapping does not create new cases: It isn't a way to collect data. In reality, constructing a sample involves genuine data acquisition, e.g., field work or lab work or using information technology systems to consolidate data. In this textbook example, we get to save all that effort and simply select at random from the population, SF. The one and only time we use the population is to draw the original sample, which, as always with a sample, we do without replacement.

Let's use bootstrapping to find the reliability of the mean arrival time calculated on a sample of size 200.

```
n <- 200
Orig_sample <- SF %>% sample_n(size = n, replace = FALSE)
```

Now, with the original sample in hand, we can draw a resample and calculate the mean arrival delay.

```
mean( ~ arr_delay,
      data = sample_n(Orig_sample, size = n, replace = TRUE))
```

```
[1] -2.2
```

By repeating this process many times, we'll be able to see how much variation there is from sample to sample:

```
Bootstrap_trials <- do(500) * mean( ~ arr_delay,
  data = sample_n(Orig_sample, size = n, replace = TRUE))
favstats( ~ mean, data = Bootstrap_trials)
```

```
   min    Q1 median     Q3  max  mean   sd  n missing
 -9.04 -3.98  -2.25 -0.564 4.57 -2.28 2.37 500       0
```

We can compare this to a (hypothetical) sample of size $n = 1,000$ from the original SF flights.

```
Trials_200 <- do(500) *
  mean( ~ arr_delay, data = sample_n(SF, size = n, replace = FALSE))
favstats( ~ mean, data = Trials_200)
```

```
   min    Q1 median   Q3  max mean   sd   n missing
 -5.64 0.241   2.29 4.51 13.3 2.47 3.11 500       0
```

Notice that the population was not used in the bootstrap, just the original sample. What's remarkable here is that the standard error calculated in this way, 2.4 minutes, is a reasonable approximation to the standard error of the sampling population calculated in the previous section (3.1 minutes).

The distribution of values in the bootstrap trials is called the *bootstrap distribution*. It's not exactly the same as the sampling distribution, but for moderate to large sample sizes it has been proven to approximate those aspects of the sampling distribution that we care most about, such as the standard error [69].

Let's return to our original example of setting a travel policy for selecting flights from New York to San Francisco. Recall that we decided to set a goal of arriving in time for the meeting 98% of the time. We can calculate the 98th percentile from our sample of size $n = 100$ flights, and use bootstrapping to see how reliable that sample statistic is.

The sample itself suggests a policy of scheduling a flight to arrive 85 minutes early.

```
qdata( ~ arr_delay, p = 0.98, data = Orig_sample)

     p quantile
  0.98    85.00
```

We can check the reliability of that estimate using bootstrapping.

```
Bootstrap_trials <- do(500) *
  qdata( ~ arr_delay, p = 0.98,
         data = sample_n(Orig_sample, size = n, replace = TRUE))
favstats( ~ quantile, data = Bootstrap_trials)

  min   Q1 median   Q3 max mean   sd   n missing
   51 79.1     85 85.4 186 87.2 16.6 500       0
```

The bootstrapped standard error is about 17 minutes. The corresponding 95% confidence interval is 87 ± 33 minutes. A policy based on this would be practically a shot in the dark: unlikely to hit the target.

One way to fix things might be to collect more data, hoping to get a more reliable estimate of the 98th percentile. Let's generate a sample with $n = 10,000$ cases.

```
  min   Q1 median   Q3  max mean sd   n missing
 24.1 40.1   47.3 54.2 97.5 47.8 11 500       0
```

Disappointing! The 95% confidence interval is still very broad, 48 ± 22 minutes. The standard error of the 98th percentile estimated from a sample of size $n = 10,000$ is not better. This is showing us that estimates of the 98th percentile are not very reliable, since it is by definition in the tail of the distribution. Having more data doesn't cure all ills. Knowing this, we might decide not to set our goal in terms of the unreliable 98th percentile, or at least to tell our boss that there is no way to guarantee that the policy based on 98 percent will come close to meeting its goal. Or, even better, we might decide to examine things more closely, as in the next section.

7.4 Outliers

One place where more data is helpful is in identifying unusual or extreme events: *outliers*. Suppose we consider any flight delayed by seven hours (420 minutes) or more as an extreme event (see Section 12.5). While an arbitrary choice, 420 minutes may be valuable as a marker for seriously delayed flights.

```
SF %>%
  filter(arr_delay >= 420) %>%
  select(month, day, dep_delay, arr_delay, carrier)
```

```
# A tibble: 7  5
  month   day dep_delay arr_delay carrier
  <int> <int>     <dbl>     <dbl> <chr>
1    12     7       374       422      UA
2     7     6       589       561      DL
3     7     7       629       676      VX
4     7     7       653       632      VX
5     7    10       453       445      B6
6     7    10       432       433      VX
7     9    20      1014      1007      AA
```

Most of the very long delays (five of seven) were in July, and Virgin America (VX) is the most frequent offender. Immediately, this suggests one possible route for improving the outcome of the business travel policy we have been asked to develop. We could tell people to arrive extra early in July and to avoid VX.

But let's not rush into this. The outliers themselves may be misleading. These outliers account for a tiny fraction of the flights into San Francisco in 2013. That's a small component of our goal of having a failure rate of 2% in getting to meetings on time. And there was an even more extremely rare event at SFO in July 2013: the crash-landing of Asiana Airlines flight 214. We might remove these points to get a better sense of the main part of the distribution.

Pro Tip: Outliers can often tell us interesting things. How they should be handled depends on their cause. Outliers due to data irregularities or errors should be fixed. Other outliers may yield important insights. Outliers should never be dropped unless there is a clear rationale. If outliers are dropped this should be clearly reported.

Figure 7.2 displays the histogram without those outliers.

Note that the large majority of flights arrive without any delay or a delay of less than 60 minutes. Might we be able to identify patterns that can presage when the longer delays are likely to occur? The 14 outliers suggested that month or carrier may be linked to long delays. Let's see how that plays out with the large majority of data.

```
SF %>%
  mutate(long_delay = arr_delay > 60) %>%
  tally(~ long_delay | month, data = .)
```

```
           month
long_delay   1    2    3    4    5    6    7    8    9   10   11   12
      TRUE  29   21   61  112   65  209  226   96   65   36   51   66
     FALSE 856  741  812  993 1128  980  966 1159 1124 1177 1107 1093
```

We see that June and July (months 6 and 7) are problem months.

```
SF %>% filter(arr_delay < 420) %>%
  ggplot(aes(arr_delay)) + geom_histogram(binwidth = 15)
```

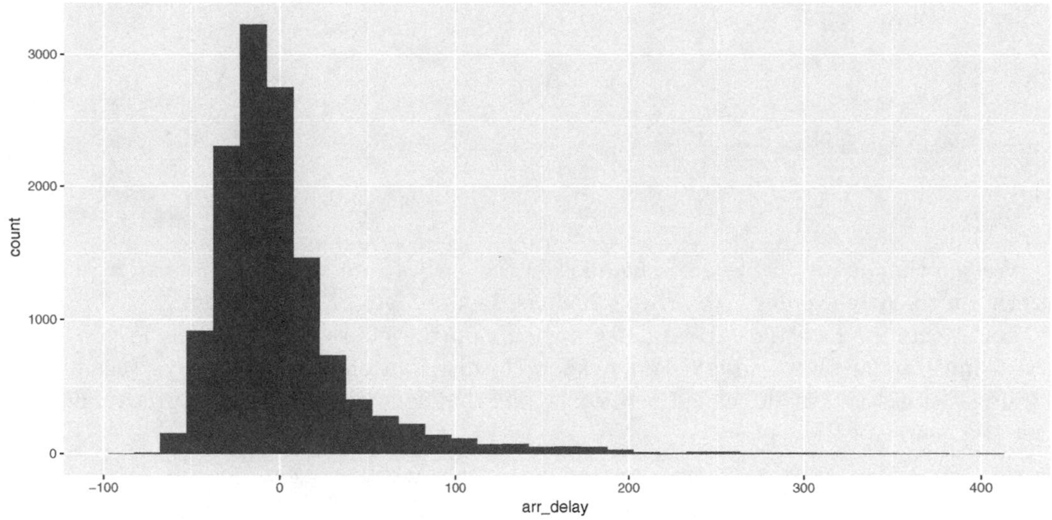

Figure 7.2: Distribution of flight arrival delays in 2013 for flights to San Francisco from NYC airports that were delayed less than seven hours. The distribution features a long right tail (even after pruning the outliers).

```
SF %>%
  mutate(long_delay = arr_delay > 60) %>%
    tally(~ long_delay | carrier, data = .)

           carrier
long_delay  AA   B6   DL   UA   VX
     TRUE  148   86   91  492  220
    FALSE 1250  934 1757 6236 1959
```

Delta Airlines (DL) has reasonable performance. These two simple analyses hint at a policy that might advise travelers to plan to arrive extra early in June and July and to consider Delta as an airline for travel to SFO (see Section 12.5 for a fuller discussion of which airlines seem to have fewer delays in general).

7.5 Statistical models: Explaining variation

In the previous section, we used month of the year and airline to narrow down the situations in which the risk of an unacceptable flight delay is large. Another way to think about this is that we are *explaining* part of the variation in arrival delay from flight to flight. *Statistical modeling* provides a way to relate variables to one another. Doing so helps us better understand the system we are studying.

To illustrate modeling, let's consider another question from the airline delays data set: What impact, if any, does scheduled time of departure have on expected flight delay? Many people think that earlier flights are less likely to be delayed, since flight delays tend to cascade over the course of the day. Is this theory supported by the data?

We first begin by considering time of day. In the nycflights13 package, the flights data frame has a variable (hour) that specifies the *scheduled* hour of departure.

```
tally( ~ hour, data = SF)

hour
    5     6     7     8     9    10    11    12    13    14    15    16    17    18    19
   55   663  1696   987   429  1744   413   504   476   528   946   897  1491  1091   731
   20    21
  465    57
```

We see that many flights are scheduled in the early to mid-morning and from the late afternoon to early evening. None are scheduled before 5 am or after 10 pm.

Let's examine how the arrival delay depends on the hour. We'll do this in two ways: first using standard box-and-whiskers to show the distribution of arrival delays; second with a kind of statistical model called a *linear model* that lets us track the mean arrival delay over the course of the day.

```
SF %>%
  ggplot(aes(x = hour, y = arr_delay)) +
  geom_boxplot(alpha = 0.1, aes(group = hour)) + geom_smooth(method = "lm") +
  xlab("Scheduled hour of departure") + ylab("Arrival delay (minutes)") +
  coord_cartesian(ylim = c(-30, 120))
```

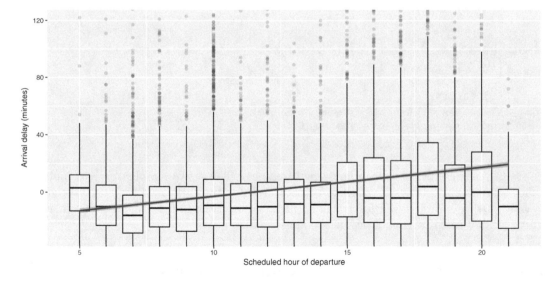

Figure 7.3: Association of flight arrival delays with scheduled departure time for flights to San Francisco from New York airports in 2013.

Figure 7.3 displays the arrival delay versus schedule departure hour. The average arrival delay increases over the course of the day. The trend line itself is created via a *regression model* (see Appendix E).

```
mod1 <- lm(arr_delay ~ hour, data = SF)
msummary(mod1)
```

```
              Estimate Std. Error t value Pr(>|t|)
(Intercept)  -22.9327     1.2328   -18.6   <2e-16 ***
hour           2.0149     0.0915    22.0   <2e-16 ***


Residual standard error: 46.8 on 13171 degrees of freedom
Multiple R-squared:  0.0355,Adjusted R-squared:  0.0354
F-statistic:  484 on 1 and 13171 DF,  p-value: <2e-16
```

The number under the "Estimate" for hour indicates that the arrival delay increases by about 2 minutes per hour. Over the 15 hours of flights, this leads to a 30-minute increase in arrival delay for flights at the end of the day. The msummary() function also calculates the standard error: 0.09 minutes per hour. Or, stated as a 95% confidence interval, this model indicates that arrival delay increases by 2.0 ± 0.18 minutes per hour. The rightmost column gives the *p-value*, a way of translating the estimate and standard error onto a scale from zero to one. By convention, p-values below 0.05 provide a kind of certificate testifying that random, accidental patterns would be unlikely to generate an estimate as large as that observed. The tiny p-value given in the report (2e-16 is 0.0000000000000002) is another way of saying that this confidence interval rules out the possibility that the two-minutes-per-hour increase in arrival delay is just an accidental pattern.

Re-read those last three sentences. Confusing? Despite an almost universal practice of presenting p-values, they are mostly misunderstood even by scientists and other professionals. The p-value conveys much less information than usually supposed: The "certificate" might not be worth the paper it's printed on (see Section 7.7).

Can we do better? What additional factors might help to explain flight delays? Let's look at departure airport, carrier (airline), month of the year, and day of the week. Some wrangling will let us extract the day of the week (dow) from the year, month, and day of month. We'll also create a variable season that summarizes what we already know about the month: that June and July are the months with long delays. These will be used as *explanatory variables* to account for the *response variable*: arrival delay.

```
library(lubridate)
SF <- SF %>%
  mutate(day = ymd(paste0(year, "-", month, "-", day)),
         dow = as.character(wday(day, label = TRUE)),
         season = ifelse(month %in% 6:7, "summer", "other month"))
```

Now we can build a model that includes variables we want to use to explain arrival delay.

```
mod2 <- lm(arr_delay ~ hour + origin + carrier + season + dow, data = SF)
msummary(mod2)
```

```
              Estimate Std. Error t value Pr(>|t|)
(Intercept)  -24.5408     2.1745  -11.29  < 2e-16 ***
hour           2.0642     0.0898   22.98  < 2e-16 ***
originJFK      4.1989     1.0044    4.18  2.9e-05 ***
carrierB6    -10.3322     1.8797   -5.50  3.9e-08 ***
carrierDL    -18.4011     1.6146  -11.40  < 2e-16 ***
carrierUA     -4.7825     1.4808   -3.23  0.00124 **
carrierVX     -5.0365     1.5979   -3.15  0.00163 **
seasonsummer  25.3272     1.0307   24.57  < 2e-16 ***
```

```
dowMon              1.4438     1.4444    1.00   0.31755
dowSat             -5.9460     1.5617   -3.81   0.00014 ***
dowSun              5.5372     1.4709    3.76   0.00017 ***
dowThurs            3.3359     1.4461    2.31   0.02108 *
dowTues            -1.8487     1.4502   -1.27   0.20241
dowWed             -0.5014     1.4491   -0.35   0.72935

Residual standard error: 45.4 on 13159 degrees of freedom
Multiple R-squared:  0.0922,Adjusted R-squared:  0.0913
F-statistic:  103 on 13 and 13159 DF,  p-value: <2e-16
```

The numbers in the "Estimate" column tell us that we should add 4.2 minutes to the average delay if departing from JFK (instead of EWR—Newark). Delta has a better average delay than the other carriers. Delays are on average longer in June and July (by 25 minutes), and on Sundays (by 6 minutes). Recall that the Aviana crash was on July 6th (a Saturday) with a number of extreme delays on the 7th (a Sunday).

The model also indicates that Sundays involve roughly five minutes of additional delays; Saturdays are six minutes less delayed on average. (Each of the days of the week is being compared to Friday.) The standard errors tell us the precision of these estimates; the p-values describe whether the individual patterns are consistent with what might be expected to occur by accident even if there were no systemic association between the variables.

In this example, we've used lm() to construct what are called *linear models*. Linear models describe how the mean of the response variable varies with the explanatory variables. They are the most widely used statistical modeling technique, but there are others. In particular, since our original motivation was to set a policy about business travel, we might want a modeling technique that lets us look at another question: What is the probability that a flight will be, say, greater than 100 minutes late? Without going into detail, we'll mention that a technique called *logistic regression* is appropriate.

7.6   Confounding and accounting for other factors

We drill the mantra "correlation does not imply causation" into students whenever statistics are discussed. While the statement is certainly true, there are times when correlations *do* imply causal relationships (beyond just in carefully conducted randomized trials). A major concern for observational data is whether *other factors* may be the determinants of the observed relationship between two factors. Such other factors may *confound* the relationship being studied.

Randomized trials in scientific experiments are considered the gold standard for evidence-based research. Such trials, sometimes called *A/B tests*, are commonly undertaken to compare the effect of a treatment (e.g., two different Web pages). By controlling who receives a new intervention and who receives a control (or standard treatment), the investigator ensures that, on average, all other factors are balanced between the two groups. This allows them to conclude that if there are differences in the outcomes measured at the end of the trial, they can be attributed to the application of the treatment.

While they are ideal, randomized trials are not practical in many settings. It is not ethical to randomize some children to smoke and the others not to smoke in order to determine whether cigarettes cause lung cancer. It is not practical to randomize adults to either drink coffee or abstain to determine whether it has long-term health impacts. Observational (or "found") data may be the only feasible way to answer important questions.

Let's consider an example using data on average teacher salaries and average total SAT scores for the 50 United States. The SAT (Scholastic Aptitude Test) is a high-stakes exam used for entry into college. Are higher teacher salaries associated with better outcomes on the test at the state level? If so, should we adjust salaries to improve test performance? Figure 7.4 displays a scatterplot of these data. We also fit a linear regression model.

```
library(mdsr)
SAT_2010 <- mutate(SAT_2010, Salary = salary/1000)
SAT_plot <- ggplot(data = SAT_2010, aes(x = Salary, y = total)) +
  geom_point() + geom_smooth(method = "lm") +
  ylab("Average total score on the SAT") +
  xlab("Average teacher salary (thousands of USD)")
SAT_plot
```

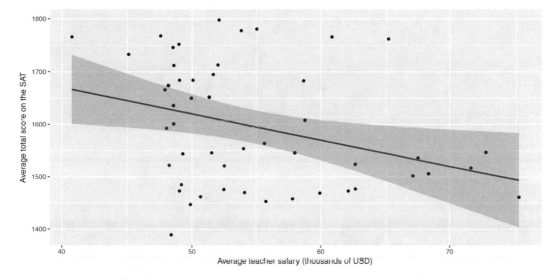

Figure 7.4: Scatterplot of average SAT scores versus average teacher salaries (in thousands of dollars) for the 50 United States in 2010.

```
SAT_mod1 <- lm(total ~ Salary, data = SAT_2010)
msummary(SAT_mod1)

            Estimate Std. Error t value Pr(>|t|)
(Intercept) 1871.10     113.14   16.54   <2e-16 ***
Salary        -5.02       2.05   -2.45    0.018 *

Residual standard error: 111 on 48 degrees of freedom
Multiple R-squared:  0.111,Adjusted R-squared:  0.0927
F-statistic: 6.01 on 1 and 48 DF,  p-value: 0.0179
```

Lurking in the background, however, is another important factor. The percentage of students who take the SAT in each state varies dramatically (from 3% to 93% in 2010). We can create a variable called SAT_grp that divides the states into two groups.

```
favstats(~ sat_pct, data = SAT_2010)

 min Q1 median Q3 max mean sd  n missing
   3  6     27 68  93 38.5 32 50       0

SAT_2010 <- SAT_2010 %>%
  mutate(SAT_grp = ifelse(sat_pct <= 27, "Low", "High"))
tally(~ SAT_grp, data = SAT_2010)

SAT_grp
High  Low
  25   25
```

Figure 7.5 displays a scatterplot of these data stratified by the grouping of percentage taking the SAT.

```
SAT_plot %+% SAT_2010 + aes(color = SAT_grp)
```

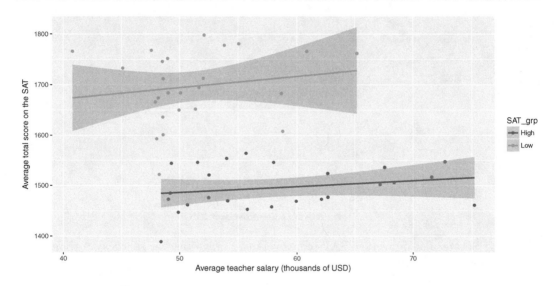

Figure 7.5: Scatterplot of average SAT scores versus average teacher salaries (in thousands of dollars) for the 50 United States in 2010, stratified by the percentage of students taking the SAT in each state.

```
coef(lm(total ~ Salary, data = filter(SAT_2010, SAT_grp == "Low")))

(Intercept)        Salary
   1583.27          2.22

coef(lm(total ~ Salary, data = filter(SAT_2010, SAT_grp == "High")))

(Intercept)        Salary
   1428.38          1.16
```

For each of the groups, average teacher salary is positively associated with average SAT score. But when we collapse over this variable, average teacher salary is negatively associated with average SAT score. This form of confounding is a quantitative version of *Simpson's paradox* and arises in many situations. It can be summarized in the following way:

1. Among states with a low percentage taking the SAT, teacher salaries and SAT scores are positively associated.

2. Among states with a high percentage taking the SAT, teacher salaries and SAT scores are positively associated.

3. Among all states, salaries and SAT scores are negatively associated.

Addressing confounding is straightforward if the confounding variables are measured. Stratification is one approach (as seen above). Multiple regression is another technique. Let's add the sat_pct variable into the model.

```
SAT_mod2 <- lm(total ~ Salary + sat_pct, data = SAT_2010)
msummary(SAT_mod2)

            Estimate Std. Error t value Pr(>|t|)
(Intercept) 1589.007     58.471    27.2   <2e-16 ***
Salary         2.637      1.149     2.3    0.026 *
sat_pct       -3.553      0.278   -12.8   <2e-16 ***

Residual standard error: 53.2 on 47 degrees of freedom
Multiple R-squared:  0.801,Adjusted R-squared:  0.792
F-statistic: 94.5 on 2 and 47 DF,  p-value: <2e-16
```

We now see that the slope for Salary is positive and statistically significant when we control for sat_pct. This is consistent with the results when the model was stratified by SAT_grp.

We still can't really conclude that teacher salaries cause improvements in SAT scores; however, the associations that we observe after accounting for the confounding are likely more reliable than those that do not take those factors into account.

Pro Tip: Data scientists spend most of their time working with observational data. When seeking to find meaning from such data, it is important to be on the lookout for potential confounding factors that could distort observed associations.

7.7 The perils of p-values

We close with a reminder of the perils of *null hypothesis statistical testing*. Recall that a *p-value* is defined as the probability of seeing a sample statistic as extreme (or more extreme) than the one that was observed if it were really the case that patterns in the data are a result of random chance (This hypothesis, that only randomness is in play, is called the *null hypothesis*.) For the SAT and salary example, the null hypothesis would be that the population regression coefficient (slope) is zero. Typically, when using *hypothesis testing*, analysts declare results with a p-value of $\alpha = 0.05$ or smaller as *statistically significant*, while values larger than 0.05 are declared non-significant.

Keep in mind that p-values are computed by simulating a world in which a null hypothesis is set to be true (see Chapter 10). The p-value indicates the quality of the concordance between the data and the simulation results. A large p-value indicates the data are concordant with the simulation. A very small p-value means otherwise: that the simulation is irrelevant to describing the mechanism behind the observed patterns. Unfortunately, that in itself tells us little about what kind of hypothesis would be relevant. Ironically, a "significant result" means that we get to reject the null hypothesis but doesn't tell us what hypothesis to accept.

Pro Tip: Always report the actual p-value (or a statement that it is less than some small value such as $p < 0.0001$) rather than just the decision (reject null vs. fail to reject the null). In addition, confidence intervals are often more interpretable and should be reported as well.

The problem with p-values is even more vexing in most real-world investigations. Analyses might involve not just a single hypothesis test but instead have dozens or more. In such a situation, even small p-values do not demonstrate discordance between the data and the null hypothesis, so the statistical analysis may tell us nothing at all.

In an attempt to restore meaning to p-values, investigators are starting to clearly delineate and pre-specify the primary and secondary outcomes for a randomized trial. Imagine that such a trial has five outcomes that are defined as being of primary interest. If the usual procedure in which a test is declared statistically significant if its p-value is less than 0.05 is used, the null hypotheses are true, and the tests are independent, we would expect that we would reject one or more of the null hypotheses more than 22% of the time (considerably more than 5% of the time we want).

```
1 - (1-0.05)^5
```

```
[1] 0.226
```

Clinical trialists have adapted to this problem by using more stringent determinations for statistical significance. A simple, albeit conservative approach is use of a *Bonferroni* correction. Consider dividing our α-level by the number of tests, and only rejecting the null hypothesis when the p-value is less than this adjusted value. In our example, the new threshold would be 0.01 (and the overall experiment-wise error rate is preserved at 0.05).

```
1 - (1-.01)^5
```

```
[1] 0.049
```

For observational analyses without pre-specified protocols, it is much harder to determine what (if any) Bonferroni correction is appropriate.

Pro Tip: For analyses that involve many hypothesis tests it is appropriate to include a note of possible limitations that some of the results may be spurious due to *multiple comparisons*.

A related problem has been called the *garden of forking paths* by Andrew Gelman of Columbia University. Most analyses involve many decisions about how to code data, determine important factors, and formulate and then revise models before the final analyses are set. This process involves looking at the data to construct a parsimonious representation. For example, a continuous predictor might be cut into some arbitrary groupings to assess the relationship between that predictor and the outcome. Or certain variables might be included or excluded from a regression model in an exploratory process.

This process tends to lead towards hypothesis tests that are biased against a null result, since decisions that yield more of a signal (or smaller p-value) might be chosen rather than other options. In clinical trials, the garden of forking paths problem may be less common, since analytic plans need to be prespecified and published. For most data science problems, however, this is a vexing issue that leads to questions about *reproducible results*.

7.8 Further resources

While this chapter raises many important issues related to the appropriate use of statistics in data science, it can only scratch the surface. A number of accessible books provide background in basic statistics [63] and statistical practice [202, 89]. Rice's excellent text [174] provides a modern introduction to the foundations of statistics (see also [148, 108, 105, 93]) along with the derivation of the sampling distribution of the median (pages 409–410). Shalizi's forthcoming *Advanced Data Analysis from an Elementary Point of View* (http://www.stat.cmu.edu/~cshalizi/ADAfaEPoV) provides a technical introduction to a wide range of important topics in statistics, including causal inference.

Null hypothesis testing and p-values are a vexing topic for many analysts. To help clarify these issues, the American Statistical Association endorsed a statement on p-values [209] that laid out six principles:

1. P-values can indicate how incompatible the data are with a specified statistical model.

2. P-values do not measure the probability that the studied hypothesis is true, or the probability that the data were produced by random chance alone.

3. Scientific conclusions and business or policy decisions should not be based only on whether a p-value passes a specific threshold.

4. Proper inference requires full reporting and transparency.

5. A p-value, or statistical significance, does not measure the size of an effect or the importance of a result.

6. By itself, a p-value does not provide a good measure of evidence regarding a model or hypothesis.

Hesterberg [102, 101] discusses the potential and perils for resampling-based inference. Hastie and Efron [68] provide an overview of modern inference techniques.

Missing data can be said to provide job security for data scientists since it arises in almost all real-world studies. A number of principled approaches have been developed to account for missing values, most notably multiple imputation. Accessible references to the extensive literature on incomplete data include [133, 171, 110].

While clinical trials are often considered a gold standard for evidence-based decision making, it is worth noting that they are almost always imperfect. Subjects may not comply with the intervention that they were randomized to. They make break the *blinding* and learn what treatment they have been assigned. Some subjects may drop out of the study. All of these issues complicate analysis and interpretation and have led to improvements in trial design and analysis along with the development of causal inference models. The CONSORT (Consolidated Standards of Reporting Trials) statement (http://www.consort-statement.org) was developed to alleviate problems with trial reporting.

Reproducibility and the perils of multiple comparisons have been the subject of much discussion in recent years. Nuzzo [150] summarizes why p-values are not as reliable as often assumed. The STROBE (Strengthening the Reporting of Observational Studies in Epidemiology, http://www.strobe-statement.org) statement discusses ways to improve the use of inferential methods (see also Appendix D).

7.9   Exercises

Exercise 7.1

Calculate and interpret a 95% confidence interval for the mean age of mothers from the classic Gestataion data set from the mosaicData package.

Exercise 7.2

Use the bootstrap to generate and interpret a 95% confidence interval for the median age of mothers for the classic Gestation data set from the mosaicData package.

Exercise 7.3

Use the bootstrap to generate a 95% confidence interval for the regression parameters in a model for weight as a function of age for the Gestation data frame from the mosaicData package.

Exercise 7.4

We saw that a 95% confidence interval for a mean was constructed by taking the estimate and adding and subtracting two standard deviations. How many standard deviations should be used if a 99% confidence interval is desired?

Exercise 7.5

Minnesota Twins: In 2010, the Minnesota Twins played their first season at Target Field. However, up through 2009, the Twins played at the Metrodome (an indoor stadium). In the Metrodome, air ventilator fans are used both to keep the roof up and to ventilate the stadium. Typically, the air is blown from all directions into the center of the stadium.

According to a retired supervisor in the Metrodome, in the late innings of some games the fans would be modified so that the ventilation air would blow out from home plate toward the outfield. The idea is that the air flow might increase the length of a fly ball. To see if manipulating the fans could possibly make any difference, a group of students at the University of Minnesota and their professor built a 'cannon' that used compressed air to shoot baseballs. They then did the following experiment.

- Shoot balls at angles around 50 degrees with velocity of around 150 feet per second.

- Shoot balls under two different settings: headwind (air blowing from outfield toward home plate) or tailwind (air blowing from home plate toward outfield).

- Record other variables: weight of the ball (in grams), diameter of the ball (in cm), and distance of the ball's flight (in feet).

Background: People who know little or nothing about baseball might find these basic facts useful. The batter stands near "home plate" and tries to hit the ball toward the outfield. A "fly ball" refers to a ball that is hit into the air. It is desirable to hit the ball as far as possible. For reasons of basic physics, the distance is maximized when the ball is hit at an intermediate angle steeper than 45 degrees from the horizontal.

The variables are described in the following table.

Cond	the wind conditions, a categorical variable with levels Headwind, Tailwind
Angle	the angle of ball's trajectory
Velocity	velocity of ball in feet per second
BallWt	weight of ball in grams
BallDia	diameter of ball in inches
Dist	distance in feet of the flight of the ball

Here is the output of several models.

```
> lm1 <- lm(Dist ~ Cond, data=ds)   # FIRST MODEL


> summary(lm1)
Coefficients:
            Estimate Std. Error t value Pr(>|t|)
(Intercept)  350.768      2.179 160.967   <2e-16
CondTail       5.865      3.281   1.788   0.0833
---
Residual standard error: 9.499 on 32 degrees of freedom
Multiple R-squared: 0.0908,     Adjusted R-squared: 0.06239
F-statistic: 3.196 on 1 and 32 DF,  p-value: 0.0833


> confint(lm1)
                2.5 %    97.5 %
(Intercept) 346.32966 355.20718
CondTail     -0.81784  12.54766


> # SECOND MODEL
> lm2 <- lm(Dist ~ Cond + Velocity + Angle + BallWt + BallDia, data=ds)
> summary(lm2)
Coefficients:
            Estimate Std. Error t value Pr(>|t|)
(Intercept) 181.7443   335.6959   0.541  0.59252
CondTail      7.6705     2.4593   3.119  0.00418
Velocity      1.7284     0.5433   3.181  0.00357
Angle        -1.6014     1.7995  -0.890  0.38110
BallWt       -3.9862     2.6697  -1.493  0.14659
BallDia     190.3715    62.5115   3.045  0.00502
---
Residual standard error: 6.805 on 28 degrees of freedom
Multiple R-squared: 0.5917,     Adjusted R-squared: 0.5188
F-statistic: 8.115 on 5 and 28 DF,  p-value: 7.81e-05


> confint(lm2)
                  2.5 %      97.5 %
(Intercept) -505.8974691  869.386165
CondTail       2.6328174   12.708166
Velocity       0.6155279    2.841188
Angle         -5.2874318    2.084713
BallWt        -9.4549432    1.482457
BallDia       62.3224999  318.420536
```

Consider the results from the model of `Dist` as a function of `Cond` (first model). Briefly summarize what this model says about the relationship between the wind conditions and the distance travelled by the ball. Make sure to say something sensible about the strength of evidence that there is any relationship at all.

Exercise 7.6

Twins, continued: Briefly summarize the model that has `Dist` as the response variable and includes the other variables as explanatory variables (second model) by reporting and interpretating the `CondTail` parameter. This second model suggests a somewhat different result for the relationship between `Dist` and `Cond`. Summarize the differences and explain in statistical terms why the inclusion of the other explanatory variables has affected the results.

Exercise 7.7

Smoking and mortality: The `Whickham` data set in the `mosaicData` package includes data on age, smoking, and mortality from a one-in-six survey of the electoral roll in Whickham, a mixed urban and rural district near Newcastle upon Tyne, in the United Kingdom. The survey was conducted in 1972–1974 to study heart disease and thyroid disease. A follow-up on those in the survey was conducted twenty years later. Describe the association between smoking status and mortality in this study. Be sure to consider the role of age as a possible confounding factor.

Exercise 7.8

A data scientist working for a company that sells mortgages for new home purchases might be interested in determining what factors might be predictive of defaulting on the loan. Some of the mortgagees have missing income in their data set. Would it be reasonable for the analyst to drop these loans from their analytic data set? Explain.

Exercise 7.9

Missing data: The `NHANES` data set in the `NHANES` package includes survey data collected by the U.S. National Center for Health Statistics (NCHS), which has conducted a series of health and nutrition surveys since the early 1960s. An investigator is interested in fitting a model to predict the probability that a female subject will have a diagnosis of diabetes. Predictors for this model include age and BMI. Imagine that only 1/10 of the data are available but that these data are sampled randomly from the full set of observations (this mechanism is called "Missing Completely at Random", or MCAR). What implications will this sampling have on the results?

Exercise 7.10

More missing data: Imagine that only 1/10 of the data are available but that these data are sampled from the full set of observations such that missingness depends on age, with older subjects less likely to be observed than younger subjects. (this mechanism is called "Covariate Dependent Missingness", or CDM). What implications will this sampling have on the results?

Exercise 7.11

More missing data: Imagine that only 1/10 of the data are available but that these data are sampled from the full set of observations such that missingness depends on diabetes status (this mechanism is called "Non-Ignorable Non-Response", or NINR). What implications will this sampling have on the results?

Chapter 8

Statistical learning and predictive analytics

Thus far, we have discussed two primary methods for investigating relationships among variables in our data: graphics and regression models. Graphics are often interpretable through intuitive inspection alone. They can be used to identify patterns and relationships in data—this is called *exploratory data analysis*. Regression models can help us quantify the magnitude and direction of relationships among variables. Thus, both are useful for helping us understand the world and then tell a coherent story about it.

However, graphics are not always the best way to explore or to present data. Graphics work well when there are two or three or even four variables involved. As we saw in Chapter 2, two variables can be represented with position on paper or on screen via a scatterplot. Ultimately, that information is processed by the eye's retina. To represent a third variable, color or size can be used. In principle, more variables can be represented by other graphical aesthetics: shape, angle, color saturation, opacity, facets, etc., but doing so raises problems for human cognition—people simply struggle to integrate so many graphical modes into a coherent whole.

While regression scales well into higher dimensions, it is a limited modeling framework. Rather, it is just one type of model, and the space of all possible models is infinite. In the next two chapters we will explore this space by considering a variety of models that exist outside of a regression framework. The idea that a general specification for a model could be tuned to a specific data set automatically has led to the field of *machine learning*.

The term machine learning was coined in the late 1950s to label a set of inter-related algorithmic techniques for extracting information from data without human intervention.

In the days before computers were invented, the dominant modeling framework was regression, which is based heavily on the mathematical disciplines of linear algebra and calculus. Many of the important concepts in machine learning emerged from the development of regression, but models that are associated with machine learning tend to be valued more for their ability to make accurate predictions and scale to large data sets, as opposed to the mathematical simplicity, ease of interpretation of the parameters, and solid inferential setting that has made regression so widespread. Nevertheless, regression and related statistical techniques from Chapter 7 provide an important foundation for understanding machine learning. Appendix E provides a brief overview of regression modeling.

There are two main branches in machine learning: *supervised learning* (modeling a specific response variable as a function of some explanatory variables) and *unsupervised learning* (approaches to finding patterns or groupings in data where there is no clear response variable).

In unsupervised learning, the outcome is unmeasured, and thus the task is often framed as a search for otherwise *unmeasured features* of the cases. For instance, assembling DNA data into an evolutionary tree is a problem in unsupervised learning. No matter how much DNA data you have, you don't have a direct measurement of where each organism fits on the "true" evolutionary tree. Instead, the problem is to create a representation that organizes the DNA data themselves.

By contrast, in supervised learning—which includes regression—the data being studied already include measurements of outcome variables. For instance, in the NHANES data, there is already a variable indicating whether or not a person has diabetes. Building a model to explore or describe how other variables are related to diabetes (weight? age? smoking?) is an exercise in supervised learning.

We discuss several types of supervised learning models in this chapter and postpone discussion of unsupervised learning to the next. It is important to understand that we cannot provide an in-depth treatment of each technique in this book. Rather, our goal is to provide a high-level overview of machine learning techniques that you are likely to come across. By reading these chapters, you will understand the general goals of machine learning, the evaluation techniques that are typically employed, and the basic models that are most commonly used. For a deeper understanding of these techniques, we strongly recommend [121] or [98].

8.1   Supervised learning

The basic goal of supervised learning is to find a *function* that accurately describes how different measured explanatory variables can be combined to make a prediction about a response variable.

A function represents a relationship between inputs and an output (see Appendix C). Outdoor temperature is a function of season: Season is the input; temperature is the output. Length of the day—i.e., how many hours of daylight—is a function of latitude and day of the year: Latitude and day of the year (e.g., March 22) are the inputs; day length is the output. For something like a person's risk of developing diabetes, we might suspect that age and obesity are likely informative, but how should they be combined?

A bit of R syntax will help with defining functions: the tilde. The tilde is used to define what the output variable (or outcome, on the left-hand side) is and what the input variables (or predictors, on the right-hand side) are. You'll see expressions like this:

```
diabetic ~ age + sex + weight + height
```

Here, the variable `diabetic` is marked as the output, simply because it is on the left of the tilde (~). The variables `age`, `sex`, `weight`, and `height` are to be the inputs to the function. You may also see the form `diabetic ~ .` in certain places. The dot to the right of the tilde is a shortcut that means: "use all the available variables (except the output)." The object above has class `formula` in R.

There are several different goals that might motivate constructing a function.

- Predict the output given an input. It is February, what will the temperature be? Or on June 15th in Northampton, Massachusetts, U.S.A. (latitude 42.3 deg N), how many hours of daylight will there be?

- Determine which variables are useful inputs. It is obvious from experience that temperature is a function of season. But in less familiar situations, e.g., predicting diabetes, the relevant inputs are uncertain or unknown.

- Generate hypotheses. For a scientist trying to figure out the causes of diabetes, it can be useful to construct a predictive model, then look to see what variables turn out to be related to the risk of developing this disorder. For instance, you might find that diet, age, and blood pressure are risk factors. Socioeconomic status is not a direct cause of diabetes, but it might be that there an association through factors related to the accessibility of health care. That "might be" is a hypothesis, and one that you probably would not have thought of before finding a function relating risk of diabetes to those inputs.

- Understand how a system works. For instance, a reasonable function relating hours of daylight to day-of-the-year and latitude reveals that the northern and southern hemisphere have reversed patterns: Long days in the southern hemisphere will be short days in the northern hemisphere.

Depending on your motivation, the kind of model and the input variables may differ. In understanding how a system works, the variables you use should be related to the actual, causal mechanisms involved, e.g., the genetics of diabetes. For predicting an output, it hardly matters what the causal mechanisms are. Instead, all that's required is that the inputs are known at a time *before* the prediction is to be made.

8.2 Classifiers

A logistic regression model (see Appendix E) takes a set of explanatory variables and converts them into a probability. In such a model the analyst specifies the form of the relationship and what variables are included. If $\mathbf{X}$ is the *matrix* of our p explanatory variables, we can think of this as a function $f : \mathbb{R}^p \to (0, 1)$ that returns a value $\pi \in (0, 1)$. However, since the actual values of the response variable y are binary (i.e., in $\{0, 1\}$), we can implement rules $g : (0, 1) \to \{0, 1\}$ that round values of p to either 0 or 1. Thus, our rounded logistic regression models are essentially functions $h : \mathbb{R}^k \to \{0, 1\}$, such that $h(\mathbf{X}) = g(f(\mathbf{X}))$ is always either 0 or 1. Such models are known as *classifiers*. More generally, whereas regression models for quantitative response variables return real numbers, models for categorical response variables are called classifiers.

Classifiers are an important complement to regression models in the fields of machine learning and predictive modeling. Whereas regression models have a quantitative response variable (and can thus be visualized as a geometric surface), classification models have a categorical response (and are often visualized as a discrete surface (i.e., a tree)). In the next section, we will discuss a particular type of *classifier* called a *decision tree. Regression trees* are analogous to decision trees, but with a quantitative response variable.[1]

8.2.1 Decision trees

A decision tree is a tree-like flowchart that assigns class labels to individual observations. Each branch of the tree separates the records in the data set into increasingly "pure" (i.e., homogeneous) subsets, in the sense that they are more likely to share the same class label.

How do we construct these trees? First, note that the number of possible decision trees grows exponentially with respect to the number of variables p. In fact, it has been proven that an efficient algorithm to determine the optimal decision tree almost certainly does not

---

[1]The oft-used acronym CART stands for "classification and regression trees."

exist [115].[2] The lack of a globally optimal algorithm means that there are several competing heuristics for building decision trees that employ greedy (i.e., locally optimal) strategies. While the differences among these algorithms can mean that they will return different results (even on the same data set), we will simplify our presentation by restricting our discussion to *recursive partitioning* decision trees. The R package that builds these decision trees is accordingly called `rpart`.

The partitioning in a decision tree follows Hunt's algorithm, which is itself recursive. Suppose that we are somewhere in the decision tree, and that $D_t = (y_t, \mathbf{X}_t)$ is the set of records that are associated with node t and that $\{y_1, y_2\}$ are the available class labels for the response variable.[3] Then:

- If all records in D_t belong to a single class, say, y_1, then t is a leaf node labeled as y_1.

- Otherwise, *split* the records into at least two child nodes, in such a way that the *purity* of the new set of nodes exceeds some threshold. That is, the records are separated more distinctly into groups corresponding to the response class. In practice, there are several competitive methods for optimizing the purity of the candidate child nodes, and—as noted above—we don't know the optimal way of doing this.

A decision tree works by running Hunt's algorithm on the full training data set.

What does it mean to say that a set of records is "purer" than another set? Two popular methods for measuring the purity of a set of candidate child nodes are the *Gini coefficient* and the *information* gain. Both are implemented in `rpart()`, which uses the Gini measurement by default. If $w_i(t)$ is the fraction of records belonging to class i at node t, then

$$Gini(t) = 1 - \sum_{i=1}^{2} (w_i(t))^2, \qquad Entropy(t) = - \sum_{i=1}^{2} w_i(t) \cdot \log_2 w_i(t)$$

The information gain is the change in entropy. The following example should help to clarify how this works in practice.

8.2.2 Example: High-earners in the 1994 United States Census

A marketing analyst might be interested in finding factors that can be used to predict whether a potential customer is a high-earner. The 1994 United States Census provides information that can inform such a model, with records from 32,561 adults that include a binary variable indicating whether each person makes greater or less than $50,000 (more than $80,000 today after accounting for inflation). This is our response variable.

```
library(mdsr)
census <- read.csv(
"http://archive.ics.uci.edu/ml/machine-learning-databases/adult/adult.data",
  header = FALSE)
names(census) <- c("age", "workclass", "fnlwgt", "education",
  "education.num", "marital.status", "occupation", "relationship",
  "race", "sex", "capital.gain", "capital.loss", "hours.per.week",
  "native.country", "income")
glimpse(census)
```

---

[2]Specifically, the problem of determining the optimal decision tree is NP-complete, meaning that it does not have a polynomial-time solution unless $P = NP$, which would be the most life-altering scientific discovery in the history of human civilization.

[3]For simplicity, we focus on a binary outcome in this chapter, but classifiers can generalize to any number of discrete response values.

```
Observations: 32,561
Variables: 15
$ age             <int> 39, 50, 38, 53, 28, 37, 49, 52, 31, 42, 37, 30,...
$ workclass       <fctr> State-gov, Self-emp-not-inc, Private, Priv...
$ fnlwgt          <int> 77516, 83311, 215646, 234721, 338409, 284582, 1...
$ education       <fctr> Bachelors, Bachelors, HS-grad, 11th, Bach...
$ education.num   <int> 13, 13, 9, 7, 13, 14, 5, 9, 14, 13, 10, 13, 13,...
$ marital.status  <fctr> Never-married, Married-civ-spouse, Divorced...
$ occupation      <fctr> Adm-clerical, Exec-managerial, Handlers-cle...
$ relationship    <fctr> Not-in-family, Husband, Not-in-family, Hus...
$ race            <fctr> White, White, White, Black, Black, White...
$ sex             <fctr> Male, Male, Male, Male, Female, Female, ...
$ capital.gain    <int> 2174, 0, 0, 0, 0, 0, 0, 0, 14084, 5178, 0, 0, 0...
$ capital.loss    <int> 0, 0, 0, 0, 0, 0, 0, 0, 0, 0, 0, 0, 0, 0, 0, 0,...
$ hours.per.week  <int> 40, 13, 40, 40, 40, 40, 16, 45, 50, 40, 80, 40,...
$ native.country  <fctr> United-States, United-States, United-States...
$ income          <fctr> <=50K, <=50K, <=50K, <=50K, <=50K, <=50K...
```

For reasons that we will discuss later, we will first separate our data set into two pieces by separating the rows at random. A sample of 80% of the rows will become the training data set, with the remaining 20% set aside as the testing (or "hold-out") data set.

```
set.seed(364)
n <- nrow(census)
test_idx <- sample.int(n, size = round(0.2 * n))
train <- census[-test_idx, ]
nrow(train)
```

```
[1] 26049
```

```
test <- census[test_idx, ]
nrow(test)
```

```
[1] 6512
```

Note that only about 24% of those in the sample make more than $50k. Thus, the *accuracy* of the *null model* is about 76%, since we can get that many right by just predicting that everyone makes less than $50k.

```
tally(~income, data = train, format = "percent")
```

```
income
 <=50K    >50K
  75.7    24.3
```

Pro Tip: Always benchmark your predictive models against a reasonable null model.

Let's consider the optimal split for income using only the variable capital.gain, which measures the amount each person paid in capital gains taxes. According to our tree, the optimal split occurs for those paying more than $5095.5 in capital gains:

```
library(rpart)
rpart(income ~ capital.gain, data = train)

n= 26049

node), split, n, loss, yval, (yprob)
      * denotes terminal node

1) root 26049 6320   <=50K (0.7575 0.2425)
  2) capital.gain< 5.1e+03 24784 5120   <=50K (0.7936 0.2064) *
  3) capital.gain>=5.1e+03 1265    63  >50K (0.0498 0.9502) *
```

Although nearly 80% of those who paid less than $5095.5 in capital gains tax made less than $50k, about 95% of those who paid more than $5095.5 in capital gains tax made *more* than $50k. Thus, splitting (partitioning) the records according to this criterion helps to divide them into relatively purer subsets. We can see this distinction geometrically as we divide the training records in Figure 8.1.

```
split <- 5095.5
train <- train %>% mutate(hi_cap_gains = capital.gain >= split)

ggplot(data = train, aes(x = capital.gain, y = income)) +
  geom_count(aes(color = hi_cap_gains),
    position = position_jitter(width = 0, height = 0.1), alpha = 0.5) +
  geom_vline(xintercept = split, color = "dodgerblue", lty = 2) +
  scale_x_log10(labels = scales::dollar)
```

Thus, this decision tree uses a single variable (`capital.gains`) to partition the data set into two parts: those who paid more than $5095.5 in capital gains, and those who did not. For the former—who make up 0.951 of all observations—we get 79.4% right by predicting that they made less than $50k. For the latter, we get 95% right by predicting that they made more than $50k. Thus, our overall accuracy jumps to 80.1%, easily besting the 75.7% in the null model.

How did the algorithm know to pick $5095.5 as the threshold value? It tried all of the sensible values, and this was the one that lowered the Gini coefficient the most. This can be done efficiently, since thresholds will always be between actual values of the splitting variable, and thus there are only $O(n)$ possible splits to consider.

So far, we have only used one variable, but we can build a decision tree for `income` in terms of all of the other variables in the data set. (We have left out `native.country` because it is a categorical variable with many levels, which can make some learning models computationally infeasible.)

```
form <- as.formula("income ~ age + workclass + education + marital.status +
  occupation + relationship + race + sex + capital.gain + capital.loss +
  hours.per.week")
mod_tree <- rpart(form, data = train)
mod_tree

n= 26049

node), split, n, loss, yval, (yprob)
```

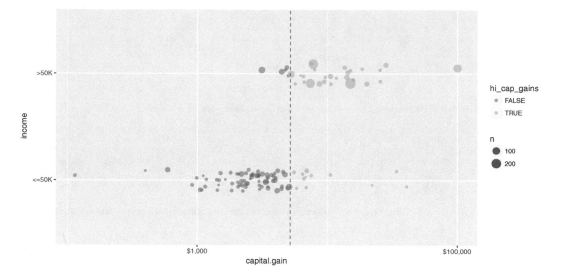

Figure 8.1: A single partition of the census data set using the `capital.gain` variable to determine the split. Color, and the vertical line at $5,095.50 in capital gains tax indicate the split. If one paid more than this amount, one almost certainly made more than $50,000 in income. On the other hand, if one paid less than this amount in capital gains, one almost certainly made less than $50,000.

```
    * denotes terminal node

1) root 26049 6320  <=50K (0.7575 0.2425)
  2) relationship= Not-in-family, Other-relative, Own-child, Unmarried
      14196  947  <=50K (0.9333 0.0667)
    4) capital.gain< 7.07e+03 13946  706  <=50K (0.9494 0.0506) *
    5) capital.gain>=7.07e+03 250    9  >50K (0.0360 0.9640) *
  3) relationship= Husband, Wife 11853 5370  <=50K (0.5470 0.4530)
    6) education= 10th, 11th, 12th, 1st-4th, 5th-6th, 7th-8th, 9th,
        Assoc-acdm, Assoc-voc, HS-grad, Preschool, Some-college
        8280 2770  <=50K (0.6656 0.3344)
     12) capital.gain< 5.1e+03 7857 2360  <=50K (0.7003 0.2997) *
     13) capital.gain>=5.1e+03 423    9  >50K (0.0213 0.9787) *
    7) education= Bachelors, Doctorate, Masters, Prof-school
        3573  972  >50K (0.2720 0.7280) *
```

In this more complicated tree, the optimal first split now does not involve `capital.gain`, but rather `relationship`. A basic visualization of the tree can be created using the `plot()` function from the `rpart` package.

```
plot(mod_tree)
text(mod_tree, use.n = TRUE, all = TRUE, cex = 0.7)
```

A much nicer-looking plot (shown in Figure 8.2) is available through the `partykit` package, which contains a series of functions for working with decision trees.

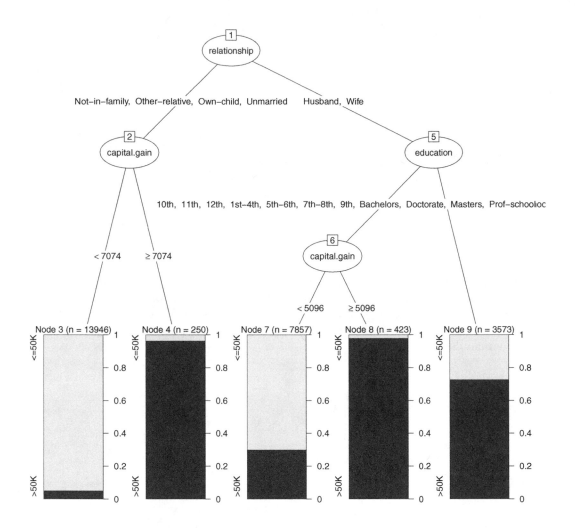

Figure 8.2: Decision tree for income using the census data.

```
library(partykit)
plot(as.party(mod_tree))
```

Figure 8.2 shows the decision tree itself, while Figure 8.3 shows how the tree recursively partitions the original data. Here, the first question is whether relationship status is Husband or Wife. If not, then a capital gains threshold of $7,073.50 is used to determine one's income. 96.4% of those who paid more than the threshold earned more than $50k, but 94.9% of those who paid less than the threshold did not. For those whose relationship status was Husband or Wife, the next question was whether you had a college degree. If so, then the model predicts with 72.8% accuracy that you made more than $50k. If not, then again we ask about capital gains tax paid, but this time the threshold is $5,095.50. 97.9% of those who were neither a husband nor a wife, and had no college degree, but paid more than that amount in capital gains tax, made more than $50k. On the other hand, 70% of those who paid below the threshold made less than $50k.

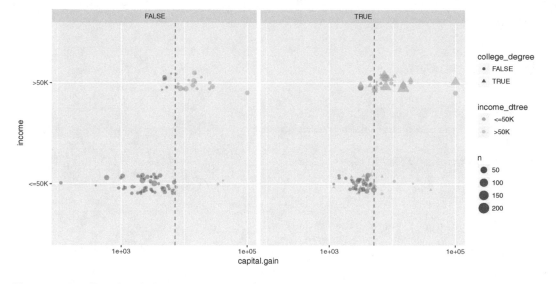

Figure 8.3: Graphical depiction of the full recursive partitioning decision tree classifier. On the left, those whose relationship status is neither "Husband" nor "Wife" are classified based on their capital gains paid. On the right, not only is the capital gains threshold different, but the decision is also predicated on whether the person has a college degree.

```
train <- train %>%
  mutate(husband_or_wife = relationship %in% c(" Husband", " Wife"),
         college_degree = husband_or_wife & education %in%
           c(" Bachelors", " Doctorate", " Masters", " Prof-school"),
         income_dtree = predict(mod_tree, type = "class"))

cg_splits <- data.frame(husband_or_wife = c(TRUE, FALSE),
                        vals = c(5095.5, 7073.5))

ggplot(data = train, aes(x = capital.gain, y = income)) +
  geom_count(aes(color = income_dtree, shape = college_degree),
             position = position_jitter(width = 0, height = 0.1),
             alpha = 0.5) +
  facet_wrap(~ husband_or_wife) +
  geom_vline(data = cg_splits, aes(xintercept = vals),
             color = "dodgerblue", lty = 2) +
  scale_x_log10()
```

Since there are exponentially many trees, how did the algorithm know to pick this one? The *complexity parameter* controls whether to keep or prune possible splits. That is, the algorithm considers many possible splits (i.e., new branches on the tree), but prunes them if they do not sufficiently improve the predictive power of the model (i.e., bear fruit). By default, each split has to decrease the error by a factor of 1%. This will help to avoid *overfitting* (more on that later). Note that as we add more splits to our model, the relative error decreases.

```
printcp(mod_tree)


Classification tree:
rpart(formula = form, data = train)

Variables actually used in tree construction:
[1] capital.gain education    relationship

Root node error: 6317/26049 = 0.243

n= 26049

      CP nsplit rel error xerror    xstd
1 0.1289      0    1.000  1.000 0.01095
2 0.0641      2    0.742  0.742 0.00982
3 0.0367      3    0.678  0.678 0.00947
4 0.0100      4    0.641  0.641 0.00926

# plotcp(mod_tree)
```

An important tool in verifying a model's accuracy is called the *confusion matrix* (really). Simply put, this is a two-way table that counts how often our model made the correct prediction. Note that there are two different types of mistakes that our model can make: predicting a high income when the income was in fact low, and predicting a low income when the income was in fact high.

```
train <- train %>%
  mutate(income_dtree = predict(mod_tree, type = "class"))
confusion <- tally(income_dtree ~ income, data = train, format = "count")
confusion

            income
income_dtree  <=50K  >50K
       <=50K  18742  3061
        >50K    990  3256

sum(diag(confusion)) / nrow(train)

[1] 0.84449
```

In this case, the accuracy of the decision tree classifier is now 84.4%, a considerable improvement over the null model.

8.2.3   Tuning parameters

The decision tree that we built above was based on the default parameters. Most notably, our tree was pruned so that only splits that decreased the overall lack of fit by 1% were retained. If we lower this threshold to 0.2%, then we get a more complex tree.

```
mod_tree2 <- rpart(form, data = train, control = rpart.control(cp = 0.002))
```

Can you find the accuracy of this more complex tree. Is it more or less accurate than our original tree?

8.2.4 Random forests

A natural extension of a decision tree is a *random forest*. A random forest is collection of decision trees that are aggregated by majority rule. In a sense, a random forest is like a collection of bootstrapped (see Chapter 7) decision trees. A random forest is constructed by:

1. Choosing the number of decision trees to grow (controlled by the `ntree` argument) and the number of variables to consider in each tree (`mtry`)

2. Randomly selecting the rows of the data frame *with replacement*

3. Randomly selecting `mtry` variables from the data frame

4. Building a decision tree on the resulting data set

5. Repeating this procedure `ntree` times

A prediction for a new observation is made by taking the majority rule from all of the decision trees in the forest. Random forests are available in R via the `randomForest` package. They can be very effective, but are sometimes computationally expensive.

```
library(randomForest)
mod_forest <- randomForest(form, data = train, ntree = 201, mtry = 3)
mod_forest
```

```
Call:
 randomForest(formula = form, data = train, ntree = 201, mtry = 3)
               Type of random forest: classification
                     Number of trees: 201
No. of variables tried at each split: 3

        OOB estimate of  error rate: 13.31%
Confusion matrix:
        <=50K  >50K class.error
<=50K   18471  1261    0.063906
>50K     2205  4112    0.349058
```

```
sum(diag(mod_forest$confusion)) / nrow(train)
```

```
[1] 0.86694
```

Because each tree in a random forest uses a different set of variables, it is possible to keep track of which variables seem to be the most consistently influential. This is captured by the notion of *importance*. While—unlike p-values in a regression model—there is no formal statistical inference here, importance plays an analogous role in that it may help to generate hypotheses. Here, we see that `capital.gain` and `age` seem to be influential, while race and sex do not.

```
library(tibble)
importance(mod_forest) %>%
  as.data.frame() %>%
  rownames_to_column() %>%
  arrange(desc(MeanDecreaseGini))
```

```
          rowname MeanDecreaseGini
1             age          1068.76
2    capital.gain          1064.39
3       education           982.65
4    relationship           951.87
5      occupation           905.27
6  marital.status           880.90
7  hours.per.week           627.88
8       workclass           337.92
9    capital.loss           326.13
10           race           145.34
11            sex           110.08
```

A model object of class `randomForest` also has a `predict()` method for making new predictions.

8.2.5   Nearest neighbor

Thus far, we have focused on using data to build models that we can then use to predict outcomes on a new set of data. A slightly different approach is offered by *lazy learners*, which seek to predict outcomes without constructing a "model." A very simple, yet widely-used approach is *k-nearest neighbor*.

Recall that data with p attributes (explanatory variables) are manifest as points in a p-dimensional space. The Euclidean distance between any two points in that space can be easily calculated in the usual way as the square root of the sum of the squared deviations. Thus, it makes sense to talk about the *distance* between two points in this p-dimensional space, and as a result, it makes sense to talk about the distance between two observations (rows of the data frame). Nearest neighbor classifiers exploit this property by assuming that observations that are "close" to each other probably have similar outcomes.

Suppose we have a set of training data $(\mathbf{X}, y) \in \mathbb{R}^{n \times p} \times \mathbb{R}^n$. For some positive integer k, a k-nearest neighbor algorithm classifies a new observation x^* by:

1. Finding the k observations in the training data $\mathbf{X}$ that are closest to x^*, according to some distance metric (usually Euclidean). Let $D(x^*) \subseteq (\mathbf{X}, y)$ denote this set of observations.

2. For some aggregate function f, computing $f(y)$ for the k values of y in $D(x^*)$ and assigning this value (y^*) as the predicted value of the response associated with x^*. The logic is that since x^* is similar to the k observations in $D(x^*)$, the response associated with x^* is likely to be similar to the responses in $D(x^*)$. In practice, simply taking the value shared by the majority (or a plurality) of the y's is enough.

Note that a k-NN classifier does not need to process the training data before making new classifications—it can do this on the fly. A simple k-NN classifier (without many options) is provided by the `knn()` function in the `class` package. Note that since the distance metric only makes sense for quantitative variables, we have to restrict our data set to those first.

```
library(class)
# distance metric only works with quantitative variables
train_q <- train %>%
  select(age, education.num, capital.gain, capital.loss, hours.per.week)
income_knn <- knn(train_q, test = train_q, cl = train$income, k = 10)

confusion <- tally(income_knn ~ income, data = train, format = "count")
confusion
```

```
          income
income_knn  <=50K  >50K
     <=50K  18875  2997
      >50K    857  3320
```

```
sum(diag(confusion)) / nrow(train)
```

```
[1] 0.85205
```

k-NN classifiers are widely used in part because they are easy to understand and code. They also don't require any pre-processing time. However, predictions can be slow, since the data must be processed at that time.

The usefulness of k-NN can depend importantly on the geometry of the data. Are the points clustered together? What is the distribution of the distances among each variable? A wider scale on one variable can dwarf a narrow scale on another variable.

An appropriate choice of k will depend on the application and the data. Cross-validation can be used to optimize the choice of k. In Figure 8.4, we show how the misclassification rate increases as k increases. That is, if one seeks to minimize the misclassification rate *on this data set*, then the optimal value of k is 1.[4] This method of optimizing the value of the parameter k is a form of *cross-validation* (see below).

```
knn_error_rate <- function(x, y, numNeighbors, z = x) {
  y_hat <- knn(train = x, test = z, cl = y, k = numNeighbors)
  return(sum(y_hat != y) / nrow(x))
}
ks <- c(1:15, 20, 30, 40, 50)
train_rates <- sapply(ks, FUN = knn_error_rate, x = train_q, y = train$income)
knn_error_rates <- data.frame(k = ks, train_rate = train_rates)
ggplot(data = knn_error_rates, aes(x = k, y = train_rate)) +
  geom_point() + geom_line() + ylab("Misclassification Rate")
```

8.2.6 Naïve Bayes

Another relatively simple classifier is based on Bayes theorem. Bayes theorem is a very useful result from probability that allows conditional probabilities to be calculated from other conditional probabilities. It states:

$$\Pr(y|x) = \frac{\Pr(xy)}{\Pr(x)} = \frac{\Pr(x|y)\Pr(y)}{\Pr(x)} .$$

---

[4]In section 8.4.5, we discuss why this particular optimization criterion might not be the wisest choice.

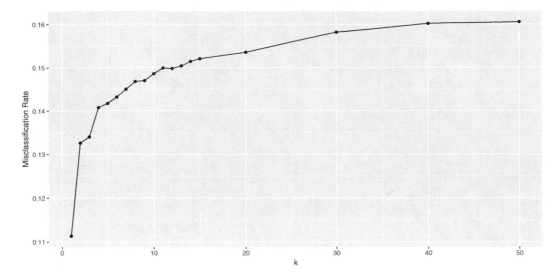

Figure 8.4: Performance of nearest neighbor classifier for different choices of k on census training data.

How does this relate to a naïve Bayes classifier? Suppose that we have a binary response variable y and we want to classify a new observation x^* (recall that x is a vector). Then if we can compute that the conditional probability $\Pr(y = 1|x^*) > \Pr(y = 0|x^*)$, we have evidence that $y = 1$ is a more likely outcome for x^* than $y = 0$. This is the crux of a naïve Bayes classifier. In practice, how we arrive at the estimates $\Pr(y = 1|x^*)$ are based on Bayes theorem and estimates of conditional probabilities derived from the training data $(\mathbf{X}, y)$.

Consider the first person in the training data set. This is a 39-year-old white male with a bachelor's degree working for a state government in a clerical role. In reality, this person made less than \$50,000.

```
head(train, 1)

  age  workclass fnlwgt  education education.num marital.status
1  39  State-gov  77516  Bachelors            13  Never-married
      occupation   relationship   race   sex capital.gain capital.loss
1  Adm-clerical  Not-in-family  White  Male         2174             0
  hours.per.week native.country income hi_cap_gains husband_or_wife
1             40  United-States  <=50K        FALSE           FALSE
  college_degree income_dtree
1          FALSE         <=50K
```

The naïve Bayes classifier would make a prediction for this person based on the probabilities observed in the data. For example, in this case the probability $\Pr(\text{male}|{>}50\text{k})$ of being male given that you had high income is 0.845, while the unconditional probability of being male is $\Pr(\text{male}) = 0.670$. We know that the overall probability of having high income is $\Pr({>}50\text{k}) = 0.243$. Bayes's rule tells us that the resulting probability of having high income given that one is male is:

$$\Pr({>}50\text{k}|\text{male}) = \frac{\Pr(\text{male}|{>}50\text{k}) \cdot \Pr({>}50\text{k})}{\Pr(\text{male})} = \frac{0.845 \cdot 0.243}{0.670} = 0.306\,.$$

This simple example illustrates the case where we have a single explanatory variable (e.g., sex), but the Naïve Bayes model extends to multiple variables by making the sometimes overly simplistic assumption that the explanatory variables are conditionally independent (hence the name "naïve").

A naïve Bayes classifier is provided in R by the `naiveBayes()` function from the e1071 package. Note that like `lm()` and `glm()`, a `naiveBayes()` object has a `predict()` method.

```
library(e1071)
mod_nb <- naiveBayes(form, data = train)
income_nb <- predict(mod_nb, newdata = train)
confusion <- tally(income_nb ~ income, data = train, format = "count")
confusion
```

```
           income
income_nb   <=50K   >50K
    <=50K   18587   3605
     >50K    1145   2712
```

```
sum(diag(confusion)) / nrow(train)
```

```
[1] 0.81765
```

8.2.7 Artificial neural networks

An *artificial neural network* is yet another classifier. While the impetus for the artificial neural network comes from a biological understanding of the brain, the implementation here is entirely mathematical.

```
library(nnet)
mod_nn <- nnet(form, data = train, size = 5)
```

```
# weights:   296
initial   value 21842.825468
iter  10 value 13198.315933
iter  20 value 11190.055832
iter  30 value 10252.441741
iter  40 value 9937.073100
iter  50 value 9591.448419
iter  60 value 9319.908227
iter  70 value 9062.177126
iter  80 value 8918.313144
iter  90 value 8826.858128
iter 100 value 8729.189597
final   value 8729.189597
stopped after 100 iterations
```

A neural network is a directed graph (see Chapter 16) that proceeds in stages. First, there is one node for each input variable. In this case, because each factor level counts as its own variable, there are 57 input variables. These are shown on the left in Figure 8.5. Next, there are a series of nodes specified as a *hidden* layer. In this case, we have specified five nodes for the hidden layer. There are shown in the middle of Figure 8.5, and each of the input variables are connected to these hidden nodes. Each of the hidden nodes is connected to the single output variable. In addition, nnet() adds two control nodes, the first of which is connected to the five hidden nodes, and the latter is connected to the output node. The total number of edges is thus $pk + k + k + 1$, where k is the number of hidden nodes. In this case, there are $57 \cdot 5 + 5 + 5 + 1 = 296$ edges.

The algorithm iteratively searches for the optimal set of weights for each edge. Once the weights are computed, the neural network can make predictions for new inputs by running these values through the network.

```
income_nn <- predict(mod_nn, newdata = train, type = "class")
confusion <- tally(income_nn ~ income, data = train, format = "count")
confusion
```

```
             income
income_nn   <=50K   >50K
    <=50K   17871   2128
     >50K    1861   4189
```

```
sum(diag(confusion)) / nrow(train)
```

```
[1] 0.84687
```

8.3   Ensemble methods

The benefit of having multiple classifiers is that they can be easily combined into a single classifier. Note that there is a real probabilistic benefit to having multiple prediction systems, especially if they are independent. For example, if you have three independent classifiers with error rates ϵ_1, ϵ_2, and ϵ_3, then the probability that all three are wrong is $\prod_{i=1}^{3} \epsilon_i$. Since $\epsilon_i < 1$ for all i, this probability is lower than any of the individual error rates. Moreover, the probability that at least one of the classifiers is correct is $1 - \prod_{i=1}^{3} \epsilon_i$, which will get closer to 1 as you add more classifiers—even if you have not improved the individual error rates!

Consider combining the k-NN, naïve Bayes, and artificial neural network classifiers that we have build previously. Suppose that we build an ensemble classifier by taking the majority vote from each. Does this ensemble classifier outperform any of the individual classifiers?

```
income_ensemble <- ifelse((income_knn == " >50K") +
                          (income_nb == " >50K") +
                          (income_nn == " >50K") >= 2, " >50K", " <=50K")
confusion <- tally(income_ensemble ~ income, data = train, format = "count")
confusion
```

```
                 income
income_ensemble   <=50K   >50K
          <=50K   18790   3039
           >50K     942   3278
```

```
sum(diag(confusion)) / nrow(train)
```

```
[1] 0.84717
```

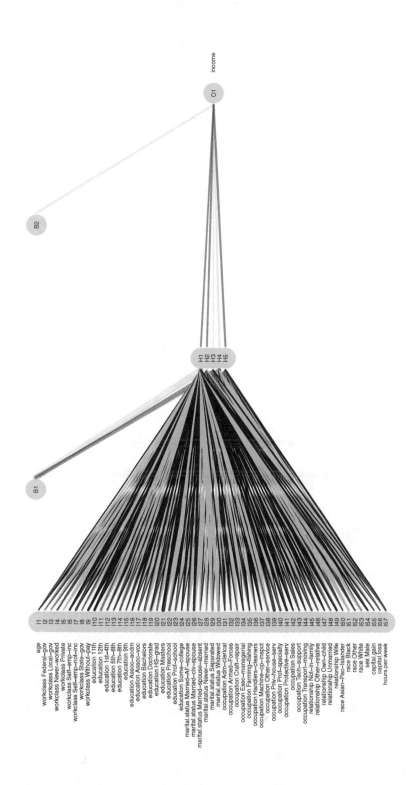

Figure 8.5: Visualization of an artificial neural network. The input 57 input variables are shown on the bottom, with the five hidden nodes in the middle, and the single output variable at the top.

In this case, it doesn't—the k-NN classifier achieves a slightly higher 85% accuracy rate. Nevertheless, ensemble methods are a simple but effective way of hedging your bets.

8.4   Evaluating models

How do you know if your model is a good one? In this section, we outline some of the key concepts in model evaluation—a critical step in predictive analytics.

8.4.1   Cross-validation

One of the most seductive traps that modelers fall into is *overfitting*. Every model discussed in this chapter is *fit* to a set of data. That is, given a set of *training* data and the specification for the type of model (e.g., decision tree, artificial neural network, etc.), each algorithm will determine the optimal set of parameters for that model and those data. However, if the model works well on those training data, but not so well on a set of *testing* data—that the model has never seen—then the model is said to be *overfit*. Perhaps the most elementary mistake in predictive analytics is to overfit your model to the training data, only to see it later perform miserably on the testing set.

In predictive analytics, data sets are often divided into two sets:

Training The set of data on which you build your model

Testing Once your model is built, you test it by evaluating it against data that it has not previously seen.

For example, in this chapter we set aside 80% of the observations to use as a training set, but held back another 20% for testing. The 80/20 scheme we have employed in this chapter is among the simplest possible schemes, but there are many more complicated schemes.

Another approach to combat this problem is *cross-validation*. To perform a 2-fold cross-validation:

1. Randomly separate your data (by rows) into two data sets with the same number of observations. Let's call them X_1 and X_2.

2. Build your model on the data in X_1, and then run the data in X_2 through your model. How well does it perform? Just because your model performs well on X_1 (this is known as *in-sample* testing), does not imply that it will perform as well on the data in X_2 (*out-of-sample* testing).

3. Now reverse the roles of X_1 and X_2, so that the data in X_2 is used for training, and the data in X_1 is used for testing.

4. If your first model is overfit, then it will likely not perform as well on the second set of data.

More complex schemes for cross-validating are possible. k-fold cross-validation is the generalization of 2-fold cross validation, in which the data are separated into k equal-sized partitions, and each of the k partitions is chosen to be the testing set once, with the other $k - 1$ partitions used for training.

8.4.2   Measuring prediction error

For evaluating models with a quantitative response, there are a variety of criteria that are commonly used. Here we outline three of the simplest and most common. The following presumes a vector of real observations denoted y and a corresponding vector of prediction $\hat{y}$:

RMSE Root mean squared error is probably the most common:

$$RMSE(y, \hat{y}) = \sqrt{\frac{1}{n} \sum_{i=1}^{n} (y - \hat{y})^2}\,.$$

The RMSE has several desirable properties. Namely, it is in the same units as the response variable y, it captures both overestimates and underestimates equally, and it penalizes large misses heavily.

MAE Mean absolute error is similar to the RMSE, but does not penalize large misses as heavily, due to the lack of a squared term:

$$MAE(y, \hat{y}) = \frac{1}{n} \sum_{i=1}^{n} |y - \hat{y}|\,.$$

Correlation The previous two methods require that the units and scale of the predictions $\hat{y}$ are the same as the response variable y. While this is of course necessary for accurate predictions, some predictive models merely want to track the trends in the response. In such cases the *correlation* between y and $\hat{y}$ may suffice.

In addition to the usual Pearson product-moment correlation, measures of rank correlation are also occasionally useful. That is, instead of trying to minimize $y - \hat{y}$, it might be enough to make sure that the $\hat{y}_i$'s are in the same relative order as the y_i's. Popular measures of rank correlation include Spearman's ρ and Kendall's τ.

Coefficient of determination (R^2) The coefficient of determination is measured on a scale of $[0, 1]$, with 1 indicating a perfect match between y and $\hat{y}$.

8.4.3   Confusion matrix

For classifiers, we have already seen the confusion matrix, which is a common way to assess the effectiveness of the model.

8.4.4   ROC curves

Recall that each of the classifiers we have discussed in this chapter are capable of producing not only a binary class label, but also the predicted probability of belonging to either class. Rounding the probabilities in the usual way (using 0.5 as a threshold) is not a good idea, since the average probability might not be anywhere near 0.5, and thus we could have far too many predictions in one class.

For example, in the census data, only about 24% of the people in the training set had income above \$50,000. Thus, a sensible predictive model should predict that about 24% of the people have incomes above \$50,000. Consider the raw probabilities returned by the naïve Bayes model.

```
income_probs <- mod_nb %>%
  predict(newdata = train, type = "raw") %>%
  as.data.frame()
head(income_probs, 3)

    <=50K        >50K
1 0.98882 0.01117538
2 0.84292 0.15707605
3 0.99973 0.00026597
```

If we round these using a threshold of 0.5, then only 15% are predicted to have high incomes. Note that here we are able to work with the unfortunate leading space in the variable names by wrapping them with backticks.

```
names(income_probs)

[1] " <=50K" " >50K"

tally(~` >50K` > 0.5, data = income_probs, format = "percent")

` >50K` > 0.5
  TRUE  FALSE
14.807 85.193
```

A better alternative would be to use the overall observed percentage (i.e., 24%) as a threshold instead:

```
tally(~` >50K` > 0.24, data = income_probs, format = "percent")

` >50K` > 0.24
  TRUE  FALSE
19.939 80.061
```

This is an improvement, but a more principled approach to assessing the quality of a classifier is a *receiver operating characteristic* curve. This considers all possible threshold values for rounding, and graphically displays the trade-off between *sensitivity* (the true positive rate) and *specificity* (the true negative rate). What is actually plotted is the true positive rate as a function of the false positive rate.

ROC curves are common in machine learning and operations research as well as assessment of test characteristics and medical imaging. They can be constructed in R using the ROCR package. Note that ROC curves operate on the fitted probabilities in $(0, 1)$.

```
pred <- ROCR::prediction(income_probs[,2], train$income)
perf <- ROCR::performance(pred, 'tpr', 'fpr')
class(perf) # can also plot(perf)

[1] "performance"
attr(,"package")
[1] "ROCR"
```

We can draw an ROC curve by directly plotting the `perf` object we computed above, but since we'd like to gussy it up a bit later, we will do it with `ggplot2` instead. However,

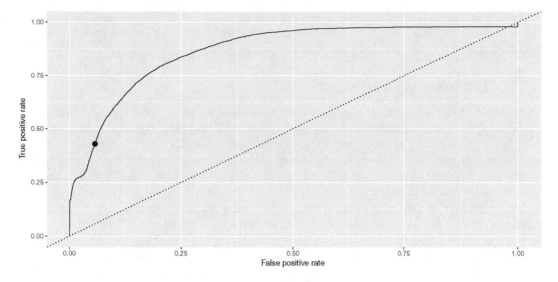

Figure 8.6: ROC curve for naive Bayes model.

to get there we will have to extract the x and y values from `perf`, which requires accessing the S4 slots using the @ notation.

```
perf_df <- data.frame(perf@x.values, perf@y.values)
names(perf_df) <- c("fpr", "tpr")
roc <- ggplot(data = perf_df, aes(x = fpr, y = tpr)) +
  geom_line(color="blue") + geom_abline(intercept=0, slope=1, lty=3) +
  ylab(perf@y.name) + xlab(perf@x.name)
```

In Figure 8.6 the upper-left corner represents a perfect classifier, which would have a true positive rate of 1 and a false positive rate of 0. On the other hand, a random classifier would lie along the diagonal, since it would be equally likely to make either kind of mistake.

The actual naïve Bayes model that we used had the following true and false positive rates, which are indicated in Figure 8.6 by the black dot.

```
confusion <- tally(income_nb ~ income, data = train, format = "count")
confusion

        income
income_nb  <=50K  >50K
   <=50K   18587  3605
    >50K    1145  2712

sum(diag(confusion)) / nrow(train)

[1] 0.81765

tpr <- confusion[" >50K", " >50K"] / sum(confusion[, " >50K"])
fpr <- confusion[" >50K", " <=50K"] / sum(confusion[, " <=50K"])
roc + geom_point(x = fpr, y = tpr, size = 3)
```

Depending on our tolerance for false positives vs. false negatives, we could modify the way that our naïve Bayes model rounds probabilities, which would have the effect of moving the black dot in Figure 8.6 along the blue curve.

8.4.5   Bias-variance trade-off

We want to have models that minimize both bias and variance, but to some extent these are mutually exclusive goals. A complicated model will have less bias, but will in general have higher variance. A simple model can reduce variance, but at the cost of increased bias. The optimal balance between bias and variance depends on the purpose for which the model is constructed (e.g. prediction vs. description of causal relationships) and the system being modeled. One helpful class of techniques—called *regularization*—provides model architectures that can balance bias and variance in a graduated way. Examples of regularization techniques are ridge regression and the lasso (see Section 8.6).

For example, in Section 8.2.5, we showed how the misclassification rate *on the training data* of the k-NN model increased as k increased. That is, as information from more neighbors—who are necessarily farther away from the target observation—was incorporated into the prediction for any given observation, those predictions got worse. This is not surprising, since the actual observation is in the training data set and that observation necessarily has distance 0 from the target observation. The error rate is not zero likely due to many points having the exact same coordinates in this five-dimensional space. However, as seen in Figure 8.7, the story is quite different when evaluating the k-NN model *on the testing set*. Here, the truth is *not* in the training set, and so pooling information across more observations leads to *better* predictions—at least for a while. Again, this should not be surprising—we saw in Chapter 7 how means are less variable than individual observations. Generally, one hopes to minimize the misclassification rate on data that the model has not seen (i.e., the testing data) without introducing too much bias. In this case that point occurs somewhere between $k = 5$ and $k = 10$. We can see this in Figure 8.7, since the accuracy on the testing data set improves rapidly up to $k = 5$, but then very slowly for larger values of k.

```
test_q <- test %>%
  select(age, education.num, capital.gain, capital.loss, hours.per.week)
test_rates <- sapply(ks, FUN = knn_error_rate, x = train_q,
                     y = train$income, z = test_q)
knn_error_rates <- knn_error_rates %>% mutate(test_rate = test_rates)
library(tidyr)
knn_error_rates_tidy <- knn_error_rates %>%
  gather(key = "type", value = "error_rate", -k)
ggplot(data = knn_error_rates_tidy, aes(x = k, y = error_rate)) +
  geom_point(aes(color = type)) + geom_line(aes(color = type)) +
  ylab("Misclassification Rate")
```

8.4.6   Example: Evaluation of income models

Recall that we separated the 32,561 observations in the census data set into a training set that contained 80% of the observations and a testing set that contained the remaining 20%. Since the separation was done by selecting rows uniformly at random, and the number of observations was fairly large, it seems likely that both the training and testing set will contain similar information. For example, the distribution of capital.gain is similar in

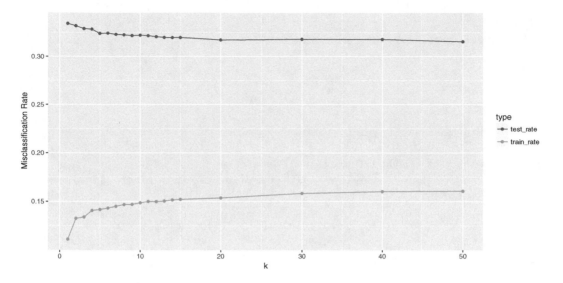

Figure 8.7: Performance of nearest neighbor classifier for different choices of k on census training and testing data.

both the testing and training sets. Nevertheless, it is worth formally testing the performance of our models on both sets.

```
favstats(~ capital.gain, data = train)

 min Q1 median Q3    max    mean     sd     n missing
   0  0      0  0  99999  1084.4  7428.6  26049       0

favstats(~ capital.gain, data = test)

 min Q1 median Q3    max    mean     sd     n missing
   0  0      0  0  99999  1050.8  7210.1  6512        0
```

First, we build the null model that simply predicts that everyone makes less than $50,000. (See Appendix E for an introduction to logistic regression.) We'll add this to the list of models that we built previously in this chapter.

```
mod_null <- glm(income ~ 1, data = train, family = binomial)
mods <- list(mod_null, mod_tree, mod_forest, mod_nn, mod_nb)
```

While each of the models we have fit have different classes in R (see B.4.6), each of those classes has a predict() method that will generate predictions.

```
lapply(mods, class)

[[1]]
[1] "glm" "lm"

[[2]]
[1] "rpart"
```

```
[[3]]
[1] "randomForest.formula" "randomForest"

[[4]]
[1] "nnet.formula" "nnet"

[[5]]
[1] "naiveBayes"

predict_methods <- methods("predict")
predict_methods[grepl(pattern = "(glm|rpart|randomForest|nnet|naive)",
  predict_methods)]

[1] "predict.glm"         "predict.glmmPQL"     "predict.glmtree"
[4] "predict.naiveBayes"  "predict.nnet"        "predict.randomForest"
[7] "predict.rpart"
```

Thus, we can iterate through the list of models and apply the appropriate `predict()` method to each object.

```
predictions_train <- data.frame(
  y = as.character(train$income),
  type = "train",
  mod_null = predict(mod_null, type = "response"),
  mod_tree = predict(mod_tree, type = "class"),
  mod_forest = predict(mod_forest, type = "class"),
  mod_nn = predict(mod_nn, type = "class"),
  mod_nb = predict(mod_nb, newdata = train, type = "class"))
predictions_test <- data.frame(
  y = as.character(test$income),
  type = "test",
  mod_null = predict(mod_null, newdata = test, type = "response"),
  mod_tree = predict(mod_tree, newdata = test, type = "class"),
  mod_forest = predict(mod_forest, newdata = test, type = "class"),
  mod_nn = predict(mod_nn, newdata = test, type = "class"),
  mod_nb = predict(mod_nb, newdata = test, type = "class"))
predictions <- bind_rows(predictions_train, predictions_test)
glimpse(predictions)

Observations: 32,561
Variables: 7
$ y          <fctr> <=50K, <=50K, <=50K, <=50K, <=50K, <=50K, <...
$ type       <chr> "train", "train", "train", "train", "train", "train...
$ mod_null   <dbl> 0.2425, 0.2425, 0.2425, 0.2425, 0.2425, 0.2425, 0.2...
$ mod_tree   <fctr> <=50K, >50K, <=50K, <=50K, >50K, >50K, <=50...
$ mod_forest <fctr> <=50K, <=50K, <=50K, <=50K, >50K, >50K, <=5...
$ mod_nn     <fctr> >50K, <=50K, <=50K, <=50K, >50K, >50K, <=50...
$ mod_nb     <fctr> <=50K, <=50K, <=50K, <=50K, <=50K, <=50K, <...
```

As you can see, while each of the models returned the right number of predictions, they describe those predictions differently. The null model returned a probability, which we want to round. The other models returned a factor of levels. We will unify this and gather() it into a tidy format (see Chapter 5).

```
predictions_tidy <- predictions %>%
  mutate(mod_null = ifelse(mod_null < 0.5, " <=50K", " >50K")) %>%
  gather(key = "model", value = "y_hat", -type, -y)
glimpse(predictions_tidy)

Observations: 162,805
Variables: 4
$ y     <fctr> <=50K, <=50K, <=50K, <=50K, <=50K, <=50K, <=50K,...
$ type  <chr> "train", "train", "train", "train", "train", "train", "t...
$ model <chr> "mod_null", "mod_null", "mod_null", "mod_null", "mod_nul...
$ y_hat <chr> " <=50K", " <=50K", " <=50K", " <=50K", " <=50K", " <=50...
```

Now that we have the predictions for each model, we just need to compare them to the truth (y), and tally the results. We can do this using some dplyr machinations (note the use of the unite() function from the tidyr package).

```
predictions_summary <- predictions_tidy %>%
  group_by(model, type) %>%
  summarize(N = n(), correct = sum(y == y_hat, 0),
            positives = sum(y == " >50K"),
            true_pos = sum(y_hat == " >50K" & y == y_hat),
            false_pos = sum(y_hat == " >50K" & y != y_hat)) %>%
  mutate(accuracy = correct / N,
         tpr = true_pos / positives,
         fpr = false_pos / (N - positives)) %>%
  ungroup() %>%
  gather(val_type, val, -model, -type) %>%
  unite(temp1, type, val_type, sep = "_") %>%    # glue variables
  spread(temp1, val) %>%
  arrange(desc(test_accuracy)) %>%
  select(model, train_accuracy, test_accuracy, test_tpr, test_fpr)
predictions_summary

# A tibble: 5  5
      model train_accuracy test_accuracy test_tpr test_fpr
      <chr>          <dbl>         <dbl>    <dbl>    <dbl>
1 mod_forest        0.86694       0.86364  0.64436 0.069366
2   mod_tree        0.84449       0.84459  0.50459 0.051524
3     mod_nn        0.84687       0.84413  0.65157 0.097033
4     mod_nb        0.81765       0.82217  0.41929 0.054731
5   mod_null        0.75750       0.76597  0.00000 0.000000
```

We note that even though the neural network slightly outperformed the decision tree on the training set, the decision tree performed slightly better on the testing set. In this case, however, the accuracy rates of all models were about the same on both the training and testing sets.

In Figure 8.8, we compare the ROC curves for all five census models on the testing data set. Some data wrangling is necessary before we can gather the information to make these curves.

```
outputs <- c("response", "prob", "prob", "raw", "raw")
roc_test <- mapply(predict, mods, type = outputs,
    MoreArgs = list(newdata = test)) %>%
  as.data.frame() %>%
  select(1,3,5,6,8)
names(roc_test) <-
  c("mod_null", "mod_tree", "mod_forest", "mod_nn", "mod_nb")
glimpse(roc_test)

Observations: 6,512
Variables: 5
$ mod_null   <dbl> 0.2425, 0.2425, 0.2425, 0.2425, 0.2425, 0.2425, 0.2...
$ mod_tree   <dbl> 0.299733, 0.727960, 0.299733, 0.299733, 0.299733, 0...
$ mod_forest <dbl> 0.5920398, 0.5771144, 0.0945274, 0.0348259, 0.59701...
$ mod_nn     <dbl> 0.669130, 0.649079, 0.249087, 0.166118, 0.695224, 0...
$ mod_nb     <dbl> 1.8642e-01, 2.6567e-01, 4.8829e-02, 3.5218e-02, 4.2...

get_roc <- function(x, y) {
  pred <- ROCR::prediction(x$y_hat, y)
  perf <- ROCR::performance(pred, 'tpr', 'fpr')
  perf_df <- data.frame(perf@x.values, perf@y.values)
  names(perf_df) <- c("fpr", "tpr")
  return(perf_df)
}

roc_tidy <- roc_test %>%
  gather(key = "model", value = "y_hat") %>%
  group_by(model) %>%
  dplyr::do(get_roc(., y = test$income))


ggplot(data = roc_tidy, aes(x = fpr, y = tpr)) +
  geom_line(aes(color = model)) +
  geom_abline(intercept = 0, slope = 1, lty = 3) +
  ylab(perf@y.name) + xlab(perf@x.name) +
  geom_point(data = predictions_summary, size = 3,
    aes(x = test_fpr, y = test_tpr, color = model))
```

8.5   Extended example: Who has diabetes?

Consider the relationship between age and diabetes mellitus, a group of metabolic diseases characterized by high blood sugar levels. As with many diseases, the risk of contracting diabetes increases with age and is associated with many other factors. Age does not suggest a way to avoid diabetes: there is no way for you to change your age. You can, however, change things like diet, physical fitness, etc. Knowing what is predictive of diabetes can be helpful in practice, for instance, to design an efficient screening program to test people for the disease.

Let's start simply. What is the relationship between age, body-mass index (BMI), and diabetes for adults surveyed in NHANES? Note that the overall rate of diabetes is relatively low.

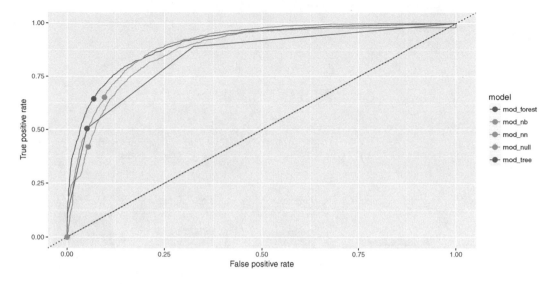

Figure 8.8: Comparison of ROC curves across five models on the Census testing data. The null model has a true positive rate of zero and lies along the diagonal. The Bayes has a lower true positive rate than the other models. The random forest may be the best overall performer, as its curve lies furthest from the diagonal.

```
library(NHANES)
people <- NHANES %>%
  select(Age, Gender, Diabetes, BMI, HHIncome, PhysActive) %>%
  na.omit()
glimpse(people)

Observations: 7,555
Variables: 6
$ Age       <int> 34, 34, 34, 49, 45, 45, 45, 66, 58, 54, 58, 50, 33,...
$ Gender    <fctr> male, male, male, female, female, female, female, ...
$ Diabetes  <fctr> No, No, No, No, No, No, No, No, No, No, No, No, No...
$ BMI       <dbl> 32.22, 32.22, 32.22, 30.57, 27.24, 27.24, 27.24, 23...
$ HHIncome  <fctr> 25000-34999, 25000-34999, 25000-34999, 35000-44999...
$ PhysActive <fctr> No, No, No, No, Yes, Yes, Yes, Yes, Yes, Yes, Yes,...

tally(~ Diabetes, data = people, format = "percent")

Diabetes
     No      Yes
90.9464   9.0536
```

We illustrate the use of a decision tree using all of the variables except for household income in Figure 8.9. From the original data shown in Figure 8.10, it appears that older people, and those with higher BMIs, are more likely to have diabetes.

```
whoIsDiabetic <- rpart(Diabetes ~ Age + BMI + Gender + PhysActive,
  data = people, control = rpart.control(cp = 0.005, minbucket = 30))
whoIsDiabetic
```

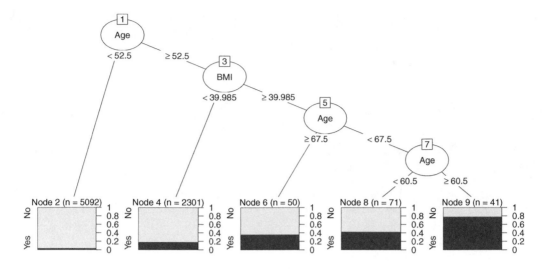

Figure 8.9: Illustration of decision tree for diabetes.

```
n= 7555

node), split, n, loss, yval, (yprob)
      * denotes terminal node

 1) root 7555 684 No (0.909464 0.090536)
   2) Age< 52.5 5092 188 No (0.963079 0.036921) *
   3) Age>=52.5 2463 496 No (0.798620 0.201380)
     6) BMI< 39.985 2301 416 No (0.819209 0.180791) *
     7) BMI>=39.985 162  80 No (0.506173 0.493827)
      14) Age>=67.5 50  18 No (0.640000 0.360000) *
      15) Age< 67.5 112  50 Yes (0.446429 0.553571)
        30) Age< 60.5 71  30 No (0.577465 0.422535) *
        31) Age>=60.5 41   9 Yes (0.219512 0.780488) *
```

```
plot(as.party(whoIsDiabetic))
```

If you are 52 or younger, then you very likely do not have diabetes. However, if you are 53 or older, your risk is higher. If your BMI is above 40—indicating obesity—then your risk increases again. Strangely—and this may be evidence of overfitting—your risk is highest if you are between 61 and 67 years old. This partition of the data is overlaid on Figure 8.10.

```
ggplot(data = people, aes(x = Age, y = BMI)) +
  geom_count(aes(color = Diabetes), alpha = 0.5) +
  geom_vline(xintercept = 52.5) +
  geom_segment(x = 52.5, xend = 100, y = 39.985, yend = 39.985) +
  geom_segment(x = 67.5, xend = 67.5, y = 39.985, yend = Inf) +
  geom_segment(x = 60.5, xend = 60.5, y = 39.985, yend = Inf) +
  annotate("rect", xmin = 60.5, xmax = 67.5, ymin = 39.985,
    ymax = Inf, fill = "blue", alpha = 0.1)
```

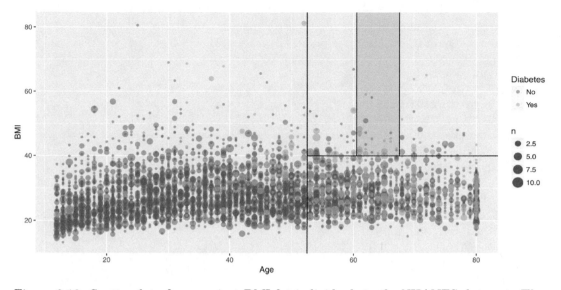

Figure 8.10: Scatterplot of age against BMI for individuals in the NHANES data set. The green dots represent a collection of people with diabetes, while the pink dots represent those without diabetes.

Figure 8.10 is a nice way to visualize a complex model. We have plotted our data in two quantitative dimensions (Age and BMI) while using color to represent our binary response variable (Diabetes). The decision tree simply partitions this two-dimensional space into axis-parallel rectangles. The model makes the same prediction for all observations within each rectangle. It is not hard to imagine—although it is hard to draw—how this recursive partitioning will scale to higher dimensions.

Note, however, that Figure 8.10 provides a clear illustration of the strengths and weaknesses of models based on recursive partitioning. These types of models can *only* produce axis-parallel rectangles in which all points in each rectangle receive the same prediction. This makes these models relatively easy to understand and apply, but it is not hard to imagine a situation in which they might perform miserably (e.g., what if the relationship was non-linear?). Here again, this underscores the importance of visualizing your model *in the data space* [233] as demonstrated in Figure 8.10.

We can visualize any model in a similar fashion. To do this, we will tile the (*Age, BMI*)-plane with a fine grid of points.

```
ages <- range(~ Age, data = people)
bmis <- range(~ BMI, data = people)
res <- 100
fake_grid <- expand.grid(
  Age = seq(from = ages[1], to = ages[2], length.out = res),
  BMI = seq(from = bmis[1], to = bmis[2], length.out = res))
```

Next, we will evaluate each of our six models on each grid point, taking care to retrieve not the classification itself, but the probability of having diabetes.

```
form <- as.formula("Diabetes ~ Age + BMI")
dmod_tree <- rpart(form, data = people,
                   control = rpart.control(cp = 0.005, minbucket = 30))
```

```
dmod_forest <- randomForest(form, data = people, ntree = 201, mtry = 3)
dmod_nnet <- nnet(form, data = people, size = 6)

# weights:  25
initial  value 3639.632154
iter  10 value 2191.855612
iter  20 value 2101.429936
iter  30 value 1900.247565
iter  40 value 1878.343889
iter  50 value 1868.729142
iter  60 value 1866.214396
iter  70 value 1865.463551
iter  80 value 1865.400693
iter  90 value 1865.123549
iter 100 value 1865.067787
final  value 1865.067787
stopped after 100 iterations

dmod_nb <- naiveBayes(form, data = people)

pred_tree <- predict(dmod_tree, newdata = fake_grid)[, "Yes"]
pred_forest <- predict(dmod_forest, newdata = fake_grid,
  type = "prob")[, "Yes"]
pred_knn <- people %>%
  select(Age, BMI) %>%
  knn(test = select(fake_grid, Age, BMI), cl = people$Diabetes, k = 5) %>%
  as.numeric() - 1
pred_nnet <- predict(dmod_nnet, newdata = fake_grid, type = "raw") %>%
  as.numeric()
pred_nb <- predict(dmod_nb, newdata = fake_grid, type = "raw")[, "Yes"]
```

To evaluate the null model, we'll need the overall percentage of those with diabetes.

```
p <- tally(~ Diabetes, data = people, format = "proportion")["Yes"]
```

We next build a data frame with these vectors, and then gather() it into a long format.

```
res <- fake_grid %>%
  mutate(
    "Null" = rep(p, nrow(fake_grid)), "Decision Tree" = pred_tree,
    "Random Forest" = pred_forest, "k-Nearest Neighbor" = pred_knn,
    "Neural Network" = pred_nnet, "Naive Bayes" = pred_nb) %>%
  gather(key = "model", value = "y_hat", -Age, -BMI)
```

Figure 8.11 illustrates each model in the data space. The differences between the models are striking. The rigidity of the decision tree is apparent, especially relative to the flexibility of the k-NN model. However, the k-NN model makes bold binary predictions, whereas the random forest has similar flexibility, but more nuance. The null model makes uniform predictions, while the naïve Bayes model produces a non-linear horizon similar to what we would expect from a logistic regression model.

```
ggplot(data = res, aes(x = Age, y = BMI)) +
  geom_tile(aes(fill = y_hat), color = NA) +
  geom_count(aes(color = Diabetes), alpha = 0.4, data = people) +
  scale_fill_gradient(low = "white", high = "dodgerblue") +
  scale_color_manual(values = c("gray", "gold")) +
  scale_size(range = c(0, 2)) +
  scale_x_continuous(expand = c(0.02,0)) +
  scale_y_continuous(expand = c(0.02,0)) +
  facet_wrap(~model)
```

8.6 Regularization

Regularization is a technique where constraints are added to a regression model to prevent overfitting. Two techniques for *regularization* include *ridge regression* and the *LASSO* (least absolute shrinkage and selection operator). Instead of fitting a model that minimizes $\sum_{i=1}^{n}(y - \hat{y})^2$ where $\hat{y} = \mathbf{X}'\beta$, ridge regression adds a constraint that $\sum_{j=1}^{p}\beta_j^2 \leq c_1$ and the LASSO imposes the constraint that $\sum_{j=1}^{p}|\beta_j| \leq c_2$, for some constants c_1 and c_2.

These methods are considered part of statistical or machine learning since they shrink coefficients (for ridge regression) or select predictors (for the LASSO) automatically. They are particularly helpful when the set of predictors is large.

8.7 Further resources

We have focused on classification in this chapter, although supervised statistical learning models can be fit to quantitative outcomes. Such extensions are included in the exercises.

All readers are encouraged to consult [121] for a fuller treatments of these topics. A free PDF of this book is available online at http://www-bcf.usc.edu/~gareth/ISL. A graduate-level version of that text (also freely downloadable at http://www-stat.stanford.edu/~tibs/ElemStatLearn) is [98]. Another helpful source is [189], which has more of a computer science flavor. Breiman [37] is a classic paper that describes two cultures in statistics: prediction and modeling.

The partykit::ctree() function builds a recursive partitioning model using conditional inference trees. The functionality is similar to rpart() but uses different criteria to determine the splits. The partykit package also includes a cforest() function. The caret package provides a number of useful functions for training and plotting classification and regression models. The glmnet and lars packages include support for regularization methods. The RWeka package provides an R interface to the comprehensive Weka machine learning library, which is written in Java.

8.8 Exercises

Exercise 8.1

The ability to get a good night's sleep is correlated with many positive health outcomes. The NHANES data set contains a binary variable SleepTrouble that indicates whether each person has trouble sleeping. For each of the following models:

1. Build a classifier for SleepTrouble

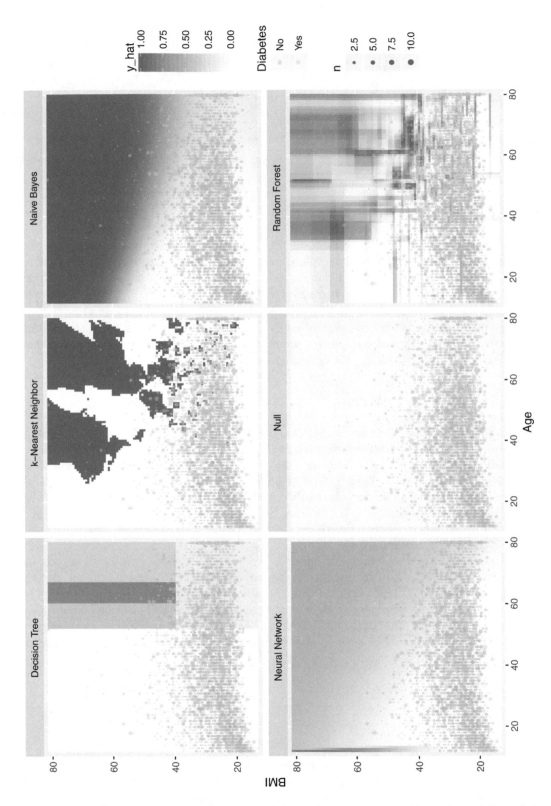

Figure 8.11: Comparison of predictive models in the data space. Note the rigidity of the decision tree, the flexibility of k-NN and the random forest, and the bold predictions of k-NN.

2. Report its effectiveness on the `NHANES` training data

3. Make an appropriate visualization of the model

4. Interpret the results. What have you learned about people's sleeping habits?

You may use whatever variable you like, except for `SleepHrsNight`.

1. Null model

2. Logistic regression

3. Decision tree

4. Random forest

5. Neural network

6. Naïve Bayes

7. *k*-NN

Exercise 8.2

Repeat the previous exercise, but now use the quantitative response variable `SleepHrsNight`. Build and interpret the following models:

1. Null model

2. Multiple regression

3. Regression tree

4. Random forest

5. Ridge regression

6. LASSO

Exercise 8.3

Repeat either of the previous exercises, but this time first separate the `NHANES` data set uniformly at random into 75% training and 25% testing sets. Compare the effectiveness of each model on training vs. testing data.

Exercise 8.4

Repeat the first exercise, but for the the variable `PregnantNow`. What did you learn about who is pregnant?

Exercise 8.5

The `nasaweather` package contains data about tropical storms from 1995–2005. Consider the scatterplot between the wind speed and pressure of these storms shown below.

```
library(mdsr)
library(nasaweather)
ggplot(data = storms, aes(x = pressure, y = wind, color = type)) +
  geom_point(alpha = 0.5)
```

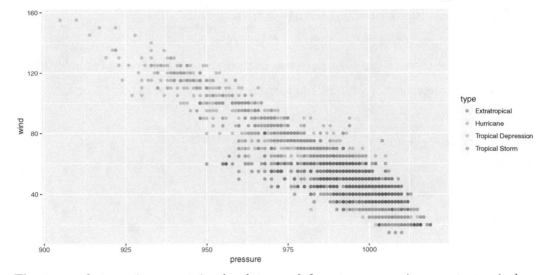

The type of storm is present in the data, and four types are given: extratropical, hurricane, tropical depression, and tropical storm. There are complicated and not terribly precises definitions for storm type. Build a classifier for the type of each storm as a function of its wind speed and pressure.

Why would a descision tree make a particularly good classifier for these data? Visualize your classifier in the data space in a manner similar to Figure 8.10 or 8.11.

Exercise 8.6

Fit a series of supervised learning models to predict arrival delays for flights from New York to SFO using the nycflights13 package. How do the conclusions change from the multiple regression model presented in the Chapter 7?

Exercise 8.7

Use the College Scorecard Data (https://collegescorecard.ed.gov/data) to model student debt as a function of institutional characteristics using the techniques described in this chapter. Compare and contrast results from at least three methods. (Note that a considerable amount of data wrangling will be needed.)

Chapter 9

Unsupervised learning

In the previous chapter, we explored models for learning about a response variable y from a set of explanatory variables $\mathbf{X}$. This process is called *supervised learning* because the response variable provides not just a clear goal for the modeling (to improve predictions about future y's), but also a guide (sometimes called the "ground truth"). In this chapter, we explore techniques in *unsupervised learning*, where there is no response variable y. Here, we simply have a set of observations $\mathbf{X}$, and we want to understand the relationships among them.

9.1   Clustering

Figure 9.1[1] shows an evolutionary tree of mammals. We humans (hominidae) are on the far left of the tree. The numbers at the branch points are estimates of how long ago—in millions of years—the branches separated. According to the diagram, rodents and primates diverged about 90 million years ago.

How do evolutionary biologists construct a tree like this? They study various traits of different kinds of mammals. Two mammals that have similar traits are deemed closely related. Animals with dissimilar traits are distantly related. By combining all of this information about the proximity of species, biologists can propose these kinds of evolutionary trees.

A tree—sometimes called a *dendrogram*—is an attractive organizing structure for relationships. Evolutionary biologists imagine that at each branch point there was an actual animal whose descendants split into groups that developed in different directions. In evolutionary biology, the inferences about branches come from comparing existing creatures to one another (as well as creatures from the fossil record). Creatures with similar traits are in nearby branches while creatures with different traits are in faraway branches. It takes considerable expertise in anatomy and morphology to know which similarities and differences are important. Note, however, that there is no outcome variable—just a construction of what is closely related or distantly related.

Trees can describe degrees of similarity between different things, regardless of how those relationships came to be. If you have a set of objects or cases, and you can measure how similar any two of the objects are, you can construct a tree. The tree may or may not reflect some deeper relationship among the objects, but it often provides a simple way to visualize relationships.

---

[1]Reprinted with permission under Creative Commons Attribution 2.0 Generic. No changes were made to this image.

MAMMALIA

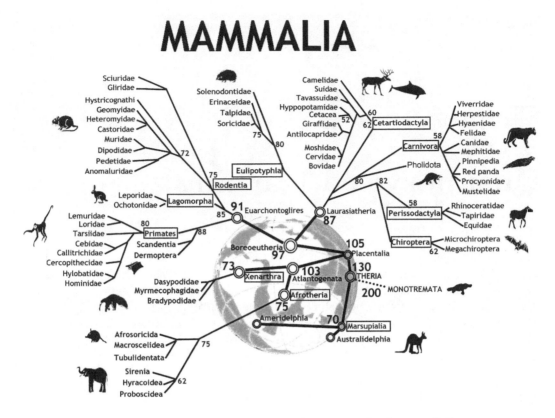

Figure 9.1: An evolutionary tree for mammals. Source: [92]

9.1.1   Hierarchical clustering

When the description of an object consists of a set of numerical variables (none of which is a *response*), there are two main steps in constructing a tree to describe the relationship among the cases in the data:

1. Represent each case as a point in a Cartesian space.

2. Make branching decisions based on how close together points or clouds of points are.

   To illustrate, consider the unsupervised learning process of identifying different types of cars. The United States Department of Energy maintains automobile characteristics for thousands of cars: miles per gallon, engine size, number of cylinders, number of gears, etc. Please see their guide for more information. Here, we download a ZIP file from their website that contains fuel economy rating for the 2016 model year.

```
download.file("https://www.fueleconomy.gov/feg/epadata/16data.zip",
              destfile = "data/fueleconomy.zip")
unzip("data/fueleconomy.zip", exdir = "data/fueleconomy/")
```

Next, we use the readxl package to read this file into R, clean up some of the resulting variable names, select a small subset of the variables, and filter for distinct models of Toyota vehicles. The resulting data set contains information about 75 different models that Toyota produces.

```
library(mdsr)
library(readxl)
filename <- list.files("data/fueleconomy", pattern = "public\\.xlsx")[1]
cars <- read_excel(paste0("data/fueleconomy/", filename)) %>% data.frame()
cars <- cars %>%
  rename(make = Mfr.Name, model = Carline, displacement = Eng.Displ,
    cylinders = X..Cyl, city_mpg = City.FE..Guide....Conventional.Fuel,
    hwy_mpg = Hwy.FE..Guide....Conventional.Fuel, gears = X..Gears) %>%
  select(make, model, displacement, cylinders, gears, city_mpg, hwy_mpg) %>%
  distinct(model, .keep_all = TRUE) %>%
  filter(make == "Toyota")
rownames(cars) <- cars$model
glimpse(cars)

Observations: 75
Variables: 7
$ make         <chr> "Toyota", "Toyota", "Toyota", "Toyota", "Toyota",...
$ model        <chr> "FR-S", "RC 200t", "RC 300 AWD", "RC 350", "RC 35...
$ displacement <dbl> 2.0, 2.0, 3.5, 3.5, 3.5, 5.0, 1.5, 1.8, 5.0, 2.0,...
$ cylinders    <dbl> 4, 4, 6, 6, 6, 8, 4, 4, 8, 4, 6, 6, 6, 4, 4, 4, 4...
$ gears        <dbl> 6, 8, 6, 8, 6, 8, 6, 1, 8, 8, 6, 8, 6, 6, 1, 4, 6...
$ city_mpg     <dbl> 25, 22, 19, 19, 19, 16, 33, 43, 16, 22, 19, 19, 1...
$ hwy_mpg      <dbl> 34, 32, 26, 28, 26, 25, 42, 40, 24, 33, 26, 28, 2...
```

As a large automaker, Toyota has a diverse lineup of cars, trucks, SUVs, and hybrid vehicles. Can we use unsupervised learning to categorize these vehicles in a sensible way with only the data we have been given?

For an individual quantitative variable, it is easy to measure how far apart any two cars are: Take the difference between the numerical values. The different variables are, however, on different scales and in different units. For example, gears ranges only from 1 to 8, while city_mpg goes from 13 to 58. This means that some decision needs to be made about rescaling the variables so that the differences along each variable reasonably reflect how different the respective cars are. There is more than one way to do this, and in fact, there is no universally "best" solution—the best solution will always depend on the data and your domain expertise. The dist() function takes a simple and pragmatic point of view: Each variable is equally important.[2]

The output of dist() gives the *distance* from each individual car to every other car.

```
car_diffs <- dist(cars)
str(car_diffs)

Class 'dist'  atomic [1:2775] 4.88 12.2 10.73 12.2 16.35 ...
  ..- attr(*, "Size")= int 75
  ..- attr(*, "Labels")= chr [1:75] "FR-S" "RC 200t" "RC 300 AWD" "RC 350" ...
  ..- attr(*, "Diag")= logi FALSE
  ..- attr(*, "Upper")= logi FALSE
  ..- attr(*, "method")= chr "euclidean"
  ..- attr(*, "call")= language dist(x = cars)
```

---

[2]The default distance metric used by dist() is the Euclidean distance. Recall that we discussed this in Chapter 8 in our explanation of k-nearest-neighbor methods.

	Atlanta	Boston	Chicago	Dallas	Denver	Houston	Las Vegas	Los Angeles
Atlanta		1095	715	805	1437	844	1920	2230
Boston	1095		983	1815	1991	1886	2500	3036
Chicago	715	983		931	1050	1092	1500	2112
Dallas	805	1815	931		801	242	1150	1425
Denver	1437	1991	1050	801		1032	885	1174
Houston	844	1886	1092	242	1032		1525	1556
Las Vegas	1920	2500	1500	1150	885	1525		289
Los Angeles	2230	3036	2112	1425	1174	1556	289	

Figure 9.2: Distances between some U.S. cities.

```
car_mat <- car_diffs %>% as.matrix()
car_mat[1:6, 1:6] %>% round(digits = 2)
```

```
            FR-S RC 200t RC 300 AWD RC 350 RC 350 AWD   RC F
FR-S        0.00    4.88      12.20  10.73      12.20  16.35
RC 200t     4.88    0.00       8.79   6.61       8.79  12.41
RC 300 AWD 12.20    8.79       0.00   3.35       0.00   5.32
RC 350     10.73    6.61       3.35   0.00       3.35   5.83
RC 350 AWD 12.20    8.79       0.00   3.35       0.00   5.32
RC F       16.35   12.41       5.32   5.83       5.32   0.00
```

This point-to-point distance matrix is analogous to the tables that used to be printed on road maps giving the distance from one city to another, like Figure 9.2, which states that it is 1,095 miles from Atlanta to Boston, or 715 miles from Atlanta to Chicago. Notice that the distances are symmetric: It is the same distance from Boston to Los Angeles as from Los Angeles to Boston (3,036 miles, according to the table).

Knowing the distances between the cities is not the same thing as knowing their locations. But the set of mutual distances is enough information to reconstruct the relative positions of the cities.

Cities, of course, lie on the surface of the earth. That need not be true for the "distance" between automobile types. Even so, the set of mutual distances provides information equivalent to knowing the relative positions of these cars in a p-dimensional space. This can be used to construct branches between nearby items, then to connect those branches, and so on until an entire tree has been constructed. The process is called *hierarchical clustering*. Figure 9.3 shows a tree constructed by hierarchical clustering that relates Toyota car models to one another.

```
library(ape)
car_diffs %>%
  hclust() %>%
  as.phylo() %>%
  plot(cex = 0.9, label.offset = 1)
```

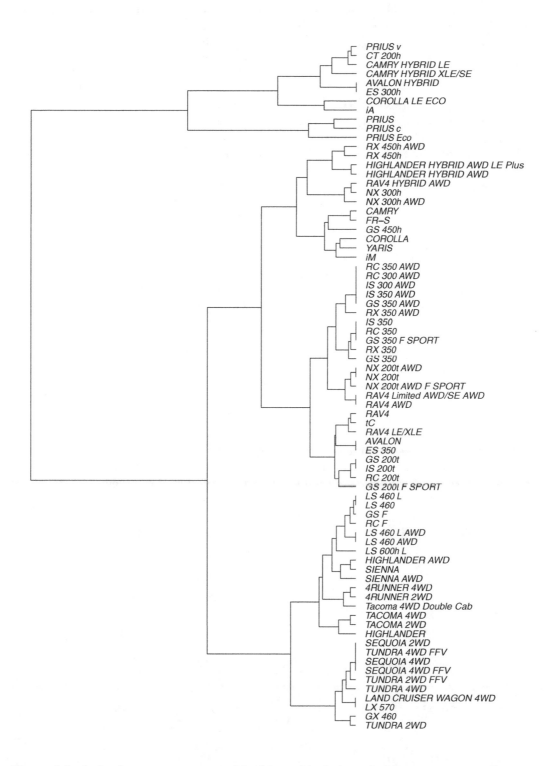

Figure 9.3: A dendrogram constructed by hierarchical clustering from car-to-car distances implied by the Toyota fuel economy data.

There are many ways to graph such trees, but here we have borrowed from biology by graphing these cars as a phylogenetic tree, similar to Figure 9.1. Careful inspection of Figure 9.3 reveals some interesting insights. The first branch in the tree is evidently between hybrid vehicles and all others. This makes sense, since hybrid vehicles use a fundamentally different type of power to achieve considerably better fuel economy. Moreover, the first branch among conventional cars divides large trucks and SUVs (e.g., Sienna, Tacoma, Sequoia, Tundra, Land Cruiser) from smaller cars and cross-over SUVs (e.g., Camry, Corolla, Yaris, RAV4). We are confident that the gearheads in the readership will identify even more subtle logic to this clustering. One could imagine that this type of analysis might help a car-buyer or marketing executive quickly decipher what might otherwise be a bewildering product line.

9.1.2 k-means

Another way to group similar cases is to assign each case to one of several distinct groups, but without constructing a hierarchy. The output is not a tree but a choice of group to which each case belongs. (There can be more detail than this; for instance, a probability for each group that a specific case belongs to the group.) This is like classification except that here there is no response variable. Thus, the definition of the groups must be inferred implicitly from the data.

As an example, consider the cities of the world (in `WorldCities`). Cities can be different and similar in many ways: population, age structure, public transportation and roads, building space per person, etc. The choice of *features* (or variables) depends on the purpose you have for making the grouping.

Our purpose is to show you that clustering via machine learning can actually identify genuine patterns in the data. We will choose features that are utterly familiar: the latitude and longitude of each city.

You already know about the location of cities. They are on land. And you know about the organization of land on earth: most land falls in one of the large clusters called continents. But the `WorldCities` data doesn't have any notion of continents. Perhaps it is possible that this feature, which you long ago internalized, can be learned by a computer that has never even taken grade-school geography.

For simplicity, consider the 4,000 biggest cities in the world and their longitudes and latitudes.

```
BigCities <- WorldCities %>%
  arrange(desc(population)) %>%
  head(4000) %>%
  select(longitude, latitude)
glimpse(BigCities)

Observations: 4,000
Variables: 2
$ longitude <dbl> 121.4581, -58.3772, 72.8826, -99.1277, 67.0822, 28.9...
$ latitude  <dbl> 31.22, -34.61, 19.07, 19.43, 24.91, 41.01, 28.65, 14...
```

Note that in these data, there is no ancillary information—not even the name of the city. However, the *k-means* clustering algorithm will separate these 4,000 points—each of which is located in a two-dimensional plane—into k clusters based on their locations alone.

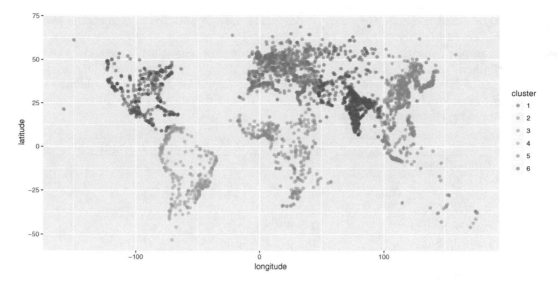

Figure 9.4: The world's 4,000 largest cities, clustered by the 6-means clustering algorithm.

```
set.seed(15)
library(mclust)
city_clusts <- BigCities %>%
   kmeans(centers = 6) %>%
   fitted("classes") %>%
   as.character()
BigCities <- BigCities %>% mutate(cluster = city_clusts)
BigCities %>% ggplot(aes(x = longitude, y = latitude)) +
   geom_point(aes(color = cluster), alpha = 0.5)
```

As shown in Figure 9.4, the clustering algorithm seems to have identified the continents. North and South America are clearly distinguished, as is most of Africa. The cities in North Africa are matched to Europe, but this is consistent with history, given the European influence in places like Morocco, Tunisia, and Egypt. Similarly, while the cluster for Europe extends into what is called Asia, the distinction between Europe and Asia is essentially historic, not geographic. East Asia and Central Asia are marked as distinct, largely because the low population areas of Tibet and Siberia look the same as the major oceans to the algorithm.

9.2   Dimension reduction

Often, a variable carries little information that is relevant to the task at hand. Even for variables that are informative, there can be redundancy or near duplication of variables. That is, two or more variables are giving essentially the same information—they have similar patterns across the cases.

Such irrelevant or redundant variables make it harder to learn from data. The irrelevant variables are simply noise that obscures actual patterns. Similarly, when two or more variables are redundant, the differences between them may represent random noise. Furthermore, for some machine learning algorithms, a large number of variables p will present computational challenges. It is usually helpful to remove irrelevant or redundant variables so that they—and the noise they carry—don't obscure the patterns that machine learning algorithms could identify.

For example, consider votes in a parliament or congress. Specifically, consider the Scottish Parliament in 2008.[3] Legislators often vote together in pre-organized blocs, and thus the pattern of "ayes" and "nays" on particular ballots may indicate which members are affiliated (i.e., members of the same political party). To test this idea, you might try clustering the members by their voting record.

name	S1M-1	S1M-1007.1	S1M-1007.2	S1M-1008
Adam, Brian	1	1	-1	0
Aitken, Bill	1	1	1	-1
Alexander, Ms Wendy	1	-1	-1	1
Baillie, Jackie	1	-1	-1	1
Barrie, Scott	-1	-1	-1	1
Boyack, Sarah	0	-1	-1	1
Brankin, Rhona	0	-1	0	1
Brown, Robert	-1	-1	-1	1
Butler, Bill	0	0	0	0
Campbell, Colin	1	1	-1	0

Table 9.1: Sample voting records data from the Scottish Parliament.

Table 9.1 shows a small part of the voting record. The names of the members of parliament are the cases. Each ballot—identified by a file number such as S1M-4.3—is a variable. A 1 means an "aye" vote, -1 is "nay", and 0 is an abstention. There are $n = 134$ members and $p = 773$ ballots—note that in this data set p far exceeds n. It is impractical to show all of the more than 100,000 votes in a table, but there are only 3 possible votes, so displaying the table as an image (as in Figure 9.5) works well.

```
Votes %>%
  mutate(Vote = factor(vote, labels = c("Nay","Abstain","Aye"))) %>%
  ggplot(aes(x = bill, y = name, fill = Vote)) +
    geom_tile() + xlab("Ballot") + ylab("Member of Parliament") +
    scale_fill_manual(values = c("darkgray", "white", "goldenrod")) +
    scale_x_discrete(breaks = NULL, labels = NULL) +
    scale_y_discrete(breaks = NULL, labels = NULL)
```

Figure 9.5 is a 134×773 grid in which each cell is color-coded based on one member of Parliament's vote on one ballot. It is hard to see much of a pattern here, although you may notice the Scottish tartan structure. The tartan pattern provides an indication to experts that the matrix can be approximated by a matrix of low-rank.

9.2.1 Intuitive approaches

As a start, Figure 9.6 shows the ballot values for all of the members of parliament for just two arbitrarily selected ballots. To give a better idea of the point count at each position, the values are jittered by adding some random noise. The red dots are the actual positions. Each point is one member of parliament. Similarly aligned members are grouped together at one of the nine possibilities marked in red: (Aye, Nay), (Aye, Abstain), (Aye, Aye),

---

[3]The Scottish Parliament example was constructed by then-student Caroline Ettinger and her faculty advisor, Andrew Beveridge, at Macalester College, and presented in Ms. Ettinger's senior capstone thesis.

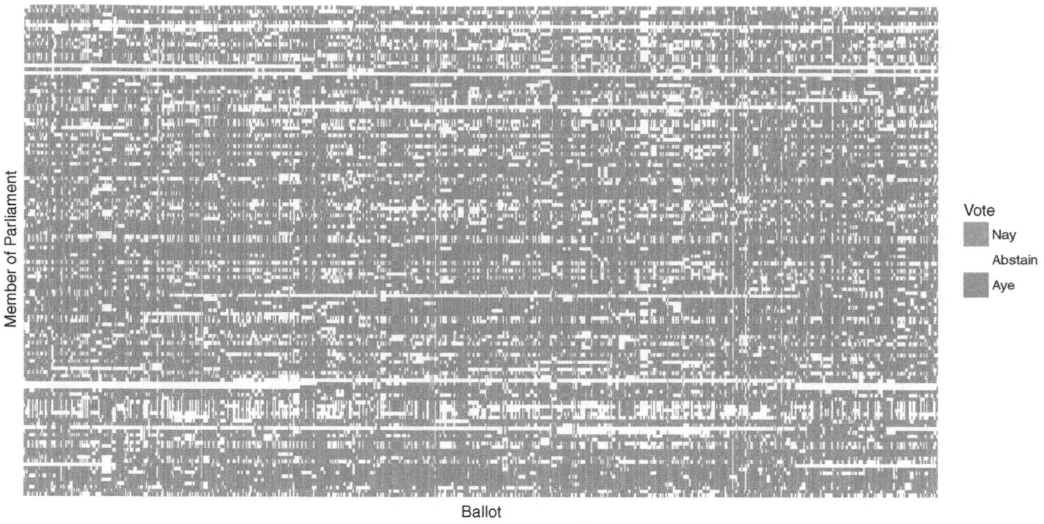

Figure 9.5: Visualization of the Scottish Parliament votes.

and so on through to (Nay, Nay). In these two ballots, eight of the nine possibilities are populated. Does this mean that there are eight clusters of members?

```
Votes %>% filter(bill %in% c("S1M-240.2", "S1M-639.1")) %>%
  tidyr::spread(key = bill, value = vote) %>%
  ggplot(aes(x = `S1M-240.2`, y = `S1M-639.1`)) +
    geom_point(alpha = 0.7,
               position = position_jitter(width = 0.1, height = 0.1)) +
    geom_point(alpha = 0.01, size = 10, color = "red" )
```

Intuition suggests that it would be better to use *all* of the ballots, rather than just two. In Figure 9.7, the first 387 ballots (half) have been added together, as have the remaining ballots. Figure 9.7 suggests that there might be two clusters of members who are aligned with each other. Using all of the data seems to give more information than using just two ballots.

```
Votes %>%
  mutate(set_num = as.numeric(factor(bill)),
    set =
      ifelse(set_num < max(set_num) / 2, "First_Half", "Second_Half")) %>%
  group_by(name, set) %>%
  summarize(Ayes = sum(vote)) %>%
  tidyr::spread(key = set, value = Ayes) %>%
  ggplot(aes(x = First_Half, y = Second_Half)) +
  geom_point(alpha = 0.7, size = 5)
```

9.2.2   Singular value decomposition

You may ask why the choice was made to add up the first half of the ballots as x and the remaining ballots as y. Perhaps there is a better choice to display the underlying patterns. Perhaps we can think of a way to add up the ballots in a more meaningful way.

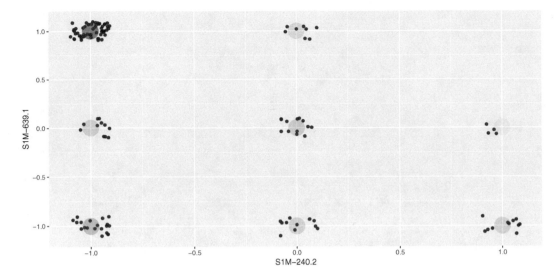

Figure 9.6: Scottish Parliament votes for two ballots.

In fact, there is a mathematical approach to finding the *best* approximation to the ballot–voter matrix using simple matrices, called *singular value decomposition* (SVD). (The statistical dimension reduction technique of Principal Component Analysis (PCA) can be accomplished using SVD.) The mathematics of SVD draw on a knowledge of matrix algebra, but the operation itself is accessible to anyone. Geometrically, SVD (or PCA) amounts to a rotation of the coordinate axes so that more of the variability can be explained using just a few variables. Figure 9.8 shows the position of each member on the two principal components that explain the most variability.

```
Votes_wide <- Votes %>% tidyr::spread(key = bill, value = vote)
vote_svd <- Votes_wide %>%
  select(-name) %>%
  svd()
voters <- vote_svd$u[ , 1:5] %>% as.data.frame()
clusts <- voters %>% kmeans(centers = 6)
voters <- voters %>% mutate(cluster = as.factor(clusts$cluster))
ggplot(data = voters, aes(x = V1, y = V2)) +
  geom_point(aes(x = 0, y = 0), color = "red", shape = 1, size = 7) +
  geom_point(size = 5, alpha = 0.6, aes(color = cluster)) +
  xlab("Best Vector from SVD") + ylab("Second Best Vector from SVD") +
  ggtitle("Political Positions of Members of Parliament")
```

Figure 9.8 shows, at a glance, that there are three main clusters. The red circle marks the *average* member. The three clusters move away from average in different directions. There are several members whose position is in-between the average and the cluster to which they are closest. These clusters may reveal the alignment of Scottish members of parliament according to party affiliation and voting history.

For a graphic, one is limited to using two variables for position. Clustering, however, can be based on many more variables. Using more SVD sums may allow the three clusters to be split up further. The color in Figure 9.8 above shows the result of asking for six clusters using the five best SVD sums. In reality, there are six national Scottish political parties. The confusion matrix below compares the actual party of each member to the

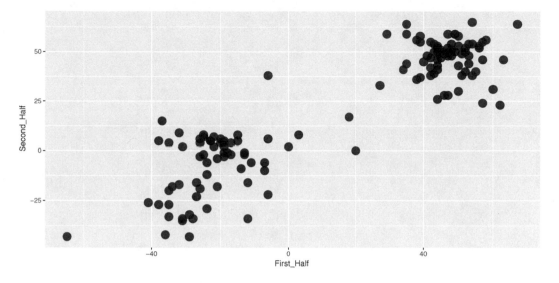

Figure 9.7: Scatterplot showing the correlation between Scottish Parliament votes in two arbitrary collections of ballots.

cluster memberships.

```
voters <- voters %>%
  mutate(name = Votes_wide$name) %>%
  left_join(Parties, by = c("name" = "name"))
tally(party ~ cluster, data = voters)
```

	cluster					
party	1	2	3	4	5	6
Member for Falkirk West	0	1	0	0	0	0
Scottish Conservative and Unionist Party	0	1	0	18	1	0
Scottish Green Party	0	0	0	0	0	1
Scottish Labour	0	2	49	0	6	1
Scottish Liberal Democrats	0	0	1	0	3	13
Scottish National Party	34	1	0	0	1	0
Scottish Socialist Party	0	0	0	0	0	1

How well did the clustering algorithm do? The party affiliation of each member of parliament is known, even though it wasn't used in finding the clusters. For each of the parties with multiple members, the large majority of members were placed into a unique cluster for that party. In other words, the technique has identified correctly nearly all of the members of the four different parties with significant representation (i.e., Conservative and Unionist, Labour, Liberal Democrats, and National).

```
ballots <- vote_svd$v[ , 1:5] %>% as.data.frame()
clust_ballots <- kmeans(ballots, centers = 16)
ballots <- ballots %>% mutate(cluster = as.factor(clust_ballots$cluster),
  bill = names(Votes_wide)[-1])
```

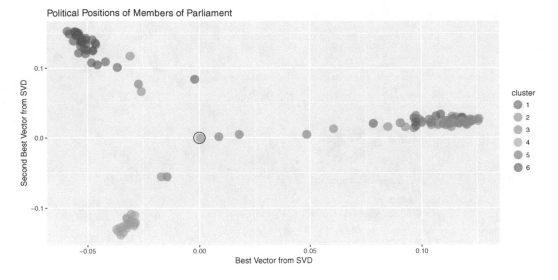

Figure 9.8: Clustering members of Scottish Parliament based on SVD along the members.

```
ggplot(data = ballots, aes(x = V1, y = V2)) +
  geom_point(aes(x = 0, y = 0), color = "red", shape = 1, size = 7) +
  geom_point(size = 5, alpha = 0.6, aes(color = cluster)) +
  xlab("Best Vector from SVD") + ylab("Second Best Vector from SVD") +
  ggtitle("Influential Ballots")
```

There is more information to be extracted from the ballot data. Just as there are clusters of political positions, there are clusters of ballots that might correspond to such factors as social effect, economic effect, etc. Figure 9.9 shows the position of ballots, using the first two principal components.

There are obvious clusters in this figure. Still, interpretation can be tricky. Remember that, on each issue, there are both "aye" and "nay" votes. This accounts for the symmetry of the dots around the center (indicated in red). The opposing dots along each angle from the center might be interpreted in terms of *socially liberal* versus *socially conservative* and *economically liberal* versus *economically conservative*. Deciding which is which likely involves reading the bill itself, as well as a nuanced understanding of Scottish politics.

Finally, the principal components can be used to re-arrange members of parliament and separately re-arrange ballots while maintaining each person's vote. This amounts simply to re-ordering the members in a way other than alphabetical and similarly with the ballots. This can bring dramatic clarity to the appearance of the data—as shown in Figure 9.10— where the large, nearly equally sized, and opposing voting blocs of the two major political parties (the National and Labour parties) become obvious. Alliances among the smaller political parties muddy the waters on the lower half of Figure 9.10.

```
Votes_svd <- Votes %>%
  mutate(Vote = factor(vote, labels = c("Nay", "Abstain", "Aye"))) %>%
  inner_join(ballots, by = "bill") %>%
  inner_join(voters, by = "name")
ggplot(data = Votes_svd,
  aes(x = reorder(bill, V1.x), y = reorder(name, V1.y), fill = Vote)) +
```

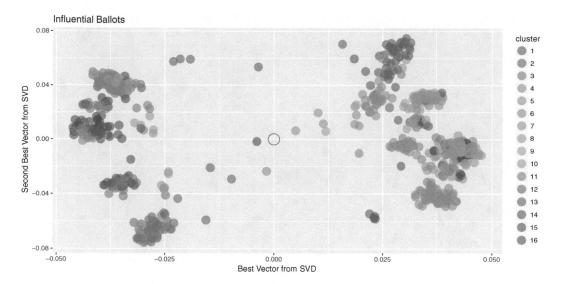

Figure 9.9: Clustering of Scottish Parliament ballots based on SVD along the ballots.

```
geom_tile() + xlab("Ballot") + ylab("Member of Parliament") +
scale_fill_manual(values = c("darkgray", "white", "goldenrod")) +
scale_x_discrete(breaks = NULL, labels = NULL) +
scale_y_discrete(breaks = NULL, labels = NULL)
```

The person represented by the top row in Figure 9.10 is Nicola Sturgeon, the leader of the Scottish National Party. Along the primary vector identified by our SVD, she is the most extreme voter. According to Wikipedia, the National Party belongs to a "mainstream European social democratic tradition."

```
Votes_svd %>%
  arrange(V1.y) %>%
  head(1)

    bill              name vote Vote      V1.x     V2.x    V3.x    V4.x
1 S1M-1 Sturgeon, Nicola    1  Aye -0.00391 -0.00167 -0.0498 -0.0734
    V5.x cluster.x  V1.y    V2.y    V3.y   V4.y     V5.y cluster.y
1 0.0137        16 -0.059  0.153 -0.0832 0.0396 -0.00198         1
                 party
1 Scottish National Party
```

Conversely, the person at the bottom of Figure 9.10 is Paul Martin, a member of the Scottish Labour Party. It is easy to see in Figure 9.10 that Martin opposed Sturgeon on most ballot votes.

```
Votes_svd %>%
  arrange(V1.y) %>%
  tail(1)

            bill         name vote Vote    V1.x     V2.x    V3.x    V4.x
103582 S1M-4064 Martin, Paul    1  Aye  0.0322 -0.00484 -0.0653 -0.0317
```

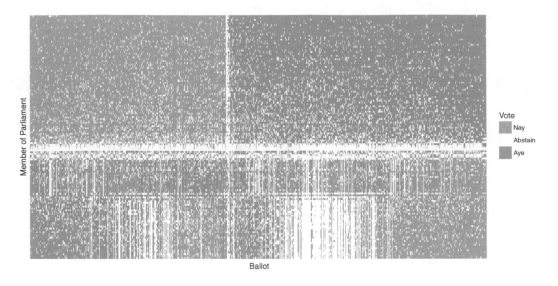

Figure 9.10: Illustration of the Scottish Parliament votes when ordered by the primary vector of the SVD.

```
           V5.x cluster.x  V1.y    V2.y    V3.y   V4.y     V5.y cluster.y
103582 0.00946          4 0.126  0.0267 -0.0425  0.056 -0.00423         3
                 party
103582 Scottish Labour
```

The beauty of Figure 9.10 is that it brings profound order to the chaos apparent in Figure 9.5. This was accomplished by simply ordering the rows (members of Parliament) and the columns (ballots) in a sensible way. In this case, the ordering was determined by the primary vector identified by the SVD of the voting matrix. This is yet another example of how machine learning techniques can identify meaningful patterns in data, but human beings are required to bring domain knowledge to bear on the problem in order to extract meaningful contextual understanding.

9.3   Further resources

The machine learning and phylogenetics CRAN task views provide guidance to functionality within R. Readers interested in learning more about unsupervised learning are encouraged to consult [121] or [98]. Kuiper [129] includes an accessible treatment of principal component analysis.

9.4   Exercises

Exercise 9.1

Consider the k-means clustering algorithm applied to the `BigCities` data and displayed in Figure 9.4. Would you expect to obtain different results if the location coordinates were *projected* (see Chapter 14)?

Exercise 9.2

Carry out and interpret a clustering of vehicles from another manufacturer using the approach outlined in Section 9.1.1.

Exercise 9.3

Project the WorldCities coordinates using the Gall–Peters projection and run the k-means algorithm again. Are the resulting clusters importantly different from those identified in Figure 9.4?

Exercise 9.4

Re-fit the k–means algorithm on the BigCities data with a different value of k (i.e., not six). Experiment with different values of k and report on the sensitivity of the algorithm to changes in this parameter.

Exercise 9.5

Baseball players are voted into the Hall of Fame by the members of the Baseball Writers of America Association. Quantitative criteria are used by the voters, but they are also allowed wide discretion. The following code identifies the position players who have been elected to the Hall of Fame and tabulates a few basic statistics, including their number of career hits (H), home runs (HR), and stolen bases (SB). Use the kmeans() function to perform a cluster analysis on these players. Describe the properties that seem common to each cluster.

```
library(mdsr)
library(Lahman)
hof <- Batting %>%
  group_by(playerID) %>%
  inner_join(HallOfFame, by = c("playerID" = "playerID")) %>%
  filter(inducted == "Y" & votedBy == "BBWAA") %>%
  summarize(tH = sum(H), tHR = sum(HR), tRBI = sum(RBI), tSB = sum(SB)) %>%
  filter(tH > 1000)
```

Exercise 9.6

Building on the previous exercise, compute new statistics and run the clustering algorithm again. Can you produce clusters that you think are more pure? Justify your choices.

Exercise 9.7

Perform the clustering on *pitchers* who have been elected to the Hall of Fame. Use wins (W), strikeouts (SO), and saves (SV) as criteria.

Exercise 9.8

Use the College Scorecard Data (https://collegescorecard.ed.gov/data) to cluster educational institutions using the techniques described in this chapter. Be sure to include variables related to student debt, number of students, graduation rate, and selectivity. (Note that a considerable amount of data wrangling will be needed.)

Chapter 10

Simulation

10.1 Reasoning in reverse

In Chapter 1 of this book we stated a simple truth: The purpose of data science is to turn data into usable information. Another way to think of this is that we use data to improve our understanding of the systems we live and work with: Data → Understanding.

This chapter is about computing techniques relating to the reverse way of thinking: Speculation → Data. In other words, this chapter is about "making up data."

Many people associate "making up data" with deception. Certainly, data can be made up for exactly that purpose. Our purpose is different. We are interested in legitimate purposes for making up data, purposes that support the proper use of data science in transforming data into understanding.

How can made-up data be legitimately useful? In order to make up data, you need to build a mechanism that contains, implicitly, an idea about how the system you are interested in works. The data you make up tell you what data generated by that system would look like. There are two main (legitimate) purposes for doing this:

- Conditional inference. If our mechanism is reflective of how the real system works, the data it generates are similar to real data. You might use these to inform tweaks to the mechanism in order to produce even more representative results. This process can help you refine your understanding in ways that are relevant to the real world.

- Winnowing out hypotheses. To "winnow" means to remove from a set the less desirable choices so that what remains is useful. Traditionally, grain was winnowed to separate the edible parts from the inedible chaff. For data science, the set is composed of hypotheses, which are ideas about how the world works. Data are generated from each hypothesis and compared to the data we collect from the real world. When the hypothesis-generated data fails to resemble the real-world data, we can remove that hypothesis from the set. What remains are hypotheses that are plausible candidates for describing the real-world mechanisms.

"Making up" data is undignified, so we will leave that term to refer to fraud and trickery. In its place we'll use use *simulation*, which derives from "similar." Simulations involve constructing mechanisms that are similar to how systems in the real world work—or at least to our belief and understanding of how such systems work.

10.2   Extended example: Grouping cancers

There are many different kinds of cancer, often given the name of the tissue in which they originate: lung cancer, ovarian cancer, prostate cancer, and so on. Different kinds of cancer are treated with different chemotherapeutic drugs. But perhaps the tissue origin of each cancer is not the best indicator of how it should be treated. Could we find a better way? Let's revisit the data introduced in Section 3.2.4.

Like all cells, cancer cells have a genome containing tens of thousands of genes. Sometimes just a few genes dictate a cell's behavior. Other times there are networks of genes that regulate one another's expression in ways that shape cell features, such as the over-rapid reproduction characteristic of cancer cells. It is now possible to examine the expression of individual genes within a cell. So-called *microarrays* are routinely used for this purpose. Each microarray has tens to hundreds of thousands of *probes* for gene activity. The result of a microarray assay is a snapshot of gene activity. By comparing snapshots of cells in different states, it's possible to identify the genes that are expressed differently in the states. This can provide insight into how specific genes govern various aspects of cell activity.

A data scientist, as part of a team of biomedical researchers, might take on the job of compiling data from many microarray assays to identify whether different types of cancer are related based on their gene expression. For instance, the NCI60 data (provided by the etl_NCI60() function in the mdsr package) contains readings from assays of $n = 60$ different cell lines of cancer of different tissue types. For each cell line, the data contain readings on over $p > 40,000$ different probes. Your job might be to find relationships between different cell lines based on the patterns of probe expression. For this purpose, you might find useful the techniques of statistical learning and unsupervised learning from Chapters 8 and 9 may be useful to you.

However, there is a problem. Even cancer cells have to carry out the routine actions that all cells use to maintain themselves. Presumably, the expression of most of the genes in the NCI60 data are irrelevant to the pecularities of cancer and the similarities and differences between different cancer types. Data interpreting methods—including those in Chapter 8—can be swamped by a wave of irrelevant data. They are more likely to be effective if the irrelevant data can be removed. Dimension reduction methods such as those described in Chapter 9 can be attractive for this purpose.

When you start down the road toward your goal of finding links among different cancer types, you don't know if you will reach your destination. If you don't, before concluding that there are no relationships, it's helpful to rule out some other possibilities. Perhaps the data reduction and data interpretation methods you used are not powerful enough. Another set of methods might be better. Or perhaps there isn't enough data to be able to detect the patterns you are looking for.

Simulations can help here. To illustrate, consider a rather simple data reduction technique for the NCI60 microarray data. If the expression of a probe is the same or very similar across all the different cancers, there's nothing that that probe can tell us about the links among cancers. One way to quantify the variation in a probe from cell line to cell line is the standard deviation of microarray readings for that probe.

It is a straightforward exercise in data wrangling to calculate this for each probe. The NCI60 data come in a wide form: a matrix that's 60 columns wide (one for each cell line) and 41,078 rows long (one row for each probe). This expression will find the standard deviation across cell lines for each probe.

```
library(mdsr)
library(tidyr)
NCI60 <- etl_NCI60()
```

```
Spreads <- NCI60 %>%
  gather(value = expression, key = cellLine, -Probe) %>%
  group_by(Probe) %>%
  summarize(N = n(), spread = sd(expression)) %>%
  arrange(desc(spread)) %>%
  mutate(order = row_number())
```

NCI60 has been rearranged into narrow format in Spreads, with columns Probe and spread for each of 32,344 probes. (A large number of the probes appear several times in the microarray, in one case as many as 14 times.) We arrange this dataset in descending order by the size of the standard deviation, so we can collect the probes that exhibit the most variation across cell lines by taking the topmost ones in Spreads. For ease in plotting, the variable order has been added to mark the order of each probe in the list.

How many of the probes with top standard deviations should we include in further data reduction and interpretation? 1? 10? 1000? 10,000? How should we go about answering this question? We'll use a simulation to help determine the number of probes that we select.

```
Sim_spreads <- NCI60 %>%
  gather(value = expression, key = cellLine, -Probe) %>%
  mutate(Probe = shuffle(Probe)) %>%
  group_by(Probe) %>%
  summarize(N = n(), spread = sd(expression)) %>%
  arrange(desc(spread)) %>%
  mutate(order = row_number())
```

What makes this a simulation is the fourth line of the expression where we call shuffle(). In that line, we replace each of the probe labels with a randomly selected label. The result is that the expression has been statistically disconnected from any other variable, particularly cellLine. The simulation creates the kind of data that would result from a system in which the probe expression data is meaningless. In other words, the simulation mechanism matches the null hypothesis that the probe labels are irrelevant. By comparing the real NCI60 data to the simulated data, we can see which probes give evidence that the null hypothesis is false. Let's compare the top 500 spread values in Spreads and Sim_spreads.

We can tell a lot from the results of the simulation shown in Figure 10.1. If we decided to use the top 500 probes, we would risk including many that were no more variable than random noise (i.e., that which could have been generated under the null hypothesis).

But if we set the threshold much lower, including, say, only those probes with a spread greater than 5.0, we would be unlikely to include any that were generated by a mechanism consistent with the null hypothesis. The simulation is telling us that it would be good to look at roughly the top 50 probes, since that is about how many in NCI60 were out of the range of the simulated results for the null hypothesis. Methods of this sort are often identified as *false discovery rate* methods.

10.3 Randomizing functions

There are as many possible simulations as there are possible hypotheses—that is, an unlimited number. Different hypotheses call for different techniques for building simulations. But there are some techniques that appear in a wide range of simulations. It's worth knowing about these.

```
Spreads %>%
  filter(order <= 500) %>%
  ggplot(aes(x = order, y = spread)) +
  geom_line(color = "blue", size = 2) +
  geom_line(data = filter(Sim_spreads, order <= 500), color = "red", size = 2)
```

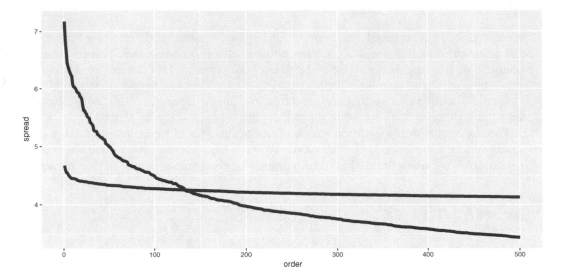

Figure 10.1: Comparing the variation in expression for individual probes across cell lines in the NCI60 data (blue) and a simulation of a null hypothesis (red).

The previous example about false discovery rates in gene expression uses an everyday method of randomization: *shuffling*. Shuffling is, of course, a way of destroying any genuine order in a sequence, leaving only those appearances of order that are due to chance. Closely related methods, *sampling* and *resampling*, were introduced in Chapter 7 when we used simulation to assess the statistical significance of patterns observed in data.

Counter-intuitively, the use of random numbers is an important component of many simulations. In simulation, we want to induce variation. For instance, the simulated probes for the cancer example do not all have the same spread. But in creating that variation, we do not want to introduce any structure other than what we specify explicitly in the simulation. Using random numbers ensures that any structure that we find in the simulation is either due to the mechanism we've built for the simulation or is purely accidental.

The workhorse of simulation is the generation of random numbers in the range from zero to one, with each possibility being equally likely. In R, the most widely used such *uniform random number generator* is runif(). For instance, here we ask for five uniform random numbers:

```
runif(5)
```

```
[1] 0.673 0.897 0.125 0.139 0.358
```

Other randomization devices can be built out of uniform random number generators. To illustrate, here is a device for selecting one value at random from a vector:

```
select_one <- function(vec) {
  n <- length(vec)
  ind <- which.max(runif(n))
  vec[ind]
}
select_one(letters) # letters are a, b, c, ..., z

[1] "k"

select_one(letters)

[1] "h"
```

The select_one() function is functionally equivalent to sample_n() with the size argument set to 1. However, random numbers are so important that you should try to use generators that have been written by experts and vetted by the community. There is a lot of sophisticated theory behind programs that generate uniform random numbers. After all, you generally don't want sequences of random numbers to repeat themselves. (An exception is described in Section 10.7.) The theory has to do with techniques for making repeated sub-sequences as rare as possible.

Perhaps the widest use of simulation in data analysis involves the randomness introduced by sampling, resampling, and shuffling. These operations are provided by the functions sample(), resample(), and shuffle(). These functions sample uniformly at random from a data frame (or vector) with or without replacement, or permute the rows of a data frame. resample() is equivalent to sample() with the replace argument set to TRUE, while shuffle() is equivalent to sample() with size equal to the number of rows in the data frame and replace equal to FALSE. Non-uniform sampling can be achieved using the prob argument.

Other important functions for building simulations are those that generate random numbers with certain important properties. We've already seen runif() for creating uniform random numbers. Very widely used are rnorm(), rexp(), and rpois() for generating numbers that are distributed normally (that is, in the bell-shaped, Gaussian distribution), exponentially, and with a Poisson pattern, respectively. These different distributions correspond to idealized descriptions of mechanisms in the real world. For instance, events that are equally likely to happen at any time (e.g., earthquakes) will tend to have a time spacing between events that is exponential. Events that have a rate that remains the same over time (e.g., the number of cars passing a point on a road in one minute) are often modeled using a Poisson distribution. There are many other forms of distributions that are considered good models of particular random processes. Functions analogous to runif() and rnorm() are available for other common probability distributions (see the Probability Distributions CRAN Task View).

10.4   Simulating variability

10.4.1   The partially planned rendezvous

Imagine a situation where Sally and Joan plan to meet to study in their college campus center [144]. They are both impatient people who will wait only ten minutes for the other before leaving. But their planning was incomplete. Sally said, "Meet me between 7 and 8 tonight at the center." When should Joan plan to arrive at the campus center? And what is the probability that they actually meet?

A simulation can help answer these questions. Joan might reasonably assume that it doesn't really matter when she arrives, and that Sally is equally likely to arrive any time between 7:00 and 8:00 pm.

So to Joan, Sally's arrival time is random and uniformly distributed between 7:00 and 8:00 pm. The same is true for Sally. Such a simulation is easy to write: generate uniform random numbers between 0 and 60 minutes after 7:00 pm. For each pair of such numbers, check whether or not the time difference between them is ten minutes or less. If so, they successfully met. Otherwise, they missed each other.

Here's an implementation in R, with 100,000 trials of the simulation being run to make sure that the possibilities are well covered.

```
n <- 100000
sim_meet <- data.frame(
  sally <- runif(n, min = 0, max = 60),
  joan <- runif(n, min = 0, max = 60)) %>%
  mutate(result = ifelse(abs(sally - joan) <= 10,
    "They meet", "They do not"))
tally(~ result, format = "percent", data = sim_meet)


result
They do not   They meet
       69.4        30.6


binom.test(~result, n, success = "They meet", data = sim_meet)

data:  sim_meet$result   [with success = They meet]
number of successes = 30000, number of trials = 1e+05, p-value
<2e-16
alternative hypothesis: true probability of success is not equal to 0.5
95 percent confidence interval:
 0.303 0.309
sample estimates:
probability of success
                0.306
```

There's about a 30% chance that they meet (the true probability is $11/36 \approx 0.3055556$). The confidence interval is narrow enough that any decision Joan might consider ("Oh, it seems unlikely we'll meet. I'll just skip it.") would be the same regardless of which end of the confidence interval is considered. So the simulation is good enough for Joan's purposes. (If the interval was not narrow enough for this, you would want to add more trials. The $1/\sqrt{n}$ rule for the width of a confidence interval described in Chapter 7 can guide your choice.)

Often, it's valuable to visualize the possibilities generated in the simulation as in Figure 10.2. The arrival times uniformly cover the rectangle of possibilities, but only those that fall into the stripe in the center of the plot are successful. Looking at the plot, Joan notices a pattern. For any arrival time she plans, the probability of success is the fraction of a vertical band of the plot that is covered in blue. For instance, if Joan chose to arrive at 7:20, the probability of success is the proportion of blue in the vertical band with boundaries of 20 minutes and 30 minutes on the horizontal axis. Joan observes that near 0 and 60 minutes, the probability goes down, since the diagonal band tapers. This observation guides an important decision: Joan will plan to arrive somewhere from 7:10 to 7:50. Following this strategy, what is the probability of success? (Hint: Repeat the simulation but replace joan <- runif(n, min = 0, max = 60)) with joan <- runif(n, min = 10, max = 50)).) If Joan had additional information about Sally ("She wouldn't arrange to meet at 7:21—most likely at 7:00, 7:15, 7:30, or 7:45.") the simulation can be easily modified, e.g., sally <- resample(c(0, 15, 30, 45), n) to incorporate that hypothesis.

10.4.2 The jobs report

One hour before the opening of the stock market on the first Friday of each month, the Bureau of Labor Statistics releases the employment report. This widely anticipated estimate of the monthly change in non-farm payroll is an economic indicator that often leads to stock market shifts.

If you read the financial blogs, you'll hear lots of speculation before the report is released, and lots to account for the change in the stock market in the minutes *after* the report comes out. And you'll hear a lot of anticipation of the consequences of that month's job report on the prospects for the economy as a whole. It happens that many financiers read a lot into the ups and downs of the jobs report. (And other people, who don't take the report so seriously, see opportunities in responding to the actions of the believers.)

You are a skeptic. You know that in the months after the jobs report, an updated number is reported that is able to take into account late-arriving data that couldn't be included in the original report. One analysis, the article "How not to be misled by the jobs report" from the May 1, 2014 *New York Times* modeled the monthly report as a random number from a Gaussian distribution, with a mean of 150,000 jobs and a standard deviation of 65,000 jobs.

You are preparing a briefing for your bosses to convince them not to take the jobs report itself seriously as an economic indicator. For many bosses, the phrases "Gaussian distribution," "standard deviation," and "confidence interval" will trigger a primitive "I'm not listening!" response, so your message won't get through in that form.

It turns out that many such people will have a better understanding of a simulation than of theoretical concepts. You decide on a strategy: Use a simulation to generate a year's worth of job reports. Ask the bosses what patterns they see and what they would look for in the next month's report. Then inform them that there are no actual patterns in the graphs you showed them.

```
jobs_true <- 150
jobs_se <- 65   # in thousands of jobs
gen_samp <- function(true_mean, true_sd,
                     num_months = 12, delta = 0, id = 1) {
  samp_year <- rep(true_mean, num_months) +
    rnorm(num_months, mean = delta * (1:num_months), sd = true_sd)
  return(data.frame(jobs_number = samp_year,
                    month = as.factor(1:num_months), id = id))
}
```

We begin by defining some constants that will be needed, along with a function to calculate a year's worth of monthly samples from this known truth. Since the default value of delta is equal to zero, the "true" value remains constant over time. When the function argument true_sd is set to 0, no random noise is added to the system.

Next, we prepare a data frame that contains the function argument values over which we want to simulate. In this case, we want our first simulation to have no random noise—thus the true_sd argument will be set to 0 and the id argument will be set to Truth. Following that, we will generate three random simulations with true_sd set to the assumed value of jobs_se. The data frame params contains complete information about the simulations we want to run.

```
n_sims <- 3
params <- data.frame(sd = c(0, rep(jobs_se, n_sims)),
                     id = c("Truth", paste("Sample", 1:n_sims)))
params

  sd        id
1  0     Truth
2 65  Sample 1
3 65  Sample 2
4 65  Sample 3
```

Finally, we will actually perform the simulation using the do() function from the dplyr package (see Chapter 5). This will iterate over the params data frame and apply the appropriate values to each simulation.

```
df <- params %>%
  group_by(id) %>%
  dplyr::do(gen_samp(true_mean = jobs_true, true_sd = .$sd, id = .$id))
```

Figure 10.3 displays the "true" number as well as three realizations from the simulation. While all of the three samples are taken from a "true" universe where the jobs number is constant, each could easily be misinterpreted to conclude that the numbers of new jobs was decreasing at some point during the series. The moral is clear: It is important to be able to understand the underlying variability of a system before making inferential conclusions.

10.4.3   Restaurant health and sanitation grades

We take our next simulation from the data set of restaurant health violations in New York City. To help ensure the safety of patrons, health inspectors make unannounced inspections at least once per year to each restaurant. Establishments are graded based on a range of criteria including food handling, personal hygiene, and vermin control. Those with a score between 0 and 13 points receive a coveted A grade, those with 14 to 27 points receive the less desirable B, and those of 28 or above receive a C. We'll display values in a subset of this range to focus on the threshold between an A and B grade.

```
minval <- 7
maxval <- 19
JustScores <- Violations %>%
  filter(score >= minval & score <= maxval) %>%
  select(dba, score) %>%
  unique()
```

Figure 10.4 displays the distribution of restaurant violation scores. Is something unusual happening at the threshold of 13 points (the highest value to still receive an A)? Or could sampling variability be the cause of the dramatic decrease in the frequency of restaurants graded between 13 and 14 points? Let's carry out a simple simulation in which a grade of 13 or 14 is equally likely. The rflip() function allows us to flip a fair coin that determines whether a grade is a 14 (heads) or 13 (tails).

```
ggplot(data = sim_meet, aes(x = joan, y = sally, color = result)) +
  geom_point(alpha = 0.3) +
  geom_abline(intercept = 10, slope = 1) +
  geom_abline(intercept = -10, slope = 1)
```

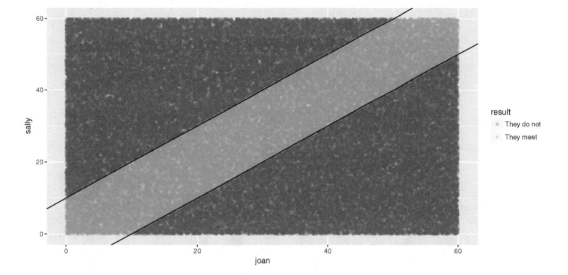

Figure 10.2: Distribution of Sally and Joan arrival times (shaded area indicates where they meet).

```
ggplot(data = df, aes(x = month, y = jobs_number)) +
  geom_hline(yintercept = jobs_true, linetype = 2) +
  geom_bar(stat = "identity") +
  facet_wrap(~ id) + ylab("Number of new jobs (in thousands)")
```

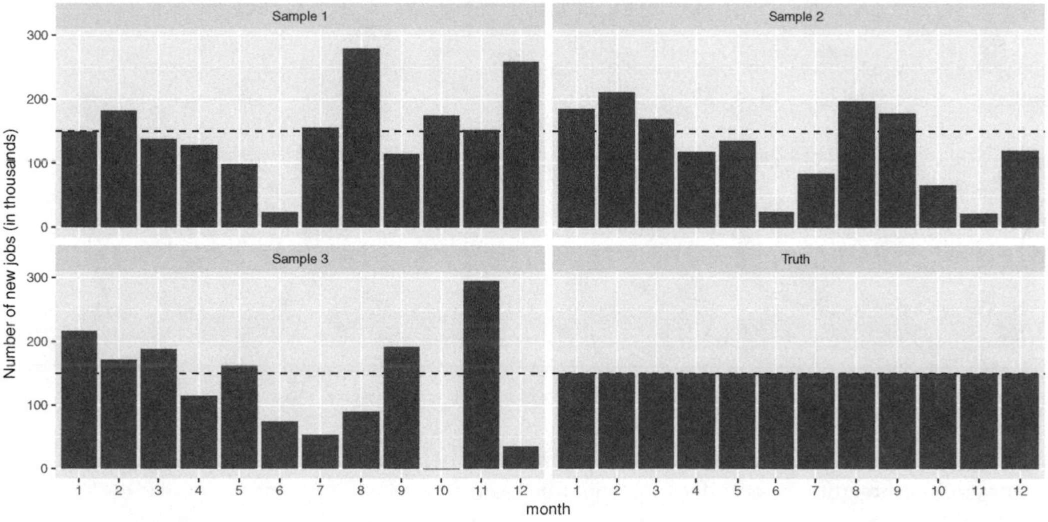

Figure 10.3: True number of new jobs from simulation as well as three realizations from a simulation.

```
ggplot(data = JustScores, aes(x = score)) +
  geom_histogram(binwidth = 0.5) + geom_vline(xintercept = 13, linetype = 2) +
  scale_x_continuous(breaks = minval:maxval) +
  annotate("text", x = 10.5, y = 10300,
           label = "A grade: score of 13 or less")
```

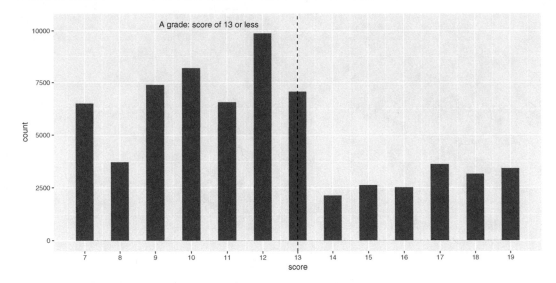

Figure 10.4: Distribution of NYC restaurant health violation scores.

```
scores <- tally(~score, data = JustScores)
scores
```

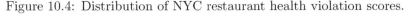

```
score
   7    8    9   10   11   12   13   14   15   16   17   18   19
6499 3709 7396 8200 6568 9858 7063 2127 2618 2513 3614 3150 3415

obs_diff <- scores["13"] - scores["14"]
mean(scores[c("13", "14")])

[1] 4595

RandomFlip <- do(1000) * rflip(scores["13"] + scores["14"])
head(RandomFlip, 3)

  n.13 heads tails prop.13
1 9190  4637  4553   0.505
2 9190  4622  4568   0.503
3 9190  4656  4534   0.507
```

Figure 10.5 demonstrates that the observed number of restaurants with a 14 are nowhere near what we would expect if there was an equal chance of receiving a score of 13 or 14. While the number of restaurants receiving a 13 might exceed the number receiving a 14 by 100 or so due to chance alone, there is essentially no chance of observing 5,000 more 13s than 14s if the two scores are truly equally likely. (It is not surprising given the large number of restaurants inspected in New York City that we wouldn't observe much sampling

```
ggplot(data = RandomFlip, aes(x = heads)) +
  geom_histogram(binwidth = 5) + xlim(c(2100, NA)) +
  geom_vline(xintercept = scores["14"], col = "red") +
  annotate("text", x = 2137, y = 45, label = "observed", hjust = "left") +
  xlab("Number of restaurants with scores of 14 (if equal probability)")
```

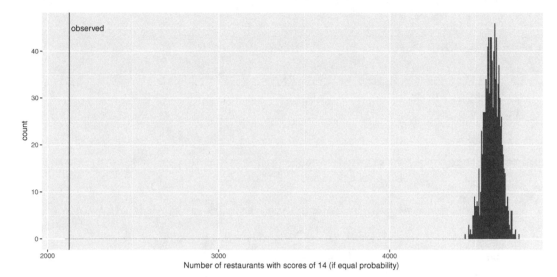

Figure 10.5: Distribution of health violation scores under a randomization procedure.

variability in terms of the proportion that are 14.) It appears as if the inspectors tend to give restaurants near the threshold the benefit of the doubt, and not drop their grade from A to B if the restaurant is on the margin between a 13 and 14 grade.

This is another situation where simulation can provide a more intuitive solution starting from first principles than an investigation using more formal statistical methods. (A more nuanced test of the "edge effect" might be considered given the drop in the numbers of restaurants with violation scores between 14 and 19.)

10.5 Simulating a complex system

Simulations can be very helpful in understanding the behavior of complex systems. As an example, consider a relatively simple system consisting of a bank with a single teller. We can make specific assumptions about the number of customers that enter the bank at any point in time, and the length of the transactions that they will conduct. This can allow the bank manager to predict typical wait times for customers.

To make this concrete, assume that one day the first customer arrives at 9:02 am and requires five minutes to be assisted. Another customer arrives at 9:05 am and requires three minutes, while a third customer arrives at 9:08 am and needs two minutes for their transaction. The first customer has a total time of five minutes, the second has a total time of five minutes (two minutes waiting plus three minutes being served), and the last customer experienced a total time of four minutes (two minutes waiting plus two minutes being served). Even though these three customers required only 10 minutes total of service time, they spent an average of 4.7 minutes at the bank due to the queuing.

To code this simulation, we employ algorithmic thinking (see Appendix C) and create some simple helper functions that can be used to break the problem down into manageable pieces.

```
any_active <- function(df) {
  # return TRUE if someone has not finished
  return(max(df$endtime) == Inf)
}

next_customer <- function(df) {
  # returns the next customer in line
  res <- filter(df, endtime == Inf) %>%
    arrange(arrival)
  return(head(res, 1))
}

update_customer <- function(df, cust_num, end_time) {
  # sets the end time of a specific customer
  return(mutate(df, endtime =
    ifelse(custnum == cust_num, end_time, endtime)))
}
```

We define a function to run the simulation with default values for the number of customers per minute, the expected length of transaction, and the number of hours that the bank is opened. We will assume that the number of customers follows a Poisson distribution (useful for modeling counts) and the transaction times follow an exponential distribution (long right tail with most transactions happening quickly but with some transactions taking a long time).

```
run_sim <- function(n = 1/2, m = 3/2, hours = 6) {
# simulation of bank where there is just one teller
# n: expected number of customers per minute
# m: expected length of transaction is m minutes
# hours: bank open for this many hours
  customers <- rpois(hours * 60, lambda = n)
  arrival <- numeric(sum(customers))
  position <- 1
  for (i in 1:length(customers)) {
    numcust <- customers[i]
    if (numcust != 0) {
      arrival[position:(position + numcust - 1)] <- rep(i, numcust)
      position <- position + numcust
    }
  }
  duration <- rexp(length(arrival), rate = 1/m)    # E[X]=m
  df <- data.frame(arrival, duration, custnum = 1:length(duration),
              endtime = Inf, stringsAsFactors = FALSE)

  endtime <- 0 # set up beginning of simulation
  while (any_active(df)) { # anyone left to serve?
    next_one <- next_customer(df)
    now <- ifelse(next_one$arrival >= endtime, next_one$arrival, endtime)
```

```
    endtime <- now + next_one$duration
    df <- update_customer(df, next_one$custnum, endtime)
  }
  df <- mutate(df, totaltime = endtime - arrival)
  return(favstats(~ totaltime, data = df))
}
```

```
sim_results <- do(3) * run_sim()
sim_results
```

	min	Q1	median	Q3	max	mean	sd	n	missing	.row	.index
1	0.000449	1.81	3.43	5.28	12.2	3.77	2.51	188	0	1	1
2	0.013429	2.34	8.17	14.25	31.2	9.72	8.47	175	0	1	2
3	0.092717	4.80	13.64	25.18	40.7	15.15	11.05	193	0	1	3

We see that the number of customers over the six-hour period at the bank ranged from 175 to 193, with the worst delays (mean of 15.15 minutes) on the day with the most customers. Several customers had to wait more than half an hour. Given enough computational time, one could run more simulations and come up with reasonable approximations for the distributions of the number of customers served and their respective waiting times. This information might lead the bank to consider adding a second teller. Consider how you might modify this simulation to model a second teller.

10.6 Random networks

As noted in Chapter 2, a network (or graph) is a collection of nodes, along with edges that connect certain pairs of those nodes. Networks are often used to model real-world systems that contain these pairwise relationships. Although these networks are often simple to describe, many of the interesting problems in the mathematical discipline of graph theory are very hard to solve analytically, and intractable computationally [83]. For this reason, simulation has become a useful technique for exploring questions in *network science*. We illustrate how simulation can be used to verify properties of *random graphs* in Chapter 16.

10.7 Key principles of simulation

Many of the key principles needed to develop the capacity to simulate come straight from computer science, including aspects of design, modularity, and reproducibility. In this section we will briefly propose guidelines for simulations.

Design

It is important to consider design issues relative to simulation. As the analyst, you control all aspects and decide what assumptions and scenarios to explore. You have the ability (and responsibility) to determine which scenarios are relevant and what assumptions are appropriate. The choice of scenarios depends on the underlying model: they should reflect plausible situations that are relevant to the problem at hand. It is often useful to start with a simple setting, then gradually add complexity as needed.

Modularity

It is very helpful to write a function to implement the simulation, which can be called repeatedly with different options and parameters (see Appendix C). Spending time planning what features the simulation might have, and how these can be split off into different functions (that might be reused in other simulations) will pay off handsomely.

Reproducibility and random number seeds

It is important that simulations are both reproducible and representative. Sampling variability is inherent in simulations: Our results will be sensitive to the number of computations that we are willing to carry out. We need to find a balance to avoid unneeded calculations while ensuring that our results aren't subject to random fluctuation. What is a reasonable number of simulations to consider? Let's revisit Sally and Joan, who will meet only if they both arrive within ten minutes of each other. How variable are our estimates if we carry out only num_sim = 100 simulations? We'll assess this by carrying out 5,000 replications, saving the results from each simulation of 100 possible meetings. Then we'll repeat the process, but with num_sim = 400 and num_sim = 1600. Note that we can do this efficiently using mosaic::do() and dplyr::do() in conjunction.

```
campus_sim <- function(num_sim = 1000, wait = 10) {
  sally <- runif(num_sim, min = 0, max = 60)
  joan <- runif(num_sim, min = 0, max = 60)
  return(sum(abs(sally - joan) <= wait) / num_sim)
}
reps <- 5000
params <- data.frame(num_sims = c(100, 400, 1600))
sim_results <- params %>%
  group_by(num_sims) %>%
  dplyr::do(mosaic::do(reps) * campus_sim(.$num_sims))
favstats(campus_sim ~ num_sims, data = sim_results)

  num_sims   min    Q1 median    Q3   max  mean     sd    n missing
1      100 0.140 0.270  0.300 0.340 0.490 0.306 0.0456 5000       0
2      400 0.228 0.290  0.305 0.320 0.395 0.306 0.0228 5000       0
3     1600 0.263 0.297  0.305 0.313 0.349 0.305 0.0116 5000       0
```

Note that each of the simulations yields an unbiased estimate of the true probability that they meet, but there is variability within each individual simulation (of size 100, 400, or 1600). The standard deviation is halved each time we increase the number of simulations by a factor of 4. We can display the results graphically (see Figure 10.6).

What would be a reasonable value for num_sim in this setting? The answer depends on how accurate we want to be. (And we can also simulate to see how variable our results are!) Carrying out 20,000 simulations yields relatively little variability and would likely be sufficient for a first pass. We could state that these results have *converged* sufficiently close to the true value since the sampling variability due to the simulation is negligible.

```
sim_results <- do(reps) * campus_sim(num_sim = 20000)
favstats(~ campus_sim, data = sim_results)

   min    Q1 median    Q3   max  mean      sd    n missing
 0.294 0.303  0.306 0.308 0.318 0.306 0.00327 5000       0
```

```
ggplot(data = sim_results, aes(x = campus_sim, color = factor(num_sims))) +
  geom_density(size = 2) +
  scale_x_continuous("Proportion of times that Sally and Joan meet")
```

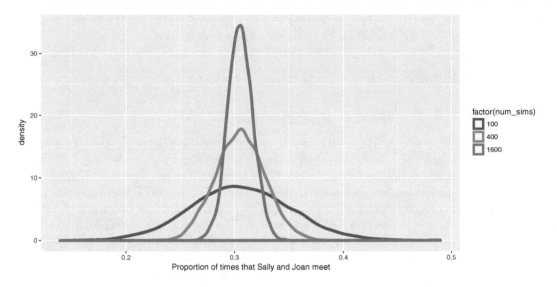

Figure 10.6: Convergence of the estimate of the proportion of times that Sally and Joan meet.

Given the inherent nature of variability due to sampling, it can be very useful to set (and save) a *seed* for the pseudo-random number generator (using the set.seed() function). This ensures that the results are the same each time the simulation is run since the simulation will use the same list of random numbers. The seed itself is arbitrary, but each seed defines a different sequence of random numbers.

```
set.seed(1974)
campus_sim()
```

```
[1] 0.308
```

```
campus_sim()
```

```
[1] 0.331
```

```
set.seed(1974)
campus_sim()
```

```
[1] 0.308
```

10.8 Further resources

This chapter has been a basic introduction to simulation. Over the last 25 years, the ability to use simulation to match observed data has become an essential component of Bayesian statistics. A central technique is called *Markov chain Monte Carlo* (MCMC). There's not enough room to give a useful introduction here, but you can expect to hear more and more about it as your career unfolds.

Rizzo [175] provides a comprehensive introduction to statistical computing in R, while [108] and [105] describe the use of R for simulation studies. The importance of simulation as part of an analyst's toolbox is enunciated in [8] and [106]. The simstudy package can be used to simplify data generation or exploration using simulation.

10.9   Exercises

Exercise 10.1

The lonely recording device: This problem demonstrates the ways that empirical simulations can complement analytic (closed-form) solutions. Consider an example where a recording device that measures remote activity is placed in a remote location. The time, T, to failure of the remote device has an exponential distribution with mean of 3 years. Since the location is so remote, the device will not be monitored during its first two years of service. As a result, the time to discovery of its failure is $X = \max(T, 2)$. The problem here is to determine the average of the time to discovery of the truncated variable (in probability parlance, the expected value of the observed variable X, E[X]).

The analytic solution is fairly straightforward, but requires calculus. We need to evaluate:

$$E[X] = \int_0^2 2 * f(u)du + \int_2^\infty u * f(u)du,$$

where $f(u) = 1/3 \exp(-1/3 * u)$ for $u > 0$.

Is calculus strictly necessary here? Conduct a simulation to estimate (or check) the value for the average time to discovery.

Exercise 10.2

More on the jobs number: In this chapter, we considered a simulation where the true jobs number remained constant over time. Modify the call to the function provided in that example so that the true situation is that there are 15,000 new jobs created every month. Set your random number seed to the value 1976. Summarize what you might conclude from these results as if you were a journalist without a background in data science.

Exercise 10.3

Simulating data from a logistic regression model: Generate $n = 5000$ observations from a logistic regression model with parameters intercept $\beta_0 = -1$, slope $\beta_1 = 0.5$, and distribution of the predictor being normal with mean 1 and standard deviation 1. Calculate and interpret the resulting parameter estimates and confidence intervals.

Exercise 10.4

The Monty Hall problem: The Monty Hall problem illustrates a simple setting where intuition is often misleading. The situation is based on the TV game show "Let's Make a Deal." First, Monty (the host) puts a prize behind one of three doors. Then the player chooses a door. Next, (without moving the prize) Monty opens an unselected door, revealing that the prize is not behind it. The player may then switch to the other nonselected door. Should the player switch?

Many people see that there are now two doors to choose between and feel that since Monty can always open a nonprize door, there is still equal probability for each door. If that were the case, the player might as well keep the original door. This intuition is so attractive that when Marilyn vos Savant asserted that the player should switch (in her *Parade* magazine column), there were reportedly 10,000 letters asserting she was wrong.

A correct intuitive route is to observe that Monty's door is fixed. The probability that the player has the right door is 1/3 before Monty opens the nonprize door, and remains 1/3 after that door is open. This means that the probability the prize is behind one of the other doors is 2/3, both before and after Monty opens the nonprize door. After Monty opens the nonprize door, the player gets a 2/3 chance of winning by switching to the remaining door. If the player wants to win, they should switch doors.

One way to prove to yourself that switching improves your chances of winning is through simulation. In fact, even deciding how to code the problem may be enough to convince yourself to switch.

In the simulation, you need to assign the prize to a door, then make an initial guess. If the guess was right, Monty can open either door. We'll switch to the other door. Rather than have Monty choose a door, we'll choose one, under the assumption that Monty opened the other one. If our initial guess was wrong, Monty will open the only remaining nonprize door, and when we switch we'll be choosing the prize door.

Exercise 10.5

Restaurant violations: Is there evidence that restaurant health inspectors in New York City also give the benefit of the doubt to those at the threshold between a B grade (14 to 27) or C grade (28 or above)?

Exercise 10.6

Equal variance assumption: What is the impact of the violation of the equal variance assumption for linear regression models? Repeatedly generate data from a "true" model given by the following code.

```
n <- 250
rmse <- 1
x1 <- rep(c(0,1), each=n/2)     # x1 resembles 0 0 0 ... 1 1 1
x2 <- runif(n, min=0, max=5)
beta0 <- -1
beta1 <- 0.5
beta2 <- 1.5
y <- beta0 + beta1*x1 + beta2*x2 + rnorm(n, mean=0, sd=rmse + x2)
```

For each simulation, fit the linear regression model and display the distribution of 1,000 estimates of the β_1 parameter (note that you need to generate the vector of outcomes each time). Does the distribution of the parameter follow a normal distribution?

Exercise 10.7

Skewed residuals: What is the impact if the residuals from a linear regression model are skewed (and not from a normal distribution)? Repeatedly generate data from a "true" model given by:

```
n <- 250
rmse <- 1
x1 <- rep(c(0,1), each=n/2)      # x1 resembles 0 0 0 ... 1 1 1
x2 <- runif(n, min=0, max=5)
beta0 <- -1
beta1 <- 0.5
beta2 <- 1.5
y <- beta0 + beta1*x1 + beta2*x2 + rexp(n, rate=1/2)
```

For each simulation, fit the linear regression model and display the distribution of 1,000 estimates of the β_1 parameter (note that you need to generate the vector of outcomes each time).

Exercise 10.8

Meeting in the campus center: Sally and Joan plan to meet to study in their college campus center. They are both impatient people who will only wait 10 minutes for the other before leaving. Rather than pick a specific time to meet, they agree to head over to the campus center sometime between 7:00 and 8:00 pm. Let both arrival times be uniformly distributed over the hour, and assume that they are independent of each other. What is the probability that they actually meet? Find the exact (analytical) solution.

Exercise 10.9

Meeting in the campus center (redux): Sally and Joan plan to meet to study in their college campus center. They are both impatient people who will only wait 10 minutes for the other before leaving. Rather than pick a specific time to meet, they agree to head over to the campus center sometime between 7:00 and 8:00 pm. Let both arrival times be normally distributed with mean 30 minutes past and a standard deviation of 10 minutes. Assume that they are independent of each other. What is the probability that they actually meet? Estimate the answer using simulation techniques introduced in this chapter, with at least 10,000 simulations.

Exercise 10.10

Consider a queueing example where customers arrive at a bank at a given minute past the hour and are served by the next available teller. Use the following data to explore wait times for a bank with one teller vs. one with two tellers, where the duration of the transaction is given below.

	arrival	duration
1	1.00	3.00
2	3.00	2.00
3	7.00	5.00
4	10.00	6.00
5	11.00	8.00
6	15.00	1.00

What is the average total time for customers in the bank with one teller? What is the average for a bank with two tellers?

Exercise 10.11

The time a manager takes to interview a job applicant has an exponential distribution with mean of half an hour, and these times are independent of each other. The applicants are scheduled at quarter-hour intervals beginning at 8:00 am, and all of the applicants arrive exactly on time (this is an excellent thing to do, by the way). When the applicant with an 8:15 am appointment arrives at the manager's office office, what is the probability that she will have to wait before seeing the manager? What is the expected time that her interview will finish?

Exercise 10.12

Tossing coins: Two people toss a fair coin 4 times each. Find the probability that they throw equal numbers of heads. Also estimate the probability that they throw equal numbers of heads using a simulation in R (with an associated 95% confidence interval for your estimate).

Part III

Topics in Data Science

Chapter 11

Interactive data graphics

As we discussed in Chapter 1, the practice of data science involves many different elements. In Part I, we laid a foundation for data science by developing a basic understanding of data wrangling, data visualization, and ethics. In Part II, we focused on building statistical models and using those models to learn from data. However, to this point we have focused mainly on traditional two-dimensional data (e.g., rows and columns) and data graphics. In this part, we tackle the heterogeneity found in many modern data: spatial, text, network, and relational data. We explore interactive data graphics that leap out of the printed page. Finally, we address the volume of data—concluding with a discussion of "big data" and the tools that you are likely to see when working with it.

In Chapter 2 we laid out a systematic framework for composing data graphics. A similar grammar of graphics employed by the `ggplot2` package provided a mechanism for creating data graphics in Chapter 3. In this chapter, we explore a few alternatives for making more complex—and in particular, dynamic—data graphics.

11.1 Rich Web content using `D3.js` and `htmlwidgets`

As Web browsers became more complex in the mid-2000s, the desire to have interactive data visualizations in the browser grew. Thus far, all of the data visualization techniques that we have discussed are based on static images. However, newer tools have made it considerably easier to create interactive data graphics.

JavaScript is a programming language that allows Web developers to create client-side Web applications. This means that computations are happening *in the client's browser*, as opposed to taking place on the host's Web servers. JavaScript applications can be more responsive to client interaction than dynamically-served Web pages that rely on a server-side scripting language, like PHP or Ruby.

The current state of the art for client-side dynamic data graphics on the Web is a JavaScript library called `D3.js`, or just `D3`, which stands for "data-driven documents." One of the lead developers of `D3` is Mike Bostock, formerly of *The New York Times* and Stanford University.

More recently, Ramnath Vaidyanathan and the developers at `RStudio` have created the `htmlwidgets` package, which provides a bridge between R and `D3`. Specifically, the `htmlwidgets` framework allows R developers to create packages that render data graphics in HTML using `D3`. Thus, R programmers can now make use of `D3` without having to learn JavaScript. Furthermore, since R Markdown documents also render as HTML, R users can easily create interactive data graphics embedded in annotated Web documents. This is a highly active area of development. In what follows we illustrate a few of the more obviously useful `htmlwidgets` packages.

11.1.1 Leaflet

Perhaps the `htmlwidget` that is getting the greatest attention is `leaflet`. This package enables dynamic geospatial maps to be drawn using the `Leaflet` JavaScript library and the OpenStreetMaps API. The use of this package requires knowledge of spatial data, and thus we postpone our illustration of its use until Chapter 14.

11.1.2 Plot.ly

Plot.ly specializes in online dynamic data visualizations, and in particular, the ability to translate code to generate data graphics between R, Python, and other data software tools. This project is based on the `plotly.js` JavaScript library, which is available under an open-source license. The functionality of Plot.ly can be accessed in R through the `plotly` package.

What makes `plotly` especially attractive is that it can convert any `ggplot2` object into a `plotly` object using the `ggplotly()` function. This enables immediate interactivity for existing data graphics. Features like *brushing* (where selected points are marked) and *mouse-over* annotations (where points display additional information when the mouse hovers over them) are automatic. For example, in Figure 11.1 we display a static plot of the frequency of the names of births in the United States of the four members of the Beatles over time (using data from the `babynames` package).

```
library(mdsr)
library(babynames)
Beatles <- babynames %>%
  filter(name %in% c("John", "Paul", "George", "Ringo") & sex == "M")
beatles_plot <- ggplot(data = Beatles, aes(x = year, y = n)) +
  geom_line(aes(color = name), size = 2)
beatles_plot
```

After running the `ggplotly()` function on that object, a plot is displayed in RStudio or in a Web browser. The exact values can be displayed by mousing-over the lines. In addition, brushing, panning, and zooming are supported. In Figure 11.2, we show a still from that dynamic image.

```
library(plotly)
ggplotly(beatles_plot)
```

11.1.3 DataTables

The datatables (`DT`) package provides a quick way to make data tables interactive. Simply put, it enables tables to be searchable, sortable, and pageable automatically. Figure 11.3 displays a screenshot of the first rows of the `Beatles` table as rendered by `DT`. Note the search box and clickable sorting arrows.

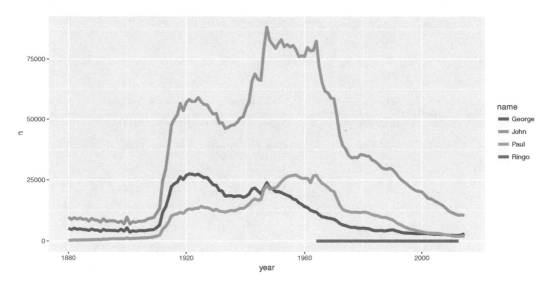

Figure 11.1: ggplot2 depiction of the frequency of Beatles names over time.

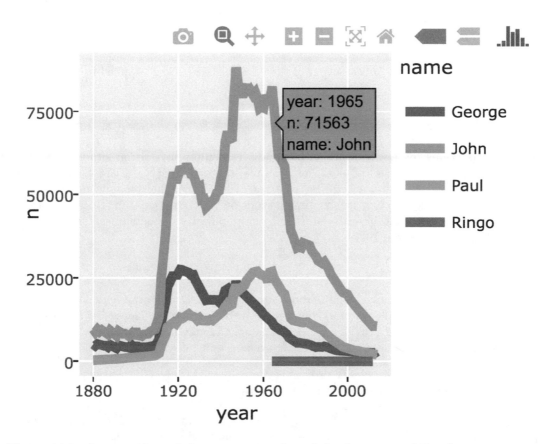

Figure 11.2: A screenshot of the interactive plot of the frequency of Beatles names over time.

```
library(DT)
datatable(Beatles, options = list(pageLength = 25))
```

Show 10 ▼ entries					Sea
	year ⇕	sex ⇕	name ⇕	n ⇕	
1	1880	M	John	9655	
2	1880	M	George	5126	
3	1880	M	Paul	301	
4	1881	M	John	8769	
5	1881	M	George	4664	
6	1881	M	Paul	291	
7	1882	M	John	9557	
8	1882	M	George	5193	
9	1882	M	Paul	397	
10	1883	M	John	8894	

Showing 1 to 10 of 430 entries   Previous   **1**   2   3   4

Figure 11.3: A screenshot of the output of the `DataTables` package applied to the `Beatles` names.

11.1.4   dygraphs

The `dygraphs` package generates interactive time series plots with the ability to brush over time intervals and zoom in and out. For example, the popularity of Beatles names could be made dynamic with just a little bit of extra code. Here, the dynamic range selector allows for the easy selection of specific time periods on which to focus. In the live version of Figure 11.4, one can zoom in on the uptick in the popularity of the names `John` and `Paul` during the first half of the 1960s.

11.1.5   streamgraphs

A *streamgraph* is a particular type of time series plot that uses area as a visual cue for quantity. Streamgraphs allow you to compare the values of several time series at once. The streamgraphs `htmlwidget` provides access to the `streamgraphs.js` D3 library. Figure 11.5 displays our `Beatles` names time series as a streamgraph.

11.2   Dynamic visualization using ggvis

The `ggvis` package provides a different set of tools to create interactive graphics for exploratory data analysis. `ggvis` uses the Vega JavaScript library, which is a visualization grammar that is *not* built on D3 or the `htmlwidgets` frameworks. In this example we demonstrate how to create a visualization of the proportion of male names that are `John` as a function of the number of names over time, where the user can mouse-over a value to see

```
library(dygraphs)
Beatles %>%
  filter(sex == "M") %>%
  select(year, name, prop) %>%
  tidyr::spread(key = name, value = prop) %>%
  dygraph(main = "Popularity of Beatles names over time") %>%
  dyRangeSelector(dateWindow = c("1940", "1980"))
```

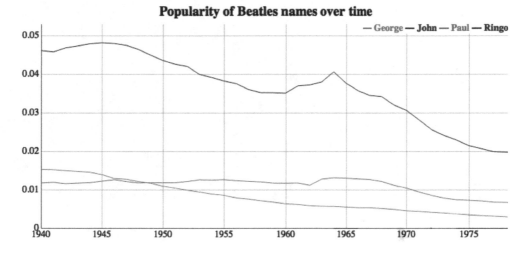

Figure 11.4: A screenshot of the dygraphs display of the popularity of Beatles names over time. In this screenshot, the years range from 1940 to 1980, but in the live version, one can expand or contract that timespan.

the year, number, and proportion. This is an alternative representation of the time series that allows the analyst to see changes in the population size over time along with name preferences.

We need a helper function to display the appropriate values as a mouse-over: This function (which we have called all_values()) is passed as an argument to the chain of commands used to display points and set up the hovering (see Figure 11.6). All columns of the selected rows are displayed. Many other capabilities are made available by modifying the function.

```
John <- filter(Beatles, name=="John")
all_values <- function(x) {
  if (is.null(x)) return(NULL)
  row <- John[John$year == x$year, ]
  paste0(names(row), ": ", format(row), collapse = "<br />")
}
```

11.3 Interactive Web apps with Shiny

Shiny is a framework for R that can be used to create interactive Web applications. It is particularly attractive because it provides a high-level structure to easily prototype and

```
# devtools::install_github("hrbrmstr/streamgraph")
library(streamgraph)
Beatles %>% streamgraph(key = "name", value = "n", date = "year") %>%
  sg_fill_brewer("Accent")
```

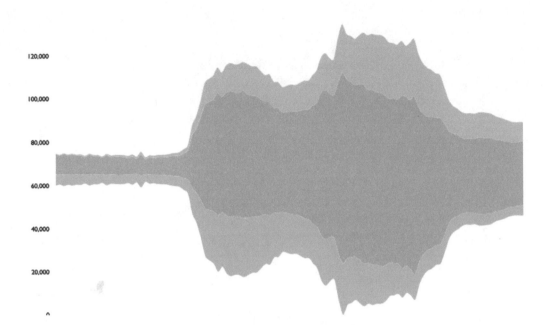

Figure 11.5: A screenshot of the streamgraph display of Beatles names over time.

deploy apps. While a full discussion of Shiny is outside the scope of the book, we will demonstrate how one might create a dynamic Web app that allows the user to explore the data set of babies with the same names as the Beatles.

One way to write a Shiny app involves creating a ui.R file that controls the user interface, and a server.R file to display the results. These files communicate with each other using *reactive objects* input and output. Reactive expressions are special constructions that use input from *widgets* to return a value. These allow the application to automatically update when the user clicks on a button, changes a slider, or provides other input. For this example we'd like to let the user pick the start and end years along with a set of checkboxes to include their favorite Beatles.

The ui.R file sets up a title, creates inputs for the start and end years (with default values), creates a set of check boxes for each of the Beatles' names, then plots the result.

```
# ui.R
beatles_names <- c("John", "Paul", "George", "Ringo")

shinyUI(bootstrapPage(
  h3("Frequency of Beatles names over time"),
  numericInput("startyear", "Enter starting year",
               value = 1960, min = 1880, max = 2014, step = 1),
  numericInput("endyear", "Enter ending year",
               value = 1970, min = 1881, max = 2014, step = 1),
```

```
library(ggvis)
John %>%
  ggvis(~n, ~prop, fill = ~year) %>%
  layer_points() %>%
  add_tooltip(all_values, "hover")
```

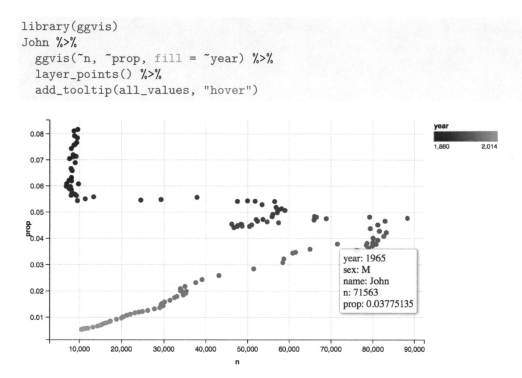

Figure 11.6: A screenshot of the ggvis display of the proportion and number of male babies named "John" over time.

```
  checkboxGroupInput('names', 'Names to display:',
                     sort(unique(beatles_names)),
                     selected = c("George", "Paul")),
  plotOutput("plot")
))
```

The server.R file loads needed packages, performs some data wrangling, extracts the reactive objects using the input object, then generates the desired plot. The renderPlot() function returns a reactive object called plot that is referenced in ui.R. Within this function, the values for the years and Beatles are used within a call to filter() to identify what to plot.

```
# server.R
library(mdsr)
library(babynames)
library(shiny)
Beatles <- babynames %>%
  filter(name %in% c("John", "Paul", "George", "Ringo") & sex == "M")

shinyServer(function(input, output) {
  output$plot <- renderPlot({
    ds <- Beatles %>%
      filter(year >= input$startyear, year <= input$endyear,
             name %in% input$names)
```

```
    ggplot(data = ds, aes(x = year, y = prop, color = name)) +
       geom_line(size = 2)
  })
})
```

Shiny Apps can be run locally within RStudio, or deployed on a Shiny App server (such as http://shinyapps.io). Figure 11.7 displays the results when only Paul and George are checked when run locally.

```
library(shiny)
runApp('.')
```

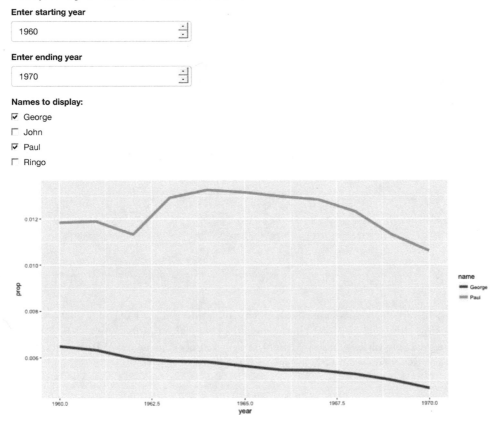

Figure 11.7: A screenshot of the Shiny app displaying babies with Beatles names.

11.4   Further customization

There are endless possibilities for customizing plots in R. One important concept is the notion of *themes*. In the next section, we will illustrate how to customize a ggplot2 theme by defining the one we are using in this book.

ggplot2 provides many different ways to change the appearance of a plot. A comprehensive system of customizations is called a *theme*. In ggplot2, a theme is a list of 57 different attributes that define how axis labels, titles, grid lines, etc. are drawn. The default theme is theme_grey().

```
length(theme_grey())
```

```
[1] 57
```

For example, the most notable features of theme_grey() are the distinctive grey background and white grid lines. The panel.background and panel.grid.major properties control these aspects of the theme.

```
theme_grey()["panel.background"]
```

```
$panel.background
List of 5
 $ fill         : chr "grey92"
 $ colour       : logi NA
 $ size         : NULL
 $ linetype     : NULL
 $ inherit.blank: logi TRUE
 - attr(*, "class")= chr [1:2] "element_rect" "element"
```

```
theme_grey()["panel.grid.major"]
```

```
$panel.grid.major
List of 6
 $ colour       : chr "white"
 $ size         : NULL
 $ linetype     : NULL
 $ lineend      : NULL
 $ arrow        : logi FALSE
 $ inherit.blank: logi TRUE
 - attr(*, "class")= chr [1:2] "element_line" "element"
```

A number of useful themes are built into ggplot2, including theme_bw() for a more traditional white background, theme_minimal(), and theme_classic(). These can be invoked using the eponymous functions. We compare theme_grey() with theme_bw() in Figure 11.8.

```
beatles_plot
beatles_plot + theme_bw()
```

We can modify a theme on-the-fly using the theme() function. In Figure 11.9 we illustrate how to change the background color and major grid lines color.

```
beatles_plot + theme(panel.background = element_rect(fill = "cornsilk"),
                     panel.grid.major = element_line(color = "dodgerblue"))
```

How did we know the names of those colors? You can display R's built-in colors using the colors() function. There are more intuitive color maps on the Web.

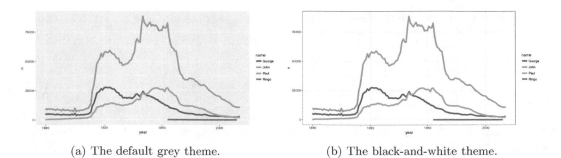

(a) The default grey theme.   (b) The black-and-white theme.

Figure 11.8: Comparison of two `ggplot2` themes.

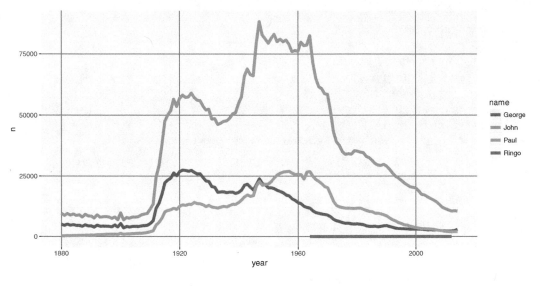

Figure 11.9: Beatles plot with custom `ggplot2` theme.

```
head(colors())
```

```
[1] "white"          "aliceblue"      "antiquewhite"   "antiquewhite1"
[5] "antiquewhite2"  "antiquewhite3"
```

To create a new theme, write a function that will return a complete `ggplot2` theme. One could write this function by completely specifying all 57 items. However, in this case we illustrate how the `%+replace%` operator can be used to modify an existing theme. We start with `theme_grey()` and change the background color, major and minor grid lines colors, and the default font.

```
theme_mdsr <- function(base_size = 12, base_family = "Bookman") {
    theme_grey(base_size = base_size, base_family = base_family) %+replace%
      theme(
          axis.text          = element_text(size = rel(0.8)),
          axis.ticks         = element_line(colour = "black"),
          legend.key         = element_rect(colour = "grey80"),
          panel.background   = element_rect(fill = "whitesmoke", colour = NA),
```

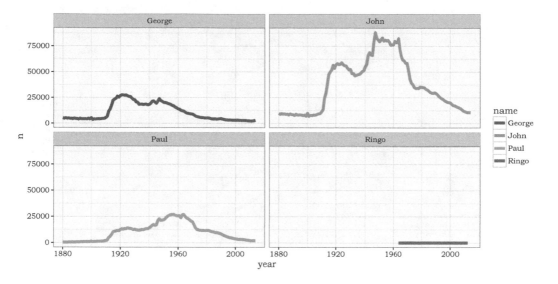

Figure 11.10: Beatles plot with customized `mdsr` theme.

```
panel.border        = element_rect(fill = NA, colour = "grey50"),
panel.grid.major    = element_line(colour = "grey80", size = 0.2),
panel.grid.minor    = element_line(colour = "grey92", size = 0.5),
strip.background    = element_rect(fill = "grey80", colour = "grey50",
    size = 0.2)
  )
}
```

With our new theme defined, we can apply it in the same way as any of the built-in themes—namely, by calling the `theme_mdsr()` function. Figure 11.10 shows how this stylizes the faceted Beatles time series plot.

```
beatles_plot + facet_wrap(~name) + theme_mdsr()
```

Many people have taken to creating their own themes for ggplot2. In particular, the ggthemes package features useful (`theme_solarized()`), humorous (`theme_tufte()`), whimsical (`theme_fivethirtyeight()`), and even derisive (`theme_excel()`) themes. Another humorous theme is `theme_xkcd()`, which attempts to mimic the popular Web comic's distinctive hand-drawn styling. This functionality is provided by the xkcd package.

```
library(xkcd)
```

To set xkcd up, we need to download the pseudo-handwritten font, import it, and then `loadfonts()`. Note that the destination for the fonts is system dependent: On Mac OS X this should be `~/Library/Fonts` instead of `~/.fonts`.

```
download.file("http://simonsoftware.se/other/xkcd.ttf",
            dest = "~/.fonts/xkcd.ttf", mode = "wb")
font_import(pattern = "[X/x]kcd", prompt = FALSE)
loadfonts()
```

Figure 11.11: Prevalence of Beatles names drawn in the style of an xkcd Web comic.

In Figure 11.11, we show the xkcd-styled plot of the popularity of the Beatles names.

```
beatles_plot + theme_xkcd()
```

11.5   Extended example: Hot dog eating

Writing in 2011, former *New York Times* data graphic intern Nathan Yau noted that "Adobe Illustrator is the industry standard. Every graphic that goes to print at *The New York Times* either was created or edited in Illustrator" [242]. To underscore his point, Yau presents the data graphic shown in Figure 11.12, created in R but modified in Illustrator.

Five years later, the *New York Times* data graphic department now produces much of their content using D3.js, an interactive JavaScript library that we discuss in Section 11.1. Nevertheless, what follows is our best attempt to recreate Figure 11.12 entirely within R using ggplot2 graphics. After saving the plot as a PDF, we can open it in Illustrator or Inkscape for further customization if necessary.

Pro Tip: Undertaking such "Copy the Master" exercises [147] is a good way to deepen your skills.

```
library(mdsr)
hd <- readr::read_csv(
  "http://datasets.flowingdata.com/hot-dog-contest-winners.csv")
names(hd) <- gsub(" ", "_", names(hd)) %>% tolower()
glimpse(hd)

Observations: 31
Variables: 5
$ year        <int> 1980, 1981, 1982, 1983, 1984, 1985, 1986, 1987, 198...
```

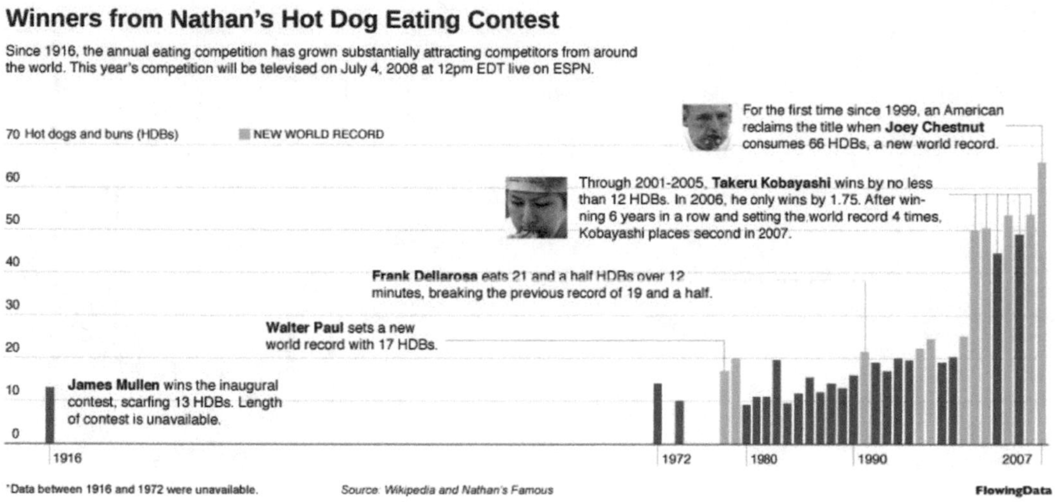

Winners from Nathan's Hot Dog Eating Contest

Since 1916, the annual eating competition has grown substantially attracting competitors from around the world. This year's competition will be televised on July 4, 2008 at 12pm EDT live on ESPN.

Figure 11.12: Nathan Yau's Hot Dog Eating data graphic (reprinted with permission from `flowingdata.com`).

```
$ winner      <chr> "Paul Siederman & Joe Baldini", "Thomas DeBerry", "...
$ dogs_eaten  <dbl> 9.1, 11.0, 11.0, 19.5, 9.5, 11.8, 15.5, 12.0, 14.0,...
$ country     <chr> "United States", "United States", "United States", ...
$ new_record  <int> 0, 0, 0, 0, 0, 0, 0, 0, 0, 0, 0, 1, 0, 0, 0, 0, 1, ...
```

The hd data table doesn't provide any data from before 1980, so we need to estimate them from Figure 11.12 and manually add these rows to our data frame.

```
new_data <- data.frame(
  year = c(1979, 1978, 1974, 1972, 1916),
  winner = c(NA, "Walter Paul", NA, NA, "James Mullen"),
  dogs_eaten = c(19.5, 17, 10, 14, 13),
  country = rep(NA, 5), new_record = c(1,1,0,0,0)
)
hd <- bind_rows(hd, new_data)
glimpse(hd)

Observations: 36
Variables: 5
$ year        <dbl> 1980, 1981, 1982, 1983, 1984, 1985, 1986, 1987, 198...
$ winner      <chr> "Paul Siederman & Joe Baldini", "Thomas DeBerry", "...
$ dogs_eaten  <dbl> 9.1, 11.0, 11.0, 19.5, 9.5, 11.8, 15.5, 12.0, 14.0,...
$ country     <chr> "United States", "United States", "United States", ...
$ new_record  <dbl> 0, 0, 0, 0, 0, 0, 0, 0, 0, 0, 0, 1, 0, 0, 0, 0, 1, ...
```

Note that we only want to draw some of the years on the horizontal axis and only every 10th value on the vertical axis.

```
xlabs <- c(1916, 1972, 1980, 1990, 2007)
ylabs <- seq(from = 0, to = 70, by = 10)
```

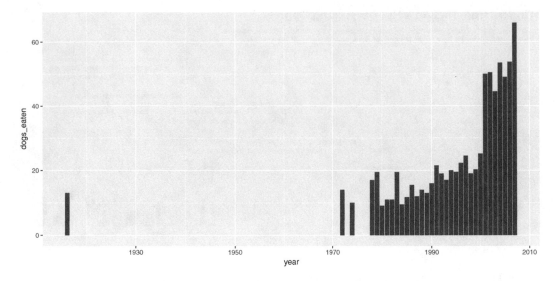

Figure 11.13: A simple bar graph of hot dog eating.

Finally, the plot only shows the data up until 2008, even though the file contains more recent information than that. Let's define a subset that we'll use for plotting.

```
hd_plot <- hd %>% filter(year < 2008)
```

Our most basic plot is shown in Figure 11.13.

```
p <- ggplot(data = hd_plot, aes(x = year, y = dogs_eaten)) +
  geom_bar(stat = "identity")
p
```

This doesn't provide the context of Figure 11.12, nor the pizzazz. Although most of the important data are already there, we still have a great deal of work to do to make this data graphic as engaging as Figure 11.12. Our recreation is shown in Figure 11.14.

We aren't actually going to draw the y-axis—instead we are going to places the labels for the y values on the plot. We'll put the locations for those values in a data frame.

```
ticks_y <- data.frame(x = 1912, y = ylabs)
```

There are many text annotations, and we will collect those into a single data frame.

```
text <- bind_rows(
  # Frank Dellarosa
  data.frame(x = 1951.5, y = 37,
    label = paste("Frank Dellarosa eats 21 and a half HDBs over 12\n",
      "minutes, breaking the previous record of 19 and a half."), adj = 0),
  # Joey Chestnut
  data.frame(x = 1976.5, y = 69,
    label = paste("For the first time since 1999, an American\n",
      "reclaims the title when Joey Chestnut\n",
      "consumes 66 HDBs, a new world record."), adj = 0),
```

```
# Kobayashi
data.frame(x = 1960.5, y = 55,
  label = paste("Through 2001-2005, Takeru Kobayashi wins by no less\n",
    "than 12 HDBs. In 2006, he only wins by 1.75. After win-\n",
    "ning 6 years in a row and setting the world record 4 times,\n",
    "Kobayashi places second in 2007."), adj = 0),
# Walter Paul
data.frame(x = 1938, y = 26, label = "Walter Paul sets a new
world record with 17 HDBs.", adj = 0),
# James Mullen
data.frame(x = 1917, y = 10, label = "James Mullen wins the inaugural
contest, scarfing 13 HDBs. Length
of contest unavailable.", adj = 0),
data.frame(x = 1935, y = 72, label = "NEW WORLD RECORD"),
data.frame(x = 1914, y = 72, label = "Hot dogs and buns (HDBs)"),
data.frame(x = 1940, y = 2,
  label = "*Data between 1916 and 1972 were unavailable"),
data.frame(x = 1922, y = 2, label = "Source: FlowingData")
)
```

The grey segments that connect the text labels to the bars in the plot must be manually specified in another data frame.

```
segments <- bind_rows(
  data.frame(x = c(1984, 1991, 1991, NA), y = c(37, 37, 21, NA)),
  data.frame(x = c(2001, 2007, 2007, NA), y = c(69, 69, 66, NA)),
  data.frame(x = c(2001, 2007, 2007, NA), y = c(69, 69, 66, NA)),
  data.frame(x = c(1995, 2006, 2006, NA), y = c(58, 58, 53.75, NA)),
  data.frame(x = c(2005, 2005, NA), y = c(58, 49, NA)),
  data.frame(x = c(2004, 2004, NA), y = c(58, 53.5, NA)),
  data.frame(x = c(2003, 2003, NA), y = c(58, 44.5, NA)),
  data.frame(x = c(2002, 2002, NA), y = c(58, 50.5, NA)),
  data.frame(x = c(2001, 2001, NA), y = c(58, 50, NA)),
  data.frame(x = c(1955, 1978, 1978), y = c(26, 26, 17)))
```

Finally, we draw the plot, layering on each of the elements that we defined above.

```
p + geom_bar(stat = "identity", aes(fill = factor(new_record))) +
  geom_hline(yintercept = 0, color = "darkgray") +
  scale_fill_manual(name = NULL,
    values = c("0" = "#006f3c", "1" = "#81c450")) +
  scale_x_continuous(name = NULL, breaks = xlabs, minor_breaks = NULL,
    limits = c(1912, 2008), expand = c(0, 1)) +
  scale_y_continuous(name = NULL, breaks = ylabs, labels = NULL,
    minor_breaks = NULL, expand = c(0.01, 1)) +
  geom_text(data = ticks_y, aes(x = x, y = y + 2, label = y), size = 3) +
  ggtitle("Winners from Nathan's hot dog eating contest") +
  geom_text(data = text, aes(x = x, y = y, label = label),
    hjust = "left", size = 3) +
  geom_path(data = segments, aes(x = x, y = y), col = "darkgray") +
  # Key
```

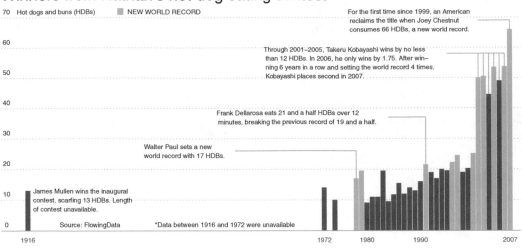

Figure 11.14: Recreating the hot dog graphic in R.

```
geom_rect(xmin = 1933, ymin = 70.75, xmax = 1934.3, ymax = 73.25,
    fill = "#81c450", color = "white") +
guides(fill = FALSE) +
theme(panel.background = element_rect(fill = "white"),
    panel.grid.major.y = element_line(color = "gray", linetype = "dotted"),
    plot.title = element_text(size = rel(2)),
    axis.ticks.length = unit(0, "cm"))
```

11.6 Further resources

The http://www.htmlwidgets.org website includes a gallery of showcased applications of JavaScript in R. Details and examples of use of the ggvis package can be found at http://ggvis.rstudio.com. The Shiny gallery (http://shiny.rstudio.com/gallery) includes a number of interactive visualizations (and associated code), many of which feature JavaScript libraries. The RStudio Shiny cheat sheet is a useful reference.

The extrafonts package makes use of the full suite of fonts that are installed on your computer, rather than the relatively small sets of fonts that R knows about. (These are often device and operating system dependent, but three fonts—sans, serif, and mono—are always available.) For a more extensive tutorial on how to use the extrafonts package, see http://tinyurl.com/fonts-rcharts.

11.7 Exercises

Exercise 11.1

The macleish package contains weather data collected every 10 minutes in 2015 from two weather stations in Whately, Massachusetts.

```
library(macleish)
head(whately_2015)

# A tibble: 6  8
                when temperature wind_speed wind_dir rel_humidity
              <dttm>       <dbl>      <dbl>    <dbl>        <dbl>
1 2015-01-01 00:00:00       -9.32       1.40      225         54.5
2 2015-01-01 00:10:00       -9.46       1.51      248         55.4
3 2015-01-01 00:20:00       -9.44       1.62      258         56.2
4 2015-01-01 00:30:00       -9.30       1.14      244         56.4
5 2015-01-01 00:40:00       -9.32       1.22      238         56.9
6 2015-01-01 00:50:00       -9.34       1.09      242         57.2
# ... with 3 more variables: pressure <int>, solar_radiation <dbl>,
#   rainfall <int>
```

Using `ggpplot2`, create a data graphic that displays the average temperature over each 10-minute interal (`temperature`) as a function of time (`when`). Create annotations to include context about the four seasons: the date of the vernal and autumnal equinoxes, and the summer and winter solstices.

Exercise 11.2

Repeat the previous question, but include context on major storms listed on the Wikipedia pages: 2014–2015 North American Winter and 2015-2016 North American Winter.

Exercise 11.3

Create the time series plot of the weather data in the first exercise using `plotly`.

Exercise 11.4

Create the time series plot of the weather data in the first exercise using `dygraphs`.

```
library(dygraphs)
```

Exercise 11.5

Create the time series plot of the weather data in the first exercise using `ggvis`.

```
library(ggvis)
```

Exercise 11.6

Create a Shiny app to display an interactive time series plot of the `macleish` weather data. Include a selection box to alternate between data from the `whately_2015` and `orchard_2015` weather stations.

Exercise 11.7

Using data from the `fec` package, create a Shiny app similar to the one at `https://beta.fec.gov/data/candidates/president/`.

Exercise 11.8

Using data from the Lahman package, create a Shiny app that displays career leaderboards similar to the one at http://www.baseball-reference.com/leaders/HR_season. shtml. Allow the user to select a statistic of their choice, and to choose between Career, Active, Progressive, and Yearly League leaderboards.

Exercise 11.9

The following code generates a scatterplot with marginal histograms.

```
library(ggplot2)
library(ggExtra)
p <- ggplot(HELPrct, aes(x = age, y = cesd)) + geom_point() +
  theme_classic() + stat_smooth(method = "loess", formula = y ~ x, size = 2)
ggExtra::ggMarginal(p, type = "histogram", binwidth = 3)
```

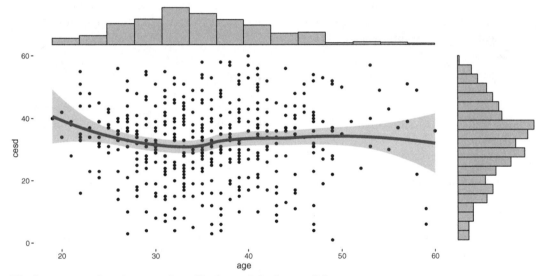

Find an example where such a display might be useful.

Exercise 11.10

Create your own ggplot2 theme. Use the material from Chapter 2 to justify the design choices you made.

Chapter 12

Database querying using SQL

Thus far, most of the data that we have encountered in this book (such as the Lahman baseball data in Chapter 4) has been small—meaning that it will fit easily in a personal computer's memory. In this chapter, we will explore approaches for working with data sets that are larger—let's call them *medium* data. These data will fit on a personal computer's hard disk, but not necessarily in its memory. Thankfully, a venerable solution for retrieving medium data from a database has been around since the 1970s: SQL (structured query language). Database management systems implementing SQL provide a ubiquitous architecture for storing and querying data that is relational in nature. While the death of SQL has been presaged many times, it continues to provide an effective solution for medium data. Its wide deployment makes it a "must-know" tool for data scientists. For those of you with bigger appetites, we will consider some extensions that move us closer to a true "Big-Data" setting in Chapter 17.

12.1 From `dplyr` to SQL

Recall the `airlines` data that we encountered in Chapter 7. Using the `dplyr` verbs that we developed in Chapter 4, consider retrieving the top on-time carriers with at least 100 flights arriving at JFK in September 1996. If the data are stored in data frames called `flights` and `carriers`, then we might write a `dplyr` pipeline like this:

```
q <- flights %>%
  filter(year == 1996 & month == 9) %>%
  filter(dest == "JFK") %>%
  inner_join(carriers, by = c("carrier" = "carrier")) %>%
  group_by(name) %>%
  summarize(N = n(),
            pct_ontime = sum(arr_delay <= 15) / n()) %>%
  filter(N >= 100) %>%
  arrange(desc(pct_ontime))
head(q, 4)

Source:   query [?? x 3]
Database: mysql 5.5.47-0ubuntu0.14.04.1 [r-user@localhost:/airlines]

               name     N pct_ontime
              <chr> <dbl>      <dbl>
```

```
1 Trans World Airways LLC   1332      0.783
2    United Air Lines Inc.   449      0.764
3     Delta Air Lines Inc.   806      0.758
4  American Airlines Inc.    833      0.688
```

However, the `flights` data frame can become very large. Going back to 1987, there are more than 169 million individual flights—each comprising a different row in this table. These data occupy nearly 20 gigabytes as CSVs, and thus are problematic to store in a personal computer's memory. Instead, we write these data to disk, and use a querying language to access only those rows that interest us. In this case, we configured `dplyr` to access the `flights` data on a MySQL server. The `src_scidb()` function from the `mdsr` package provides a connection to the `airlines` database that lives on a remote MySQL server and stores it as the object `db`. The `tbl()` function from `dplyr` maps the `flights` (carriers) table in that `airlines` database to an object in R, in this case also called `flights` (carriers).

```
db <- src_scidb("airlines")
flights <- tbl(db, "flights")
carriers <- tbl(db, "carriers")
```

Note that while we can use the `flights` and `carriers` objects *as if* they were data frames, they are not, in fact, `data.frames`. Rather, they have class `tbl_mysql`, and more generally, `tbl`. A `tbl` is a special kind of object created by `dplyr` that behaves similarly to a `data.frame`.

```
class(flights)
```

```
[1] "tbl_mysql" "tbl_sql"    "tbl_lazy"   "tbl"
```

Note also that in the output of our pipeline above, there is an explicit mention of a MySQL database. We set up this database ahead of time (see Chapter 13 for instructions on doing this), but `dplyr` allows us to interact with these `tbls` as if they were `data.frames` in our R session. This is a powerful and convenient illusion!

What is actually happening is that `dplyr` translates our pipeline into SQL. We can see the translation by passing the pipeline through the `show_query()` function.

```
show_query(q)
```

```
<SQL>
SELECT *
FROM (SELECT *
FROM (SELECT 'name', count(*) AS 'N', SUM('arr_delay' <= 15.0) / count(*) AS
'pct_ontime'
FROM (SELECT * FROM (SELECT *
FROM (SELECT *
FROM 'flights'
WHERE ('year' = 1996.0 AND 'month' = 9.0)) 'npewebtdhn'
WHERE ('dest' = 'JFK')) 'ybazwpwszb'
    INNER JOIN
    'carriers'
    USING ('carrier')) 'uiveflwkzu'
GROUP BY 'name') 'eodsdkrnpg'
```

```
WHERE ('N' >= 100.0)) 'uwsdequxik'
ORDER BY 'pct_ontime' DESC
```

Understanding this output is not important—the translator here is creating temporary tables with unintelligible names—but it should convince you that even though we wrote our pipeline in R, it was translated to SQL. dplyr will do this automatically any time you are working with objects of class tbl_sql. If we were to write an SQL query equivalent to our pipeline, we would write it in a more readable format:

```
SELECT
  c.name,
  sum(1) as N,
  sum(arr_delay <= 15) / sum(1) as pct_ontime
FROM flights f
JOIN carriers c ON f.carrier = c.carrier
WHERE year = 1996 AND month = 9
  AND dest = 'JFK'
GROUP BY name
HAVING N >= 100
ORDER BY pct_ontime desc
LIMIT 0,4;
```

How did dplyr perform this translation?[1] As we learn SQL, the parallels will become clear (e.g., the dplyr verb filter() corresponds to the SQL WHERE clause). But what about the formulas we put in our summarize() command? Notice that the R command n()() was converted into count(*) in SQL. This is not magic either: the translate_sql() function provides translation between R commands and SQL commands. For example, it will translate basic mathematical expressions.

```
translate_sql(mean(arr_delay))
```

```
<SQL> avg("arr_delay") OVER ()
```

However, it only recognizes a small set of the most common operations—it cannot magically translate any R function into SQL. So for example, the very common R function paste0(), which concatenates strings, is not translated.

```
translate_sql(paste0("this", "is", "a", "string"))
```

```
<SQL> PASTE0('this', 'is', 'a', 'string')
```

This is a good thing—since it allows you to pass arbitrary SQL code through. But you have to know what you are doing. Since there is no SQL function called paste0(), this will throw an error, even though it is a perfectly valid R expression.

```
carriers %>%
  mutate(name_code = paste0(name, "(", carrier, ")"))
```

```
Source:   query [?? x 3]
Database: mysql 5.5.47-0ubuntu0.14.04.1 [r-user@localhost:/airlines]
```

---

[1] The difference between the SQL query that we wrote and the translated SQL query that dplyr generated from our pipeline is a consequence of the syntactic logic of dplyr and needn't concern us.

```
Error in .local(conn, statement, ...):
could not run statement:  FUNCTION airlines.PASTE0 does not exist

class(carriers)

[1] "tbl_mysql" "tbl_sql"   "tbl_lazy"  "tbl"
```

Because carriers is a tbl_sql and not a data.frame, the MySQL server is actually doing the computations here. The dplyr pipeline is simply translated into SQL and submitted to the server. To make this work, we need to replace paste0() with its MySQL equivalent command, which is CONCAT.

```
carriers %>%
  mutate(name_code = CONCAT(name, "(", carrier, ")"))

Source:   query [?? x 3]
Database: mysql 5.5.47-0ubuntu0.14.04.1 [r-user@localhost:/airlines]

      carrier                                       name
       <chr>                                       <chr>
1       02Q                                Titan Airways
2       04Q                           Tradewind Aviation
3       05Q                          Comlux Aviation, AG
4       06Q                Master Top Linhas Aereas Ltd.
5       07Q                          Flair Airlines Ltd.
6       09Q                              Swift Air, LLC
7       0BQ                                          DCA
8       0CQ                            ACM AIR CHARTER GmbH
9       0GQ Inter Island Airways, d/b/a Inter Island Air
10      0HQ       Polar Airlines de Mexico d/b/a Nova Air
# ... with more rows, and 1 more variables: name_code <chr>
```

The syntax of this looks a bit strange, since CONCAT is not a valid R expression—but it works.

Another alternative is to pull the carriers data into R using the collect() function first, and then use paste0() as before.[2] The collect() function breaks the connection to the MySQL server and returns a data.frame (which is also a tbl_df).

```
carriers %>%
  collect() %>%
  mutate(name_code = paste0(name, "(", carrier, ")"))

# A tibble: 1,610   3
    carrier                                 name
     <chr>                                 <chr>
1     02Q                          Titan Airways
2     04Q                     Tradewind Aviation
3     05Q                    Comlux Aviation, AG
4     06Q          Master Top Linhas Aereas Ltd.
```

---

[2]Of course, this will work well when the carriers table is not too large, but could become problematic if it is.

```
5      07Q                          Flair Airlines Ltd.
6      09Q                            Swift Air, LLC
7      0BQ                                      DCA
8      0CQ                     ACM AIR CHARTER GmbH
9      0GQ Inter Island Airways, d/b/a Inter Island Air
10     0HQ       Polar Airlines de Mexico d/b/a Nova Air
# ... with 1,600 more rows, and 1 more variables: name_code <chr>
```

This example illustrates that when using dplyr with a src_sql backend, one must be careful to use expressions that SQL can understand. This is just one more reason why it is important to know SQL on its own, and not rely entirely on the dplyr front-end (as wonderful as it is).

For querying a database, the choice of whether to use dplyr or SQL is largely a question of convenience. If you want to work with the result of your query in R, then use dplyr. If, on the other hand, you are pulling data into a Web application, you likely have no alternative other than writing the SQL query yourself. dplyr is just one SQL client that only works in R, but there are SQL servers all over the world, in countless environments. Furthermore, as we will see in Chapter 17, even the big data tools that supersede SQL assume prior knowledge of SQL. Thus, in this chapter we will learn how to write SQL queries.

12.2   Flat-file databases

It may be the case that all of the data that you have encountered thus far has been in a proprietary format (e.g., R, Minitab, SPSS, Stata) or has taken the form of a single CSV (comma-separated value) file. This file consists of nothing more than rows and columns of data, usually with a header row providing names for each of the columns. Such a file is known as known as a *flat file*, since it consists of just one flat (e.g., two-dimensional) file. A *spreadsheet* application—like Excel or Google Spreadsheets—allows a user to open a flat file, edit it, and also provides a slew of features for generating additional columns, formatting cells, etc. In R, the read_csv command from the readr package converts a flat file database into a data.frame.

These flat-file databases are both extremely common and extremely useful, so why do we need anything else? One set of limitations comes from computer hardware. A personal computer has two main options for storing data:

- Memory (RAM): the amount of data that a computer can work on at once. Modern computers typically have a few gigabytes of memory. A computer can access data in memory extremely quickly (tens of GBs per second).

- Hard Disk: the amount of data that a computer can store permanently. Modern computers typically have hundreds or even thousands of gigabytes (terabytes) of storage space. However, accessing data on disk is orders of magnitude slower than accessing data in memory (hundreds of MBs per second).

Thus, there is a trade-off between storage space (disks have more room) and speed (memory is much faster to access). It is important to recognize that these are *physical* limitations—if you only have 4 Gb of RAM on your computer, you simply can't read more than 4 Gb of data into memory.[3]

---

[3]In practice, the limit is much lower than that, since the operating system occupies a fair amount of memory. Virtual memory, which uses the hard disk to allocate extra memory, can be another workaround, but cannot sidestep the throughput issue given the inherent limitations of hard drives or solid state devices.

In general, all objects in your R workspace are stored in memory. Note that the `carriers` object that we created earlier occupies very little memory (since the data still lives on the SQL server), whereas `collect(carriers)` pulls the data into R and occupies much more memory.

Pro Tip: You can find out how much memory an object occupies in R using the `object.size()` function and the `print()` method.

```
print(object.size(carriers), units = "Kb")
```

```
7.5 Kb
```

```
print(object.size(collect(carriers)), units = "Kb")
```

```
209.6 Kb
```

For a typical R user, this means that it can be difficult or impossible to work with a data set stored as a `data.frame` that is larger than a few Gb. The following bit of code will illustrate that a data set of random numbers with 100 columns and 1 million rows occupies more than three-quarters of a Gb of memory on this computer.

```
n <- 100 * 1000000
x <- matrix(runif(n), ncol = 100)
dim(x)
```

```
[1] 1000000      100
```

```
print(object.size(x), units = "Mb")
```

```
762.9 Mb
```

Thus, by the time that `data.frame` reached 10 million rows, it would be problematic for most personal computers—probably making your machine sluggish and unresponsive—and it could never reach 100 million rows. But Google processes over 3.5 *billion* search queries per day! We know that they get stored somewhere—where do they all go?

To work effectively with larger data, we need a system that stores *all* of the data on disk, but allows us to access a portion of the data in memory easily. A relational database—which stores data in a collection of linkable tables—provides a powerful solution to this problem. While more sophisticated approaches are available to address big data challenges, databases are a venerable solution for "medium data."

12.3   The SQL universe

SQL (Structured Query Language) is a programming language for *relational database management systems*. Originally developed in the 1970s, it is a mature, powerful, and widely used storage and retrieval solution for data of many sizes. Google, Facebook, Twitter, Reddit, LinkedIn, Instagram, and countless other companies all access large datastores using SQL.

Relational database management systems (RDBMS) are very efficient for data that is naturally broken into a series of *tables* that are linked together by *keys*. A table is a two-dimensional array of data that has *records* (rows) and *fields* (columns). It is very much like a `data.frame` in R, but there are some important differences that make SQL more efficient under certain conditions.

The theoretical foundation for SQL is based on *relational algebra* and *tuple relational calculus*. These ideas were developed by mathematicians and computer scientists, and while they are not required knowledge for our purposes, they help to solidify SQL's standing as a data storage and retrieval system.

SQL has been an American National Standards Institute (ANSI) standard since 1986, but that standard is only loosely followed by its implementing developers. Unfortunately, this means that there are many different dialects of SQL, and translating between them is not always trivial. However, the broad strokes of the SQL language are common to all, and by learning one dialect, you will be able to easily understand any other [126].

Major implementations of SQL include:

Oracle: corporation that claims #1 market share by revenue—now owns MySQL.

Microsoft SQL Server: another widespread corporate SQL product.

SQLite: a lightweight, open-source version of SQL that has recently become the most widely used implementation of SQL, in part due to its being embedded in Android, the world's most popular mobile operating system. SQLite is an excellent choice for relatively simple applications—like storing data associated with a particular mobile app—but has neither the features nor the scalability for persistent, multi-user, multi-purpose applications.

MySQL: the most popular client-server RDBMS. It is open source, but is now owned by Oracle Corporation, and that has caused some tension in the open-source community. One of the original developers of MySQL, Monty Widenius, now maintains MariaDB as a community fork. MySQL is used by Facebook, Google, LinkedIn, and Twitter.

PostgreSQL: a feature-rich, standards-compliant, open-source implementation growing in popularity. PostgreSQL hews closer to the ANSI standard than MySQL, supports more functions and data types, and provides powerful procedural languages that can extend its base functionality. It is used by Reddit and Instagram, among others.

MonetDB and MonetDBLite: open source implementations that are column-based, rather than the traditional row-based systems. Column-based RDBMSs scale better for big data. `MonetDBLite` is an R package that provides a local experience similar to SQLite.

Vertica: a commercial column-based implementation founded by Postgres originator Michael Stonebraker and now owned by Hewlett Packard.

We will focus on MySQL, but most aspects are similar in PostgreSQL or SQLite (see Appendix F for setup instructions).

12.4 The SQL data manipulation language

MySQL is based on a client-server model. This means that there is a *database server* that stores the data and executes queries. It can be located on the user's local computer or on a remote server. We will be connecting to the server located at `scidb.smith.edu`. To retrieve data from the server, one can connect to it via any number of client programs. One can of course use the command-line `mysql` program, or the official GUI application: MySQL Workbench. While we encourage the reader to explore both options—we most often use the

Workbench for MySQL development—the output you will see in this presentation comes directly from the MySQL command line client.

Pro Tip: Even though `dplyr` enables one to execute most queries using R syntax, and without even worrying so much *where* the data are stored, learning SQL is valuable in its own right due to its ubiquity.

Pro Tip: If you are just learning SQL for the first time, use the command-line client and/or one of the GUI applications. The former provides the most direct feedback, and the latter will provide lots of helpful information.

Information about setting up a MySQL database can be found in Appendix F: we assume that this has been done on a local or remote machine. In what follows, you will see SQL commands and their results in chunks of text. These are the results as returned from the command line client. To run these on your computer, please see section F.4 for information about connecting to a MySQL server.

As noted in Chapter 1, the `airlines` package streamlines construction an SQL database containing over 169 million flights. These data come directly from the U.S. Bureau of Transportation Statistics. In what follows, we access a remote SQL database that we have already set up using the `airlines` package. Note that this database is relational, and thus it consists of many tables. We can list the tables with:

```
SHOW TABLES;
```

```
+--------------------+
| Tables_in_airlines |
+--------------------+
| airports           |
| carriers           |
| flights            |
| planes             |
| summary            |
| weather            |
+--------------------+
```

Note that every SQL statement must end with a semicolon. To see what columns are present in the `airports` table, we ask for a description The output of `DESCRIBE` tells us the names of the field (or variables) in the table, as well as their data type, and what kind of keys might be present (we will learn more about keys in Chapter 13).

```
DESCRIBE airports;
```

```
+---------+---------------+------+-----+---------+-------+
| Field   | Type          | Null | Key | Default | Extra |
+---------+---------------+------+-----+---------+-------+
| faa     | varchar(3)    | NO   | PRI |         |       |
| name    | varchar(255)  | YES  |     | NULL    |       |
| lat     | decimal(10,7) | YES  |     | NULL    |       |
| lon     | decimal(10,7) | YES  |     | NULL    |       |
| alt     | int(11)       | YES  |     | NULL    |       |
```

```
| tz      | smallint(4)  | YES |   | NULL   |   |   |
| dst     | char(1)      | YES |   | NULL   |   |   |
| city    | varchar(255) | YES |   | NULL   |   |   |
| country | varchar(255) | YES |   | NULL   |   |   |
+---------+--------------+-----+-----+--------+-------+
```

Next, we want to build a *query*. Queries in SQL start with the SELECT keyword, and consist of several clauses, which have to be written in this order:

SELECT allows you to list the columns, or functions operating on columns, that you want to retrieve. This is an analogous operation to the select() verb in dplyr, potentially combined with mutate().

FROM specifies the table where the data are.

JOIN allows you to stitch together two or more tables using a key. This is analogous to the join() commands in dplyr.

WHERE allows you to filter the records according to some criteria. This is an analogous operation to the filter() verb in dplyr.

GROUP BY allows you to aggregate the records according to some shared value. This is an analogous operation to the group_by() verb in dplyr.

HAVING is like a WHERE clause that operates on the result set—not the records themselves. This is analogous to applying a second filter() command in dplyr, after the rows have already been aggregated.

ORDER BY is exactly what it sounds like—it specifies a condition for ordering the rows of the result set. This is analogous to the arrange() verb in dplyr.

LIMIT restricts the number of rows in the output. This is similar to the R command head(), but somewhat more versatile.

Only the SELECT and FROM clauses are required. Thus, the simplest query one can write is:

```
SELECT * FROM flights;
```

DO NOT EXECUTE THIS QUERY! This will cause all 169 million records to be dumped! This will not only crash your machine, but also tie up the server for everyone else!
A safe query is:

```
SELECT * FROM flights LIMIT 0,10;
```

We can specify a subset of variables.

```
SELECT year, month, day, dep_time, sched_dep_time, dep_delay, orig
FROM flights LIMIT 0,10;

ERROR 1054 (42S22) at line 1: Unknown column 'orig' in 'field list'
```

Concept	SQL	R
Filter by rows & columns	`SELECT col1, col2` `FROM a` `WHERE col3 = 'x'`	`a %>%` `filter(col3 == "x")` `%>%` `select(col1, col2)`
Aggregate by rows	`SELECT id, sum(col1)` `FROM a` `GROUP BY id`	`a %>%` `group_by(id) %>%` `summarize(sum(col1))`
Combine two tables	`SELECT *` `FROM a` `JOIN b ON a.id = b.id`	`a %>%` `inner_join(b, by =` `c("id" = "id"))`

Table 12.1: Equivalent commands in SQL and R, where *a* and *b* are SQL tables and R `data.frames`.

The astute reader will recognize the similarities between the five idioms for single table analysis and the join operations discussed in Chapter 4 and the SQL syntax. This is not a coincidence! In the contrary, `dplyr` represents a concerted effort to bring the almost natural language SQL syntax to R. In this book, we have presented the R syntax first, since much of our content is predicated on the basic data wrangling skills developed in Chapter 4. But historically, SQL predated the `dplyr` by decades. In Table 12.1, we illustrate the functional equivalence of SQL and `dplyr` commands.

12.4.1   `SELECT...FROM`

As noted above, every SQL `SELECT` query must contain `SELECT` and `FROM`. The analyst may specify columns to be retrieved. We saw above that the `airports` table contains seven columns. If we only wanted to retrieve the FAA `code` and `name` of each airport, we could write the query:

```
SELECT code, name FROM airports;
```

In addition to columns that are present in the database, one can retrieve columns that are functions of other columns. For example, if we wanted to return the geographic coordinates of each airport as an (x, y) pair, we could combine those fields.

```
SELECT
  name,
  concat('(', lat, ', ', lon, ')')
FROM airports
LIMIT 0,6;
```

```
+--------------------------------+----------------------------------+
| name                           | concat('(', lat, ', ', lon, ')') |
+--------------------------------+----------------------------------+
| Lansdowne Airport              | (41.1304722, -80.6195833)        |
| Moton Field Municipal Airport  | (32.4605722, -85.6800278)        |
| Schaumburg Regional            | (41.9893408, -88.1012428)        |
| Randall Airport                | (41.4319120, -74.3915611)        |
| Jekyll Island Airport          | (31.0744722, -81.4277778)        |
| Elizabethton Municipal Airport | (36.3712222, -82.1734167)        |
```

```
+------------------------------------+------------------------------------+
```

Note that the column header for the derived column is ungainly, since it consists of the entire formula that we used to construct it! This is difficult to read, and would be cumbersome to work with. An easy fix is to give this derived column an *alias*. We can do this using the keyword AS.

```
SELECT
  name,
  concat('(', lat, ', ', lon, ')') AS coords
FROM airports
LIMIT 0,6;
```

```
+------------------------------+----------------------------+
| name                         | coords                     |
+------------------------------+----------------------------+
| Lansdowne Airport            | (41.1304722, -80.6195833)  |
| Moton Field Municipal Airport| (32.4605722, -85.6800278)  |
| Schaumburg Regional          | (41.9893408, -88.1012428)  |
| Randall Airport              | (41.4319120, -74.3915611)  |
| Jekyll Island Airport        | (31.0744722, -81.4277778)  |
| Elizabethton Municipal Airport| (36.3712222, -82.1734167) |
+------------------------------+----------------------------+
```

We can also use AS to refer to a column in the table by a different name in the result set.

```
SELECT
  name AS airportName,
  concat('(', lat, ', ', lon, ')') AS coords
FROM airports
LIMIT 0,6;
```

```
+------------------------------+----------------------------+
| airportName                  | coords                     |
+------------------------------+----------------------------+
| Lansdowne Airport            | (41.1304722, -80.6195833)  |
| Moton Field Municipal Airport| (32.4605722, -85.6800278)  |
| Schaumburg Regional          | (41.9893408, -88.1012428)  |
| Randall Airport              | (41.4319120, -74.3915611)  |
| Jekyll Island Airport        | (31.0744722, -81.4277778)  |
| Elizabethton Municipal Airport| (36.3712222, -82.1734167) |
+------------------------------+----------------------------+
```

This brings an important distinction to the fore: In SQL, it is crucial to distinguish between clauses that operate *on the rows of the original table* versus those that operate *on the rows of the result set*. Here, name, lat, and lon are columns in the original table—they are written to the disk on the SQL server. On the other hand, airportName and coords exist only in the result set—which is passed from the server to the client and is not written to the disk.

The preceding examples show the SQL equivalents of the dplyr commands select, mutate, and rename.

12.4.2 WHERE

The WHERE clause is analogous to the `filter` command in `dplyr`—it allows you to restrict the set of rows that are retrieved to only those rows that match a certain condition. Thus, while there are several million rows in the `flights` table in each year—each corresponding to a single flight—there were only a few dozen flights that left Bradley International Airport on June 26th, 2013.

```
SELECT
  year, month, day, origin, dest,
  flight, carrier
FROM flights
WHERE year = 2013 AND month = 6 AND day = 26
AND origin = 'BDL'
LIMIT 0,6;
```

```
+------+-------+------+--------+------+--------+---------+
| year | month | day  | origin | dest | flight | carrier |
+------+-------+------+--------+------+--------+---------+
| 2013 |     6 |   26 | BDL    | EWR  |   4714 | EV      |
| 2013 |     6 |   26 | BDL    | MIA  |   2015 | AA      |
| 2013 |     6 |   26 | BDL    | DTW  |   1644 | DL      |
| 2013 |     6 |   26 | BDL    | BWI  |   2584 | WN      |
| 2013 |     6 |   26 | BDL    | ATL  |   1065 | DL      |
| 2013 |     6 |   26 | BDL    | DCA  |   1077 | US      |
+------+-------+------+--------+------+--------+---------+
```

It would be convenient to search for flights in a date range. Unfortunately, there is no date field in this table—but rather separate columns for the year, month, and day. Nevertheless, we can tell SQL to interpret these columns as a date, using the `str_to_date` function.[4] To do this, we first need to collect these columns as a string, and then tell SQL how to parse that string into a date.

Pro Tip: Dates and times can be challenging to wrangle. To learn more about these date tokens, see the MySQL documentation for `str_to_date`.

```
SELECT
  str_to_date(concat(year, '-', month, '-', day), '%Y-%m-%d') as theDate,
  origin,
  flight, carrier
FROM flights
WHERE year = 2013 AND month = 6 AND day = 26
  AND origin = 'BDL'
LIMIT 0,6;
```

```
+------------+--------+--------+---------+
| theDate    | origin | flight | carrier |
+------------+--------+--------+---------+
| 2013-06-26 | BDL    |   4714 | EV      |
```

---

[4]The analogous function in PostgreSQL is called `to_date`.

```
| 2013-06-26 | BDL    |    2015 | AA      |
| 2013-06-26 | BDL    |    1644 | DL      |
| 2013-06-26 | BDL    |    2584 | WN      |
| 2013-06-26 | BDL    |    1065 | DL      |
| 2013-06-26 | BDL    |    1077 | US      |
+------------+--------+--------+---------+
```

Note that here we have used a WHERE clause on columns that are not present in the result set. We can do this because WHERE operates only on the rows of the original table. Conversely, if we were to try and use a WHERE clause on theDate, it would not work, because (as the error suggests), theDate is not the name of a column in the flights table.

```
SELECT
  str_to_date(concat(year, '-', month, '-', day), '%Y-%m-%d') as theDate,
  origin, flight, carrier
FROM flights
WHERE theDate = '2013-06-26'
  AND origin = 'BDL'
LIMIT 0,6;

ERROR 1054 (42S22) at line 1: Unknown column 'theDate' in 'where clause'
```

A workaround is to copy and paste the definition of theDate into the WHERE clause, since WHERE *can* operate on functions of columns in the original table.

```
SELECT
  str_to_date(concat(year, '-', month, '-', day), '%Y-%m-%d') as theDate,
  origin, flight, carrier
FROM flights
WHERE str_to_date(concat(year, '-', month, '-', day), '%Y-%m-%d') =
  '2013-06-26'
  AND origin = 'BDL'
LIMIT 0,6;
```

This query will work, but here we have stumbled onto another wrinkle that exposes subtleties in how SQL executes queries. The previous query was able to make use of indices defined on the year, month, and day columns. However, the latter query is not able to make use of these indices, because it is trying to filter on functions of a combination of those columns. This makes the latter query very slow. We will return to a fuller discussion of indices in Section 13.1.

Finally, we can use the BETWEEN syntax to filter through a range of dates.

```
SELECT
  str_to_date(concat(year, '-', month, '-', day), '%Y-%m-%d') as theDate,
  origin,
  flight, carrier
FROM flights
WHERE year = 2013 AND month = 6 AND day BETWEEN 26 and 30
  AND origin = 'BDL'
LIMIT 0,6;
```

```
+-------------+--------+--------+---------+
| theDate     | origin | flight | carrier |
+-------------+--------+--------+---------+
| 2013-06-26  | BDL    |   4714 | EV      |
| 2013-06-26  | BDL    |   2015 | AA      |
| 2013-06-26  | BDL    |   1644 | DL      |
| 2013-06-26  | BDL    |   2584 | WN      |
| 2013-06-26  | BDL    |   1065 | DL      |
| 2013-06-26  | BDL    |   1077 | US      |
+-------------+--------+--------+---------+
```

Similarly, we can use the IN syntax to search for items in a specified list. Note that flights on the 27th, 28th, and 29th of June are retrieved in the query using BETWEEN but not in the query using IN.

```
SELECT
    str_to_date(concat(year, '-', month, '-', day), '%Y-%m-%d') as theDate,
    origin,
    flight, carrier
FROM flights
WHERE year = 2013 AND month = 6 AND day IN (26, 30)
    AND origin = 'BDL'
LIMIT 0,6;
```

```
+-------------+--------+--------+---------+
| theDate     | origin | flight | carrier |
+-------------+--------+--------+---------+
| 2013-06-26  | BDL    |   4714 | EV      |
| 2013-06-26  | BDL    |   2015 | AA      |
| 2013-06-26  | BDL    |   1644 | DL      |
| 2013-06-26  | BDL    |   2584 | WN      |
| 2013-06-26  | BDL    |   1065 | DL      |
| 2013-06-26  | BDL    |   1077 | US      |
+-------------+--------+--------+---------+
```

SQL also supports OR clauses in addition to AND clauses, but one must always be careful with parentheses when using OR. Note the difference in the numbers of rows returned by the following two queries. The count function simply counts the number of rows. The criteria in the WHERE clause are not evaluated left to right, but rather the ANDs are evaluated first. This means that in the first query below, all flights on the 26th day of any month, regardless of year or month, are returned.

```
SELECT
    count(*) as N
FROM flights
WHERE year = 2013 AND month = 6 OR day = 26
    AND origin = 'BDL';
```

```
+--------+
| N      |
+--------+
```

```
| 581366 |
+--------+
```

```
SELECT
  count(*) as N
FROM flights
WHERE year = 2013 AND (month = 6 OR day = 26)
  AND origin = 'BDL';
```

```
+------+
| N    |
+------+
| 2542 |
+------+
```

12.4.3 GROUP BY

The GROUP BY clause allows one to *aggregate* multiple rows according to some criteria. The challenge when using GROUP BY is specifying *how* multiple rows of data should be reduced into a single value. Aggregate functions (e.g., count, sum, max, and avg) are necessary.

We know that there were 65 flights that left Bradley Airport on June 26th, 2013, but how many belonged to each airline carrier? To get this information we need to aggregate the individual flights, based on who the carrier was.

```
SELECT
  carrier,
  count(*) AS numFlights,
  sum(1) AS numFlightsAlso
FROM flights
WHERE year = 2013 AND month = 6 AND day = 26
  AND origin = 'BDL'
GROUP BY carrier
LIMIT 0,6;
```

```
+---------+------------+----------------+
| carrier | numFlights | numFlightsAlso |
+---------+------------+----------------+
| 9E      |          5 |              5 |
| AA      |          4 |              4 |
| B6      |          5 |              5 |
| DL      |         11 |             11 |
| EV      |          5 |              5 |
| MQ      |          5 |              5 |
+---------+------------+----------------+
```

For each of these airlines, which flight left the earliest in the morning?

```
SELECT
  carrier,
  count(*) AS numFlights,
```

```
   min(dep_time)
FROM flights
WHERE year = 2013 AND month = 6 AND day = 26
   AND origin = 'BDL'
GROUP BY carrier
LIMIT 0,6;
```

```
+----------+------------+----------------+
| carrier  | numFlights | min(dep_time)  |
+----------+------------+----------------+
| 9E       |          5 |              0 |
| AA       |          4 |            559 |
| B6       |          5 |            719 |
| DL       |         11 |            559 |
| EV       |          5 |            555 |
| MQ       |          5 |              0 |
+----------+------------+----------------+
```

This is a bit tricky to figure out because the dep_time variable is stored as an integer, but would be better represented as a time data type. If it is a three-digit integer, then the first digit is the hour, but if it is a four-digit integer, then the first two digits are the hour. In either case, the last two digits are the minutes, and there are no seconds recorded. The if(condition, value if true, value if false) statement can help us with this.

```
SELECT
   carrier,
   count(*) AS numFlights,
   maketime(
      if(length(min(dep_time)) = 3,
         left(min(dep_time), 1), left(min(dep_time), 2)),
      right(min(dep_time), 2),
      0
      ) as firstDepartureTime
FROM flights
WHERE year = 2013 AND month = 6 AND day = 26
   AND origin = 'BDL'
GROUP BY carrier
LIMIT 0,6;
```

```
+----------+------------+--------------------+
| carrier  | numFlights | firstDepartureTime |
+----------+------------+--------------------+
| 9E       |          5 | 00:00:00           |
| AA       |          4 | 05:59:00           |
| B6       |          5 | 07:19:00           |
| DL       |         11 | 05:59:00           |
| EV       |          5 | 05:55:00           |
| MQ       |          5 | 00:00:00           |
+----------+------------+--------------------+
```

We can also group by more than one column, but need to be careful to specify that we apply an aggregate function to each column that we are *not* grouping by. In this case, every time we access dep_time, we apply the min function, since there may be many different values of dep_time associated with each unique combination of carrier and dest. Applying the min function returns the smallest such value unambiguously.

```
SELECT
  carrier, dest,
  count(*) AS numFlights,
  maketime(
    if(length(min(dep_time)) = 3,
      left(min(dep_time), 1), left(min(dep_time), 2)),
    right(min(dep_time), 2),
    0
    ) as firstDepartureTime
FROM flights
WHERE year = 2013 AND month = 6 AND day = 26
  AND origin = 'BDL'
GROUP BY carrier, dest
LIMIT 0,6;
```

```
+---------+------+------------+-------------------+
| carrier | dest | numFlights | firstDepartureTime |
+---------+------+------------+-------------------+
| 9E      | CVG  |          2 | 00:00:00          |
| 9E      | DTW  |          1 | 18:20:00          |
| 9E      | MSP  |          1 | 11:25:00          |
| 9E      | RDU  |          1 | 09:38:00          |
| AA      | DFW  |          3 | 07:04:00          |
| AA      | MIA  |          1 | 05:59:00          |
+---------+------+------------+-------------------+
```

12.4.4 ORDER BY

The use of aggregate function allows us to answer some very basic exploratory questions. Combining this with an ORDER BY clause will bring the most interesting results to the top. For example, which destinations are most common from Bradley in 2013?

```
SELECT
  dest, sum(1) as numFlights
FROM flights
WHERE year = 2013
  AND origin = 'BDL'
GROUP BY dest
ORDER BY numFlights desc
LIMIT 0,6;
```

```
+------+------------+
| dest | numFlights |
+------+------------+
| ORD  |       2657 |
| BWI  |       2613 |
| ATL  |       2277 |
```

```
| CLT  |         1842 |
| MCO  |         1789 |
| DTW  |         1523 |
+------+--------------+
```

Pro Tip: Note that since the ORDER BY clause cannot be executed until all of the data are retrieved, it operates on the result set, and not the rows of the original data. Thus, derived columns *can* be referenced in the ORDER BY clause.

Which of those destinations had the lowest average arrival delay time?

```
SELECT
  dest, sum(1) as numFlights,
  avg(arr_delay) as avg_arr_delay
FROM flights
WHERE year = 2013
  AND origin = 'BDL'
GROUP BY dest
ORDER BY avg_arr_delay asc
LIMIT 0,6;
```

```
+------+------------+----------------+
| dest | numFlights | avg_arr_delay  |
+------+------------+----------------+
| CLE  |         57 |       -13.0702 |
| LAX  |        127 |       -10.3071 |
| CVG  |        708 |        -7.3701 |
| MSP  |        981 |        -3.6636 |
| MIA  |        404 |        -3.2723 |
| DCA  |        204 |        -2.8971 |
+------+------------+----------------+
```

12.4.5 HAVING

Although flights to Cleveland had the lowest average arrival delay—more than 13 minutes ahead of schedule—there were only 57 flights that went to from Bradley to Cleveland in all of 2013. It probably makes more sense to consider only those destinations that had, say, at least two flights per day. We can filter our result set using a HAVING clause.

```
SELECT
  dest, sum(1) as numFlights,
  avg(arr_delay) as avg_arr_delay
FROM flights
WHERE year = 2013
  AND origin = 'BDL'
GROUP BY dest
HAVING numFlights > 365*2
ORDER BY avg_arr_delay asc
LIMIT 0,6;
```

```
+------+-----------+----------------+
| dest | numFlights | avg_arr_delay |
+------+-----------+----------------+
| MSP  |       981 |        -3.6636 |
| DTW  |      1523 |        -2.1477 |
| CLT  |      1842 |        -0.1205 |
| FLL  |      1011 |         0.2770 |
| DFW  |      1062 |         0.7495 |
| ATL  |      2277 |         4.4704 |
+------+-----------+----------------+
```

We can see now that among the airports that are common destinations from Bradley, Minneapolis–St. Paul has the lowest average arrival delay time, at nearly 4 minutes ahead of schedule, on average.[5]

It is important to understand that the HAVING clause operates on the result set. While WHERE and HAVING are similar in spirit and syntax (and indeed, in dplyr they are both masked by the filter() function), they are different, because WHERE operates on the original data in the table and HAVING operates on the result set. Moving the HAVING condition to the WHERE clause will not work.

```
SELECT
  dest, sum(1) as numFlights,
  avg(arr_delay) as avg_arr_delay
FROM flights
WHERE year = 2013
  AND origin = 'BDL'
  AND numFlights > 365*2
GROUP BY dest
ORDER BY avg_arr_delay asc
LIMIT 0,6;

ERROR 1054 (42S22) at line 1: Unknown column 'numFlights' in 'where clause'
```

On the other hand, moving the WHERE conditions to the HAVING clause will work, but could result in a major loss of efficiency. The following query will return the same result as the one we considered previously.

```
SELECT
  origin, dest, sum(1) as numFlights,
  avg(arr_delay) as avg_arr_delay
FROM flights
WHERE year = 2013
GROUP BY origin, dest
HAVING numFlights > 365*2
  AND origin = 'BDL'
ORDER BY avg_arr_delay asc
LIMIT 0,6;
```

But moving the origin = 'BDL' condition to the HAVING clause means that *all* airport destinations had to be considered. Thus, with this condition in the WHERE clause, the server

---

[5]Note: MySQL and SQLite support the use of derived column aliases in HAVING clauses, but PostgreSQL does not.

can quickly identify only those flights that left Bradley, perform the aggregation, and then filter this relatively small result set for those entries with a sufficient number of flights. Conversely, with this condition in the HAVING clause, the server is forced to consider *all* three million flights from 2013, perform the aggregation for all pairs of airports, and then filter this much larger result set for those entries with a sufficient number of flights from Bradley. The filtering of the result set is not importantly slower, but the aggregation over three million rows as opposed to a few thousand is.

Pro Tip: To maximize query efficiency, put conditions in a WHERE clause as opposed to a HAVING clause whenever possible.

12.4.6   LIMIT

A LIMIT clause simply allows you to truncate the output to a specified number of rows. This achieves an effect analogous to the R command head().

```
SELECT
   dest, sum(1) as numFlights,
   avg(arr_delay) as avg_arr_delay
FROM flights
WHERE year = 2013
   AND origin = 'BDL'
GROUP BY dest
HAVING numFlights > 365*2
ORDER BY avg_arr_delay asc
LIMIT 0,6;
```

```
+-------+------------+---------------+
| dest  | numFlights | avg_arr_delay |
+-------+------------+---------------+
| MSP   |        981 |       -3.6636 |
| DTW   |       1523 |       -2.1477 |
| CLT   |       1842 |       -0.1205 |
| FLL   |       1011 |        0.2770 |
| DFW   |       1062 |        0.7495 |
| ATL   |       2277 |        4.4704 |
+-------+------------+---------------+
```

Note, however, that it is also possible to retrieve rows not at the beginning. The first number in the LIMIT clause indicates the number of rows to skip, and the latter indicates the number of rows to retrieve. Thus, this query will return the 4th–7th airports in the previous list.

```
SELECT
   dest, sum(1) as numFlights,
   avg(arr_delay) as avg_arr_delay
FROM flights
WHERE year = 2013
   AND origin = 'BDL'
GROUP BY dest
```

```
HAVING numFlights > 365*2
ORDER BY avg_arr_delay asc
LIMIT 3,4;
```

```
+------+------------+---------------+
| dest | numFlights | avg_arr_delay |
+------+------------+---------------+
| FLL  |       1011 |        0.2770 |
| DFW  |       1062 |        0.7495 |
| ATL  |       2277 |        4.4704 |
| BWI  |       2613 |        5.0325 |
+------+------------+---------------+
```

12.4.7 JOIN

In Section 4.3 we presented the dplyr join operators: inner_join(), left_join(), and semi_join(). As you now probably expect, these operations are fundamental to SQL—and moreover, the massive success of the RDBMS paradigm is predicated on the ability to efficiently join tables together. Recall that SQL is a *relational* database management system—the relations between the tables allow you to write queries that efficiently tie together information from multiple sources. The syntax for performing these operations in SQL requires the JOIN keyword.

In general, there are four pieces of information that you need to specify in order to join two tables:

- The name of the first table that you want to join

- (optional) The *type* of join that you want to use

- The name of the second table that you want to join

- The *condition(s)* under which you want the records in the first table to match the records in the second table

There are many possible permutations of how two tables can be joined, and in many cases, a single query may involve several or even dozens of tables. In practice, the JOIN syntax varies among SQL implementations. In MySQL, OUTER JOINs are not available, but the following join types are:

- JOIN: includes all of the rows that are present in *both* tables and match.

- LEFT JOIN: includes all of the rows that are present in the first table. Rows in the first table that have no match in the second are filled with NULLs.

- RIGHT JOIN: include all of the rows that are present in the second table. This is the opposite of a LEFT JOIN.

- CROSS JOIN: the Cartesian product of the two tables. Thus, all possible combinations of rows matching the joining condition are returned.

Recall that in the flights table, the origin and destination of each flight are recorded.

```
SELECT
  origin, dest
  flight, carrier
FROM flights
WHERE year = 2013 AND month = 6 AND day = 26
  AND origin = 'BDL'
LIMIT 0,6;
```

```
+---------+--------+---------+
| origin  | flight | carrier |
+---------+--------+---------+
| BDL     | EWR    | EV      |
| BDL     | MIA    | AA      |
| BDL     | DTW    | DL      |
| BDL     | BWI    | WN      |
| BDL     | ATL    | DL      |
| BDL     | DCA    | US      |
+---------+--------+---------+
```

However, the `flights` table contains only the three-character FAA airport codes for both airports—not the full name of the airport. These cryptic abbreviations are not easily understood by humans. Which airport is `ORD`? Wouldn't it be more convenient to have the airport name in the table? It would be more convenient, but it would also be significantly less efficient from a storage and retrieval point of view, as well as more problematic from a database integrity point of view. Thus, the solution is to store information *about airports* in the `airports` table, along with these cryptic codes—which we will now call *keys*—and to only store these keys in the `flights` table—which is about *flights*, not airports. However, we can use these keys to join the two tables together in our query. In this manner we can have our cake and eat it too: The data are stored in separate tables for efficiency, but we can still have the full names in the result set if we choose. Note how once again, the distinction between the rows of the original table and the result set is critical. To write our query, we simply have to specify the table we want to join onto `flights` (e.g., `airports`) and the condition by which we want to match rows in `flights` with rows in `airports`. In this case, we want the airport code listed in `flights.dest` to be matched to the airport code in `airports.faa`. We also have to specify that we want to see the `name` column from the `airports` table in the result set.

```
SELECT
  origin, dest,
  airports.name as destAirportName,
  flight, carrier
FROM flights
JOIN airports ON flights.dest = airports.faa
WHERE year = 2013 AND month = 6 AND day = 26
  AND origin = 'BDL'
LIMIT 0,6;
```

```
+---------+------+------------------------------------+--------+---------+
| origin  | dest | destAirportName                    | flight | carrier |
+---------+------+------------------------------------+--------+---------+
| BDL     | EWR  | Newark Liberty Intl                |   4714 | EV      |
```

```
| BDL      | MIA   | Miami Intl                      |   2015 | AA       |
| BDL      | DTW   | Detroit Metro Wayne Co          |   1644 | DL       |
| BDL      | BWI   | Baltimore Washington Intl       |   2584 | WN       |
| BDL      | ATL   | Hartsfield Jackson Atlanta Intl |   1065 | DL       |
| BDL      | DCA   | Ronald Reagan Washington Natl   |   1077 | US       |
+---------+------+---------------------------------+--------+---------+
```

This is much easier to read for humans. One quick improvement to the readability of this query is to use *table aliases*. This will save us some typing now, but a considerable amount later on. A table alias is usually just a single letter after the specification of each table in the FROM and JOIN clauses. Note that these aliases can be referenced anywhere else in the query.

```
SELECT
  origin, dest,
  a.name as destAirportName,
  flight, carrier
FROM flights o
JOIN airports a ON o.dest = a.faa
WHERE year = 2013 AND month = 6 AND day = 26
  AND origin = 'BDL'
LIMIT 0,6;
```

```
+---------+------+---------------------------------+--------+---------+
| origin | dest | destAirportName                 | flight | carrier |
+---------+------+---------------------------------+--------+---------+
| BDL    | EWR  | Newark Liberty Intl             |   4714 | EV      |
| BDL    | MIA  | Miami Intl                      |   2015 | AA      |
| BDL    | DTW  | Detroit Metro Wayne Co          |   1644 | DL      |
| BDL    | BWI  | Baltimore Washington Intl       |   2584 | WN      |
| BDL    | ATL  | Hartsfield Jackson Atlanta Intl |   1065 | DL      |
| BDL    | DCA  | Ronald Reagan Washington Natl   |   1077 | US      |
+---------+------+---------------------------------+--------+---------+
```

In the same manner, there are cryptic codes in flights for the airline carriers. The full name of each carrier is stored in the carriers table, since that is the place where information about carriers are stored. We can join this table to our result set to retrieve the name of each carrier.

```
SELECT
  dest, a.name as destAirportName,
  o.carrier, c.name as carrierName
FROM flights o
JOIN airports a ON o.dest = a.faa
JOIN carriers c ON o.carrier = c.carrier
WHERE year = 2013 AND month = 6 AND day = 26
  AND origin = 'BDL'
LIMIT 0,6;
```

```
+------+---------------------------------+---------+----------------------+
| dest | destAirportName                 | carrier | carrierName          |
```

```
+-------+------------------------------+---------+--------------------------+
| EWR   | Newark Liberty Intl          | EV      | ExpressJet Airlines      |
| MIA   | Miami Intl                   | AA      | American Airlines        |
| DTW   | Detroit Metro Wayne Co       | DL      | Delta Air Lines Inc.     |
| BWI   | Baltimore Washington Intl    | WN      | Southwest Airlines       |
| ATL   | Hartsfield Jackson Atlanta Intl | DL   | Delta Air Lines Inc.     |
| DCA   | Ronald Reagan Washington Natl | US     | US Airways Inc.          |
+-------+------------------------------+---------+--------------------------+
```

Finally, to retrieve the name of the originating airport, we can join onto the same table more than once. Here the table aliases are necessary.

```
SELECT
  a2.name as origAirport,
  a1.name as destAirportName,
  c.name as carrierName
FROM flights o
JOIN airports a1 ON o.dest = a1.faa
JOIN airports a2 ON o.origin = a2.faa
JOIN carriers c ON o.carrier = c.carrier
WHERE year = 2013 AND month = 6 AND day = 26
  AND origin = 'BDL'
LIMIT 0,6;
```

```
+--------------+--------------------------------+-------------------------+
| origAirport  | destAirportName                | carrierName             |
+--------------+--------------------------------+-------------------------+
| Bradley Intl | Newark Liberty Intl            | ExpressJet Airlines Inc.|
| Bradley Intl | Miami Intl                     | American Airlines Inc.  |
| Bradley Intl | Detroit Metro Wayne Co         | Delta Air Lines Inc.    |
| Bradley Intl | Baltimore Washington Intl      | Southwest Airlines Co.  |
| Bradley Intl | Hartsfield Jackson Atlanta Intl| Delta Air Lines Inc.    |
| Bradley Intl | Ronald Reagan Washington Natl  | US Airways Inc.         |
+--------------+--------------------------------+-------------------------+
```

Now it is perfectly clear that American Eagle flight 3127 flew from Bradley International airport to Chicago O'Hare International airport on June 26th, 2013. However, in order to put this together, we had to join four tables. Wouldn't it be easier to store these data in a single table that looks like the result set? For a variety of reasons, the answer is no.

First, there are very literal storage considerations. The airports.name field has room for 50 characters.

```
DESCRIBE airports;
```

```
+----------+---------------+------+-----+---------+-------+
| Field    | Type          | Null | Key | Default | Extra |
+----------+---------------+------+-----+---------+-------+
| faa      | varchar(3)    | NO   | PRI |         |       |
| name     | varchar(255)  | YES  |     | NULL    |       |
| lat      | decimal(10,7) | YES  |     | NULL    |       |
| lon      | decimal(10,7) | YES  |     | NULL    |       |
```

alt	int(11)	YES		NULL		
tz	smallint(4)	YES		NULL		
dst	char(1)	YES		NULL		
city	varchar(255)	YES		NULL		
country	varchar(255)	YES		NULL		

This takes up considerably more space on disk that the four-character abbreviation stored in airports.faa. For small data sets, this overhead might not matter, but the flights table contains 169 million rows, so replacing the four-character origin field with a 255-character field would result in a noticeable difference in space on disk. (Plus, we'd have to do this twice, since the same would apply to dest.) We'd suffer a similar penalty for including the full name of each carrier in the flights table. Other things being equal, tables that take up less room on disk are faster to search.

Second, it would be logically inefficient to store the full name of each airport in the flights table. The name of the airport doesn't change for each flight. It doesn't make sense to store the full name of the airport any more than it would make sense to store the full name of the month, instead of just the integer corresponding to each month.

Third, what if the name of the airport *did* change? For example, in 1998 the airport with code DCA was renamed from Washington National to Ronald Reagan Washington National. It is still the same airport in the same location, and it still has code DCA—only the full name has changed. With separate tables, we only need to update a single field: the name column in the airports table for the DCA row. Had we stored the full name in the flights table we would have to make millions of substitutions, and would risk ending up in a situation in which both "Washington National" and "Reagan National" were present in the table.

When designing a database, how do you know whether to create a separate table for pieces of information? The short answer is that if you are designing a persistent, scalable database for speed and efficiency, then every *entity* should have its own table. In practice, very often it is not worth the time and effort to set this up if we are simply doing some quick analysis. But for permanent systems—like a database backend to a website—proper curation is necessary. The notions of normal forms, and specifically third normal form (3NF), provide guidance for how to properly design a database. A full discussion of this is beyond the scope of this book, but the basic idea is to "keep like with like."

Pro Tip: If you are designing a database that will be used for a long time or by a lot of people, take the extra time to design it well.

LEFT JOIN

Recall that in a JOIN—also known as an *inner* or *natural* or *regular* JOIN—all possible matching pairs of rows from the two tables are included. Thus, if the first table has n rows and the second table has m, as many as nm rows could be returned. However, in the airports table, each row has a unique airport code, and thus every row in flights will match the destination field to *at most* one row in the airports table. But what happens if no such entry is present in airports? That is, what happens if there is a destination airport in flights that has no corresponding entry in airports? If you are using a JOIN, then the offending row in flights is simply not returned. On the other hand, if you are using a LEFT JOIN, then every row in the first table is returned, and the corresponding entries from the second table are left blank. In this example, no airport names were found for several airports.

```
SELECT
  year, month, day, origin, dest,
  a.name as destAirportName,
  flight, carrier
FROM flights o
LEFT JOIN airports a ON o.dest = a.faa
WHERE year = 2013 AND month = 6 AND day = 26
  AND a.name is null
LIMIT 0,6;
```

```
+------+-------+------+--------+------+-----------------+--------+---------+
| year | month | day  | origin | dest | destAirportName | flight | carrier |
+------+-------+------+--------+------+-----------------+--------+---------+
| 2013 |     6 |   26 | BOS    | SJU  | NULL            |    261 | B6      |
| 2013 |     6 |   26 | JFK    | SJU  | NULL            |   1203 | B6      |
| 2013 |     6 |   26 | JFK    | PSE  | NULL            |    745 | B6      |
| 2013 |     6 |   26 | JFK    | SJU  | NULL            |   1503 | B6      |
| 2013 |     6 |   26 | JFK    | BQN  | NULL            |    839 | B6      |
| 2013 |     6 |   26 | JFK    | BQN  | NULL            |    939 | B6      |
+------+-------+------+--------+------+-----------------+--------+---------+
```

These airports are all in Puerto Rico: SJU is in San Juan, BQN is in Aguadilla, and PSE is in Ponce.

```
SELECT * FROM airports WHERE faa = 'SJU';
```

The result set from a LEFT JOIN is always a superset of the result set from the same query with a regular JOIN. A RIGHT JOIN is simply the opposite of a LEFT JOIN—that is, the tables have simply been specified in the opposite order. This can be useful in certain cases, especially when you are joining more than two tables.

12.4.8 UNION

Two separate queries can be combined using a UNION clause.

```
(SELECT
  year, month, day, origin, dest,
  flight, carrier
FROM flights
WHERE year = 2013 AND month = 6 AND day = 26
  AND origin = 'BDL' AND dest = 'MSP')
UNION
(SELECT
  year, month, day, origin, dest,
  flight, carrier
FROM flights
WHERE year = 2013 AND month = 6 AND day = 26
AND origin = 'JFK' AND dest = 'ORD')
LIMIT 0,10;
```

```
+------+-------+------+--------+------+--------+---------+
| year | month | day  | origin | dest | flight | carrier |
+------+-------+------+--------+------+--------+---------+
| 2013 |     6 |   26 | BDL    | MSP  |    797 | DL      |
| 2013 |     6 |   26 | BDL    | MSP  |   3338 | 9E      |
| 2013 |     6 |   26 | BDL    | MSP  |   1226 | DL      |
| 2013 |     6 |   26 | JFK    | ORD  |    905 | B6      |
| 2013 |     6 |   26 | JFK    | ORD  |   1105 | B6      |
| 2013 |     6 |   26 | JFK    | ORD  |   3523 | 9E      |
| 2013 |     6 |   26 | JFK    | ORD  |   1711 | AA      |
| 2013 |     6 |   26 | JFK    | ORD  |    105 | B6      |
| 2013 |     6 |   26 | JFK    | ORD  |   3521 | 9E      |
| 2013 |     6 |   26 | JFK    | ORD  |   3525 | 9E      |
+------+-------+------+--------+------+--------+---------+
```

This is analogous to the R operation rbind() or the dplyr operation bind_rows().

12.4.9 Subqueries

It is also possible to use a result set as if it were a table. That is, you can write one query to generate a result set, and then use that result set in a larger query as if it were a table, or even just a list of values. The initial query is called a subquery.

For example, Bradley is listed as an "international" airport, but with the exception of trips to Montreal and Toronto and occasional flights to Mexico and Europe, it is more of a regional airport. Does it have any flights coming from or going to Alaska and Hawaii? We can retrieve the list of airports outside the lower 48 states by filtering the airports table using the time zone tz column.

```
SELECT faa, name, tz, city
FROM airports a
WHERE tz < -8
LIMIT 0,6;
```

```
+------+----------------------+------+-------------+
| faa  | name                 | tz   | city        |
+------+----------------------+------+-------------+
| 369  | Atmautluak Airport   |   -9 | Atmautluak  |
| 6K8  | Tok Junction Airport |   -9 | Tok         |
| ABL  | Ambler Airport       |   -9 | Ambler      |
| ADK  | Adak Airport         |   -9 | Adak Island |
| ADQ  | Kodiak               |   -9 | Kodiak      |
| AET  | Allakaket Airport    |   -9 | Allakaket   |
+------+----------------------+------+-------------+
```

Now, let's use the airport codes generated by that query as a list to filter the flights leaving from Bradley in 2013. Note the subquery in parentheses in the query below.

```
SELECT
  dest, a.name as destAirportName,
  sum(1) as N, count(distinct carrier) as numCarriers
FROM flights o
```

```
LEFT JOIN airports a ON o.dest = a.faa
WHERE year = 2013
  AND origin = 'BDL'
  AND dest IN
    (SELECT faa
      FROM airports
      WHERE tz < -8)
GROUP BY dest;
```

No results are returned. As it turns out, Bradley did not have any outgoing flights to Alaska or Hawaii. However, it did have some flights to and from airports in the Pacific time zone.

```
SELECT
  dest, a.name as destAirportName,
  sum(1) as N, count(distinct carrier) as numCarriers
FROM flights o
LEFT JOIN airports a ON o.origin = a.faa
WHERE year = 2013
  AND dest = 'BDL'
  AND origin IN
  (SELECT faa
    FROM airports
    WHERE tz < -7)
GROUP BY origin;
```

```
+------+------------------+------+-------------+
| dest | destAirportName  | N    | numCarriers |
+------+------------------+------+-------------+
| BDL  | Mc Carran Intl   | 262  |           1 |
| BDL  | Los Angeles Intl | 127  |           1 |
+------+------------------+------+-------------+
```

We could also employ a similar subquery to create an ephemeral table.

```
SELECT
  dest, a.name as destAirportName,
  sum(1) as N, count(distinct carrier) as numCarriers
FROM flights o
JOIN (SELECT *
        FROM airports
        WHERE tz < -7) a
  ON o.origin = a.faa
WHERE year = 2013 AND dest = 'BDL'
GROUP BY origin;
```

```
+------+------------------+------+-------------+
| dest | destAirportName  | N    | numCarriers |
+------+------------------+------+-------------+
| BDL  | Mc Carran Intl   | 262  |           1 |
| BDL  | Los Angeles Intl | 127  |           1 |
+------+------------------+------+-------------+
```

Of course, we could have achieved the same result with a `JOIN` and `WHERE`:

```
SELECT
  dest, a.name as destAirportName,
  sum(1) as N, count(distinct carrier) as numCarriers
FROM flights o
LEFT JOIN airports a ON o.origin = a.faa
WHERE year = 2013
  AND dest = 'BDL'
  AND tz < -7
GROUP BY origin;
```

```
+------+------------------+------+-------------+
| dest | destAirportName  | N    | numCarriers |
+------+------------------+------+-------------+
| BDL  | Mc Carran Intl   | 262  |           1 |
| BDL  | Los Angeles Intl | 127  |           1 |
+------+------------------+------+-------------+
```

In is important to note that while subqueries are often convenient, they cannot make use of indices. Thus, in most cases it is preferable to write the query using joins as opposed to subqueries.

12.5   Extended example: FiveThirtyEight flights

Over at `FiveThirtyEight.com`, Nate Silver wrote an article about airline delays using the same Bureau of Transportation Statistics data that we have in our database. We can use this article as an exercise in querying our airlines database.

The article makes a number of claims. We'll walk through some of these. First, the article states:

> In 2014, the 6 million domestic flights the U.S. government tracked required an extra 80 million minutes to reach their destinations.
>
> The majority of flights—54 percent—arrived ahead of schedule in 2014. (The 80 million minutes figure cited earlier is a net number. It consists of about 115 million minutes of delays minus 35 million minutes saved from early arrivals.)

Although there are a number of claims here, we can verify them with a single query. Here, we compute the total number of flights, the percentage of those that were on time and ahead of schedule, and the total number of minutes of delays.

```
SELECT
  sum(1) as numFlights,
  sum(if(arr_delay < 15, 1, 0)) / sum(1) as ontimePct,
  sum(if(arr_delay < 0, 1, 0)) / sum(1) as earlyPct,
  sum(arr_delay) / 1000000 as netMinLate,
  sum(if(arr_delay > 0, arr_delay, 0)) / 1000000 as minLate,
  sum(if(arr_delay < 0, arr_delay, 0)) / 1000000 as minEarly
FROM flights o
WHERE year = 2014
LIMIT 0,6;
```

```
+------------+-----------+----------+------------+----------+----------+
| numFlights | ontimePct | earlyPct | netMinLate | minLate  | minEarly |
+------------+-----------+----------+------------+----------+----------+
|    5819811 |    0.7868 |   0.5424 |    41.6116 | 77.6157  | -36.0042 |
+------------+-----------+----------+------------+----------+----------+
```

We see the right number of flights (about 6 million), and the percentage of flights that were early (about 54%) is also about right. The total number of minutes early (about 36 million) is also about right. However, the total number of minutes late is way off (about 78 million vs. 115 million), and as a consequence, so is the net number of minutes late (about 42 million vs. 80 million). In this case, you have to read the fine print. A description of the methodology used in this analysis contains some information about the *estimates*[6] of the arrival delay for cancelled flights. The problem is that cancelled flights have an arr_delay value of 0, yet in the real-world experience of travelers, the practical delay is much longer. The FiveThirtyEight data scientists concocted an estimate of the actual delay experienced by travelers due to cancelled flights.

> A quick-and-dirty answer is that canceled flights are associated with a delay of four or five hours, on average. However, the calculation varies based on the particular circumstances of each flight.

Unfortunately, reproducing the estimates made by FiveThirtyEight is likely impossible, and certainly beyond the scope of what we can accomplish here. But since we only care about the aggregate number of minutes, we can amend our computation to add, say, 270 minutes of delay time for each cancelled flight.

```
SELECT
  sum(1) as numFlights,
  sum(if(arr_delay < 15, 1, 0)) / sum(1) as ontimePct,
  sum(if(arr_delay < 0, 1, 0)) / sum(1) as earlyPct,
  sum(if(cancelled = 1, 270, arr_delay)) / 1000000 as netMinLate,
  sum(if(cancelled = 1, 270,
    if(arr_delay > 0, arr_delay, 0))) / 1000000 as minLate,
  sum(if(arr_delay < 0, arr_delay, 0)) / 1000000 as minEarly
FROM flights o
WHERE year = 2014
LIMIT 0,6;
```

```
+------------+-----------+----------+------------+----------+----------+
| numFlights | ontimePct | earlyPct | netMinLate | minLate  | minEarly |
+------------+-----------+----------+------------+----------+----------+
|    5819811 |    0.7868 |   0.5424 |    75.8972 | 111.9014 | -36.0042 |
+------------+-----------+----------+------------+----------+----------+
```

This again puts us in the neighborhood of the estimates from the article. One has to read the fine print to properly vet these estimates. The problem is not that the estimates reported by Silver are inaccurate—on the contrary, they seem plausible and are certainly better than not correcting for cancelled flights at all. However, it is not immediately clear from reading the article (you have to read the separate methodology article) that these estimates—which account for roughly 25% of the total minutes late reported—are in fact estimates and not hard data.

---

[6]Somehow, the word "estimate" is not used to describe what is being calculated.

Southwest's Delays Are Short; United's Are Long
As share of scheduled flights, 2014

- **FLIGHTS DELAYED 15-119 MINUTES**
- **FLIGHTS DELAYED 120+ MINUTES, CANCELED OR DIVERTED**

Southwest	23.0% 3.6%
Frontier	22.5 3.5
American	21.6 5.5
United	21.3 6.7
JetBlue	19.5 5.3
US Airways	15.8 2.9
Virgin America	15.1 3.4
Delta	14.2 2.2
Alaska	11.9 1.7
Hawaiian	7.4 0.7

FIVETHIRTYEIGHT · BASED ON DATA FROM THE BUREAU OF TRANSPORTATION STATISTICS

Figure 12.1: FiveThirtyEight data graphic summarizing airline delays by carrier. Reproduced with permission.

Later in the article, Silver presents a figure (reproduced below as Fig. 12.1) that breaks down the percentage of flights that were on time, had a delay of 15 to 119 minutes, or were delayed longer than two hours. We can pull the data for this figure with the following query. Here, in order to plot these results, we need to actually bring them back into R. To do this, we will use the functionality provided by the DBI package (see Section F.4.3 for more information about connecting to a MySQL server from within R).

```
query <-
  "SELECT o.carrier, c.name,
    sum(1) as numFlights,
    sum(if(arr_delay > 15 AND arr_delay <= 119, 1, 0)) as shortDelay,
    sum(if(arr_delay >= 120 OR
           cancelled = 1 OR diverted = 1, 1, 0)) as longDelay
  FROM
    flights o
  LEFT JOIN
    carriers c ON o.carrier = c.carrier
  WHERE year = 2014
  GROUP BY carrier
  ORDER BY shortDelay desc"
```

```
res <- DBI::dbGetQuery(db$con, query)
res
```

	carrier	name	numFlights	shortDelay	longDelay
1	WN	Southwest Airlines Co.	1174633	263237	42205
2	EV	ExpressJet Airlines Inc.	686021	136207	59663
3	OO	SkyWest Airlines Inc.	613030	107192	33114
4	DL	Delta Air Lines Inc.	800375	105194	19818
5	AA	American Airlines Inc.	537697	103360	22447
6	UA	United Air Lines Inc.	493528	93721	20923
7	MQ	Envoy Air	392701	87711	31194
8	US	US Airways Inc.	414665	64505	12328
9	B6	JetBlue Airways	249693	46618	12789
10	F9	Frontier Airlines Inc.	85474	18410	2959
11	AS	Alaska Airlines Inc.	160257	18366	2613
12	FL	AirTran Airways Corporation	79495	11918	2702
13	VX	Virgin America	57510	8356	1976
14	HA	Hawaiian Airlines Inc.	74732	5098	514

Reproducing the figure requires a little bit of work. We begin by stripping the names of the airlines of uninformative labels.

```
res <- res %>%
  mutate(name = gsub("Air(lines|ways| Lines)", "", name),
         name = gsub("(Inc\\.|Co\\.|Corporation)", "", name),
         name = gsub("\\(.*\\)", "", name),
         name = gsub(" *$", "", name))
res
```

	carrier	name	numFlights	shortDelay	longDelay
1	WN	Southwest	1174633	263237	42205
2	EV	ExpressJet	686021	136207	59663
3	OO	SkyWest	613030	107192	33114
4	DL	Delta	800375	105194	19818
5	AA	American	537697	103360	22447
6	UA	United	493528	93721	20923
7	MQ	Envoy Air	392701	87711	31194
8	US	US	414665	64505	12328
9	B6	JetBlue	249693	46618	12789
10	F9	Frontier	85474	18410	2959
11	AS	Alaska	160257	18366	2613
12	FL	AirTran	79495	11918	2702
13	VX	Virgin America	57510	8356	1976
14	HA	Hawaiian	74732	5098	514

Next, it is now clear that FiveThirtyEight has considered airline mergers and regional carriers that are not captured in our data. Specifically: "We classify all remaining AirTran flights as Southwest flights." Envoy Air serves American Airlines. However, there is a bewildering network of alliances among the other regional carriers. Greatly complicating matters, ExpressJet and Skywest serve multiple national carriers (primarily United, American, and Delta) under different flight numbers. FiveThirtyEight provides a footnote detailing how they have assigned flights carried by these regional carriers, but we have chosen to ignore that here and include ExpressJet and SkyWest as independent carriers. Thus, the data that we show in Figure 12.2 does not match the data shown in Figure 12.1 exactly, but we hope you will agree that it gets the broad strokes correct.

```
carriers2014 <- res %>%
  mutate(groupName = ifelse(name %in%
    c("Envoy Air", "American Eagle"), "American", name)) %>%
  mutate(groupName =
    ifelse(groupName == "AirTran", "Southwest", groupName)) %>%
  group_by(groupName) %>%
  summarize(numFlights = sum(numFlights),
            wShortDelay = sum(shortDelay),
            wLongDelay = sum(longDelay)) %>%
  mutate(wShortDelayPct = wShortDelay / numFlights,
         wLongDelayPct = wLongDelay / numFlights,
         delayed = wShortDelayPct + wLongDelayPct,
         ontime = 1 - delayed)
carriers2014
```

```
# A tibble: 12  8
       groupName numFlights wShortDelay wLongDelay wShortDelayPct
         <chr>      <dbl>      <dbl>      <dbl>      <dbl>
1        Alaska     160257      18366       2613     0.1146
2      American     930398     191071      53641     0.2054
3         Delta     800375     105194      19818     0.1314
4     ExpressJet    686021     136207      59663     0.1985
5       Frontier     85474      18410       2959     0.2154
6       Hawaiian     74732       5098        514     0.0682
7        JetBlue     249693      46618      12789     0.1867
8        SkyWest     613030     107192      33114     0.1749
9      Southwest    1254128     275155      44907     0.2194
10        United     493528      93721      20923     0.1899
11            US     414665      64505      12328     0.1556
12 Virgin America     57510       8356       1976     0.1453
# ... with 3 more variables: wLongDelayPct <dbl>, delayed <dbl>,
#   ontime <dbl>
```

After tidying this data frame using the gather() function (see Chapter 5), we can draw the figure as a stacked bar chart.

```
carriers_tidy <- carriers2014 %>%
  select(groupName, wShortDelayPct, wLongDelayPct, delayed) %>%
  tidyr::gather(key = "delay_type", value = "pct", -groupName, -delayed)
delay_chart <- ggplot(data = carriers_tidy,
                      aes(x = reorder(groupName, pct, max), y = pct)) +
  geom_bar(stat = "identity", aes(fill = delay_type)) +
  scale_fill_manual(name = NULL, values = c("red", "gold"),
        labels = c("Flights Delayed 120+ Minutes, Canceled or Diverted",
                   "Flights Delayed 15-119 Minutes")) +
  scale_y_continuous(limits = c(0, 1)) +
```

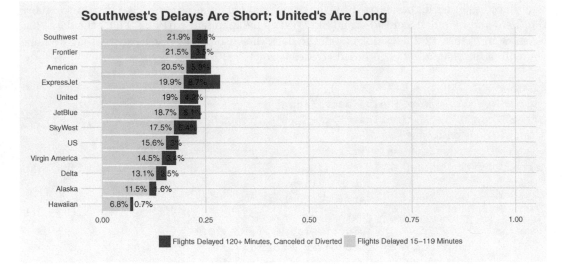

Figure 12.2: Re-creation of the FiveThirtyEight plot on flight delays.

```
coord_flip() +
ggtitle("Southwest's Delays Are Short; United's Are Long") +
ylab(NULL) + xlab(NULL) +
ggthemes::theme_fivethirtyeight()
```

Getting the right text labels in the right places to mimic Figure 12.1 requires additional wrangling. We show our best effort in Figure 12.2. In fact, by comparing the two figures, it becomes clear that many of the long delays suffered by United and American passengers occur on flights operated by ExpressJet and Skywest.

```
delay_chart +
  geom_text(data = filter(carriers_tidy, delay_type == "wShortDelayPct"),
    aes(label = paste0(round(pct * 100, 1), "% ")), hjust = "right") +
  geom_text(data = filter(carriers_tidy, delay_type == "wLongDelayPct"),
    aes(y = delayed - pct, label = paste0(round(pct * 100, 1), "% ")),
          hjust = "left", nudge_y = 0.01)
```

The rest of the analysis is predicated on FiveThirtyEight's definition of *target time*, which is different than the scheduled time in the database. To compute it would take us far astray. In another graphic in the article, FiveThirtyEight reports the slowest and fastest airports among the 30 largest airports.

Using arrival delay time instead of the FiveThirtyEight-defined target time, we can produce a similar table by joining the results of two queries together.

```
queryDest <- "SELECT
          dest,
          sum(1) as numFlights,
          avg(arr_delay) as avgArrivalDelay
        FROM
          flights o
        WHERE year = 2014
        GROUP BY dest
        ORDER BY numFlights desc
        LIMIT 0, 30"
dests <- DBI::dbGetQuery(db$con, queryDest)
queryArr <- "SELECT
          origin,
          sum(1) as numFlights,
          avg(arr_delay) as avgDepartDelay
        FROM
          flights o
        WHERE year = 2014
        GROUP BY origin
        ORDER BY numFlights desc
        LIMIT 0, 30"
origins <- DBI::dbGetQuery(db$con, queryArr)
dests %>%
  left_join(origins, by = c("dest" = "origin")) %>%
  select(dest, avgDepartDelay, avgArrivalDelay) %>%
  arrange(desc(avgDepartDelay))
```

```
   dest avgDepartDelay avgArrivalDelay
1   ORD         14.301          13.148
2   MDW         12.801           7.399
3   DEN         11.350           7.595
4   IAD         11.338           7.453
5   HOU         11.282           8.066
6   DFW         10.687           8.999
7   BWI         10.186           6.044
8   BNA          9.472           8.943
9   EWR          8.704           9.612
10  IAH          8.405           6.750
11  MCO          8.298           7.178
12  SFO          8.124          12.185
13  PHL          6.977           6.225
14  LAS          6.947           6.316
15  FLL          6.485           7.829
16  JFK          6.475           6.969
17  TPA          6.474           7.060
18  PHX          6.392           4.180
19  LAX          6.084           6.504
20  LGA          5.831           8.129
21  SAN          5.674           6.526
22  CLT          5.479           2.772
23  MIA          4.436           3.234
```

24	BOS	4.417	4.668
25	DCA	4.313	4.159
26	ATL	4.195	4.346
27	DTW	4.082	2.404
28	MSP	3.890	3.798
29	SEA	2.485	3.113
30	SLC	0.496	0.884

Finally, FiveThirtyEight produces a simple table ranking the airlines by the amount of time added versus *typical*—another of their creations—and target time.

What we can do instead is compute a similar table for the average arrival delay time by carrier, *after controlling for the routes*. First, we compute the average arrival delay time for each route.

```
query <- "SELECT
            origin, dest,
            sum(1) as numFlights,
            avg(arr_delay) as avgDelay
          FROM
            flights o
          WHERE year = 2014
          GROUP BY origin, dest"
routes <- dbGetQuery(db$con, query)
head(routes)
```

```
  origin dest numFlights avgDelay
1    ABE  ATL        829     5.43
2    ABE  DTW        665     3.23
3    ABE  ORD        144    19.51
4    ABI  DFW       2832    10.70
5    ABQ  ATL        893     1.92
6    ABQ  BWI        559     6.60
```

Next, we perform the same calculation, but this time, we add `carrier` to the `GROUP BY` clause.

```
query <- "SELECT
            origin, dest,
            o.carrier, c.name,
            sum(1) as numFlights,
            avg(arr_delay) as avgDelay
          FROM
            flights o
          LEFT JOIN
            carriers c ON o.carrier = c.carrier
          WHERE year = 2014
          GROUP BY origin, dest, o.carrier"
routes_carriers <- dbGetQuery(db$con, query)
```

Next, we merge these two data sets, matching the routes traveled by each carrier with the route averages across all carriers

```
routes_aug <- left_join(routes_carriers, routes,
  by = c("origin" = "origin", "dest" = "dest"))
head(routes_aug)
```

```
  origin dest carrier                       name numFlights.x avgDelay.x
1    ABE  ATL      DL      Delta Air Lines Inc.          186       1.67
2    ABE  ATL      EV  ExpressJet Airlines Inc.          643       6.52
3    ABE  DTW      EV  ExpressJet Airlines Inc.          665       3.23
4    ABE  ORD      EV  ExpressJet Airlines Inc.          144      19.51
5    ABI  DFW      EV  ExpressJet Airlines Inc.          219       7.00
6    ABI  DFW      MQ                 Envoy Air         2613      11.01
  numFlights.y avgDelay.y
1          829       5.43
2          829       5.43
3          665       3.23
4          144      19.51
5         2832      10.70
6         2832      10.70
```

Note that routes_aug contains both the average arrival delay time for each carrier on each route that it flies (avgDelay.x) as well as the average arrival delay time for each route across all carriers (avgDelay.y). We can then compute the difference between these times, and aggregate the weighted average for each carrier.

```
routes_aug %>%
  group_by(carrier) %>%
  # use gsub to remove parentheses
  summarise(carrier_name = gsub("\\(.*\\)", "", first(name)),
            numRoutes = n(), numFlights = sum(numFlights.x),
            wAvgDelay = sum(numFlights.x * (avgDelay.x - avgDelay.y),
                na.rm = TRUE) / sum(numFlights.x)) %>%
  arrange(wAvgDelay)
```

```
# A tibble: 14  5
   carrier              carrier_name numRoutes numFlights wAvgDelay
     <chr>                     <chr>     <int>      <dbl>     <dbl>
1       VX            Virgin America        72      57510    -2.694
2       FL AirTran Airways Corporation      170      79495    -1.552
3       AS       Alaska Airlines Inc.       242     160257    -1.445
4       US           US Airways Inc.        378     414665    -1.306
5       DL      Delta Air Lines Inc.        900     800375    -1.005
6       UA     United Air Lines Inc.        621     493528    -0.982
7       MQ                 Envoy Air        442     392701    -0.455
8       AA    American Airlines Inc.        390     537697    -0.034
9       HA     Hawaiian Airlines Inc.        56      74732     0.272
10      OO     SkyWest Airlines Inc.       1250     613030     0.358
11      B6           JetBlue Airways        316     249693     0.767
12      EV  ExpressJet Airlines Inc.       1534     686021     0.845
13      WN     Southwest Airlines Co.      1284    1174633     1.133
14      F9    Frontier Airlines Inc.        326      85474     2.289
```

12.6   SQL vs. R

This chapter contains an introduction to the database querying language SQL. However, along the way we have highlighted the similarities and differences between the way certain things are done in R versus how they are done in SQL. While the rapid development of `dplyr` has brought fusion to the most common data management operations shared by both R and SQL, while at the same time shielding the user from concerns about where certain operations are being performed, it is important for a practicing data scientist to understand the relative strengths and weaknesses of each of their tools.

Thus, while the process of slicing and dicing data can generally be performed in either R or SQL, we have already seen tasks for which one is more appropriate (e.g., faster, simpler, or more logically structured) than the other. R is a statistical computing environment that is developed for the purpose of data analysis. If the data are small enough to be read into memory, then R puts a vast array of data analysis functions at your fingertips. However, if the data are large enough to be problematic in memory, then SQL provides a robust, parallelizable, and scalable solution for data storage and retrieval. The SQL query language, or the `dplyr` interface, enable one to efficiently perform basic data management operations on smaller pieces of the data. However, there is an upfront cost to creating a well-designed SQL database. Moreover, the analytic capabilities of SQL are very limited, offering only a few simple statistical functions (e.g., `avg`, `sd`, etc.—although user-defined extensions are possible)). Thus, while SQL is usually a more robust solution for *data management*, it is a poor substitute for R when it comes to *data analysis*.

12.7   Further resources

The documentation for MySQL, PostgreSQL, and SQLite are the authoritative sources for complete information on their respective syntaxes. We have also found [126] to be a useful reference.

12.8   Exercises

Each of the following exercises can be solved via a single SQL query. Equivalently, each can be solved via a single pipeline of `dplyr` commands. Write valid solutions using both methods.

The exercises about flights assume that you have access to an SQL database that has been populated with the appropriate flight delay data. Please see the `src_scidb()` function in the `mdsr` package for access to these data on a pre-populated server. To create your own database, use the `airlines` package.

Exercise 12.1

How many domestic flights flew into Dallas-Fort Worth (`DFW`) on May 14, 1998?

Exercise 12.2

Find all flights between `JFK` and `SFO` in 1994. How many were cancelled? What percentage of the total number of flights were cancelled?

Exercise 12.3

Of all the destinations from Chicago O'Hare (`ORD`), which were the most common in 1997?

Exercise 12.4

Which airport had the highest average arrival delay time in 2008?

Exercise 12.5

How many domestic flights came into or flew out of Bradley Airport (BDL) in 2012?

Exercise 12.6

List the airline and flight number for all flights between LAX and JFK on September 26th, 1990.

The following questions require use of the Lahman package and reference basic baseball terminology. Please see https://en.wikipedia.org/wiki/Baseball_statistics for comprehensive explanations of any acronyms.

Exercise 12.7

List the names of all batters who have at least 300 home runs (HR) and 300 stolen bases (SB) in their careers and rank them by career batting average (H/AB).

Exercise 12.8

List the names of all pitchers who have at least 300 wins (W) and 3,000 strikeouts (SO) in their careers and rank them by career winning percentage ($W/(W + L)$).

Exercise 12.9

The attainment of either 500 home runs (HR) or 3,000 hits (H) in a career is considered to be among the greatest achievements to which a batter can aspire. These milestones are thought to guarantee induction into the Baseball Hall of Fame, and yet several players who have attained either milestone have not been inducted into the Hall of Fame. Identify them.

The following question may require more than one query, and a more thoughtful response.

Exercise 12.10

Based on data from 2012 only, and assuming that transportation to the airport is not an issue, would you rather fly out of JFK, LaGuardia (LGA), or Newark (EWR)? Why or why not?

Chapter 13

Database administration

In Chapter 12, we learned how to write SELECT queries to retrieve data from an existing SQL server. Of course, these queries depend on that server being configured, and the proper data loaded into it. In this chapter, we provide the tools necessary to set up a new database and populate it. Furthermore, we present concepts that will help you construct efficient databases that enable faster query performance. While the treatment herein is not sufficient to make you a seasoned database administrator, it should be enough to allow you to start experimenting with SQL databases on your own.

As in Chapter 12, the code that you see in this chapter illustrates exchanges between a MySQL server and the command line client. In places where R is involved, we will make that explicit. We assume that you are able to log in to a MySQL server. (See Appendix F for instructions on how to install, configure, and log in to an SQL server.)

13.1 Constructing efficient SQL databases

While it is often helpful to think about SQL tables as being analogous to data.frames in R, there are some important differences. In R, a data.frame is a list of vectors that have the same length. Each of those vectors has a specific data type (e.g., integers, character strings, etc.), but those data types can vary across the columns. The same is true of tables in SQL, but there are additional constraints that we can impose on SQL tables that can improve both the logical integrity of our data, as well as the performance we can achieve when searching it.

13.1.1 Creating new databases

Once you have logged into MySQL, you can see what databases are available to you by running the SHOW DATABASES command at the mysql> prompt:

```
SHOW DATABASES;

+--------------------+
| Database           |
+--------------------+
| information_schema |
| airlines           |
| imdb               |
| lahman             |
```

```
| math              |
| retrosheet        |
| yelp              |
+-------------------+
```

In this case, the airlines database already exists. But if it didn't, we could create it using the CREATE DATABASE command.

CREATE DATABASE airlines;

Since we will continue to work with the airlines database, we can save ourselves some typing by using the USE command to make that connection explicit.

USE airlines;

Now that we are confined to the airlines database, there is no ambiguity in asking what tables are present.

SHOW TABLES;

```
+---------------------+
| Tables_in_airlines  |
+---------------------+
| airports            |
| carriers            |
| flights             |
| planes              |
| summary             |
| weather             |
+---------------------+
```

13.1.2   CREATE TABLE

Recall that in Chapter 12 we used the DESCRIBE statement to display the definition of each table. This lists each field, its data type, whether there are keys or indices defined on it, and whether NULL values are allowed. For example, the airports table has the following definition.

DESCRIBE airports;

```
+----------+---------------+------+-----+---------+-------+
| Field    | Type          | Null | Key | Default | Extra |
+----------+---------------+------+-----+---------+-------+
| faa      | varchar(3)    | NO   | PRI |         |       |
| name     | varchar(255)  | YES  |     | NULL    |       |
| lat      | decimal(10,7) | YES  |     | NULL    |       |
| lon      | decimal(10,7) | YES  |     | NULL    |       |
| alt      | int(11)       | YES  |     | NULL    |       |
| tz       | smallint(4)   | YES  |     | NULL    |       |
| dst      | char(1)       | YES  |     | NULL    |       |
| city     | varchar(255)  | YES  |     | NULL    |       |
| country  | varchar(255)  | YES  |     | NULL    |       |
+----------+---------------+------+-----+---------+-------+
```

We can see from this that the `faa`, `name`, `city`, and `country` fields are defined as `varchar` (or variable character) fields. These fields contain character strings, but the length of the strings allowed varies. We know that the `faa` code is restricted to three characters, and so we have codified that in the table definition. The `dst` field contains only a single character, indicating whether daylight saving time is observed at each airport. The `lat` and `lon` fields contain geographic coordinates, which can be three-digit numbers (i.e., the maximum value is 180) with up to seven decimal places. The `tz` field can be up to a four-digit integer, while the `alt` field is allowed eleven digits. In this case, `NULL` values are allowed, and are the default, in all of the fields except for `faa`, which is the primary key.

These definitions did not come out of thin air, nor were they automatically generated. In this case, we wrote them by hand, in the following `CREATE TABLE` statement:

```
SHOW CREATE TABLE airports;

+-----------+---------------------------+
| Table     | Create Table              |
+-----------+---------------------------+
| airports  | CREATE TABLE `airports` (
  `faa` varchar(3) NOT NULL DEFAULT '',
  `name` varchar(255) DEFAULT NULL,
  `lat` decimal(10,7) DEFAULT NULL,
  `lon` decimal(10,7) DEFAULT NULL,
  `alt` int(11) DEFAULT NULL,
  `tz` smallint(4) DEFAULT NULL,
  `dst` char(1) DEFAULT NULL,
  `city` varchar(255) DEFAULT NULL,
  `country` varchar(255) DEFAULT NULL,
  PRIMARY KEY (`faa`)
) ENGINE=MyISAM DEFAULT CHARSET=latin1 |
+-----------+---------------------------+
```

As you can see, the `CREATE TABLE` command starts by defining the name of the table, and then proceeds to list the field definitions in a comma-separated list. If you want to build a base from scratch—as we do in Section 13.3—you will have to write these definitions for each table[1]. Tables that are already created can be modified using the `ALTER TABLE` command. For example, the following will change the `tz` field to two digits and change the default value to zero.

```
ALTER TABLE airports CHANGE tz tz smallint(2) DEFAULT 0;
```

13.1.3 Keys

Two related but different concepts are *keys* and *indices*. The former offers some performance advantages but is primarily useful for imposing constraints on possible entries in the database, while the latter is purely about improving the speed of retrieval.

Different RDBMSs may implement a variety of different kinds of keys, but three types are most common. In each case, suppose that we have a table with n rows and p columns.

---

[1]There are ways of automatically generating table schemas, but in many cases some manual tweaking is recommended.

PRIMARY KEY: a column or set of columns in a table that uniquely identifies each row. By convention, this column is often called id. A table can have at most one primary key, and in general it is considered good practice to define a primary key on every table (although there are exceptions to this rule). If the index spans $k < p$ columns, then even though the primary key must by definition have n rows itself, it only requires nk pieces of data, rather than the np that the full table occupies. Thus, the primary key is always smaller than the table itself, and is thus faster to search. A second critically important role of the primary key is enforcement of non-duplication. If you try to insert a row into a table that would result in a duplicate entry for the primary key, you will get an error.

UNIQUE KEY: a column or set of columns in a table that uniquely identifies each row, except for rows that contain NULL in some of those attributes. Unlike primary keys, a single table may have many unique keys. A typical use for these are in a lookup table. For example, Ted Turocy maintains a register of player ids for professional baseball players across multiple data providers. Each row in this table is a different player, and the primary key is a randomly-generated hash—each player gets exactly one value. However, each row also contains that same player's id in systems designed by ML-BAM, Baseball-Reference, Baseball Prospectus, Fangraphs, etc. This is tremendously useful for researchers working with multiple data providers, since they can easily link a player's statistics in one system to his information in another. However, this ability is predicated on the *uniqueness* of each player's id in *each* system. Moreover, many players may not have an id in every system, since data providers track minor league baseball, or even the Japanese and Korean professional leagues. Thus, the imposition of a unique key—which allows NULLs—is necessary to maintain the integrity of these data.

FOREIGN KEY: a column or set of columns that reference a primary key in another table. For example, the primary key in the carriers table is code. The carrier column in the flights table, which consists of carrier ids, is a foreign key that references carriers.code. Foreign keys don't offer any performance enhancements, but they are important for maintaining *referential integrity*, especially in transactional databases that have many insertions and deletions.

You can use the SHOW KEYS command to identify the keys in a table. Note that the carriers table has only one key defined: a primary key on code.

SHOW KEYS FROM carriers;

```
+----------+------------+----------+--------------+-------------+----------+
| Table    | Non_unique | Key_name | Seq_in_index | Column_name | Collation|
+----------+------------+----------+--------------+-------------+----------+
| carriers |          0 | PRIMARY  |            1 | carrier     | A        |
+----------+------------+----------+--------------+-------------+----------+
```

13.1.4   Indices

While keys help maintain the integrity of the data, indices impose no constraints—they simply enable faster retrieval. An index is a lookup table that helps SQL keep track of which records contain certain values. Judicious use of indices can dramatically speed up retrieval times. The technical implementation of efficient indices is an active area of research among computer scientists, and fast indices are one of the primary advantages that differentiate SQL tables from R data frames.

Indices have to be built by the database in advance, and they are then written to the disk. Thus, indices take up space on the disk (this is one of the reasons that they aren't implemented in R). For some tables with many indices, the size of the indices can even exceed the size of the raw data. Thus, when building indices, there is a trade-off to consider: You want just enough indices but not too many.

Consider the task of locating all of the rows in the `flights` table that contain the `origin` value BDL. These rows are strewn about the table in no particular order. How would you find them? A simple approach would be to start with the first row, examine the `origin` field, grab it if it contains BDL, and otherwise move to the second row. In order to ensure that all of the matching rows are returned, this algorithm must check every single one of the $n = 169$ million rows in this table! So its speed is $O(n)$. However, we have built an index on the `origin` column, and this index contains only 6,674 rows. Each row in the index corresponds to exactly one value of `origin`, and contains a lookup for the exact rows in the table that are specific to that value. Thus, when we ask for the rows for which `origin` is equal to BDL, the database will use the index to deliver those rows very quickly. In practice, the retrieval speed for indexed columns is $O(\ln n)$ (or better)—which can be a tremendous advantage when n is large.

The speed-up that indices can provide is often especially apparent when joining two large tables. To see why, consider the following toy example. Suppose we want to merge two tables on the columns whose values are listed below. To merge these records correctly, we have to do a lot of work to find the appropriate value in the second list that matches each value in the first list.

```
[1]   5 18  2  3  4  2  1
[1]   5  6  3 18  4  7  1  2
```

On the other hand, consider performing the same task on the same set of values, but having the values sorted ahead of time. Now, the merging task is very fast, because we can quickly locate the matching records. In effect, by keeping the records sorted, we have off-loaded the sorting task when we do a merge, resulting in much faster merging performance. However, this requires that we sort the records in the first place and then keep them sorted. This may slow down other operations—such as inserting new records—which now have to be done more carefully.

```
[1]   1  2  2  3  4  5 18
[1]   1  2  3  4  5  6  7 18
```

```
SHOW INDEXES FROM flights;
```

Table	Non_unique	Key_name	Seq_in_index	Column_name	Collation
flights	1	Year	1	year	A
flights	1	Date	1	year	A
flights	1	Date	2	month	A
flights	1	Date	3	day	A
flights	1	Origin	1	origin	A
flights	1	Dest	1	dest	A

```
| flights |           1 | Carrier  |          1 | carrier  | A        |
| flights |           1 | tailNum  |          1 | tailnum  | A        |
+---------+------------+----------+------------+----------+----------+
```

Since all keys are indices, MySQL does not distinguish between them, and thus the SHOW
INDEXES command is equivalent to SHOW KEYS. Note that the flights table has several keys
defined, but no primary key. The key Date spans the three columns year, month, and day.

13.1.5   EXPLAIN

It is important to have the right indices built for your specific data and the queries that
are likely to be run on it. Unfortunately, there is not always a straightforward answer to
the question of which indices to build. For the flights table, it seems likely to us that
many queries will involve searching for flights from a particular origin, or to a particular
destination, or during a particular year (or range of years), or on a specific carrier, and so
we have built indices on each of these columns. We have also built the Date index, since it
seems likely that people would want to searching for flights on a certain date. However, it
does not seems so likely that people would be search for flights in a specific month across
all years, and thus we have not built an index on month alone. The Date index contains
the month column, but this index can only be used if year is also part of the query.

You can ask MySQL for information about how it is going to perform a query using the
EXPLAIN syntax. This will help you understand how onerous your query is, without actually
running it—saving you the time of having to wait for it to execute.

If we were to run a query for long flights using the distance column, then since this
column is not indexed, the server will have to inspect each of the 169 million rows.

```
EXPLAIN SELECT * FROM flights WHERE distance > 3000;

+----+-------------+---------+------+---------------+------+---------+
| id | select_type | table   | type | possible_keys | key  | key_len |
+----+-------------+---------+------+---------------+------+---------+
|  1 | SIMPLE      | flights | ALL  | NULL          | NULL | NULL    |
+----+-------------+---------+------+---------------+------+---------+
```

On the other hand, if we search for recent flights using the year column, which has an
index built on it, then we only need to consider a fraction of those rows.

```
EXPLAIN SELECT * FROM flights WHERE year = 2013;

+----+-------------+---------+------+---------------+------+---------+
| id | select_type | table   | type | possible_keys | key  | key_len |
+----+-------------+---------+------+---------------+------+---------+
|  1 | SIMPLE      | flights | ALL  | Year,Date     | NULL | NULL    |
+----+-------------+---------+------+---------------+------+---------+
```

Note that in this case the server could have used either the index Year or the index
Date (which contains the column year). Because of the index, only the 6.3 million flights
from 2013 were consulted. Similarly, if we search by year and month, we can use the Date
index.

```
EXPLAIN SELECT * FROM flights WHERE year = 2013 AND month = 6;
```

id	select_type	table	type	possible_keys	key	key_len
1	SIMPLE	flights	ref	Year,Date	Date	6

But if we search for months across all years, we can't!

```
EXPLAIN SELECT * FROM flights WHERE month = 6;
```

id	select_type	table	type	possible_keys	key	key_len
1	SIMPLE	flights	ALL	NULL	NULL	NULL

This is because although month is part of the Date index, it is the *second* column in the index, and thus it doesn't help us when we aren't filtering on year. Thus, if it were common for our users to search on month without year, it would probably be worth building an index on month. Were we to actually run these queries, there would be a significant difference in computational time.

Using indices is especially important when performing JOIN operations on large tables. Note again how the use of the index on year speeds up the query by considering far fewer rows.

```
EXPLAIN
    SELECT * FROM planes p
    LEFT JOIN flights o ON p.tailnum = o.TailNum
    WHERE manufacturer = 'BOEING';
EXPLAIN
    SELECT * FROM planes p
    LEFT JOIN flights o ON p.Year = o.Year
    WHERE manufacturer = 'BOEING';
```

id	select_type	table	type	possible_keys	key	key_len
1	SIMPLE	p	ALL	NULL	NULL	NULL
1	SIMPLE	o	ref	tailNum	tailNum	9

id	select_type	table	type	possible_keys	key	key_len
1	SIMPLE	p	ALL	NULL	NULL	NULL
1	SIMPLE	o	ref	Year,Date	Year	3

13.1.6   Partitioning

Another approach to speeding up queries on large tables (like `flights`) is *partitioning*. Here, we could create partitions based on the `year`. For `flights` this would instruct the server to physically write the `flights` table as a series of smaller tables, each one specific to a single value of `year`. At the same time, the server would create a logical supertable, so that to the user, the appearance of `flights` would be unchanged. This acts like a preemptive index on the `year` column.

If most of the queries to the `flights` table were for a specific year or range of years, then partitioning could significantly improve performance, since most of the rows would never be consulted. For example, if most of the queries to the `flights` database were for the past three years, then partitioning would reduce the search space of most queries to the roughly 20 million flights in the last three years instead of the 169 million rows in the last 20 years. But here again, if most of the queries to the `flights` table were about carriers across years, then this type of partitioning would not help at all. It is the job of the database designer to tailor the database structure to the pattern of queries coming from the users. As a data scientist, this may mean that you have to tailor the database structure to the queries that you are running.

13.2   Changing SQL data

In Chapter 12, we described how to query an SQL database using the `SELECT` command. Thus far in this chapter, we have discussed how to set up an SQL database, and how to optimize it for speed. None of these operations actually change data in an existing database. In this section, we will briefly touch upon the `UPDATE` and `INSERT` commands, which allow you to do exactly that.

13.2.1   UPDATE

The `UPDATE` command allows you to reset values in a table across all rows that match a certain criteria. For example, in Chapter 12 we discussed the possibility that airports could change names over time. The airport in Washington, D.C. with code `DCA` is now called Ronald Reagan Washington National.

```
SELECT faa, name FROM airports WHERE faa = 'DCA';

+-----+-------------------------------+
| faa | name                          |
+-----+-------------------------------+
| DCA | Ronald Reagan Washington Natl |
+-----+-------------------------------+
```

However, the "Ronald Reagan" prefix was added in 1998. If—for whatever reason—we wanted to go back to the old name, we could use an `UPDATE` command to change that information in the `airports` table.

```
UPDATE airports
  SET name = 'Washington National'
  WHERE faa = 'DCA';
```

An UPDATE operation can be very useful when you have to apply wholesale changes over a large number of rows. However, extreme caution is necessary, since an imprecise UPDATE query can wipe out large quantities of data, and there is no "undo" operation!

Pro Tip: Exercise extraordinary caution when performing UPDATEs.

13.2.2 INSERT

New data can be appended to an existing table with the INSERT commands. There are actually three things that can happen, depending on what you want to do when you have a primary key conflict. This occurs when one of the new rows that you are trying to insert has the same primary key value as one of the existing rows in the table.

INSERT Try to insert the new rows. If there is a primary key conflict, quit and return an error.

INSERT IGNORE Try to insert the new rows. If there is a primary key conflict, skip inserting the conflicting rows and leave the existing rows untouched.

REPLACE Try to insert the new rows. If there is a primary key key conflict, overwrite the existing rows with the new ones.

Recall that in Chapter 12 we found that the airports in Puerto Rico were not present in the `airports` table. If we wanted to add these manually, we could use INSERT.

```
INSERT INTO airports (faa, name)
  VALUES ('SJU', 'Luis Munoz Marin International Airport');
```

Since `faa` is the primary key on this table, we can insert this row without contributing values for all of the other fields. In this case, the new row corresponding to SJU would have the `faa` and `name` fields as noted above, and the default values for all of the other fields. If we were to run this operation a second time, we would get an error, because of the primary key collision on SJU. We could avoid the error by choosing to INSERT INGORE or REPLACE instead of INSERT.

13.2.3 LOAD DATA

In practice, we rarely add new data manually in this manner. Instead, new data are most often added using the LOAD DATA command. This allows a file containing new data—usually a CSV—to be inserted in bulk. This is very common, when, for example, your data comes to you daily in a CSV file and you want to keep your database up to date. The primary key collision concepts described above also apply to the LOAD DATA syntax, and are important to understand for proper database maintenance. We illustrate the use of LOAD DATA in Section 13.3.

13.3 Extended example: Building a database

The *extract-transform-load* (ETL) paradigm is common among data professionals. The idea is that many data sources need to be extracted from some external source, transformed into a different format, and finally loaded into a database system. Often, this is an iterative process that needs to be done every day, or even every hour. In such cases developing the infrastructure to automate these steps can result in dramatically increased productivity.

In this example, we will illustrate how to set up a MySQL database for the `babynames` data using the command line and SQL, but not R. As noted previously, while the `dplyr` package has made R a viable interface for querying and populating SQL databases, its functionality is not nearly complete. It is occasionally necessary to get "under the hood" with SQL. The files that correspond to this example can be found on the book website at `http://mdsr-book.github.io/`.

13.3.1 Extract

In this case, our data already lives in an R package, but in most cases, your data will live on a website, or be available in a different format. Our goal is to take that data from wherever it is and download it. For the `babynames` data, there isn't much to do, since we already have the data in an R package. We will simply load it.

```
library(babynames)
```

13.3.2 Transform

Since SQL tables conform to a row-and-column paradigm, our goal during the transform phase is to create CSV files (see Chapter 5) for each of the tables. In this example we will create tables for the `babynames` and `births` tables. You can try to add the `applicants` and `lifetables` tables on your own. We will simply write these data to CSV files using the `write.csv()` command. Since the babynames table is very long (nearly 1.8 million rows), we will just use the more recent data.

```
babynames %>%
  filter(year > 1975) %>%
  write.csv(file = "babynames.csv", row.names = FALSE)
births %>%
  write.csv(file = "births.csv", row.names = FALSE)
list.files(".", pattern = ".csv")
```

```
[1] "babynames.csv" "births.csv"
```

This raises an important question: what should we call these objects? The babynames package includes a data frame called `babynames` with one row per sex per year per name. Having both the database and a table with the same name may be confusing. To clarify which is which we will call the database `babynamedb` and the table `babynames`.

Pro Tip: Spending time thinking about the naming of databases, tables, and fields before you create them can help avoid confusion later on.

13.3.3 Load into MySQL database

Next, we need to write a script that will define the table structure for these two tables in a MySQL database (instructions for creation of a database in SQLite can be found in Section F.4.4). This script will have four parts:

1. a USE statement that ensures we are in the right schema/database

2. a series of DROP TABLE statements that drop any old tables with the same names as the ones we are going to create

3. a series of CREATE TABLE statements that specify the table structures

4. a series of LOAD DATA statements that read the data from the CSVs into the appropriate tables

The first part is easy:

```
USE babynamedb;
```

This assumes that we have a local database called babynamedata—we will create this later. The second part is easy in this case, since we only have two tables. These ensure that we can run this script as many times as we want.

```
DROP TABLE IF EXISTS babynames;
DROP TABLE IF EXISTS births;
```

Pro Tip: Be careful with the DROP TABLE statement. It destroys data.

The third step is the trickiest part—we need to define the columns precisely. The use of str(), summary(), and glimpse() are particularly useful for matching up R data types with MySQL data types. Please see the MySQL documentation for more information about what data types are supported.

```
glimpse(babynames)
```

```
Observations: 1,825,433
Variables: 5
$ year <dbl> 1880, 1880, 1880, 1880, 1880, 1880, 1880, 1880, 1880, 188...
$ sex  <chr> "F", "F", "F", "F", "F", "F", "F", "F", "F", "F", "F", "F...
$ name <chr> "Mary", "Anna", "Emma", "Elizabeth", "Minnie", "Margaret"...
$ n    <int> 7065, 2604, 2003, 1939, 1746, 1578, 1472, 1414, 1320, 128...
$ prop <dbl> 0.07238, 0.02668, 0.02052, 0.01987, 0.01789, 0.01617, 0.0...
```

In this case, we know that the year variable will only contain four-digit integers, so we can specify that this column take up only that much room in SQL. Similarly, the sex variable is just a single character, so we can restrict the width of that column as well. These savings probably won't matter much in this example, but for large tables they can make a noticeable difference.

```
CREATE TABLE `babynames` (
  `year` smallint(4) NOT NULL DEFAULT 0,
  `sex` char(1) NOT NULL DEFAULT 'F',
  `name` varchar(255) NOT NULL DEFAULT '',
  `n` mediumint(7) NOT NULL DEFAULT 0,
  `prop` decimal(21,20) NOT NULL DEFAULT 0,
  PRIMARY KEY (`year`, `sex`, `name`)
) ENGINE=MyISAM DEFAULT CHARSET=latin1;
```

In this table, each row contains the information about one name for one sex in one year. Thus, each row contains a unique combination of those three variables, and we can therefore define a primary key across those three fields. Note the use of backquotes (to denote tables and variables) and the use of regular quotes (for default values).

```
glimpse(births)

Observations: 104
Variables: 2
$ year   <int> 1909, 1910, 1911, 1912, 1913, 1914, 1915, 1916, 1917, 1...
$ births <dbl> 2718000, 2777000, 2809000, 2840000, 2869000, 2966000, 2...
```

```
CREATE TABLE `births` (
  `year` smallint(4) NOT NULL DEFAULT 0,
  `births` mediumint(8) NOT NULL DEFAULT 0,
  PRIMARY KEY (`year`)
) ENGINE=MyISAM DEFAULT CHARSET=latin1;
```

Finally, we have to tell MySQL where to find the CSV files and where to put the data it finds in them. This is accomplished using the LOAD DATA command. You may also need to add a LINES TERMINATED BY \r\n clause, but we have omitted that for clarity. Please be aware that lines terminate using different characters in different operating systems, so Windows, Mac, and Linux users may have to tweak these commands to suit their needs. The SHOW WARNINGS commands are not necessary, but they will help with debugging.

```
LOAD DATA LOCAL INFILE './babynames.csv' INTO TABLE `babynames`
  FIELDS TERMINATED BY ',' OPTIONALLY ENCLOSED BY '"' IGNORE 1 LINES;
SHOW WARNINGS;
LOAD DATA LOCAL INFILE './births.csv' INTO TABLE `births`
  FIELDS TERMINATED BY ',' OPTIONALLY ENCLOSED BY '"' IGNORE 1 LINES;
SHOW WARNINGS;
```

Putting this all together, we have the following script:

```
USE babynamedata;

DROP TABLE IF EXISTS babynames;
DROP TABLE IF EXISTS births;

CREATE TABLE `babynames` (
  `year` smallint(4) NOT NULL DEFAULT 0,
  `sex` char(1) NOT NULL DEFAULT 'F',
  `name` varchar(255) NOT NULL DEFAULT '',
  `n` mediumint(7) NOT NULL DEFAULT 0,
  `prop` decimal(21,20) NOT NULL DEFAULT 0,
  PRIMARY KEY (`year`, `sex`, `name`)
) ENGINE=MyISAM DEFAULT CHARSET=latin1;

CREATE TABLE `births` (
  `year` smallint(4) NOT NULL DEFAULT 0,
```

```
  `births` mediumint(8) NOT NULL DEFAULT 0,
  PRIMARY KEY (`year`)
) ENGINE=MyISAM DEFAULT CHARSET=latin1;

LOAD DATA LOCAL INFILE './babynames.csv' INTO TABLE `babynames`
  FIELDS TERMINATED BY ',' OPTIONALLY ENCLOSED BY '"' IGNORE 1 LINES;
LOAD DATA LOCAL INFILE './births.csv' INTO TABLE `births`
  FIELDS TERMINATED BY ',' OPTIONALLY ENCLOSED BY '"' IGNORE 1 LINES;

SELECT year, count(distinct name) as numNames
  , sum(n) as numBirths
  FROM babynames
  GROUP BY year
  ORDER BY numBirths desc
  LIMIT 0,10;
```

Note that we have added a SELECT query just to verify that our table is populated. To load this into MySQL, we must first make sure that the babynamedb database exists, and if not, we must create it.

First, we check to see if babynamedata exists. We can do this from the command line using shell commands:

```
mysql -e "SHOW DATABASES;"
```

If it doesn't exist, then we must create it:

```
mysql -e "CREATE DATABASE babynamedb;"
```

Finally, we run our script. The --show-warnings and -v flags are optional, but will help with debugging.

```
mysql --local-infile --show-warnings -v babynamedb
  < babynamedata.mysql
```

In practice, this will often result in errors or warnings the first time you try this. But by iterating this process, you will eventually refine your script such that it works as desired. If you get an 1148 error, make sure that you are using the --local-infile flag.

```
ERROR 1148 (42000): The used command is not allowed with this MySQL version
```

If you get a 29 error, make sure that the file exists in this location and that the mysql user has permission to read and execute it.

```
ERROR 29 (HY000): File './babynames.csv' not found (Errcode: 13)
```

Once the MySQL database has been created, the following commands can be used to access it from R using dplyr:

```
db <- src_mysql(dbname = "babynamedb", default.file = "~/.my.cnf",
                user = NULL, password = NULL)
babynames <- tbl(db, "babynames")
babynames %>% filter(name == "Benjamin")
```

13.4   Scalability

With the exception of SQLite, RBDMSs scale very well on a single computer to databases that take up dozens of gigabytes. For a dedicated server, even terabytes are workable on a single machine. Beyond this, many companies employ distributed solutions called *clusters*. A cluster is simply more than one machine (i.e., a node) linked together running the same RDBMS. One machine is designated as the head node, and this machine controls all of the other nodes. The actual data are distributed across the various nodes, and the head node manages queries—parceling them to the appropriate cluster nodes.

A full discussion of clusters and other distributed architectures (including *replication*) are beyond the scope of this book. In Chapter 17, we discuss alternatives to SQL that may provide higher-end solutions for bigger data.

13.5   Further resources

The *SQL in a Nutshell* book [126] is a useful reference for all things SQL.

13.6   Exercises

The exercises about flights assume that you have access to a SQL database that has been populated with the appropriate flight delay data. Please see the `src_scidb()` function in the `mdsr` package for access to these data on a pre-populated server. To create your own database, use see the `airlines` package.

Exercise 13.1

Consider the following queries:

```
SELECT * FROM flights WHERE cancelled = 1;
SELECT * FROM flights WHERE carrier = "DL";
```

Which query will execute faster? Justify your answer.

Exercise 13.2

Alice is searching for cancelled flights in the `flights` table and her query is running very slowly. She decides to build an index on `cancelled` in the hopes of speeding things up. Discuss the relative merits of her plan. What are the trade-offs? Will her query be any faster?

Exercise 13.3

The `Master` table of the `Lahman` database contains biographical information about baseball players. The primary key is the `playerID` variable. There are also variables for `retroID` and `bbrefID`, which correspond to the player's identifier in other baseball databases. Discuss the ramifications of placing a primary, unique, or foreign key on `retroID`.

Exercise 13.4

Bob wants to analyze the on-time performance of United Airlines flights across the decade of the 1990s. Discuss how the partitioning scheme of the `flights` table based on `year` will affect the performance of Bob's queries, relative to an unpartitioned table.

Exercise 13.5

Write a full table schema for the `mtcars` data set and import it into the database server of your choice.

Exercise 13.6

Write a full table schema for the two tables in the `fueleconomy` package and import them into the database server of your choice.

Exercise 13.7

Write a full table schema for the five tables in the `nasaweather` package and import them into the database server of your choice.

Exercise 13.8

Write a full table schema for the ten tables in the `usdanutrients` package and import them into the database server of your choice.

Exercise 13.9

Use the `macleish` package to download the weather data at the MacLeish Field Station. Write your own table schema from scratch and import these data into the database server of your choice.

Exercise 13.10

Use the `fec` package to download and unzip the federal election data for 2012 that were used in Chapter 2. Write your own table schema from scratch and import these data into the database server of your choice.

Chapter 14

Working with spatial data

When data contain geographic coordinates, they can be considered a type of *spatial data*. Like the "text as data" that we explore in Chapter 15, spatial data are fundamentally different than the numerical data with which we most often work. While spatial coordinates are often encoded as numbers, these numbers have special meaning, and our ability to understand them will suffer if we do not recognize their spatial nature.

The field of *spatial statistics* concerns building and interpreting models that include spatial coordinates. For example, consider a model for airport traffic using the `airlines` data. These data contain the geographic coordinates of each airport, so they are spatially-aware. But simply including the coordinates for latitude and longitude as covariates in a multiple regression model does not take advantage of the special meaning that these coordinates encode. In such a model we might be led towards the meaningless conclusion that airports at higher latitudes are associated with greater airplane traffic—simply due to the limited nature of the model and our careless use of these spatial data.

Unfortunately, a full treatment of spatial statistics is beyond the scope of this book, but there are many excellent resources for such material [34, 60]. While we won't be building spatial models in this chapter, we will learn how to manage and visualize spatial data in R. We will learn about how to work with *shapefiles*, which are a *de facto* open specification data structure for encoding spatial information. We will learn about projections (from three-dimensional space into two-dimensional space), colors (again), and how to create informative, but not misleading, spatially-aware visualizations. Our goal—as always—is to provide the reader with the technical ability and intellectual know-how to derive meaning from spatial data.

14.1   Motivation: What's so great about spatial data?

The most famous early analysis of spatial data was done by physician John Snow in 1854. In a certain London neighborhood, an outbreak of cholera killed 127 people in three days, resulting in a mass exodus of the local residents. At the time it was thought that cholera was an airborne disease caused by breathing foul air. Snow was critical of this theory, and set about discovering the true transmission mechanism.

Consider how you might use data to approach this problem. At the hospital, they might have a list of all of the patients that died of cholera. Those data might look like what is presented in Table 14.1.

Snow's genius was in focusing his analysis on the `Address` column. In a literal sense, the `Address` variable is a character vector—it stores text. This text has no obvious medical

Date	Last Name	First Name	Address	Age	Cause of death
Aug 31, 1854	Jones	Thomas	26 Broad St.	37	cholera
Aug 31, 1854	Jones	Mary	26 Broad St.	11	cholera
Oct 1, 1854	Warwick	Martin	14 Broad St.	23	cholera
⋮					

Table 14.1: Hypothetical data from 1854 cholera outbreak.

significance with respect to cholera. But we as human beings recognize that these strings of text encode *geographic locations*—they are *spatial* data. Snow's insight into this outbreak involved simply plotting these data in a geographically relevant way (see Figure 14.1).

The CholeraDeaths data are included in the mdsr package. When you plot the address of each person who died from cholera, you get something similar to what is shown in Figure 14.2.

```
library(mdsr)
library(sp)
plot(CholeraDeaths)
```

While you might see certain patterns in these data, there is no *context* provided. The map that Snow actually drew is presented in Figure 14.1. The underlying map of the London streets provides helpful context that makes the information in Figure 14.2 intelligible.

However, Snow's insight was driven by another set of data—the locations of the street-side water pumps. It may be difficult to see in the reproduction, but in addition to the lines indicating cholera deaths, there are labeled circles indicating the water pumps. A quick study of the map reveals that nearly all of the cholera cases are clustered around a single pump on the center of Broad St. Snow was able to convince local officials that this pump was the probable cause of the epidemic.

While the story presented above is factual, it may be more legend than spatial data analysts would like to believe. Much of the causality is dubious: Snow himself believed that the outbreak petered out more or less on its own, and he did not create his famous map until afterwards. Nevertheless, his map was influential in the realization among doctors that cholera is a waterborne—rather than airborne—disease.

Our idealized conception of Snow's use of spatial analysis typifies a successful episode in data science. First, the key insight was made by combining three sources of data: the cholera deaths, the locations of the water pumps, and the London street map. Second, while we now have the capability to create a spatial model directly from the data that might have led to the same conclusion, constructing such a model is considerably more difficult than simply plotting the data in the proper context. Moreover, the plot itself— properly contextualized—is probably more convincing to most people than a statistical model anyway. Human beings have a very strong intuitive ability to see spatial patterns in data, but computers have no such sense. Third, the problem was only resolved when the data-based evidence was combined with a plausible model that explained the physical phenomenon. That is, Snow *was a doctor* and his knowledge of disease transmission was sufficient to convince his colleagues that cholera was not transmitted via the air.[1]

---

[1]Unfortunately, the theory of germs and bacteria was still nearly a decade away.

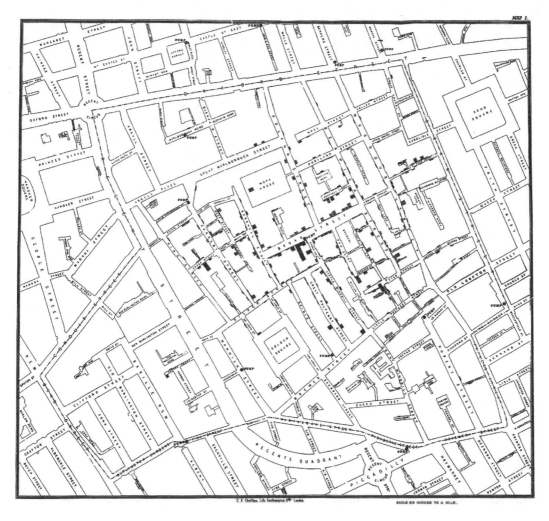

Figure 14.1: John Snow's original map of the 1854 Broad Street cholera outbreak. Source: Wikipedia

14.2 Spatial data structures

Spatial data are often stored in special data structures (i.e., not just `data.frames`). The most commonly used format for spatial data is called a shapefile. Another common format is KML. There are many other formats, and while mastering the details of any of these formats is not realistic in this treatment, there are some important basic notions that one must have in order to work with spatial data.

Shapefiles evolved as the native file format of the ArcView program developed by the Environmental Systems Research Institute (Esri), and have since become an open specification. They can be downloaded from many different government websites and other locations that publish spatial data. Spatial data consists not of rows and columns, but of geometric objects like points, lines, and polygons. Shapefiles contain vector-based instructions for drawing the boundaries of countries, counties, and towns, etc. As such, shapefiles are richer—and more complicated—data containers than simple data frames. Working with shapefiles in R can be challenging, but the major benefit is that shapefiles allow you to provide your data with a geographic context. The results can be stunning.

First, the term "shapefile" is somewhat of a misnomer, as there are several files that you must have in order to read spatial data. These files have extensions like .shp, .shx, and .dbf, and they are typically stored in a common directory.

There are *many* packages for R that specialize in working with spatial data, but two are of primary importance: sp and rgdal. The former provides class definitions for spatial objects in R. These will have the class Spatial*DataFrame, where * can be any of Pixels, Grid, Polygons, Lines, Points. The rgdal package provides access to the Geospatial Data Abstraction Library that computes map projections, as well as a series of import functions.[2]

To get a sense of how these work, we will make a re-creation of Snow's cholera map. First, download and unzip this file: http://rtwilson.com/downloads/SnowGIS_SHP.zip. After loading the rgdal package, we explore the directory that contains our shapefiles.

```
library(rgdal)
dsn <- paste0(root, "snow/SnowGIS_SHP/")
list.files(dsn)
```

```
 [1] "Cholera_Deaths.dbf"         "Cholera_Deaths.prj"
 [3] "Cholera_Deaths.sbn"         "Cholera_Deaths.sbx"
 [5] "Cholera_Deaths.shp"         "Cholera_Deaths.shx"
 [7] "OSMap_Grayscale.tfw"        "OSMap_Grayscale.tif"
 [9] "OSMap_Grayscale.tif.aux.xml" "OSMap_Grayscale.tif.ovr"
[11] "OSMap.tfw"                  "OSMap.tif"
[13] "Pumps.dbf"                  "Pumps.prj"
[15] "Pumps.sbx"                  "Pumps.shp"
[17] "Pumps.shx"                  "README.txt"
[19] "SnowMap.tfw"                "SnowMap.tif"
[21] "SnowMap.tif.aux.xml"        "SnowMap.tif.ovr"
```

Note that there are six files with the name Cholera_Deaths and another five with the name Pumps. These correspond to two different sets of shapefiles called *layers*.

```
ogrListLayers(dsn)
```

```
[1] "Cholera_Deaths" "Pumps"
attr(,"driver")
[1] "ESRI Shapefile"
attr(,"nlayers")
[1] 2
```

We'll begin by loading the Cholera_Deaths layer. Note that these shapefiles are in the ESRI format, and contain 250 "rows" of data. We will return to discussion of the mysterious CRS projection information later, but for now simply note that a specific geographic projection is encoded in these files.

```
ogrInfo(dsn, layer = "Cholera_Deaths")
```

```
Source: "data/shp/snow/SnowGIS_SHP/", layer: "Cholera_Deaths"
Driver: ESRI Shapefile; number of rows: 250
```

---

[2]Note that rgdal may require external dependencies. On Ubuntu, it requires the libgdal-dev and libproj-dev packages. On Mac OS X, it requires GDAL. Also, loading rgdal loads sp.

```
Feature type: wkbPoint with 2 dimensions
Extent: (529160 180858) - (529656 181306)
CRS: +proj=tmerc +lat_0=49 +lon_0=-2 +k=0.9996012717 +x_0=400000
     +y_0=-100000 +ellps=airy +units=m +no_defs
LDID: 87
Number of fields: 2
   name type length typeName
1    Id    0      6 Integer
2 Count    0      4 Integer
```

To load these data into R, we use the readOGR() function.

```
CholeraDeaths <- readOGR(dsn, layer = "Cholera_Deaths")

OGR data source with driver: ESRI Shapefile
Source: "data/shp/snow/SnowGIS_SHP/", layer: "Cholera_Deaths"
with 250 features
It has 2 fields

summary(CholeraDeaths)

Object of class SpatialPointsDataFrame
Coordinates:
              min     max
coords.x1 529160 529656
coords.x2 180858 181306
Is projected: TRUE
proj4string :
[+proj=tmerc +lat_0=49 +lon_0=-2 +k=0.9996012717 +x_0=400000
+y_0=-100000 +ellps=airy +units=m +no_defs]
Number of points: 250
Data attributes:
       Id          Count
 Min.   :0   Min.   : 1.00
 1st Qu.:0   1st Qu.: 1.00
 Median :0   Median : 1.00
 Mean   :0   Mean   : 1.96
 3rd Qu.:0   3rd Qu.: 2.00
 Max.   :0   Max.   :15.00
```

From the summary() command, we can see that we have loaded 250 spatial points. An important feature is that there is a data attribute associated with each of these points. This is a data.frame of values that correspond to each observation. Because SpatialPointsDataFrame is an S4 class,[3] the data slot is accessible using the @ notation.

```
str(CholeraDeaths@data)

'data.frame': 250 obs. of  2 variables:
 $ Id   : int  0 0 0 0 0 0 0 0 0 0 ...
 $ Count: int  3 2 1 1 4 2 2 2 3 2 ...
```

---

[3]For more information about S4 objects, please see [220]. These subtleties will arise rarely in this book—the only other occurrence is in Chapter 8.

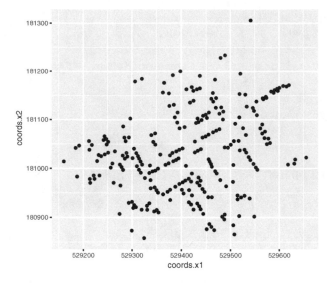

Figure 14.2: A simple `ggplot2` of the cholera deaths, with no context provided.

In this case, for each of the points, we have an associated `Id` number and a `Count` of the number of deaths at that location. To plot these data, simply use the `plot()` generic function. A sensible re-creation of Snow's map can be done in `sp`, but it will be far easier using the `ggmap` package, which we introduce next.

14.3 Making maps

While `sp` and `rgdal` do the heavy lifting, the best interface for actually drawing static maps in R is `ggmap`. The syntax employed by `ggmap` [122] is an extension of the grammar of graphics embedded in `ggplot2` that we explored in Chapter 3. Thus, we are only a few steps away from having some powerful mapping functionality.

14.3.1 Static maps with `ggmap`

Consider for a moment how you would plot the cholera deaths using `ggplot2`. One approach would be to bind the x coordinate to the longitudinal coordinate and the y coordinate to the latitude. Your map would look like this:

```
cholera_coords <- as.data.frame(coordinates(CholeraDeaths))
ggplot(cholera_coords) +
  geom_point(aes(x = coords.x1, y = coords.x2)) + coord_quickmap()
```

Figure 14.2 is not much better than what you would get from `plot()`. It is not clear what the coordinates along the axes are telling us (the units are in fact meters), so we still don't have any context for what we are seeing. What we really want is to overlay these points on the London street map—and this is exactly what `ggmap` lets us do.

`ggmap` is designed to work seamlessly with `ggplot2`. In fact, every `ggmap` object is a `ggplot2` object. The `get_map()` function returns a `ggmap` object from the result of a query to Google Maps. One can control the `zoom` level, as well as the `maptype`. Here, we note that *John Snow* is now the name of a pub on the corner of Broadwick (formerly Broad) Street and Lexington Street.

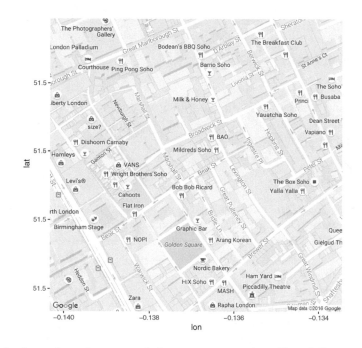

Figure 14.3: A modern-day map of the area surrounding Broad Street in London.

```
library(ggmap)
m <- get_map("John Snow, London, England", zoom = 17, maptype = "roadmap")
ggmap(m)
```

Figure 14.3 provides the context we need, but we have yet to add the layer containing the data points. Since every ggmap object is a ggplot2 object, we can use the familiar syntax that we developed in Chapter 3. The following command will cast the SpatialPointsDataFrame deaths to a data.frame, and use that data frame to map the coordinates. It will also map the number of deaths at each location to the size of the dot.

```
ggmap(m) + geom_point(data = as.data.frame(CholeraDeaths),
                   aes(x = coords.x1, y = coords.x2, size = Count))
```

If you try this, you will not see any points on the plot. Why? Note that the coordinates in the deaths object look like this:

```
head(as.data.frame(CholeraDeaths))
```

```
  Id Count coords.x1 coords.x2
1  0     3    529309    181031
2  0     2    529312    181025
3  0     1    529314    181020
4  0     1    529317    181014
5  0     4    529321    181008
6  0     2    529337    181006
```

But the coordinates in the map object (m) are (lat, long) pairs, as we can see by accessing the *bounding box* (bb) attribute of the ggmap object.

```
attr(m, "bb")
```

```
    ll.lat ll.lon ur.lat ur.lon
1   51.5  -0.14   51.5 -0.133
```

Both `deaths` and `m` have geospatial coordinates, but those coordinates are not in the same units. To understand how to get these two spatial data sources to work together, we have to understand projections.

14.3.2   Projections

The Earth happens to be an oblate spheroid—a three-dimensional flattened sphere. Yet we would like to create two-dimensional representations of the Earth that fit on pages or computer screens. The process of converting locations in a three-dimensional *geographic coordinate system* to a two-dimensional representation is called *projection*.

Once people figured out that the world was not flat, the question of how to project it followed. Since people have been making nautical maps for centuries, it would be nice if the study of map projection had resulted in a simple, accurate, universally-accepted projection system. Unfortunately, that is not the case. It is simply not possible to faithfully preserve all properties present in a three-dimensional space in a two-dimensional space. Thus there is no one best projection system—each has its own advantages and disadvantages. Further complicating matters is the fact that the Earth is not a perfect sphere, but a flattened sphere (i.e., an oblate spheroid). This means that even the mathematics behind many of these projections are non-trivial.

Two properties that a projection system might preserve—though not simultaneously—are *shape/angle* and *area*. That is, a projection system may be constructed in such a way that it faithfully represents the relative sizes of land masses in two dimensions. The Mercator projection shown at left in Figure 14.4 is a famous example of a projection system that does *not* preserve area. Its popularity is a result of its *angle*-preserving nature, which makes it useful for navigation. Unfortunately, it also greatly distorts the size of features near the poles, where land masses become infinitely large.

```
library(maps)
map("world", projection = "mercator", wrap = TRUE)
map("world", projection = "cylequalarea", param = 45, wrap = TRUE)
```

The Gall–Peters projection shown at right in Figure 14.4 does preserve area. Note the difference between the two projections when comparing the size of Greenland to Africa. In reality (as shown in the Gall–Peters projection) Africa is 14 times larger than Greenland. However, because Greenland is much closer to the North Pole, its area is greatly distorted in the Mercator projection, making it appear to be larger than Africa.

This particular example—while illustrative—became famous because of the socio-political controversy in which these projections became embroiled. Beginning in the 1960s, a German filmmaker named Arno Peters alleged that the commonly used Mercator projection was an instrument of cartographic imperialism, in that it falsely focused attention on Northern and Southern countries at the expense of those in Africa and South America closer to the equator. Peters had a point—the Mercator projection has many shortcomings—but unfortunately his claims about the virtues of the Gall–Peters projection (particularly its originality) were mostly false. Peters either ignored or was not aware that cartographers had long campaigned against the Mercator.

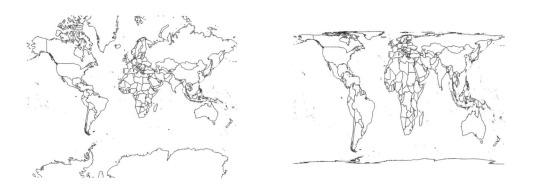

(a) Mercator   (b) Gall-Peters

Figure 14.4: The world according to the Mercator (left) and Gall–Peters (right) projections.

Nevertheless, you should be aware that the "default" projection can be very misleading. As a data scientist, your choice of how to project your data can have a direct influence on what viewers will take away from your data maps. Simply ignoring the implications of projections is not an ethically tenable position! While we can't offer a comprehensive list of map projections here, two common general-purpose map projections are the Lambert conformal conic projection and the Albers equal-area conic projection (see Figure 14.5). In the former, angles are preserved, while in the latter neither scale nor shape are preserved, but gross distortions of both are minimized.

```
map("state", projection = "lambert",
    parameters = c(lat0 = 20, lat1 = 50), wrap = TRUE)
map("state", projection = "albers",
    parameters = c(lat0 = 20, lat1 = 50), wrap = TRUE)
```

Pro Tip: Always think about how your data are projected when making a map.

A *coordinate reference system* (CRS) is needed to keep track of geographic locations. Every spatially-aware object in R can have a projection string, encoded using the PROJ.4 map projection library. These can be retrieved (or set) using the proj4string() command.

```
proj4string(CholeraDeaths) %>% strwrap()
```

```
[1] "+proj=tmerc +lat_0=49 +lon_0=-2 +k=0.9996012717 +x_0=400000"
[2] "+y_0=-100000 +ellps=airy +units=m +no_defs"
```

It should be clear by now that the science of map projection is complicated, and it is likely unclear how to decipher this cryptic list of symbols. What we can say is that +proj=tmerc indicates that these data are encoded using a Transverse Mercator projection. The Airy ellipsoid is being used (+ellps=airy), and the units are meters (+units=m). The rest of the

(a) Lambert conformal conic   (b) Albers equal area

Figure 14.5: The contiguous United States according to the Lambert conformal conic (left) and Albers equal area (right) projections. We have specified that the scales are true on the 20th and 50th parallels.

terms in the string are parameters that specify properties of that projection. The unfamiliar coordinates that we saw earlier for the CholeraDeaths data set were relative to this CRS.

There are many CRSs, but a few are most common. A set of EPSG (European Petroleum Survey Group) codes provides a shorthand for the full PROJ.4 strings (like the one shown above). The most commonly-used are:

EPSG:4326 Also known as WGS84, this is the standard for GPS systems and Google Earth.

EPSG:3857 A Mercator projection used in maps tiles[4] by Google Maps, Open Street Maps, etc.

EPSG:27700 Also known as OSGB 1936, or the British National Grid: United Kingdom Ordnance Survey. It is commonly used in Britain.

The CRS() function will translate from the shorthand EPSG code to the full-text PROJ.4 string.

```
CRS("+init=epsg:4326")

CRS arguments:
 +init=epsg:4326 +proj=longlat +datum=WGS84 +no_defs +ellps=WGS84
+towgs84=0,0,0

CRS("+init=epsg:3857")

CRS arguments:
```

---

[4]Google Maps and other online maps are composed of a series of square static images called *tiles*. These are pre-fetched and loaded as you scroll, creating the appearance of a larger image.

```
 +init=epsg:3857 +proj=merc +a=6378137 +b=6378137 +lat_ts=0.0
+lon_0=0.0 +x_0=0.0 +y_0=0 +k=1.0 +units=m +nadgrids=@null
+no_defs

CRS("+init=epsg:27700")

CRS arguments:
 +init=epsg:27700 +proj=tmerc +lat_0=49 +lon_0=-2 +k=0.9996012717
+x_0=400000 +y_0=-100000 +datum=OSGB36 +units=m +no_defs
+ellps=airy
+towgs84=446.448,-125.157,542.060,0.1502,0.2470,0.8421,-20.4894
```

The CholeraDeaths points did not show up on our earlier map because we did not project them into the same coordinate system as the map data. Since we can't project the ggmap image, we had better project the points in the cholera CholeraDeaths data. As noted above, Google Maps tiles are projected in the espg:3857 system. However, they are confusingly returned with coordinates in the epsg:4326 system. Thus, we use the spTransform() function in the rgdal package to project our CholeraDeaths data to epsg:4326.

```
cholera_latlong <- CholeraDeaths %>% spTransform(CRS("+init=epsg:4326"))
```

Note that the *bounding box* in our new coordinates are in the same familiar units as our map object.

```
bbox(cholera_latlong)

             min     max
coords.x1 -0.138 -0.131
coords.x2 51.511 51.515
```

Finally, we can see some points on our map.

```
ggmap(m) + geom_point(aes(x = coords.x1, y = coords.x2,
  size = Count), data = as.data.frame(cholera_latlong))
```

However, in Figure 14.6 the points don't seem to be in the right places. The center of the cluster is not on Broadwick Street, and some of the points are in the middle of the street (where there are no residences). A careful reading of the help file for spTransform() gives some clues to our mistake.

```
help("spTransform-methods", package = "rgdal")
```

Not providing the appropriate +datum and +towgs84 tags may lead to coordinates being out by hundreds of meters. Unfortunately, there is no easy way to provide this information: The user has to know the correct metadata for the data being used, even if this can be hard to discover.

That seems like our problem! Note that the +datum and +towgs84 arguments were missing from our PROJ.4 string. We can try to recover the EPSG code from the PROJ.4 string using the showEPSG() function.

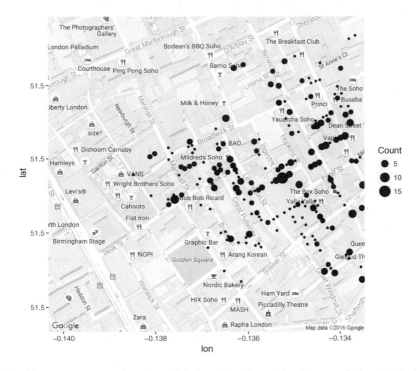

Figure 14.6: Erroneous reproduction of John Snow's original map of the 1854 cholera outbreak. The dots representing the deaths from cholera are off by hundreds of meters.

```
CholeraDeaths %>% proj4string() %>% showEPSG()

[1] "OGRERR_UNSUPPORTED_SRS"
```

Unfortunately, the `CholeraDeaths` data set is not projected in a known EPSG format. However, it has all of the same specifications as `epsg:27700`, but without the missing `+datum` and `+towgs84` tags. Furthermore, the documentation for the original data source suggests using `epsg:27700`. Thus, we first assert that the `CholeraDeaths` data is in `epsg:27700`.

```
proj4string(CholeraDeaths) <- CRS("+init=epsg:27700")
```

Now, projecting to `epsg:4326` works as intended.

```
cholera_latlong <- CholeraDeaths %>%
spTransform(CRS("+init=epsg:4326"))
snow <- ggmap(m) +
geom_point(aes(x = coords.x1, y = coords.x2,
               size = Count), data = as.data.frame(cholera_latlong))
```

All that remains is to add the locations of the pumps.

```
pumps <- readOGR(dsn, layer = "Pumps")

OGR data source with driver: ESRI Shapefile
Source: "data/shp/snow/SnowGIS_SHP/", layer: "Pumps"
```

```
with 8 features
It has 1 fields

proj4string(pumps) <- CRS("+init=epsg:27700")
pumps_latlong <- pumps %>% spTransform(CRS("+init=epsg:4326"))
snow + geom_point(data = as.data.frame(pumps_latlong),
            aes(x = coords.x1, y = coords.x2), size = 3, color = "red")
```

Figure 14.7: Reproduction of John Snow's original map of the 1854 cholera outbreak. The size of each black dot is proportional to the number of people who died from cholera at that location. The red dots indicate the location of public water pumps. The strong clustering of deaths around the water pump on Broad(wick) Street suggests that perhaps the cholera was spread through water obtained at that pump.

In Figure 14.7, we finally see the clarity that judicious uses of spatial data in the proper context can provide. It is not necessary to fit a statistical model to these data to see that nearly all of the cholera deaths occurred in people closest to the Broad Street water pump, which was later found to be drawing fecal bacteria from a nearby cesspit.

14.3.3   Geocoding, routes, and distances

The process of converting a human-readable address into geographic coordinates is called *geocoding*. While there are numerous APIs available online that will do this for you, this functionality is provided in ggmap by the geocode() function.

```
smith <- "Smith College, Northampton, MA 01063"
geocode(smith)

    lon  lat
1 -72.6 42.3
```

Note that Google will limit you to 2500 queries per day. Alternatively, the RgoogleMaps package provides similar functionality that is not capped via the getGeoCode() function.

```
library(RgoogleMaps)
amherst <- "Amherst College, Amherst, MA"
getGeoCode(amherst)

  lat   lon
42.4 -72.5
```

Distances can also be retrieved using the Google Map API accessible through ggmap. Here, we compute the distance between two of the Five Colleges[5] using the mapdist() function.

```
mapdist(from = smith, to = amherst, mode = "driving")

                                 from                        to     m
1 Smith College, Northampton, MA 01063 Amherst College, Amherst, MA 12496
    km miles seconds minutes hours
1 12.5  7.77    1424    23.7 0.396

mapdist(from = smith, to = amherst, mode = "bicycling")

                                 from                        to     m
1 Smith College, Northampton, MA 01063 Amherst College, Amherst, MA 13615
    km miles seconds minutes hours
1 13.6  8.46    2914    48.6 0.809
```

As you might suspect, you can also find routes between multiple locations using the route() command. This returns a data frame with the segments that make up individual routes.

```
legs_df <- route(smith, amherst, alternatives = TRUE)
head(legs_df) %>%
  select(m, km, miles, miles, seconds, minutes, hours, startLon, startLat)

        m    km  miles seconds minutes   hours startLon startLat
1      30 0.030 0.0186       5  0.0833 0.00139    -72.6     42.3
```

---

[5]The Five College Consortium consists of Amherst, Hampshire, Mount Holyoke, and Smith Colleges, as well as the University of Massachusetts-Amherst.

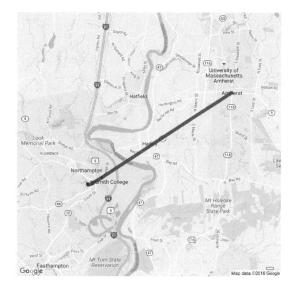

Figure 14.8: The fastest route from Smith College to Amherst College.

2	80	0.080	0.0497	29	0.4833	0.00806	-72.6	42.3
3	289	0.289	0.1796	83	1.3833	0.02306	-72.6	42.3
4	165	0.165	0.1025	40	0.6667	0.01111	-72.6	42.3
5	659	0.659	0.4095	213	3.5500	0.05917	-72.6	42.3
6	11274	11.274	7.0057	1053	17.5500	0.29250	-72.6	42.3

The qmap() (quick map) is a wrapper to ggmap() and get_map(). Since the Coolidge Bridge is the only reasonable way to get from Northampton to Amherst, there is only one possibility returned for the shortest route between Smith and Amherst, as shown in Figure 14.8.

```
qmap("The Quarters, Hadley, MA", zoom = 12, maptype = 'roadmap') +
  geom_leg(aes(x = startLon, y = startLat, xend = endLon, yend = endLat),
           alpha = 3/4, size = 2, color = "blue", data = legs_df)
```

However, shortest paths in a network are not unique (see Chapter 16). Ben's daily commute to Citi Field from his apartment in Brooklyn presented two distinct alternatives: One could take the Brooklyn-Queens Expressway (I-278 E) to the Grand Central Parkway E, or continue on the Long Island Expressway (I-495 E) and then approach from the opposite direction on the Grand Central Parkway W. The latter route is shorter, but often will take longer due to traffic. The former route is also more convenient to the Citi Field employee parking lot, as opposed to the lot by the now-demolished Shea Stadium. These routes are overlaid on the map in Figure 14.9.

```
legs_df <- route(from = "736 Leonard St, Brooklyn, NY",
                 to = "Citi Field, Roosevelt Ave, Flushing, NY",
                 alternatives = TRUE, structure = "legs")

qmap("74th St and Broadway, Queens, NY", zoom = 12, maptype = 'roadmap') +
  geom_leg(aes(x = startLon, y = startLat, xend = endLon, yend = endLat,
               color = route), alpha = 0.7, size = 2, data = legs_df)
```

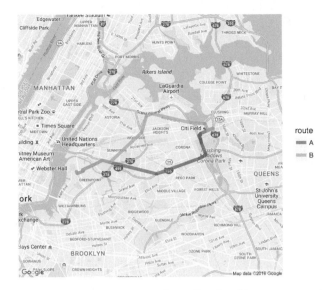

Figure 14.9: Alternative commuting routes from Ben's old apartment in Brooklyn to Citi Field. Note that the Google API only returns the endpoints of each segment, so they appear on the map as straight lines even when the actual road curves.

14.3.4   Dynamic maps with `leaflet`

Leaflet is a powerful open source JavaScript library for building interactive maps in HTML. The corresponding R package `leaflet` brings this functionality to R using the `htmlwidgets` platform.

The `leaflet` package is another part of the tidyverse, so if you are comfortable working with `dplyr` and `ggplot2`, then you already understand how `leaflet` works. Although the commands are different, the architecture is very similar to `ggmap`. However, instead of putting data-based layers on top of a static map, `leaflet` allows you to put data-based layers on top of an interactive map.

Because `leaflet` renders as HTML, you won't see any of our plots in this book (except as screen shots). However, we encourage you to run this code on your own and explore interactively.

A `leaflet` map widget is created with the `leaflet()` command. We will subsequently add layers to this widget. The first layer that we will add is a *tile* layer containing all of the static map information, which by default comes from OpenStreetMap. The second layer we will add here is a marker, which designates a point location. Note how the `addMarkers()` function can take a `data` argument, just like a `geom_*()` layer in `ggplot2` would.

```
white_house <- geocode("The White House, Washington, DC")
library(leaflet)
map <- leaflet() %>%
  addTiles() %>%
  addMarkers(lng = ~lon, lat = ~lat, data = white_house)
```

When you render this in RStudio, or in an R Markdown document with HTML output, or in a Web browser using Shiny, you will be able to scroll and zoom on the fly. In Figure 14.10 we display a static image from that plot.

We can also add a pop-up to provide more information about a particular location.

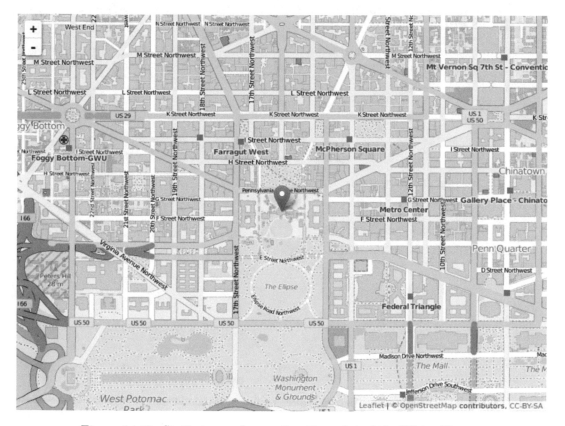

Figure 14.10: Static image from a `leaflet` plot of the White House.

```
white_house <- white_house %>%
  mutate(title = "The White House", address = "2600 Pennsylvania Ave")
map %>%
  addPopups(lng = ~lon, lat = ~lat, data = white_house,
            popup = ~paste0("<b>", title, "</b></br>", address))
```

Although `leaflet` and `ggmap` are not syntactically equivalent, they are conceptually similar. In many cases, the dynamic, zoomable, scrollable maps created by `leaflet` can be more informative than the static maps created by `ggmap`.

14.4 Extended example: Congressional districts

In the 2012 presidential election, the Republican challenger Mitt Romney narrowly defeated President Barack Obama in the state of North Carolina, winning 50.4% of the popular votes, but thereby earning all 15 electoral votes. Obama had won North Carolina in 2008—becoming the first Democrat to do so since 1976. As a swing state, North Carolina has voting patterns that are particularly interesting, and—as we will see—contentious.

The roughly 50/50 split in the popular vote suggests that there are about the same number of Democratic and Republican votes in the state. However, 10 of North Carolina's 13 congressional representatives are Republican. How can this be? In this case, spatial data can help us understand.

14.4.1   Election results

Our first step is to download the results of the 2012 congressional elections from the Federal Election Commission. These data are available through the `fec` package. Please see Appendix A.2 for more detail on how to set this up.

```
library(fec)
db <- src_mysql(default.file = "~/.my.cnf", groups = "rs-dbi",
                dbname = "fec", user = NULL, password = NULL)
fec <- etl("fec", db, dir = "~/dumps/fec")
us_elections <- tbl(fec, "house_elections") %>%
  collect()
```

Note that we have slightly more than 435 elections, since these data include U.S. territories like Puerto Rico and the Virgin Islands.

```
us_elections %>%
  group_by(state, district) %>%
  summarize(N = n()) %>%
  nrow()
```

```
[1] 445
```

According to the U.S. Constitution, congressional districts are apportioned according to population from the 2010 U.S. Census. In practice we see that this is not quite the case. These are the ten candidates who earned the most votes in the general election.

```
us_elections %>%
  select(state, district, candidate_name, party, general_votes) %>%
  arrange(desc(general_votes))
```

```
# A tibble: 2,178   5
    state              district              candidate_name party
    <chr>                 <chr>                      <chr> <chr>
1      PR                    00 Pierluisi Urrutia, Pedro R.     N
2      PR                    00      Cox Alomar, Roberto         P
3      WA 01 - UNEXPIRED TERM             DelBene, Suzan          D
4      NJ 10 - UNEXPIRED TERM      Payne, Donald M., Jr.          D
5      KY 04 - UNEXPIRED TERM            Massie, Thomas           R
6      MI 11 - UNEXPIRED TERM          Bentivolio, Kerry          R
7      PA                    02           Fattah, Chaka           D
8      WA                    07         McDermott, Jim            D
9      WA 01 - UNEXPIRED TERM            Koster, John             R
10     MI                    14           Peters, Gary            D
# ... with 2,168 more rows, and 1 more variables: general_votes <int>
```

We are interested in the results from North Carolina. Thus, we create a data frame specific to that state, with the votes aggregated by congressional district. As there are 13 districts, the nc_results data frame will have exactly 13 rows.

```
district_elections <- us_elections %>%
  mutate(district = stringr::str_sub(district, 1, 2)) %>%
  group_by(state, district) %>%
  summarize(N = n(), total_votes = sum(general_votes, na.rm = TRUE),
    d_votes = sum(ifelse(party == "D", general_votes, 0)),
    r_votes = sum(ifelse(party == "R", general_votes, 0))) %>%
  mutate(other_votes = total_votes - d_votes - r_votes,
         r_pct = r_votes / total_votes,
         r_win = r_votes > d_votes)
nc_results <- district_elections %>% filter(state == "NC")
nc_results
```

Source: local data frame [13 x 9]
Groups: state [1]

	state	district	N	total_votes	d_votes	r_votes	other_votes	r_pct
	<chr>	<chr>	<int>	<int>	<dbl>	<dbl>	<dbl>	<dbl>
1	NC	01	4	338066	254644	77288	6134	0.229
2	NC	02	8	311397	128973	174066	8358	0.559
3	NC	03	3	309885	114314	195571	0	0.631
4	NC	04	4	348485	259534	88951	0	0.255
5	NC	05	3	349197	148252	200945	0	0.575
6	NC	06	4	364583	142467	222116	0	0.609
7	NC	07	4	336736	168695	168041	0	0.499
8	NC	08	8	301824	137139	160695	3990	0.532
9	NC	09	13	375690	171503	194537	9650	0.518
10	NC	10	6	334849	144023	190826	0	0.570
11	NC	11	10	331426	141107	190319	0	0.574
12	NC	12	3	310908	247591	63317	0	0.204
13	NC	13	5	370610	160115	210495	0	0.568

... with 1 more variables: r_win <lgl>

Note that the distribution of the number of votes cast across congressional districts in North Carolina is very narrow—all of the districts had between 301 and 376 thousand votes cast.

```
favstats(~ total_votes, data = nc_results)
```

min	Q1	median	Q3	max	mean	sd	n	missing
301824	311397	336736	349197	375690	337204	24175	13	0

However, as the close presidential election suggests, the votes of North Carolinans were roughly evenly divided among Democratic and Republican congressional candidates. In fact, state Democrats earned a narrow majority—50.6%—of the votes. Yet the Republicans won nine of the 13 races.[6]

---

[6]The 7th district was the closest race in the entire country, with Democratic incumbent Mike McIntyre winning by just 655 votes. After McIntyre's retirement, Republican challenger David Rouzer won the seat easily in 2014.

```
nc_results %>%
  summarize(N = n(), repub_win = sum(r_win),
            state_votes = sum(total_votes), state_d = sum(d_votes),
            state_r = sum(r_votes)) %>%
  mutate(d_pct = state_d / state_votes, r_pct = state_r / state_votes)

# A tibble: 1  8
  state      N repub_win state_votes state_d state_r d_pct r_pct
  <chr> <int>     <int>       <int>   <dbl>   <dbl> <dbl> <dbl>
1   NC    13         9     4383656 2218357 2137167 0.506 0.488
```

One clue is to look at the distribution of the percentage of Republican votes in each district.

```
nc_results %>%
  select(district, r_pct) %>%
  arrange(desc(r_pct))

Source: local data frame [13 x 3]
Groups: state [1]

   state district r_pct
   <chr>    <chr> <dbl>
1     NC       03 0.631
2     NC       06 0.609
3     NC       05 0.575
4     NC       11 0.574
5     NC       10 0.570
6     NC       13 0.568
7     NC       02 0.559
8     NC       08 0.532
9     NC       09 0.518
10    NC       07 0.499
11    NC       04 0.255
12    NC       01 0.229
13    NC       12 0.204
```

In the nine districts that Republicans won, their share of the vote ranged from a narrow (51.8%) to a comfortable (63.1%) majority. With the exception of the essentially even 7th district, the three districts that Democrats won were routs, with the Democratic candidate winning between 75% and 80% of the vote. Thus, although Democrats won more votes across the state, most of their votes were clustered within three overwhelmingly Democratic districts, allowing Republicans to prevail with moderate majorities across the remaining nine districts.

Democratic voters tend to live in cities, so perhaps they were simply clustered in three cities, while Republican voters were spread out across the state in more rural areas. There is some truth to this. Let's look at the districts.

14.4.2   Congressional districts

To do this, we first download the congressional district shapefiles for the 113th Congress.

```
src <- "http://cdmaps.polisci.ucla.edu/shp/districts113.zip"
lcl <- paste0(root, "districts113.zip")
download.file(src, destfile = lcl)
unzip(zipfile = lcl, exdir = root)
```

Next, we read these shapefiles into R as a SpatialPolygonsDataFrame.

```
library(rgdal)
dsn_districts <- paste0(root, "districtShapes/")
ogrListLayers(dsn_districts)
```

```
[1] "districts113"
attr(,"driver")
[1] "ESRI Shapefile"
attr(,"nlayers")
[1] 1
```

```
districts <- readOGR(dsn_districts, layer = "districts113")
```

```
OGR data source with driver: ESRI Shapefile
Source: "data/shp/districtShapes/", layer: "districts113"
with 436 features
It has 15 fields
```

```
glimpse(districts@data)
```

```
Observations: 436
Variables: 15
$ STATENAME  <fctr> Arizona, Arizona, California, District Of Columbia...
$ ID         <fctr> 004113113005, 004113113001, 006113113037, 01111311...
$ DISTRICT   <fctr> 05, 01, 37, 98, 01, 02, 15, 04, 03, 02, 01, 04, 01...
$ STARTCONG  <fctr> 113, 113, 113, 113, 113, 113, 113, 113, 113, 113, ...
$ ENDCONG    <fctr> 113, 113, 113, 113, 113, 113, 113, 113, 113, 113, ...
$ DISTRICTSI <fctr> NA, NA, NA, NA, NA, NA, NA, NA, NA, NA, NA, NA, NA...
$ COUNTY     <fctr> NA, NA, NA, NA, NA, NA, NA, NA, NA, NA, NA, NA, NA...
$ PAGE       <fctr> NA, NA, NA, NA, NA, NA, NA, NA, NA, NA, NA, NA, NA...
$ LAW        <fctr> NA, NA, NA, NA, NA, NA, NA, NA, NA, NA, NA, NA, NA...
$ NOTE       <fctr> NA, NA, NA, NA, NA, NA, NA, NA, NA, NA, NA, NA, NA...
$ BESTDEC    <fctr> NA, NA, NA, NA, NA, NA, NA, NA, NA, NA, NA, NA, NA...
$ FINALNOTE  <fctr> {"From US Census website"}, {"From US Census websi...
$ RNOTE      <fctr> NA, NA, NA, NA, NA, NA, NA, NA, NA, NA, NA, NA, NA...
$ LASTCHANGE <fctr> 2014-02-14 17:40:40.110145, 2014-02-14 17:40:40.11...
$ FROMCOUNTY <fctr> F, F, F, F, F, F, F, F, F, F, F, F, F, F, F, F, F,...
```

We are investigating North Carolina, so we will create a smaller object with only those shapes using the generic subset() function (subset() behaves very much like filter()).

```
nc_shp <- subset(districts, STATENAME == "North Carolina")
plot(nc_shp, col = gray.colors(nrow(nc_shp)))
```

It is hard to see exactly what is going on here, but it appears as though there are some traditionally shaped districts, as well as some very strange and narrow districts. Unfortunately the map in Figure 14.11 is devoid of context, so it is not very informative. We need

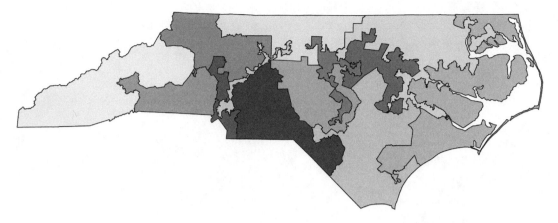

Figure 14.11: A basic map of the North Carolina congressional districts.

the `nc_results` data to provide that context, but unfortunately, these two objects are of very different classes.

```
class(nc_shp)
```

```
[1] "SpatialPolygonsDataFrame"
attr(,"package")
[1] "sp"
```

```
class(nc_results)
```

```
[1] "grouped_df" "tbl_df"       "tbl"          "data.frame"
```

14.4.3   Putting it all together

How to merge these two together? The simplest way is to use the generic merge() function from the sp package. (The merge() function as used here is conceptually equivalent to the inner_join() function from dplyr (see Chapter 4).) If the first argument is a Spatial object, then this function will combine the geometric objects with the rows of the data frame supplied in the second argument. Here, we merge() the nc_shp polygons with the nc_results election data frame using the district as the key. Note that there are 13 polygons and 13 rows.

```
nc_merged <- merge(nc_shp, as.data.frame(nc_results),
  by.x = c("DISTRICT"), by.y = c("district"))
glimpse(nc_merged@data)
```

```
Observations: 13
Variables: 23
$ DISTRICT    <fctr> 08, 09, 13, 04, 05, 10, 02, 03, 07, 12, 01, 06, 11
$ STATENAME   <fctr> North Carolina, North Carolina, North Carolina, N...
$ ID          <fctr> 037113113008, 037113113009, 037113113013, 0371131...
$ STARTCONG   <fctr> 113, 113, 113, 113, 113, 113, 113, 113, 113, 113,...
```

```
$ ENDCONG      <fctr> 113, 113, 113, 113, 113, 113, 113, 113, 113, 113,...
$ DISTRICTSI   <fctr> NA, NA, NA, NA, NA, NA, NA, NA, NA, NA, NA, NA, NA
$ COUNTY       <fctr> NA, NA, NA, NA, NA, NA, NA, NA, NA, NA, NA, NA, NA
$ PAGE         <fctr> NA, NA, NA, NA, NA, NA, NA, NA, NA, NA, NA, NA, NA
$ LAW          <fctr> NA, NA, NA, NA, NA, NA, NA, NA, NA, NA, NA, NA, NA
$ NOTE         <fctr> NA, NA, NA, NA, NA, NA, NA, NA, NA, NA, NA, NA, NA
$ BESTDEC      <fctr> NA, NA, NA, NA, NA, NA, NA, NA, NA, NA, NA, NA, NA
$ FINALNOTE    <fctr> {"From US Census website"}, {"From US Census webs...
$ RNOTE        <fctr> NA, NA, NA, NA, NA, NA, NA, NA, NA, NA, NA, NA, NA
$ LASTCHANGE   <fctr> 2014-02-14 17:40:40.110145, 2014-02-14 17:40:40.1...
$ FROMCOUNTY   <fctr> F, F, F, F, F, F, F, F, F, F, F, F
$ state        <chr> "NC", "NC", "NC", "NC", "NC", "NC", "NC", "NC", "N...
$ N            <int> 8, 13, 5, 4, 3, 6, 8, 3, 4, 3, 4, 4, 10
$ total_votes  <int> 301824, 375690, 370610, 348485, 349197, 334849, 31...
$ d_votes      <dbl> 137139, 171503, 160115, 259534, 148252, 144023, 12...
$ r_votes      <dbl> 160695, 194537, 210495, 88951, 200945, 190826, 174...
$ other_votes  <dbl> 3990, 9650, 0, 0, 0, 0, 8358, 0, 0, 0, 6134, 0, 0
$ r_pct        <dbl> 0.532, 0.518, 0.568, 0.255, 0.575, 0.570, 0.559, 0...
$ r_win        <lgl> TRUE, TRUE, TRUE, FALSE, TRUE, TRUE, TRUE, TRUE, F...
```

However, while `leaflet` understands spatial objects, `ggmap` and `ggplot2` do not. The broom package contains a series of functions that convert different kinds of objects into a tidy format. Here, we use it to tidy the `nc_merged` spatial data, and then merge it with its associated data attributes. The resulting `nc_full` data frame contains everything we know about these districts in a tidy format.

```
library(broom)
library(maptools)
nc_tidy <- tidy(nc_merged, region = "ID")
nc_full <- nc_tidy %>% left_join(nc_merged@data, by = c("id" = "ID"))
glimpse(nc_full)
```

```
Observations: 28,172
Variables: 29
$ long         <dbl> -79.0, -79.0, -79.0, -79.0, -79.0, -78.9, -78.9, -...
$ lat          <dbl> 36, 36, 36, 36, 36, 36, 36, 36, 36, 36, 36, 36, 36...
$ order        <int> 1, 2, 3, 4, 5, 6, 7, 8, 9, 10, 11, 12, 13, 14, 15,...
$ hole         <lgl> FALSE, FALSE, FALSE, FALSE, FALSE, FALSE, FALSE, F...
$ piece        <fctr> 1, 1, 1, 1, 1, 1, 1, 1, 1, 1, 1, 1, 1, 1, 1, 1...
$ group        <fctr> 037113113001.1, 037113113001.1, 037113113001.1, 0...
$ id           <chr> "037113113001", "037113113001", "037113113001", "0...
$ DISTRICT     <fctr> 01, 01, 01, 01, 01, 01, 01, 01, 01, 01, 01, 01, 0...
$ STATENAME    <fctr> North Carolina, North Carolina, North Carolina, N...
$ STARTCONG    <fctr> 113, 113, 113, 113, 113, 113, 113, 113, 113, 113,...
$ ENDCONG      <fctr> 113, 113, 113, 113, 113, 113, 113, 113, 113, 113,...
$ DISTRICTSI   <fctr> NA, NA, NA, NA, NA, NA, NA, NA, NA, NA, NA, NA, N...
$ COUNTY       <fctr> NA, NA, NA, NA, NA, NA, NA, NA, NA, NA, NA, NA, N...
$ PAGE         <fctr> NA, NA, NA, NA, NA, NA, NA, NA, NA, NA, NA, NA, N...
$ LAW          <fctr> NA, NA, NA, NA, NA, NA, NA, NA, NA, NA, NA, NA, N...
$ NOTE         <fctr> NA, NA, NA, NA, NA, NA, NA, NA, NA, NA, NA, NA, N...
$ BESTDEC      <fctr> NA, NA, NA, NA, NA, NA, NA, NA, NA, NA, NA, NA, N...
```

```
$ FINALNOTE   <fctr> {"From US Census website"}, {"From US Census webs...
$ RNOTE       <fctr> NA, NA, NA, NA, NA, NA, NA, NA, NA, NA, NA, NA, N...
$ LASTCHANGE  <fctr> 2014-02-14 17:40:40.110145, 2014-02-14 17:40:40.1...
$ FROMCOUNTY  <fctr> F, F, F, F, F, F, F, F, F, F, F, F, F, F, F, F, F...
$ state       <chr> "NC", "NC", "NC", "NC", "NC", "NC", "NC", "NC", "N...
$ N           <int> 4, 4, 4, 4, 4, 4, 4, 4, 4, 4, 4, 4, 4, 4, 4, 4, 4,...
$ total_votes <int> 338066, 338066, 338066, 338066, 338066, 338066, 33...
$ d_votes     <dbl> 254644, 254644, 254644, 254644, 254644, 254644, 25...
$ r_votes     <dbl> 77288, 77288, 77288, 77288, 77288, 77288, 77288, 7...
$ other_votes <dbl> 6134, 6134, 6134, 6134, 6134, 6134, 6134, 6134, 61...
$ r_pct       <dbl> 0.229, 0.229, 0.229, 0.229, 0.229, 0.229, 0.229, 0...
$ r_win       <lgl> FALSE, FALSE, FALSE, FALSE, FALSE, FALSE, FALSE, F...
```

Before we draw the map, we'll want to overlay the names of the districts. But where should those labels go? Since each district is represented as a polygon, it makes sense to put the label in the "center" of each polygon. But where is that "center"? One answer is the *centroid*. The rgeos package contains functionality for computing the centroids of SpatialPolygons objects and returning a SpatialPoints object.

```
library(rgeos)
nc_centroids <- gCentroid(nc_shp, byid = TRUE)
class(nc_centroids)
```

```
[1] "SpatialPoints"
attr(,"package")
[1] "sp"
```

Since we also want to associate the number of the district with each of these points, we need to convert these centroids into a SpatialPointsDataFrame by adding the map data.

```
nc_centroids <- SpatialPointsDataFrame(nc_centroids, nc_shp@data)
```

Finally, we convert the SpatialPointsDataFrame object into a tidy format for use with ggmap, and merge in the election results data.

```
nc_centroids_tidy <- as.data.frame(nc_centroids)
nc_centroids_full <- nc_centroids_tidy %>%
  inner_join(nc_results,
    by = c("STATENAME" = "state", "DISTRICT" = "district"))
```

14.4.4   Using ggmap

We are now ready to plot our map of North Carolina's congressional districts. We start by using a simple red–blue color scheme for the districts.

```
library(ggmap)
nc <- get_map("charlotte, north carolina", zoom = 6, maptype = "roadmap")
ggmap(nc) +
  geom_polygon(aes(x = long, y = lat, group = group, fill = r_win),
               alpha = 0.8, data = nc_full) +
```

```
scale_fill_manual(values = c("blue", "red")) +
geom_text(aes(x = x, y = y, label = DISTRICT), data = nc_centroids_full) +
theme_map()
```

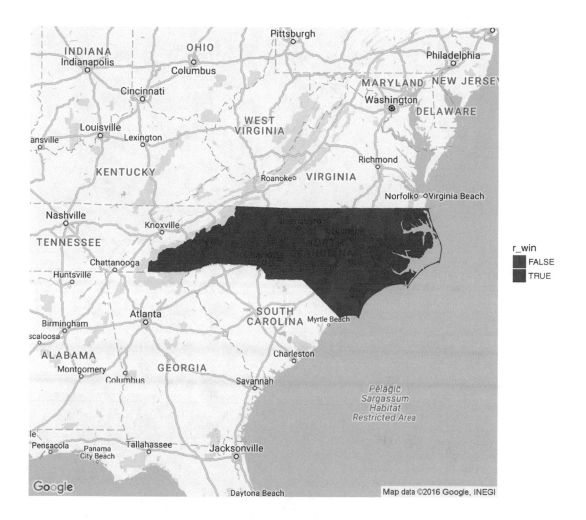

Figure 14.12: Bichromatic choropleth map of the results of the 2012 congressional elections in North Carolina.

Figure 14.12 shows that it was the Democratic districts that tended to be irregularly shaped. Districts 12 and 4 have narrow, tortured shapes—both were heavily Democratic. This plot tells us who won, but it doesn't convey the subtleties we observed about the *margins* of victory. In the next plot, we use a continuous color scale to to indicate the percentage of votes in each district. The RdBu diverging color palette comes from RColorBrewer (see Chapter 2).

```
ggmap(nc) +
  geom_polygon(aes(x = long, y = lat, group = group, fill = r_pct),
               alpha = 0.8, data = nc_full) +
  scale_fill_distiller(palette = "RdBu", limits = c(0.2,0.8)) +
  geom_text(aes(x = x, y = y, label = DISTRICT), data = nc_centroids_full) +
  theme_map()
```

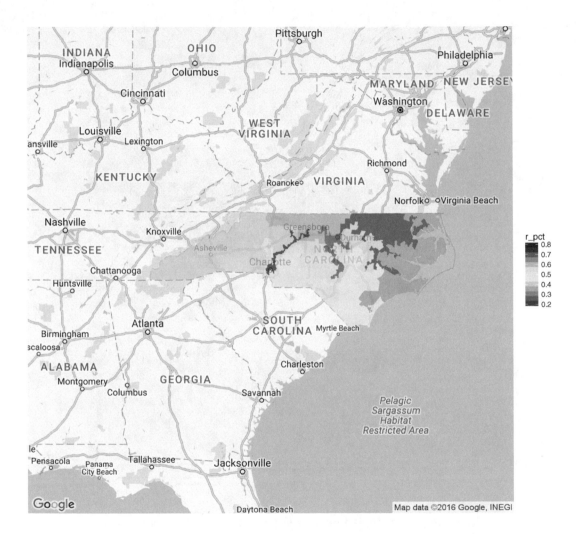

Figure 14.13: Full color choropleth of the results of the 2012 congressional elections in North Carolina. The clustering of Democratic voters is evident from the deeper blue in Democratic districts, versus the pale red in the more numerous Republican districts.

The `limits` argument to `scale_fill_distiller()` is important. This forces *red* to be the color associated with 80% Republican votes and *blue* to be associated with 80% Democratic votes. Without this argument, red would be associated with the maximum value in that data (about 63%) and blue with the minimum (about 20%). This would result in the neutral color of white not being at exactly 50%. When choosing color scales, it is critically important to make choices that reflect the data.

Pro Tip: Choose colors and scales carefully when making maps.

In Figure 14.13, we can see that the three Democratic districts are "bluer" than the nine Republican counties are "red." This reflects the clustering that we observed earlier. North Carolina has become one of the more egregious examples of *gerrymandering*, the phenomenon of when legislators of one party use their re-districting power for political gain. This is evident in Figure 14.13, where Democratic votes are concentrated in three curiously-drawn congressional districts. This enables Republican lawmakers to have 69% (9/13) of the voting power in Congress despite earning only 48.8% of the votes.

14.4.5   Using `leaflet`

Was it true that the Democratic districts were weaved together to contain many of the biggest cities in the state? A similar map made in `leaflet` would allow us to zoom in and pan out, making it easier to survey the districts.

First, we will define a color palette over the values $[0, 1]$ that ranges from red to blue.

```
library(leaflet)
pal <- colorNumeric(palette = "RdBu", domain = c(0, 1))
```

To make our plot in `leaflet`, we have to add the tiles, and then the polygons defined by the `SpatialPolygonsDataFrame` `nc_merged`. Since we want red to be associated with the percentage of Republican votes, we will map $1 - \texttt{r\_pct}$ to color. Note that we also add popups with the actual percentages, so that if you click on the map, it will show the district number and the percentage of Republican votes. A static image from the resulting `leaflet` map is shown in Figure 14.14.

```
nc_dynamic <- leaflet() %>%
  addTiles() %>%
  addPolygons(
    data = nc_merged, weight = 1, fillOpacity = 0.7, color = ~pal(1- r_pct),
    popup = ~paste("District", DISTRICT, "</br>", round(r_pct, 4))) %>%
  setView(lng = -80, lat = 35, zoom = 7)
```

14.5   Effective maps: How (not) to lie

The map shown in Figure 14.13 is an example of a *choropleth* map. This is a very common type of map where coloring and/or shading is used to differentiate a region of the map based on the value of a variable. These maps are popular, and can be very persuasive, but you should be aware of some challenges when making and interpreting choropleth maps and other data maps. Three common map types include:

- Choropleth: color or shade regions based on the value of a variable

- Proportional symbol: associate a symbol with each location, but scale its size to reflect the value of a variable

- Dot density: place dots for each data point, and view their accumulation

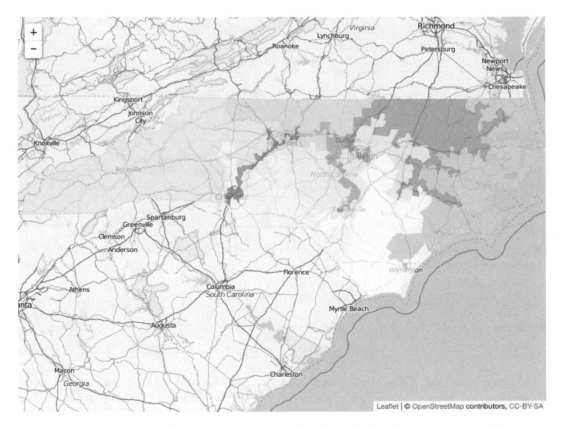

Figure 14.14: Static image from a `leaflet` plot of the North Carolina congressional districts.

In the next section, we will create a proportional symbol map. We note that in these situations the symbol placed on the map is usually two-dimensional. Thus, its size—*in area*—should be scaled in proportion to the quantity being mapped. Be aware that often the size of the symbol is defined by its *radius*. If the *radius* is in direct proportion to the quantity being mapped, then the area will be disproportionately large.

Pro Tip: Always scale the size of proportional symbols in terms of their area.

As noted in Chapter 2, the choice of scale is also important, and often done poorly. The relationship between various quantities can be altered by scale. In Chapter 2, we showed how the use of logarithmic scale can be used to improve the readability of a scatterplot. In Figure 14.13 we illustrated the importance of properly setting the scale of a proportion so that 0.5 was exactly in the middle. Try making Figure 14.13 without doing this, and see if the results are as easily interpretable.

Decisions about colors are also crucial to making an effective map. In Chapter 2, we mentioned the color palettes available through `RColorBrewer`. When making maps, categorical variables should be displayed using a *qualitative* palette, while quantitative variables should be displayed using a *sequential* or *diverging* palette. In Figure 14.13 we employed a diverging palette, because Republicans and Democrats are on two opposite ends of the scale, with the neutral white color representing 0.5.

Finally, the concept of *normalization* is fundamental. Plotting raw data values on maps can easily distort the truth. This is particularly true in the case of data maps, because area is an implied variable. Thus, on choropleth maps, we almost always want to show some sort of density or ratio rather than raw values (i.e., counts).

14.6 Extended example: Historical airline route maps

One of the more juvenile pleasures of flying is reading the material in the seat pocket in front of you. The amount of information that the airline is willing to tell you about their business never ceases to amaze. In addition to the layout of the terminals for the airports that the airlines serves, they always show a domestic airlines route map. But while those old route maps are probably long gone, the `airlines` data gives us the ability to resurrect *historical* airline route maps—for any airline.

To start, let's specify a carrier and a year. In this case we will work with Delta Airlines before their merger with Northwest Airlines in 2008.

```
my_carrier <- "DL"
my_year <- 2006
```

Next, we can use these values as parameters to a query to our `airlines` database. To make an informative map, we will need two pieces of information: a list of airports and the number of flights that it handled that year, and a list of all the segments that the carrier flew that year. First we will make connections to the `flights` and `airports` tables, respectively.

```
db <- src_scidb("airlines")
airports <- tbl(db, "airports")
flights <- tbl(db, "flights")
```

To find the airports that the airline services, we'll query the `flights` table, but join on the `airports` table to retrieve the name and location of the airport.[7]

```
destinations <- flights %>%
  filter(year == my_year, carrier == my_carrier) %>%
  left_join(airports, by = c("dest" = "faa")) %>%
  group_by(dest) %>%
  summarize(N = n(), lon = max(lon), lat = max(lat),
            # note use of MySQL syntax instead of dplyr
            name = min(CONCAT("(", dest, ") ",
              REPLACE(name, " Airport", ""))))) %>%
  collect() %>%
  na.omit()
glimpse(destinations)

Observations: 108
Variables: 5
$ dest <chr> "ABQ", "ALB", "ANC", "ATL", "AUS", "BDL", "BHM", "BNA", "...
$ N    <dbl> 1842, 444, 641, 165743, 1401, 6443, 2029, 2175, 917, 1617...
$ lon  <dbl> -106.6, -73.8, -150.0, -84.4, -97.7, -72.7, -86.8, -86.7,...
$ lat  <dbl> 35.0, 42.7, 61.2, 33.6, 30.2, 41.9, 33.6, 36.1, 43.6, 42....
$ name <chr> "(ABQ) Albuquerque International Sunport", "(ALB) Albany ...
```

---

[7]Note the use of MySQL syntax in defining the `name` field. This was necessary because both `flights` and `airports` are src_mysql objects. See Section 12.1 for a further explanation.

Next, we need to know about the flights between each airport that will make up the segments in our map. How many flights went between each pair of airports?

```
segments <- flights %>%
  filter(year == my_year, carrier == my_carrier) %>%
  group_by(origin, dest) %>%
  summarize(N = n()) %>%
  left_join(airports, by = c("origin" = "faa")) %>%
  left_join(airports, by = c("dest" = "faa")) %>%
  collect() %>%
  na.omit()
dim(segments)
```

```
[1] 489   20
```

Note that there were 108 in service, but only 489 unique pairs of airports (in either direction). Thus, Delta served only 4% of the possible flight routes among these airports.

14.6.1   Using ggmap

Since we have the geographic coordinates of the airports, we can make a map of the Delta hubs using ggmap. We will plot the airports as semi-transparent gray dots, with the area of each dot proportional to the number of flights that it served. Note that ggplot2 automatically scales points by area.

```
library(ggmap)
route_map <- qmap("junction city, kansas", zoom = 4, maptype = "roadmap") +
  geom_point(data = destinations, alpha = 0.5,
             aes(x = lon, y = lat, size = N)) +
  scale_size() +
  theme_map()
route_map
```

Note that the Delta hubs in Atlanta, Salt Lake City, Cincinnati, and New York are immediately obvious in Figure 14.15. However, the additional hubs in Minneapolis–St. Paul and Detroit, are not present—these were acquired through the merger with Northwest. At the time, Atlanta served more than five times as many flights as Salt Lake City.

```
destinations %>% arrange(desc(N))
```

```
# A tibble: 108   5
    dest      N     lon    lat                                         name
   <chr>  <dbl>   <dbl>  <dbl>                                        <chr>
1    ATL 165743  -84.4   33.6      (ATL) Hartsfield Jackson Atlanta Intl
2    SLC  30835 -112.0   40.8                 (SLC) Salt Lake City Intl
3    CVG  24274  -84.7   39.0  (CVG) Cincinnati Northern Kentucky Intl
4    LGA  21525  -73.9   40.8                         (LGA) La Guardia
5    BOS  16176  -71.0   42.4 (BOS) General Edward Lawrence Logan Intl
6    MCO  13302  -81.3   28.4                       (MCO) Orlando Intl
7    LAX  13277 -118.4   33.9                   (LAX) Los Angeles Intl
8    JFK  12756  -73.8   40.6               (JFK) John F Kennedy Intl
```

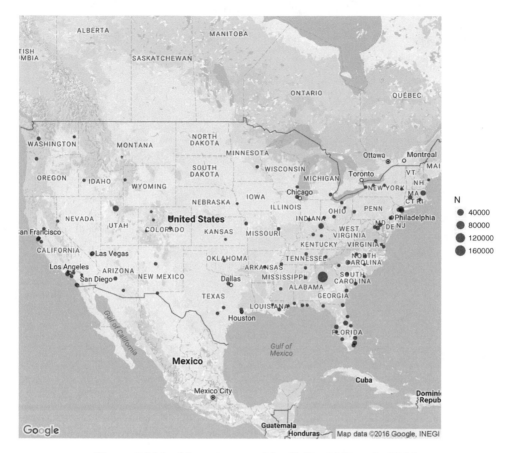

Figure 14.15: Airports served by Delta Airlines in 2006.

```
9    DCA  11625  -77.0  38.9    (DCA) Ronald Reagan Washington Natl
10   FLL  10567  -80.2  26.1    (FLL) Fort Lauderdale Hollywood Intl
# ... with 98 more rows
```

However, it remains to draw the segments connecting each airport. Here again, we will use the frequency of such segments to change the color of the lines. Figure 14.16 shows the Delta Airlines network in 2006.

```
route_map + geom_segment(
  aes(x = lon.x, y = lat.x, xend = lon.y, yend = lat.y, color = N),
  size = 0.05, arrow = arrow(length = unit(0.3, "cm")), data = segments)
```

14.6.2   Using `leaflet`

To plot our segments using `leaflet`, we have to convert them to a `SpatialLines` object. This is unfortunately a bit cumbersome. First, we will create a data frame called `lines` that has one row for each pair of airports, and contains a column of corresponding `Line` objects.

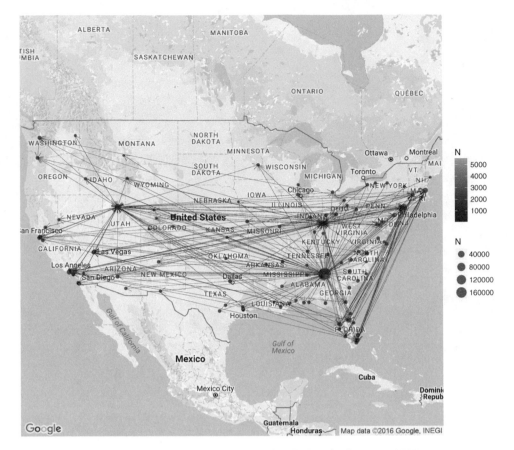

Figure 14.16: Full route map for Delta Airlines in 2006.

```
lines <- bind_rows(
  segments %>%
    select(origin, dest, lat.x, lon.x) %>%
    rename(lat = lat.x, lon = lon.x),
  segments %>%
    select(origin, dest, lat.y, lon.y) %>%
    rename(lat = lat.y, lon = lon.y)) %>%
  arrange(origin, dest) %>%
  na.omit() %>%
  group_by(origin, dest) %>%
  do(line = Line(as.data.frame(select(., lon, lat))))
)
head(lines, 3)

# A tibble: 3  3
  origin  dest        line
  <chr>  <chr>       <list>
1    ABQ    ATL  <S4: Line>
2    ABQ    CVG  <S4: Line>
3    ALB    ATL  <S4: Line>
```

Next, we write a function that will take each row of `lines` as an input, and return an object of class Lines. We use `apply()` to iterate that function over each row of the `lines` data set, returning a `list` of Lines objects.

```
make_line <- function(x) {
  Lines(list(x[["line"]]), ID = paste0(x$origin, "-", x$dest))
}
lines_list <- apply(lines, MARGIN = 1, make_line)
```

Finally, we define these as `SpatialLines` and project them into the correct coordinate system for use with `leaflet`.

```
segments_sp <- SpatialLines(lines_list, CRS("+proj=longlat"))
summary(segments_sp)

Object of class SpatialLines
Coordinates:
     min    max
x -157.9 -70.3
y   19.7  64.8
Is projected: FALSE
proj4string : [+proj=longlat +ellps=WGS84]

segments_sp <- segments_sp %>% spTransform(CRS("+init=epsg:4326"))
```

To make our map in `leaflet`, we simply have to use the `addCircles()` function to add the circle markers for each airport, and the `addPolylines()` function to add the lines for each segment. A static image of the interactive plot is shown in Figure 14.17.

```
library(leaflet)
l_map <- leaflet() %>%
  addTiles() %>%
  addCircles(lng = ~lon, lat = ~lat, weight = 1,
    radius = ~sqrt(N) * 500, popup = ~name, data = destinations) %>%
  addPolylines(weight = 0.4, data = segments_sp) %>%
  setView(lng = -80, lat = 38, zoom = 6)
l_map
```

14.7 Projecting polygons

It is worth briefly illustrating the hazards of mapping unprojected data. Consider the congressional district map for the entire country. To plot this, we follow the same steps as before, but omit the step of restricting to North Carolina. There is one additional step here for creating a mapping between state names and their abbreviations. Thankfully, these data are built into R.

```
districts_tidy <- tidy(districts, region = "ID")
districts_full <- districts_tidy %>%
  left_join(districts@data, by = c("id" = "ID")) %>%
```

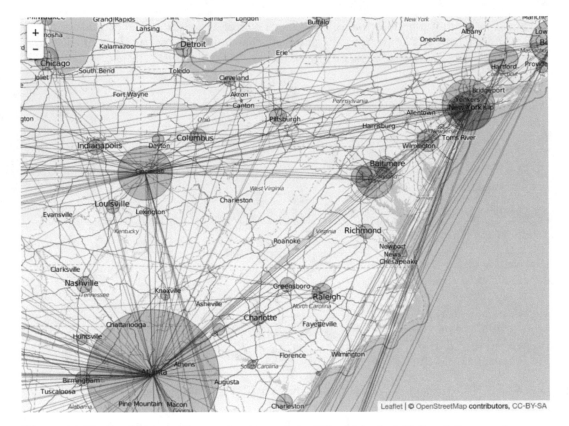

Figure 14.17: Static image from a `leaflet` plot of the historical Delta airlines route map.

```
left_join(data.frame(state.abb, state.name),
          by = c("STATENAME" = "state.name")) %>%
left_join(district_elections, by = c("state.abb" = "state",
                                     "DISTRICT" = "district"))
```

We can make the map by adding white polygons for the generic map data and then adding colored polygons for each congressional district. Some clipping will make this easier to see.

```
box <- bbox(districts)
us_map <- ggplot(data = map_data("world"),
                 aes(x = long, y = lat, group = group)) +
  geom_path(color = "black", size = 0.1) +
  geom_polygon(aes(fill = r_pct), data = districts_full) +
  scale_fill_distiller(palette = "RdBu", limits = c(0,1)) +
  theme_map() + xlim(-180, -50) + ylim(box[2,])
```

We display the Mercator projection of this base map in Figure 14.18. Note how massive Alaska appears to be in relation to the other states! Alaska is big, but it is not that big! This is a distortion of reality due to the projection.

```
us_map + coord_map()
```

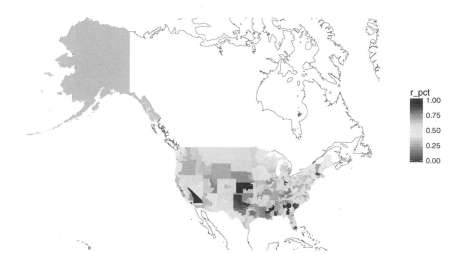

Figure 14.18: U.S. congressional election results, 2012 (Mercator projection).

We can use the Albers equal area projection to make a more representative picture, as shown in Figure 14.19. Note how Alaska is still the biggest state (and district) by area, but it is not much closer in size to Texas.

```
us_map + coord_map(projection = "albers", lat0 = 20, lat1 = 50)
```

14.8 Playing well with others

There are many technologies outside of R that allow you to work with spatial data. ArcGIS is a proprietary Geographic Information System software that is considered by many to be the industry state-of-the-art. QGIS is its open-source competitior. Both have graphical user interfaces.

Keyhole Markup Language (KML) is an XML file format for storing geographic data. KML files can be read by Google Earth and other GIS applications. A Spatial*DataFrame object in R can be written to KML using functions from either the maptools or plotKML packages. These files can then be read by ArcGIS, Google Maps, or Google Earth. Here, we illustrate how to create a KML file for the North Carolina congressional districts data frame that we defined earlier. A screenshot of the resulting output in Google Earth is shown in Figure 14.20.

```
nc_merged %>% spTransform(CRS("+init=epsg:4326")) %>%
  plotKML::kml(file = "nc_congress113.kml",
               folder.name = "113th Congress (NC)",
               colour = r_pct, fill = c("red", "blue", "white"),
               labels = DISTRICT, alpha = 0.5, kmz = TRUE)
```

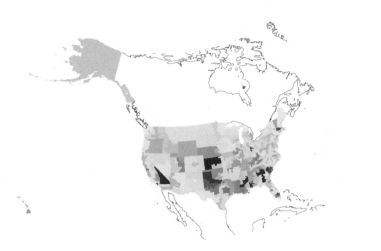

Figure 14.19: U.S. congressional election results, 2012 (Albers equal area projection).

14.9   Further resources

A helpful pocket guide to CRS systems in R contains information about projections, ellipsoids, and datums (reference points). Bivand et al. [34] discuss the mechanics of how to work with spatial data in R in addition to introducing spatial modeling. The tigris package provides access to shapefiles and demographic data from the United States Census Bureau.

Quantitative measures of gerrymandering have been a subject of interest to political scientists for some time [146, 71, 104, 139].

14.10   Exercises

Exercise 14.1

Use the spatial data in the macleish package and ggmap to make an informative static map of the MacLeish Field Station property. You may want to consult with https://www.smith.edu/ceeds/macleish_maps.php for inspiration and context.

Exercise 14.2

Use the spatial data in the macleish package and leaflet to make an informative interactive map of the MacLeish Field Station property.

Exercise 14.3

The Violations data frame in the mdsr contains information on Board of Health violations by New York City restaurants. These data contain spatial information in the form of addresses and zip codes. Use the geocode() function in ggmap to obtain spatial coordinates for these restaurants.

Exercise 14.4

Using the spatial coordinates you obtained in the previous exercise, create an informative static map using ggmap that illustrates the nature and extent of restaurant violations in New York City.

Exercise 14.5

Using the spatial coordinates you obtained in the previous exercises, create an informative interactive map using leaflet that illustrates the nature and extent of restaurant violations in New York City.

Exercise 14.6

Use the tigris package to make the congressional election district map for your home state. Do you see evidence of gerrymandering? Why or why not?

Exercise 14.7

Use the tigris package to conduct a spatial analysis of the Census data it contains for your home state. Can you illustrate how the demography of your state varies spatially?

Exercise 14.8

Use the airlines data to make the airline route map for another carrier in another year.

Exercise 14.9

Compare the airline route map for Delta Airlines in 2013 to the same map for Delta in 2003 and 1993. Discuss the history of Delta's use of hub airports. Quantify changes over time. Reflect on the more general westward expansion of air travel in the United States.

Exercise 14.10

Researchers at UCLA maintain historical congressional district shapefiles (see http://cdmaps.polisci.ucla.edu/shp). Use these data to discuss the history of gerrymandering in the United States. Is the problem better or worse today?

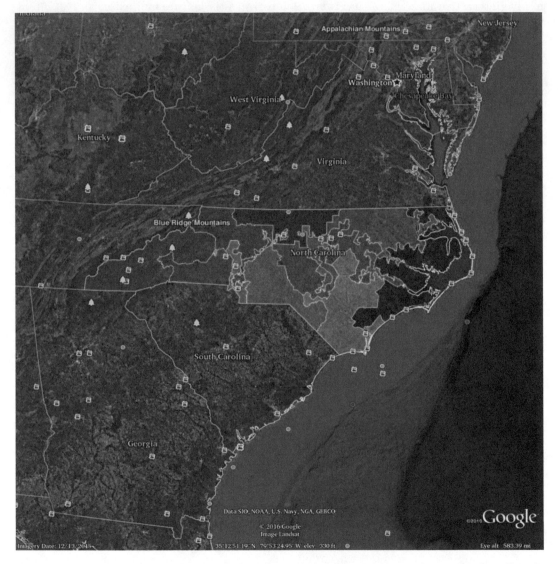

Figure 14.20: Screenshot of the North Carolina congressional districts as rendered in Google Earth, after exporting to KML. Compare with Figure 14.13.

Chapter 15

Text as data

So far, we have focused primarily on numerical data, but there is a whole field of research that focuses on textual data. Fields such as *natural language processing* and *computational linguistics* work directly with text documents to extract meaning algorithmically. Not surprisingly, the fact that computers are really good at storing text, but not very good at understanding it, whereas humans are really good at understanding text, but not very good at storing it, is a fundamental challenge.

Processing text data requires an additional set of wrangling skills. In this chapter we will introduce how text can be ingested, how *corpora* (collections of text documents) can be created, and how *regular expressions* can be used to automate searches that would otherwise be excruciatingly labor-intensive.

15.1 Tools for working with text

As noted previously, working with textual data requires new tools. In this section we introduce the powerful grammar of regular expressions.

15.1.1 Regular expressions using *Macbeth*

Project Gutenberg contains the full-text for all of Shakespeare's plays. In this example we will use text mining techniques to explore *The Tragedy of Macbeth*. The text can be downloaded directly from Project Gutenberg. Alternatively, the Macbeth_raw object is also included in the mdsr package.

```
library(mdsr)
macbeth_url <- "http://www.gutenberg.org/cache/epub/1129/pg1129.txt"
Macbeth_raw <- RCurl::getURL(macbeth_url)
```

```
data(Macbeth_raw)
```

Note that Macbeth_raw is a *single* string of text (i.e., a character vector of length 1) that contains the entire play. In order to work with this, we want to split this single string into a vector of strings using the strsplit() function. To do this, we just have to specify the end of line character(s), which in this case are: \r\n.

```
# strsplit returns a list: we only want the first element
macbeth <- strsplit(Macbeth_raw, "\r\n")[[1]]
length(macbeth)
```

[1] 3193

Now let's examine the text. Note that each speaking line begins with two spaces, followed by the speaker's name in capital letters.

```
macbeth[300:310]
```

```
 [1] "meeting a bleeding Sergeant."
 [2] ""
 [3] "  DUNCAN. What bloody man is that? He can report,"
 [4] "    As seemeth by his plight, of the revolt"
 [5] "    The newest state."
 [6] "  MALCOLM. This is the sergeant"
 [7] "    Who like a good and hardy soldier fought"
 [8] "    'Gainst my captivity. Hail, brave friend!"
 [9] "    Say to the King the knowledge of the broil"
[10] "    As thou didst leave it."
[11] "  SERGEANT. Doubtful it stood,"
```

The power of text mining comes from quantifying ideas embedded in the text. For example, how many times does the character Macbeth speak in the play? Think about this question for a moment. If you were holding a physical copy of the play, how would you compute this number? Would you flip through the book and mark down each speaking line on a separate piece of paper? Is your algorithm scalable? What if you had to do it for *all* characters in the play, and not just Macbeth? What if you had to do it for *all 37* of Shakespeare's plays? What if you had to do it for all plays written in English?

Naturally, a computer cannot read the play and figure this out, but we can find all instances of Macbeth's speaking lines by cleverly counting patterns in the text.

```
macbeth_lines <- grep("  MACBETH", macbeth, value = TRUE)
length(macbeth_lines)
```

[1] 147

```
head(macbeth_lines)
```

```
[1] "  MACBETH, Thane of Glamis and Cawdor, a general in the King's"
[2] "  MACBETH. So foul and fair a day I have not seen."
[3] "  MACBETH. Speak, if you can. What are you?"
[4] "  MACBETH. Stay, you imperfect speakers, tell me more."
[5] "  MACBETH. Into the air, and what seem'd corporal melted"
[6] "  MACBETH. Your children shall be kings."
```

The grep() function works using a *needle* in a *haystack* paradigm, wherein the first argument is the *regular expression* (or pattern) you want to find (i.e., the needle) and the second argument is the character vector in which you want to find patterns (i.e., the haystack). Note that unless the value is set to TRUE, grep() returns the *indices* of the haystack in which the needles were found. By changing the needle, we find different results:

```
length(grep(" MACDUFF", macbeth))
```

```
[1] 60
```

The grep() function—which we use in the example in the next section—uses the same syntax but returns a logical vector as long as the haystack. Thus, while the length of the vector returned by grep() is the number of matches, the length of the vector returned by grepl() is always the same as the length of the haystack vector.

```
length(grep(" MACBETH", macbeth))
```

```
[1] 147
```

```
length(grepl(" MACBETH", macbeth))
```

```
[1] 3193
```

However, both will subset the original vector in the same way, and thus in this respect they are functionally equivalent.

```
identical(macbeth[grep(" MACBETH", macbeth)],
          macbeth[grepl(" MACBETH", macbeth)])
```

```
[1] TRUE
```

To extract the piece of each matching line that actually matched, use the str_extract() function from the stringr package.

```
library(stringr)
pattern <- " MACBETH"
grep(pattern, macbeth, value = TRUE) %>%
  str_extract(pattern) %>%
  head()
```

```
[1] " MACBETH" " MACBETH" " MACBETH" " MACBETH" " MACBETH" " MACBETH"
```

Above, we use a literal string (e.g., " MACBETH") as our needle to find exact matches in our haystack. This is the simplest type of pattern for which we could have searched, but the needle that grep() searches for can be any regular expression.

Regular expression syntax is very powerful and as a result, can become very complicated. Still, regular expressions are a grammar, so that learning a few basic concepts will allow you to build more efficient searches.

Pro Tip: Regular expressions are a powerful and commonly used tool. They are implemented in many programming languages. Developing a deep understanding of regular expressions will pay off in terms of text manipulations.

- Metacharacters: . is a *metacharacter* that matches any character. Note that if you want to search for the literal value of a metacharacter (e.g., a period), you have to escape it with a backslash. To use the pattern in R, two backslashes are needed. Note the difference in the results below.

```
head(grep("MAC.", macbeth, value = TRUE))

[1] "MACHINE READABLE COPIES MAY BE DISTRIBUTED SO LONG AS SUCH COPIES"
[2] "MACHINE READABLE COPIES OF THIS ETEXT, SO LONG AS SUCH COPIES"
[3] "WITH PERMISSION.  ELECTRONIC AND MACHINE READABLE COPIES MAY BE"
[4] "THE TRAGEDY OF MACBETH"
[5] "  MACBETH, Thane of Glamis and Cawdor, a general in the King's"
[6] "  LADY MACBETH, his wife"

head(grep("MACBETH\\.", macbeth, value = TRUE))

[1] "  MACBETH. So foul and fair a day I have not seen."
[2] "  MACBETH. Speak, if you can. What are you?"
[3] "  MACBETH. Stay, you imperfect speakers, tell me more."
[4] "  MACBETH. Into the air, and what seem'd corporal melted"
[5] "  MACBETH. Your children shall be kings."
[6] "  MACBETH. And Thane of Cawdor too. Went it not so?"
```

- Character sets: Use brackets to define sets of characters to match. This pattern will match any lines that contain MAC followed by any capital letter other than A. It will match MACBETH but not MACALESTER.

```
head(grep("MAC[B-Z]", macbeth, value = TRUE))

[1] "MACHINE READABLE COPIES MAY BE DISTRIBUTED SO LONG AS SUCH COPIES"
[2] "MACHINE READABLE COPIES OF THIS ETEXT, SO LONG AS SUCH COPIES"
[3] "WITH PERMISSION.  ELECTRONIC AND MACHINE READABLE COPIES MAY BE"
[4] "THE TRAGEDY OF MACBETH"
[5] "  MACBETH, Thane of Glamis and Cawdor, a general in the King's"
[6] "  LADY MACBETH, his wife"
```

- Alternation: To search for a few specific alternatives, use the | wrapped in parentheses. This pattern will match any lines that contain either MACB or MACD.

```
head(grep("MAC(B|D)", macbeth, value = TRUE))

[1] "THE TRAGEDY OF MACBETH"
[2] "  MACBETH, Thane of Glamis and Cawdor, a general in the King's"
[3] "  LADY MACBETH, his wife"
[4] "  MACDUFF, Thane of Fife, a nobleman of Scotland"
[5] "  LADY MACDUFF, his wife"
[6] "  MACBETH. So foul and fair a day I have not seen."
```

- Anchors: Use ^ to anchor a pattern to the beginning of a piece of text, and $ to anchor it to the end.

```
head(grep("^ MAC[B-Z]", macbeth, value = TRUE))
```

```
[1] "  MACBETH, Thane of Glamis and Cawdor, a general in the King's"
[2] "  MACDUFF, Thane of Fife, a nobleman of Scotland"
[3] "  MACBETH. So foul and fair a day I have not seen."
[4] "  MACBETH. Speak, if you can. What are you?"
[5] "  MACBETH. Stay, you imperfect speakers, tell me more."
[6] "  MACBETH. Into the air, and what seem'd corporal melted"
```

- Repetitions: We can also specify the number of times that we want certain patterns to occur: ? indicates zero or one time, * indicates zero or more times, and + indicates one or more times. This quantification is applied to the previous element in the pattern—in this case, a space.

```
head(grep("^ ?MAC[B-Z]", macbeth, value = TRUE))
```

```
[1] "MACHINE READABLE COPIES MAY BE DISTRIBUTED SO LONG AS SUCH COPIES"
[2] "MACHINE READABLE COPIES OF THIS ETEXT, SO LONG AS SUCH COPIES"
```

```
head(grep("^ *MAC[B-Z]", macbeth, value = TRUE))
```

```
[1] "MACHINE READABLE COPIES MAY BE DISTRIBUTED SO LONG AS SUCH COPIES"
[2] "MACHINE READABLE COPIES OF THIS ETEXT, SO LONG AS SUCH COPIES"
[3] "  MACBETH, Thane of Glamis and Cawdor, a general in the King's"
[4] "  MACDUFF, Thane of Fife, a nobleman of Scotland"
[5] "  MACBETH. So foul and fair a day I have not seen."
[6] "  MACBETH. Speak, if you can. What are you?"
```

```
head(grep("^ +MAC[B-Z]", macbeth, value = TRUE))
```

```
[1] "  MACBETH, Thane of Glamis and Cawdor, a general in the King's"
[2] "  MACDUFF, Thane of Fife, a nobleman of Scotland"
[3] "  MACBETH. So foul and fair a day I have not seen."
[4] "  MACBETH. Speak, if you can. What are you?"
[5] "  MACBETH. Stay, you imperfect speakers, tell me more."
[6] "  MACBETH. Into the air, and what seem'd corporal melted"
```

Combining these basic rules can automate incredibly powerful and sophisticated searches, and are an increasingly necessary tool in every data scientist's toolbox.

15.1.2 Example: Life and death in *Macbeth*

Can we use these techniques to analyze the speaking patterns in Macbeth? Are there things we can learn about the play simply by noting who speaks when? Four of the major characters in *Macbeth* are the titular character, his wife Lady Macbeth, his friend Banquo, and Duncan, the King of Scotland.

We might learn something about the play by knowing when each character speaks as a function of the line number in the play. We can retrieve this information using `grepl()`.

```
Macbeth <- grepl("  MACBETH\\.", macbeth)
LadyMacbeth <- grepl("  LADY MACBETH\\.", macbeth)
Banquo <- grepl("  BANQUO\\.", macbeth)
Duncan <- grepl("  DUNCAN\\.", macbeth)
```

However, for plotting purposes we will want to convert these `logical` vectors into numeric vectors, and tidy up the data. Since there is unwanted text at the beginning and the end of the play text, we will also restrict our analysis to the actual contents of the play (which occurs from line 218 to line 3172).

```
library(tidyr)
speaker_freq <- data.frame(Macbeth, LadyMacbeth, Banquo, Duncan) %>%
  mutate(line = 1:length(macbeth)) %>%
  gather(key = "character", value = "speak", -line) %>%
  mutate(speak = as.numeric(speak)) %>%
  filter(line > 218 & line < 3172)
glimpse(speaker_freq)

Observations: 11,812
Variables: 3
$ line      <int> 219, 220, 221, 222, 223, 224, 225, 226, 227, 228, 22...
$ character <chr> "Macbeth", "Macbeth", "Macbeth", "Macbeth", "Macbeth...
$ speak     <dbl> 0, 0, 0, 0, 0, 0, 0, 0, 0, 0, 0, 0, 0, 0, 0, 0, 0, 0...
```

Before we create the plot, we will gather some helpful contextual information about when each Act begins.

```
acts_idx <- grep("^ACT [I|V]+", macbeth)
acts_labels <- str_extract(macbeth[acts_idx], "^ACT [I|V]+")
acts <- data.frame(line = acts_idx, labels = acts_labels)
```

Finally, Figure 15.1 illustrates how King Duncan of Scotland is killed early in Act II (never to speak again), with Banquo to follow in Act III. Soon afterwards in Act IV, Lady Macbeth—overcome by guilt over the role she played in Duncan's murder—kills herself. The play and Act V conclude with a battle in which Macbeth is killed.

```
ggplot(data = speaker_freq, aes(x = line, y = speak)) +
  geom_smooth(aes(color = character), method = "loess", se = 0, span = 0.4) +
  geom_vline(xintercept = acts_idx, color = "darkgray", lty = 3) +
  geom_text(data = acts, aes(y = 0.085, label = labels),
            hjust = "left", color = "darkgray") +
  ylim(c(0, NA)) + xlab("Line Number") + ylab("Proportion of Speeches")
```

15.2 Analyzing textual data

The *arXiv* (pronounced "archive") is a fast-growing electronic repository of preprints of scientific papers from many disciplines. The `aRxiv` package provides an application programming interface (API) to the files and metadata available at `arxiv.org`. We will explore 95 papers that matched the search term "`data science`" in the repository as of December

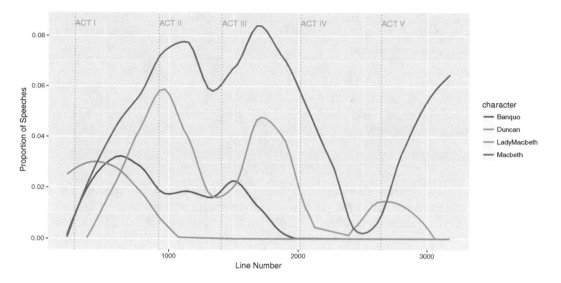

Figure 15.1: Speaking parts in *Macbeth* for four major characters. Duncan is killed early in the play and never speaks again.

2015, and see if we can crowd-source a definition of "data science." The following code was used to generate this file.

```
library(aRxiv)
DataSciencePapers <- arxiv_search(query = '"Data Science"', limit = 200)
```

We have also included the data frame DataSciencePapers in the mdsr package, so to use this selection of papers downloaded from the archive, you can simply load it.

```
data(DataSciencePapers)
```

Note that there are two columns in this data set (submitted and updated) that are clearly storing dates, but they are stored as character vectors.

```
head(DataSciencePapers)
```

```
# A tibble: 6  15
                id           submitted           updated
             <chr>              <chr>             <chr>
1 astro-ph/0701361v1 2007-01-12 03:28:11 2007-01-12 03:28:11
2         0901.2805v1 2009-01-19 10:38:33 2009-01-19 10:38:33
3         0901.3118v2 2009-01-20 18:48:59 2009-01-24 19:23:47
4         0909.3895v1 2009-09-22 02:55:14 2009-09-22 02:55:14
5         1106.2503v5 2011-06-13 17:42:32 2013-06-23 21:21:41
6         1106.3305v1 2011-06-16 18:45:32 2011-06-16 18:45:32
# ... with 12 more variables: title <chr>, abstract <chr>, authors <chr>,
#   affiliations <chr>, link_abstract <chr>, link_pdf <chr>,
#   link_doi <chr>, comment <chr>, journal_ref <chr>, doi <chr>,
#   primary_category <chr>, categories <chr>
```

To make sure that R understands those variables as dates, we will once again use the lubridate package (see Chapter 5). After this conversion, R understands that these two columns are measurements of time.

```
library(lubridate)
DataSciencePapers <- DataSciencePapers %>%
  mutate(submitted = ymd_hms(submitted), updated = ymd_hms(updated))
glimpse(DataSciencePapers)

Observations: 95
Variables: 15
$ id                <chr> "astro-ph/0701361v1", "0901.2805v1", "0901.31...
$ submitted         <dttm> 2007-01-12 03:28:11, 2009-01-19 10:38:33, 20...
$ updated           <dttm> 2007-01-12 03:28:11, 2009-01-19 10:38:33, 20...
$ title             <chr> "How to Make the Dream Come True: The Astrono...
$ abstract          <chr> "  Astronomy is one of the most data-intensiv...
$ authors           <chr> "Ray P Norris", "Heinz Andernach", "O. V. Ver...
$ affiliations      <chr> "", "", "Special Astrophysical Observatory, N...
$ link_abstract     <chr> "http://arxiv.org/abs/astro-ph/0701361v1", "h...
$ link_pdf          <chr> "http://arxiv.org/pdf/astro-ph/0701361v1", "h...
$ link_doi          <chr> "", "http://dx.doi.org/10.2481/dsj.8.41", "ht...
$ comment           <chr> "Submitted to Data Science Journal Presented ...
$ journal_ref       <chr> "", "", "", "", "EPJ Data Science, 1:9, 2012"...
$ doi               <chr> "", "10.2481/dsj.8.41", "10.2481/dsj.8.34", "...
$ primary_category  <chr> "astro-ph", "astro-ph.IM", "astro-ph.IM", "as...
$ categories        <chr> "astro-ph", "astro-ph.IM|astro-ph.CO", "astro...
```

We will begin by examining the distribution of submission years. Is there more interest in data science in more recent years?

```
tally(~ year(submitted), data = DataSciencePapers)

year(submitted)
2007 2009 2011 2012 2013 2014 2015
   1    3    3    6   13   25   44
```

We see that the first paper was submitted in 2007, but that submissions have increased almost exponentially since then—nearly doubling in each of the last five years. Let's take a closer look at that first paper.

```
DataSciencePapers %>%
  filter(year(submitted) == 2007) %>%
  glimpse()

Observations: 1
Variables: 15
$ id                <chr> "astro-ph/0701361v1"
$ submitted         <dttm> 2007-01-12 03:28:11
$ updated           <dttm> 2007-01-12 03:28:11
$ title             <chr> "How to Make the Dream Come True: The Astrono...
$ abstract          <chr> "  Astronomy is one of the most data-intensiv...
```

```
$ authors          <chr> "Ray P Norris"
$ affiliations     <chr> ""
$ link_abstract    <chr> "http://arxiv.org/abs/astro-ph/0701361v1"
$ link_pdf         <chr> "http://arxiv.org/pdf/astro-ph/0701361v1"
$ link_doi         <chr> ""
$ comment          <chr> "Submitted to Data Science Journal Presented ...
$ journal_ref      <chr> ""
$ doi              <chr> ""
$ primary_category <chr> "astro-ph"
$ categories       <chr> "astro-ph"
```

This manifesto (entitled "How to Make the Dream Come True") discussed the data-intensive field of astronomy and was submitted to the *Data Science Journal* (which helps explain why it was included in our search but doesn't include "data science" in the abstract).

What fields are generating these papers? A quick glance at the primary_category variable reveals a cryptic list of fields and sub-fields. It would be more helpful to focus simply on the primary field.

```
tally(~ primary_category, data = DataSciencePapers)
```

```
primary_category
       astro-ph       astro-ph.EP       astro-ph.GA       astro-ph.IM
              1                 1                 1                 6
   cond-mat.str-el            cs.AI             cs.CG             cs.CL
              1                 5                 1                 2
          cs.CR             cs.CY             cs.DB             cs.DC
              1                 6                 7                 2
          cs.DL             cs.DS             cs.GT             cs.IR
              1                 2                 1                 2
          cs.LG             cs.NA             cs.NI             cs.OH
              2                 1                 1                 1
          cs.SE             cs.SI           math.HO           math.OC
              2                 9                 1                 1
        math.ST  physics.chem-ph  physics.comp-ph   physics.ed-ph
              6                 1                 1                 1
   physics.geo-ph   physics.soc-ph           q-bio.PE          q-fin.GN
              1                 9                 1                 1
         stat.AP           stat.CO           stat.ME           stat.ML
              5                 3                 2                 2
         stat.OT
              4
```

Thankfully, we can use a regular expression to extract only the primary field, which may contain a dash (-), but otherwise is all lower case characters. Once we have this information extracted, we can tally() those primary fields.

```
DataSciencePapers %>%
  mutate(field = str_extract(primary_category, "^[a-z,-]+")) %>%
  tally(x = ~field) %>%
  sort()
```

```
field
cond-mat    q-bio    q-fin    math astro-ph  physics      stat       cs
       1        1        1       8        9       13        16       46
```

It appears that nearly half ($46/95 = 48\%$) of these papers come from computer science, while roughly one quarter come from physics and astrophysics, and another quarter comes from mathematics and statistics.

15.2.1   Corpora

Text mining is often performed not just on one text document, but on a collection of many text documents, called a *corpus*. Can we use these papers to craft a working definition of data science? We will begin by creating a text corpus of the arXiv abstracts using the tm (text mining) package.

```
library(tm)
Corpus <- with(DataSciencePapers, VCorpus(VectorSource(abstract)))
Corpus[[1]] %>%
  as.character() %>%
  strwrap()
```

```
 [1] "Astronomy is one of the most data-intensive of the sciences. Data"
 [2] "technology is accelerating the quality and effectiveness of its"
 [3] "research, and the rate of astronomical discovery is higher than"
 [4] "ever. As a result, many view astronomy as being in a 'Golden Age',"
 [5] "and projects such as the Virtual Observatory are amongst the most"
 [6] "ambitious data projects in any field of science. But these"
 [7] "powerful tools will be impotent unless the data on which they"
 [8] "operate are of matching quality. Astronomy, like other fields of"
 [9] "science, therefore needs to establish and agree on a set of"
[10] "guiding principles for the management of astronomical data. To"
[11] "focus this process, we are constructing a 'data manifesto', which"
[12] "proposes guidelines to maximise the rate and cost-effectiveness of"
[13] "scientific discovery."
```

In order to concentrate on the words that are important, we will find it useful to strip extraneous whitespace, remove numbers and punctuation, convert everything to lower case, and remove common English words (i.e., *stop words*). As these are common operations in text analysis, functionality is provided by tm.

```
Corpus <- Corpus %>%
  tm_map(stripWhitespace) %>%
  tm_map(removeNumbers) %>%
  tm_map(removePunctuation) %>%
  tm_map(content_transformer(tolower)) %>%
  tm_map(removeWords, stopwords("english"))
strwrap(as.character(Corpus[[1]]))
```

```
[1] "astronomy one dataintensive sciences data technology accelerating"
[2] "quality effectiveness research rate astronomical discovery higher"
[3] "ever result many view astronomy golden age projects virtual"
```

```
[4] "observatory amongst ambitious data projects field science powerful"
[5] "tools will impotent unless data operate matching quality astronomy"
[6] "like fields science therefore needs establish agree set guiding"
[7] "principles management astronomical data focus process constructing"
[8] "data manifesto proposes guidelines maximise rate costeffectiveness"
[9] "scientific discovery"
```

The removal of stop words is particularly helpful when performing text analysis. Of the words that are left, which are the most common?

15.2.2   Word clouds

At this stage, we have taken what was a coherent English paragraph and reduced it to a collection of individual, non-trivial English words. We have transformed something that was easy for humans to read into *data*. Unfortunately, it is not obvious how we can learn from these data. One rudimentary approach is to construct a *word cloud*—a kind of multivariate histogram for words. The wordcloud package can generate these graphical depictions of word frequencies.

```
library(wordcloud)
wordcloud(Corpus, max.words = 30, scale = c(8, 1),
          colors = topo.colors(n = 30), random.color = TRUE)
```

Although word clouds such as the one shown in Figure 15.2 have a somewhat dubious reputation for conveying meaning, they can be useful for quickly visualizing the prevalence of words in large corpora.

15.2.3   Document term matrices

Another important technique in text mining involves the calculation of a *term frequency-inverse document frequency (tf-idf)*, or *document term matrix*. The term frequency of a term t in a document d is denoted $tf(t, d)$ and is simply equal to the number of times that the term t appears in document d. On the other hand, the inverse document frequency measures the prevalence of a term across a set of documents D. In particular,

$$idf(t, D) = \log \frac{|D|}{|\{d \in D : t \in d\}|} .$$

Finally, $tf.idf(t, d, D) = tf(t, d) \cdot idf(t, D)$. The $tf.idf$ is commonly used in search engines, when the relevance of a particular word is needed across a body of documents.

Note that commonly used words like the will appear in every document. Thus, their inverse document frequency score will be zero, and thus their $tf.idf$ will also be zero regardless of the term frequency. This is a desired result, since words like the are never important in full-text searches. Rather, documents with high $tf.idf$ scores for a particular term will contain that particular term many times relative to its appearance across many documents. Such documents are likely to be more relevant to the search term being used.

The DocumentTermMatrix() function will create a document term matrix with one row per document and one column per term. By default, each entry in that matrix records the *term frequency* (i.e., the number of times that each word appeared in each document). However, in this case we will specify that the entries record the normalized $tf.idf$ as defined above. Note that the DTM matrix is very sparse—in this case 98% of the entries are 0. This makes sense, since most words do not appear in most documents.

Figure 15.2: A word cloud of terms that appear in the abstracts of arXiv papers on data science.

```
DTM <- DocumentTermMatrix(Corpus, control = list(weighting = weightTfIdf))
DTM

<<DocumentTermMatrix (documents: 95, terms: 3289)>>
Non-/sparse entries: 7350/305105
Sparsity           : 98%
Maximal term length: 29
Weighting          : term frequency - inverse document frequency
                     (normalized) (tf-idf)
```

We can now use the findFreqTerms() function with the DTM object to find the words with the highest $tf.idf$ scores. Note how these results differ from the word cloud in Figure 15.2. By term frequency, the word data is by far the most common, but this gives it a low idf score that brings down its $tf.idf$.

```
findFreqTerms(DTM, lowfreq = 0.8)

[1] "big"          "information" "model"        "modern"       "network"
[6] "science"      "social"       "statistical" "students"
```

Since the DTM contains all of the $tf.idf$ scores for each word, we can extract those values and calculate the score of each word across all of the abstracts.

```
DTM %>% as.matrix() %>%
  apply(MARGIN = 2, sum) %>%
  sort(decreasing = TRUE) %>%
  head(9)
```

```
      social         big    students statistical       model information
       1.112       0.991       0.971       0.904       0.891       0.884
     science      modern     network
       0.857       0.823       0.808
```

Moreover, we can identify which terms tend to show up in the same documents as the word "statistics" using the `findAssocs()` function. In this case, we compare the words that have a correlation of at least 0.5 with the terms `statistics` and `mathematics`. It is amusing that `think` and `conceptual` rise to the top of these rankings, respectively.

```
findAssocs(DTM, terms = "statistics", corlimit = 0.5)
```

```
$statistics
       think       courses      capacity      students  introductory
        0.58          0.56          0.53          0.53          0.51
```

```
findAssocs(DTM, terms = "mathematics", corlimit = 0.5)
```

```
$mathematics
  conceptual         light        review    historical   perspective          role
        0.99          0.99          0.97          0.96          0.95          0.90
      modern
        0.85
```

15.3 Ingesting text

In Chapter 5 (see Section 5.5.1) we illustrated how the `rvest` package can be used to convert tabular data presented on the Web in HTML format into a proper R data table. Here, we present another example of how this process can bring text data into R.

15.3.1 Example: Scraping the songs of the Beatles

In Chapter 11 we explored the popularity of the names for the four members of the Beatles. During their heyday from 1962–1970, the Beatles were prolific—recording 310 singles. In this example we explore some of their song titles and authorship. We begin by downloading the contents of the Wikipedia page that lists the Beatles' songs.

```
library(rvest)
library(tidyr)
library(methods)
url <- "http://en.wikipedia.org/wiki/List_of_songs_recorded_by_the_Beatles"
tables <- url %>%
  read_html() %>%
  html_nodes(css = "table")
songs <- html_table(tables[[5]])
```

```
glimpse(songs)

Observations: 310
Variables: 8
$ Title             <chr> "\"12-Bar Original\"", "\"Across the Univers...
$ Year              <chr> "1965", "1968", "1965", "1961", "1963", "196...
$ Album debut       <chr> "Anthology 2", "Let It Be", "UK: Help!\nUS: ...
$ Songwriter(s)     <chr> "Lennon, McCartney, Harrison and Starkey", "...
$ Lead vocal(s)     <chr> "", "Lennon", "Starkey", "Lennon", "Lennon",...
$ Chart position UK <chr> "", "", "", "", "", "", "", "", "700...
$ Chart position US <chr> "", "", "7001470000000000000047", "7001190...
$ Notes             <chr> "", "", "Cover, B-side", "Cover. A 1969 reco...
```

We need to clean these data a bit. Note that the `Title` variable contains quotation marks, and the `Year` variable is of type `character` (`chr`). The `Songwriters(s)` variable also contains parentheses in its name, which will make it cumbersome to work with.

```
songs <- songs %>%
  mutate(Title = gsub('\\"', "", Title), Year = as.numeric(Year)) %>%
  rename(songwriters = `Songwriter(s)`)
```

Most of the Beatles' songs were written by some combination of John Lennon and Paul McCartney. While their productive but occasionally contentious working relationship is well-documented, we might be interested in determining how many songs each person wrote. Unfortunately, a simple `tally()` of these data does not provide much clarity.

```
tally(~songwriters, data = songs) %>%
  sort(decreasing = TRUE) %>%
  head()
```

```
songwriters
            McCartney                 Lennon              Harrison
                   68                     65                    26
McCartney, with Lennon   Lennon and McCartney Lennon, with McCartney
                   23                     16                    14
```

Both Lennon and McCartney wrote songs on their own, together, and—it also appears—primarily on their own but with help from the other. Regular expressions can help us parse these inconsistent data. We already saw the number of songs written by each person individually, and it isn't hard to figure out the number of songs that each person contributed to in some form.

```
length(grep("McCartney", songs$songwriters))
```

```
[1] 139
```

```
length(grep("Lennon", songs$songwriters))
```

```
[1] 136
```

How many of these songs were the product of some type of Lennon-McCartney collaboration? Given the inconsistency in how the songwriters are attributed, it requires some ingenuity to extract these data. We can search the `songwriters` variable for either `McCartney` or `Lennon` (or both), and count these instances.

```
length(grep("(McCartney|Lennon)", songs$songwriters))
```

```
[1] 207
```

At this point, we need another regular expression to figure out how many songs they collaborated on. The following will find the pattern consisting of either `McCartney` or `Lennon`, followed by a possibly empty string of characters, followed by another instance of either `McCartney` or `Lennon`.

```
length(grep("(McCartney|Lennon).*(McCartney|Lennon)", songs$songwriters))
```

```
[1] 68
```

Note also that we can use `grepl()` in a `filter()` command to retrieve the list of songs upon which Lennon and McCartney collaborated.

```
songs %>%
  filter(grepl("(McCartney|Lennon).*(McCartney|Lennon)", songwriters)) %>%
  select(Title) %>%
  head()
```

```
                   Title
1         12-Bar Original
2         All Together Now
3             And I Love Her
4 And Your Bird Can Sing
5           Any Time at All
6                Ask Me Why
```

The Beatles have had such a profound influence upon musicians of all stripes that it might be worth investigating the titles of their songs. What were they singing about?

First, we create a corpus from the vector of song titles, remove the English stop words, and build a document term matrix using the *tf.idf* criteria. Once this is done, we find the words with the highest *tf.idf* scores.

```
song_titles <- VCorpus(VectorSource(songs$Title)) %>%
  tm_map(removeWords, stopwords("english")) %>%
  DocumentTermMatrix(control = list(weighting = weightTfIdf))
findFreqTerms(song_titles, 25)
```

```
[1] "love" "you"
```

15.3.2 Scraping data from Twitter

The micro-blogging service Twitter has a mature application programming interface (API). The `twitteR` package can be used to access these data. To use the API, an account and private key need to be set up using the `setup_twitter_oauth()` function.

```
library(twitteR)
setup_twitter_oauth(consumer_key = "u2UthjbK6YHyQSp4sPk6yjsuV",
  consumer_secret = "sC4mjd2WME5nH1FoWeSTuSy7JCP5DHjNtTYU1X6BwQ1vPZOj3v",
  access_token = "1365606414-7vPfPxStYNq6kWEATQlT8HZBd4G83BBcX4VoS9T",
  access_secret = "0hJq9KYC3eBRuZzJqSacmtJ4PNJ7tNLkGrQrVlOOJHirs")
```

```
[1] "Using direct authentication"
```

Now we can retrieve a list of tweets using a hashtag. Here, we will search for the 1000
most recent English tweets that contain the hashtag #datascience.

```
tweets <- searchTwitter("#datascience", lang = "en", n = 1000,
  retryOnRateLimit = 100)
class(tweets)
class(tweets[[1]])
```

Note that the Twitter API serves tweets as JSON objects, which are then stored as a
list of status objects, but twitteR provides the twListToDF function to collapse those
into a data frame.

```
tweet_df <- twListToDF(tweets) %>% as.tbl()
tweet_df %>%
  select(text) %>%
  head()
```

```
# A tibble: 6  1
                                                                     text
                                                                    <chr>
We have a postdoc available on #cancer #Bioinformatics and applied #machine
RT @BigDataReport_: #Java is the go to language for #IoT applications https
RT @knime: #KNIME 3.3 to offer #cloud connectors to #Amazon #S3 and #Micros
#Java is the go to language for #IoT applications https://t.co/gC2Fc5YVpJ #
        How to Become a Data Scientist https://t.co/OKFQb9zp8k #DataScience
RT @knime: #KNIME 3.3 to offer #cloud connectors to #Amazon #S3 and #Micros
```

Note that there is a rate limit on the numbers of tweets that each user can download at
one time. See https://dev.twitter.com/docs/rate-limiting/1.1 for more information
about rate limits.

We can start to analyze these tweets using some simple statistical methods. For example,
what is the distribution of the number of characters in these tweets?

```
ggplot(data = tweet_df, aes(x = nchar(text))) +
  geom_density(size = 2) +
  geom_vline(xintercept = 140) +
  scale_x_continuous("Number of Characters")
```

We can clearly see the famous 140 character limit in Figure 15.3, although a few tweets
have exceeded that limit. How is that possible?

```
tweet_df %>%
  filter(nchar(text) > 140) %>%
  select(text)
```

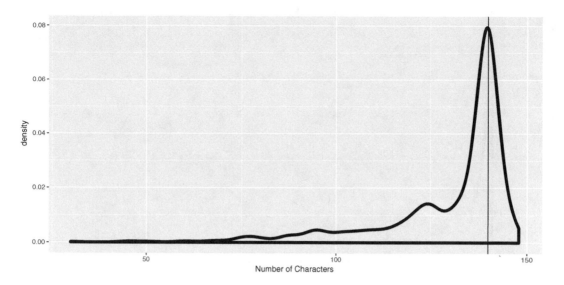

Figure 15.3: Distribution of the number of characters in a sample of tweets.

```
# A tibble: 83   1
                                                                    text
                                                                   <chr>
RT @esthermeadDev: "The NonTechnical Guide to #MachineLearning &amp; Artifi
RT @esthermeadDev: "The NonTechnical Guide to #MachineLearning &amp; Artifi
RT @kshameer: Excited to be part of the @Philips family as Director of #Bio
#ibmwow highlight: jameskobielus talks #DataScience &amp; more with an expe
RT @IBMBigData: #ibmwow highlight: @jameskobielus talks #DataScience &amp;
RT @IBMBigData: #ibmwow highlight: @jameskobielus talks #DataScience &amp;
RT @IBMBigData: #ibmwow highlight: @jameskobielus talks #DataScience &amp;
#bigdata #ibmwow highlight: jameskobielus talks #DataScience &amp; more wit
RT @IBMBigData: #ibmwow highlight: @jameskobielus talks #DataScience &amp;
#ibmwow highlight: jameskobielus talks #DataScience &amp; more with an expe
# ... with 73 more rows
```

Our best guess is that special characters like ampersands (&) only count as one character, but come through as their HTML equivalent (&), which has four characters. The three characters RT that precede a retweet might also not count.

What does the distribution of retweet counts look like? As Twitter has grown in popularity, this question has been of interest to scholars in recent years [135, 30].

```
ggplot(data = tweet_df, aes(x = retweetCount)) +
  geom_density(size = 2)
```

The distribution displayed in Figure 15.4 is highly right-skewed—a few tweets get retweeted a lot, but most don't. This behavior suggests a power-law distribution that is commonly observed in real-world networks. We describe this phenomenon in greater depth in Chapter 16.

How many of those tweets are geolocated (have latitude and longitude of the tweet location)? Unfortunately, very few tweets are actually geolocated.

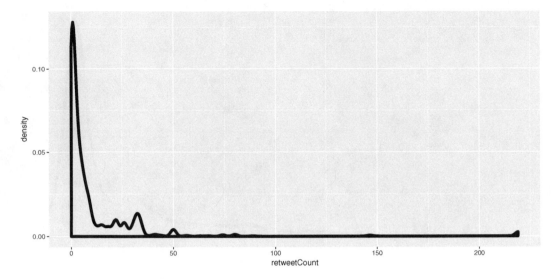

Figure 15.4: Distribution of the number of retweets in a sample of tweets.

```
tweet_df %>% filter(!is.na(longitude))

# A tibble: 2   16
                                                                    text
                                                                   <chr>
Step into #analytics for #business #digitaltrasformation #datascience @ Pal
Join your #datascience team this Friday 13:00 to #denguehack in Brussels. @
# ... with 15 more variables: favorited <lgl>, favoriteCount <dbl>,
#    replyToSN <chr>, created <dttm>, truncated <lgl>, replyToSID <lgl>,
#    id <chr>, replyToUID <chr>, statusSource <chr>, screenName <chr>,
#    retweetCount <dbl>, isRetweet <lgl>, retweeted <lgl>, longitude <chr>,
#    latitude <chr>
```

Building a database of tweets

This is not a large sample of tweets—naturally we want to see more. Unfortunately, the Twitter API does not allow you to search back in time. However, we can start to build a library of tweets by running our query every so often. twitteR also provides a mechanism for storing tweets in a SQLite database (see Chapter 12).

```
tweet_db <- tempfile()
register_sqlite_backend(tweet_db)
store_tweets_db(tweets)

[1] TRUE
```

By running our query every so often (keeping in mind the rate limit), we can slowly build up a database of tweets.

```
tweets_src <- src_sqlite(tweet_db)
old_tweets <- tweets_src %>% tbl("tweets")
glimpse(old_tweets)
```

```
Observations: NA
Variables: 16
$ text          <chr> "We have a postdoc available on #cancer #Bioinfo...
$ favorited     <int> 0, 0, 0, 0, 0, 0, 0, 0, 0, 0, 0, 0, 0, 0, 0, 0, ...
$ favoriteCount <dbl> 0, 0, 0, 1, 0, 0, 0, 0, 0, 0, 0, 0, 1, 0, 1, 0, ...
$ replyToSN     <chr> NA, NA, NA, NA, NA, NA, NA, NA, NA, NA, NA, NA, ...
$ created       <dbl> 1.48e+09, 1.48e+09, 1.48e+09, 1.48e+09, 1.48e+09...
$ truncated     <int> 0, 0, 0, 0, 0, 0, 0, 0, 0, 0, 0, 0, 0, 0, 0, 0, ...
$ replyToSID    <int> NA, NA, NA, NA, NA, NA, NA, NA, NA, NA, NA, NA, ...
$ id            <chr> "801385084842635264", "801384921432473600", "801...
$ replyToUID    <chr> NA, NA, NA, NA, NA, NA, NA, NA, NA, NA, NA, NA, ...
$ statusSource  <chr> "<a href=\"http://twitter.com\" rel=\"nofollow\"...
$ screenName    <chr> "pjballester", "RafaEntarch", "mannitan", "BigDa...
$ retweetCount  <dbl> 1, 1, 6, 1, 0, 6, 2, 9, 8, 0, 9, 2, 0, 5, 0, 2, ...
$ isRetweet     <int> 0, 1, 1, 0, 0, 1, 1, 1, 1, 0, 1, 1, 0, 1, 0, 1, ...
$ retweeted     <int> 0, 0, 0, 0, 0, 0, 0, 0, 0, 0, 0, 0, 0, 0, 0, 0, ...
$ longitude     <chr> NA, NA, NA, NA, NA, NA, NA, NA, NA, NA, NA, NA, ...
$ latitude      <chr> NA, NA, NA, NA, NA, NA, NA, NA, NA, NA, NA, NA, ...
```

```
big_data_tweets <- old_tweets %>%
  collect() %>%
  filter(grepl("#bigdata", text))
nrow(big_data_tweets) / nrow(collect(old_tweets))
```

```
[1] 0.111
```

In this sample, only a fraction of the tweets containing the hashtag `#datascience` also contained the hashtag `#bigdata`.

Trends

Twitter keeps track of which hash tags or phrases are popular in real-time—these are known as *trending topics*. Trending topics are available in many major cities and might be used to study how certain populations respond to news or world events. Here, we examine the trending topics closest to us.

First, we need to find the latitude and longitude coordinates for Smith College. We can do this using the `geocode()` function from the `ggmap` package (see Chapter 14).

```
library(ggmap)
smith <- geocode("44 College Lane, 01063")
smith
```

```
   lon  lat
1 -72.6 42.3
```

Next, we use the `closestTrendLocations()` function to retrieve the cities with trending topics that are closest to Smith.

```
with(smith, closestTrendLocations(lat = lat, long = lon))
```

```
         name         country    woeid
1 New Haven United States 2458410
```

In our case, the only nearby city with trends is New Haven. What's happening there?

```
head(getTrends(2458410))
```

```
                           name
1 #HowToAvoidPoliticsAtDinner
2       Happy Thanksgiving Eve
3           #MyProtestWouldBe
4               #LatelyIveBeen
5                  Nick Young
6        #MakeAMiserableMovie
                                                    url
1 http://twitter.com/search?q=%23HowToAvoidPoliticsAtDinner
2  http://twitter.com/search?q=%22Happy+Thanksgiving+Eve%22
3          http://twitter.com/search?q=%23MyProtestWouldBe
4            http://twitter.com/search?q=%23LatelyIveBeen
5              http://twitter.com/search?q=%22Nick+Young%22
6         http://twitter.com/search?q=%23MakeAMiserableMovie
                         query  woeid
1 %23HowToAvoidPoliticsAtDinner 2458410
2   %22Happy+Thanksgiving+Eve%22 2458410
3           %23MyProtestWouldBe 2458410
4             %23LatelyIveBeen 2458410
5               %22Nick+Young%22 2458410
6         %23MakeAMiserableMovie 2458410
```

15.4   Further resources

There are many sources to find text data online. Project Gutenberg is a massive free online library. Project Gutenberg collects the full-text of more than 50,000 books whose copyrights have expired. It is great for older, classic books. You won't find anything by Stephen King (but there is one by Stephen King-Hall). Direct access to Project Gutenberg is available in R through the gutenbergr package.

An *n*-gram is a contiguous sequence of *n* "words." Thus, a 1-gram is a single word (e.g., "text"), while a 2-gram is a pair of words (e.g. "text mining"). Google has collected *n*-grams for many books and provides an interface to these data.

Wikipedia provides a clear overview of syntax for sophisticated pattern-matching within strings using regular expressions.

The forthcoming *Tidy Text Mining in R* book by Silge and Robinson (https://github. com/dgrtwo/tidy-text-mining) has an extensive set of examples of text mining and sentiment analysis. The same authors have also written a tidytext package [186].

15.5   Exercises

Exercise 15.1

Speaking lines in Shakespeare's plays are identified by a line that starts with two spaces, then a string of capital letters and spaces (the character's name) followed by a period. Use grep() to find all of the speaking lines in *Macbeth*. How many are there?

Exercise 15.2

Find all the hyphenated words in one of Shakespeare's plays.

Exercise 15.3

Use the babynames data table from the babynames package to find the ten most popular:

1. Boys' names ending in a vowel.

2. Names ending with "joe", "jo", "Joe", or "Jo" (e.g., Billyjoe).

Exercise 15.4

Find all of the adjectives in one of Shakespeare's plays that end in more or less.

Exercise 15.5

Find all of the lines containing the stage direction Exit or Exeunt in one of Shakespeare's plays.

Exercise 15.6

Use regular expressions to determine the number of speaking lines *The Complete Works of William Shakespeare*. Here, we care only about how many times a character speaks—not what they say or for how long they speak.

Exercise 15.7

Make a bar chart displaying the top 100 characters with the greatest number of lines. *Hint*: you may want to use either the stringr::str_extract() or strsplit() function here.

Exercise 15.8

In this problem, you will do much of the work to recreate Mark Hansen's *Shakespeare Machine*. Start by watching a video clip (http://vimeo.com/54858820) of the exhibit. Use *The Complete Works of William Shakespeare* and regular expressions to find all of the hyphenated words in Shakespeare Machine. How many are there? Use %in% to verify that your list contains the following hyphenated words pictured at 00:46 of the clip.

Exercise 15.9

Find an interesting Wikipedia page with a table, scrape the data from it, and generate a figure that tells an interesting story. Include sentences interpreting the figure.

Exercise 15.10

The site stackexchange.com displays questions and answers on technical topics. The following code downloads the most recent R questions related to the dplyr package.

```
library(httr)
# Find the most recent R questions on stackoverflow
getresult <- GET("http://api.stackexchange.com",
  path = "questions",
  query = list(site = "stackoverflow.com", tagged = "dplyr"))
stop_for_status(getresult) # Ensure returned without error
questions <- content(getresult)  # Grab content
names(questions$items[[1]])     # What does the returned data look like?
```

```
 [1] "tags"                "owner"               "is_answered"
 [4] "view_count"          "answer_count"        "score"
 [7] "last_activity_date"  "creation_date"       "question_id"
[10] "link"                "title"
```

```
length(questions$items)
```

```
[1] 30
```

```
substr(questions$items[[1]]$title, 1, 68)
```

```
[1] "Dplyr: how to loop over specific columns whose names are in a list?"
```

```
substr(questions$items[[2]]$title, 1, 68)
```

```
[1] "k-fold cross-validation in dplyr?"
```

```
substr(questions$items[[3]]$title, 1, 68)
```

```
[1] "Creating a function with multiple arguments using dplyr"
```

How many questions were returned? Without using jargon, describe in words what is being displayed and how it might be used.

Exercise 15.11

Repeat the process of downloading the content from stackexchange.com related to the dplyr package and summarize the results.

Chapter 16

Network science

Network science is an emerging interdisciplinary field that studies the properties of large and complex networks. Network scientists are interested in both theoretical properties of networks (e.g., mathematical models for degree distribution) and data-based discoveries in real networks (e.g., the distribution of the number of friends on Facebook).

16.1 Introduction to network science

16.1.1 Definitions

The roots of network science are in the mathematical discipline of graph theory. There are a few basic definitions that we need before we can proceed.

- A *graph* $G = (V, E)$ is simply a set of *vertices* (or nodes) V, and a set of *edges* (or links, or even ties) E between those nodes. It may be more convenient to think about a graph as being a *network*. For example, in a network model of Facebook, each user is a vertex and each friend relation is an edge connecting two users. Thus, one can think of Facebook as a *social network*, but the underlying mathematical structure is just a graph. Discrete mathematicians have been studying graphs since Leonhard Euler posed the Seven Bridges of Königsberg problem in 1736 [73].

- Edges in graphs can be *directed* or *undirected*. The difference is whether the relationship is mutual or one-sided. For example, edges in the Facebook social network are undirected, because friendship is a mutual relationship. Conversely, edges in Twitter are directed, since you may follow someone who does not necessarily follow you.

- Edges (or less commonly, vertices) may be *weighted*. The value of the weight represents some quantitative measure. For example, an airline may envision its flight network as a graph, in which each airport is a node, and edges are weighted according to the distance (in miles) from one airport to another. (If edges are unweighted, this is equivalent to setting all weights to 1.)

- A *path* is a non-self-intersecting sequence of edges that connect two vertices. More formally, a path is a special case of a *walk*, which does allow self-intersections (i.e., a vertex may appear in the walk more than once). There may be many paths, or no paths, between two vertices in a graph, but if there are any paths, then there is at least one *shortest path* (or *geodesic*). The notion of a shortest path is dependent upon a distance measure in the graph (usually, just the number of edges, or the sum of the edge weights). A graph is *connected* if there is a path between all pairs of vertices.

- The *diameter* of a graph is the length of the longest geodesic (i.e., the longest shortest [sic] path) between any two pairs of vertices. The *eccentricity* of a vertex v in a graph is the length of the longest geodesic starting at that vertex. Thus, in some sense a vertex with a low eccentricity is more central to the graph.

- In general, graphs do not have coordinates. Thus, there is no right way to draw a graph. Visualizing a graph is more art than science, but several graph layout algorithms are popular.

- Centrality: Since graphs don't have coordinates, there is no obvious measure of *centrality*. That is, it is frequently of interest to determine which nodes are most "central" to the network, but there are many different notions of centrality. We will discuss three:

  - Degree centrality: The *degree* of a vertex within a graph is the number of edges incident to it. Thus, the degree of a node is a simple measure of centrality in which more highly connected nodes rank higher. President Obama has almost 10 million followers on Twitter, whereas the vast majority of users have fewer than a thousand. Thus, the degree of the vertex representing President Obama in the Twitter network is in the millions, and he is more central to the network in terms of degree centrality.

  - Betweenness centrality: If a vertex v is more central to a graph, then you would suspect that more shortest paths between vertices would pass through v. This is the notion of *betweenness centrality*. Specifically, let $\sigma(s,t)$ be the number of geodesics between vertices s and t in a graph. Let $\sigma_v(s,t)$ be the number of shortest paths between s and t that pass through v. Then the betweenness centrality for v is the sum of the fractions $\sigma_v(s,t)/\sigma(s,t)$ over all possible pairs (s,t). This figure ($C_B(v)$) is often normalized by dividing by the number of pairs of vertices that do not include v in the graph.

    $$C_B(v) = \frac{2}{(n-1)(n-2)} \sum_{s,t \in V \setminus \{v\}} \frac{\sigma_v(s,t)}{\sigma(s,t)},$$

    where n is the number of vertices in the graph. Note that President Obama's high degree centrality would not necessarily translate into a high betweenness centrality.

  - *Eigenvector centrality*: This is the essence of Google's PageRank algorithm, which we will discuss in Section 16.3.

  Note that there are also notions of *edge centrality* that we will not discuss further.

- In a social network, it is usually believed that if Alice and Bob are friends, and Alice and Carol are friends, then it is more likely than it otherwise would be that Bob and Carol are friends. This is the notion of *triadic closure* and it leads to measurements of *clusters* in real-world networks.

16.1.2   A brief history of network science

As noted above, the study of graph theory began in the 1700s, but the inception of the field of network science was a paper published in 1959 by the legendary Paul Erdős and Alfréd Rényi [72]. Erdős and Rényi proposed a model for a *random graph*, where the number of vertices n is fixed, but the probability of an edge connecting any two vertices is p. What do such graphs look like? What properties do they have? It is obvious that if p is very close to 0, then the graph will be almost empty, while conversely, if p is very close to 1, then the

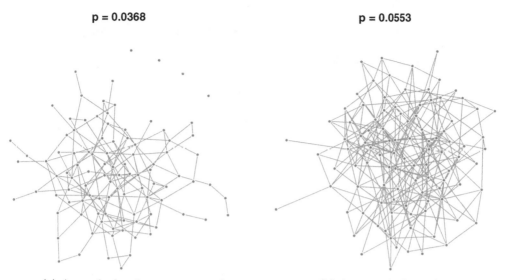

(a) A graph that is not connected.   (b) A connected graph.

Figure 16.1: Two Erdős–Rényi random graphs on 100 vertices with different values of p. The graph at left is not connected, but the graph at right is. The value of p hasn't changed by much.

graph will be almost complete. Erdős and Rényi unexpectedly proved that for many graph properties c (e.g., connectedness, the existence of a cycle of a certain size, etc.), there is a threshold function $p_c(n)$ around which the structure of the graph seems to change rapidly. That is, for values of p slightly less than $p_c(n)$, the probability that a random graph is connected is close to zero, while for values of p just a bit larger than $p_c(n)$, the probability that a random graph is connected is close to one (see Figure 16.1. This bizarre behavior has been called the *phase transition* in allusion to physics, because it evokes at a molecular level how solids turn to liquids and liquids turn to gasses. When temperatures are just above 32 degrees Fahrenheit, water is a liquid, but at just below 32 degrees, it becomes a solid.

```
library(mdsr)
library(igraph)
n <- 100
p_star <- log(n)/n

plot_er <- function(n, p, ...) {
  g <- erdos.renyi.game(n, p)
  plot(g, main = paste("p =", round(p, 4)), vertex.frame.color = "white",
    vertex.size = 3, vertex.label = NA, ...)
}
plot_er(n, p = 0.8 * p_star)
plot_er(n, p = 1.2 * p_star)
```

While many properties of the phase transition have been proven mathematically, they can often be illustrated using simulation (see Chapter 10). The `igraph` package provides the `erdos.renyi.game()` function for simulating Erdős–Rényi random graphs. In Figure 16.2, we show how the phase transition for connectedness appears around the threshold value of $p(n) = \log n/n$. With $n = 1,000$, we have $p(n) = 0.007$. Note how quickly the probability of being connected increases near the value of the threshold function.

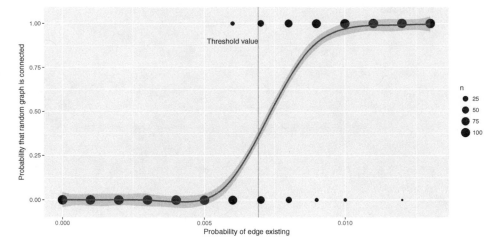

Figure 16.2: Simulation of connectedness of ER random graphs on 1,000 vertices.

```
n <- 1000
p_star <- log(n)/n
ps <- rep(seq(from = 0, to = 2 * p_star, by = 0.001), each = 100)
er_connected <- function(n, p, ...) {
  c(n = n, p = p, connected = is.connected(erdos.renyi.game(n, p)))
}
sims <- as.data.frame(t(sapply(ps, er_connected, n = n)))
ggplot(data = sims, aes(x = p, y = connected)) +
  geom_vline(xintercept = p_star, color = "darkgray") +
  geom_text(x = p_star, y = 0.9, label = "Threshold value", hjust="right") +
  labs(x = "Probability of edge existing",
       y = "Probability that random graph is connected") +
  geom_count() + geom_smooth()
```

This surprising discovery demonstrated that random graphs had interesting properties. Yet it was less clear whether the Erdős–Rényi random graph model could produce graphs whose properties were similar to those that we observe in reality. That is, while the Erdős–Rényi random graph model was interesting in its own right, did it model reality well?

The answer turned out to be "no," or at least, "not really." In particular, Watts and Strogatz identified two properties present in real-world networks that were not present in Erdős–Rényi random graphs: triadic closure and large hubs [210]. As we saw above, triadic closure is the idea that two people with a friend in common are likely to be friends themselves. Real-world (not necessarily social) networks tend to have this property, but Erdős–Rényi random graphs do not. Similarly, real-world networks tend to have large hubs—individual nodes with many edges. More specifically, whereas the distribution of the degrees of vertices in Erdős–Rényi random graphs can be shown to follow a Poisson distribution, in real-world networks the distribution tends to be flatter. The Watts–Strogatz model provides a second random graph model that produces graphs more similar to those we observe in reality.

```
g <- watts.strogatz.game(n)
```

In particular, many real-world networks, including not only social networks but also the World Wide Web, citation networks, and many others, have a degree distribution that

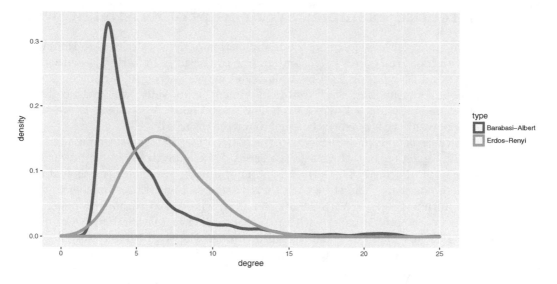

Figure 16.3: Degree distribution for two random graphs.

follows a *power-law*. These are known as *scale-free* networks and were popularized by Albert-László Barabási in two widely-cited papers [13, 3] and his highly readable book [14]. Barabási and Albert proposed a third random graph model based on the notion of *preferential attachment*. Here, new nodes are connected to old nodes based on the existing degree distribution of the old nodes. Their model produces the power-law degree distribution that has been observed in many different real-world networks.

Here again, we can illustrate these properties using simulation. The `barabasi.game()` function in `igraph` will allow us to simulate a Barabási–Albert random graph. Figure 16.3 compares the degree distribution between an Erdős–Rényi random graph and a Barabási–Albert random graph.

```
g1 <- erdos.renyi.game(n, p = log(n)/n)
g2 <- barabasi.game(n, m = 3, directed = FALSE)
summary(g1)

IGRAPH U--- 1000 3498 -- Erdos renyi (gnp) graph
+ attr: name (g/c), type (g/c), loops (g/l), p (g/n)

summary(g2)

IGRAPH U--- 1000 2994 -- Barabasi graph
+ attr: name (g/c), power (g/n), m (g/n), zero.appeal (g/n),
| algorithm (g/c)

d <- data.frame(type = rep(c("Erdos-Renyi", "Barabasi-Albert"), each = n),
                degree = c(degree(g1), degree(g2)))
ggplot(data = d, aes(x = degree, color = type)) +
  geom_density(size = 2) +
  scale_x_continuous(limits = c(0, 25))
```

Network science is a very active area of research, with interesting unsolved problems for mathematicians, computer scientists, and statisticians to investigate.

16.2   Extended example: Six degrees of Kristen Stewart

In this extended example we will explore a fun application of network science to Hollywood movies. The notion of *Six Degrees of Separation* was conjectured by a Hungarian network theorist in 1929, and later popularized by a play (and movie starring Will Smith). Stanley Milgram's famous letter-mailing *small-world* experiment supposedly lent credence to the idea that all people are connected by relatively few "social hops" [193]. That is, we are all part of a social network with a relatively small diameter (as small as 6).

Two popular incarnations of these ideas are the notion of an *Erdős number* and the Kevin Bacon game. The question in each case is the same: How many hops are you away from Paul Erdős (or Kevin Bacon)? The former is popular among academics (mathematicians especially), where edges are defined by co-authored papers. Ben's Erdős number is three, since he has co-authored a paper with Amotz Bar–Noy, who has co-authored a paper with Noga Alon, who co-authored a paper with Erdős. According to MathSciNet, Nick's Erdős number is four (through Ben given [23]; but also through Nan Laird, Fred Mosteller, and Persi Diaconis), and as of this writing, Danny's is five (through Nick). Danny's Erdős number will become four when this book is published. These data reflect the fact that Ben's research is "closer" to Erdős's, since he has written about network science [35, 25, 15, 17] and graph theory [26]. Similarly, the idea is that every actor in Hollywood can be connected to Kevin Bacon in at most six movie hops. We'll explore this idea using the IMDb (Internet Movie Database [117]).

16.2.1   Collecting Hollywood data

We will populate a Hollywood network using actors and actresses in the IMDb. In this network, each actor or actress is a node, and two actors share an edge if they have ever appeared in a movie together. Our goal will be to determine the centrality of Kevin Bacon.

First, we want to determine the edges, since we can then look up the node information based on the edges that are present. One caveat is that these networks can grow very rapidly (since the number of edges is $O(n^2)$, where n is the number of vertices). Thus, for this example, we will be conservative by including popular (at least 100,000 ratings) feature films (i.e., `kind_id` equal to 1) from 2012, and we will consider only the top 20 credited roles in each film.

To retrieve the list of edges, we need to consider all possible cast assignment pairs. To get this list, we start by forming all total pairs using the `CROSS JOIN` operation in MySQL (see Chapter 12), which has no `dplyr` equivalent. Thus, in this case we will have to actually write the SQL code and use the `DBI` interface to execute it. We will subsequently need to filter this list down to the unique pairs, which we can do by only including pairs where `person_id` from the first table is strictly less than `person_id` from the second table.

```
library(mdsr)
db <- src_scidb("imdb")
```

```
sql <-
  "SELECT a.person_id as src, b.person_id as dest,
    a.movie_id,
    a.nr_order * b.nr_order as weight,
    t.title, idx.info as ratings
  FROM imdb.cast_info a
    CROSS JOIN imdb.cast_info b USING (movie_id)
    LEFT JOIN imdb.title t ON a.movie_id = t.id
    LEFT JOIN imdb.movie_info_idx idx ON idx.movie_id = a.movie_id
  WHERE t.production_year = 2012 AND t.kind_id = 1
    AND info_type_id = 100 AND idx.info > 125000
    AND a.nr_order <= 20 AND b.nr_order <= 20
    AND a.role_id IN (1,2) AND b.role_id IN (1,2)
    AND a.person_id < b.person_id
  GROUP BY src, dest, movie_id"
E <- DBI::dbGetQuery(db$con, sql) %>%
  mutate(ratings = as.numeric(ratings))
glimpse(E)
```

```
Observations: 10,603
Variables: 6
$ src      <int> 2720, 2720, 2720, 2720, 2720, 2720, 2720, 2720, 2720,...
$ dest     <int> 113645, 363432, 799414, 906453, 1221633, 1238138, 135...
$ movie_id <int> 3164088, 3164088, 3164088, 3164088, 3164088, 3164088,...
$ weight   <dbl> 153, 255, 204, 272, 340, 68, 17, 306, 102, 136, 85, 5...
$ title    <chr> "Mud", "Mud", "Mud", "Mud", "Mud", "Mud", "Mud", "Mud...
$ ratings  <dbl> 129918, 129918, 129918, 129918, 129918, 129918, 12991...
```

We have also computed a `weight` variable that we can use to weight the edges in the resulting graph. In this case, the `weight` is based on the order in which each actor appears in the credits. So a ranking of 1 means that the actor/actress had top billing. These weights will be useful because a higher order in the credits usually means more screen time.

```
nrow(E)
```

```
[1] 10603
```

```
length(unique(E$title))
```

```
[1] 57
```

Our query resulted in 10,603 connections between 57 films. We can see that *Batman: The Dark Knight Rises* received the most user ratings on IMDb.

```
E %>%
  group_by(movie_id) %>%
  summarize(title = max(title), N = n(), numRatings = max(ratings)) %>%
  arrange(desc(numRatings))
```

```
# A tibble: 57  4
  movie_id                                    title    N numRatings
```

	<int>		<chr>	<int>	<dbl>
1	3470749	The Dark Knight Rises		190	1091382
2	3451543	The Avengers		190	944436
3	2769192	Django Unchained		190	908419
4	3496103	The Hunger Games		190	681918
5	3494301	The Hobbit: An Unexpected Journey		190	616722
6	3373175	Silver Linings Playbook		190	516618
7	3379118	Skyfall		190	505192
8	3437250	Ted		190	457339
9	3281286	Prometheus		190	444429
10	2580175	Argo		190	436948

```
# ... with 47 more rows
```

Next, we should gather some information about the vertices in this graph. We could have done this with another JOIN in the original query, but doing it now will be more efficient. (Why? See Exercise 16.1.) In this case, all we need is each actor's name and IMDb identifier.

```
actor_ids <- unique(c(E$src, E$dest))
V <- db %>%
  tbl("name") %>%
  filter(id %in% actor_ids) %>%
  select(id, name) %>%
  rename(actor_name = name) %>%
  collect() %>%
  arrange(id)
glimpse(V)
```

```
Observations: 1,047
Variables: 2
$ id         <int> 2720, 5511, 5943, 7312, 14451, 14779, 16297, 24692,...
$ actor_name <chr> "Abbott Jr., Michael", "Abkarian, Simon", "Aboutbou...
```

16.2.2   Building the Hollywood network

There are two popular R packages for network analysis: igraph and sna. Both have large user bases and are actively developed, but we will use igraph (which also has bindings for Python and C, see Chapter 17). To build a graph, we specify the edges, whether we want them to be directed, and in this case, we add the information about the vertices.

```
library(igraph)
g <- graph_from_data_frame(E, directed = FALSE, vertices = V)
summary(g)
```

```
IGRAPH UNW- 1047 10603 --
+ attr: name (v/c), actor_name (v/c), movie_id (e/n), weight
| (e/n), title (e/c), ratings (e/n)
```

From the summary() command above, we can see that we have 1,047 actors and actresses and 10,603 edges between them. Note that we have associated metadata with each edge:

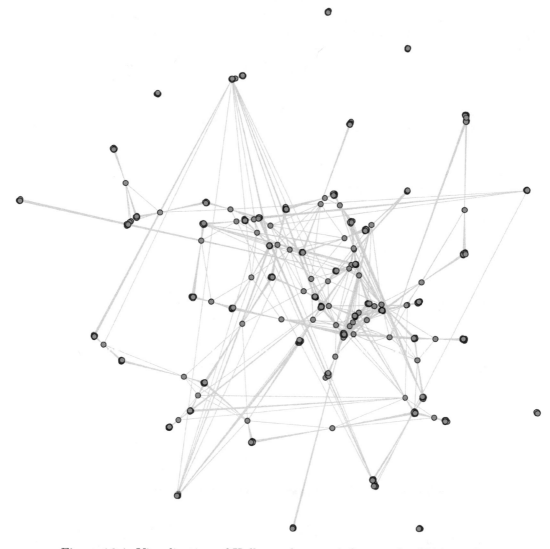

Figure 16.4: Visualization of Hollywood network for popular 2012 movies.

namely, information about the movie that gave rise to the edge, and the aforementioned `weight` metric based on the order in the credits where each actor appeared. (The idea is that top-billed stars are likely to appear on screen longer, and thus have more meaningful interactions with more of the cast.) By default, the first vertex attribute is called `name`, but we would like to keep the more informative `imdbId` label.

```
g <- set_vertex_attr(g, "imdbId", value = V(g)$name)
```

With our network intact, we can visualize it. There are *many* graphical parameters that you may wish to set, and the default choices are not always good. In this case we have 1,047 vertices, so we'll make them small, and omit labels. Figure 16.4 displays the results.

```
plot(g, edge.color = "lightgray", vertex.size = 2, vertex.label = NA)
```

It is easy to see the clusters based on movies, but you can also see a few actors who
have appeared in multiple movies, and how they tend to be more "central" to the network.
If an actor has appeared in multiple movies, then it stands to reason that they will have
more connections to other actors. This is captured by degree centrality.

```
g <- set_vertex_attr(g, "degree", value = degree(g))
as_data_frame(g, what = "vertices") %>%
  arrange(desc(degree)) %>%
  head()
```

```
     name             actor_name  imdbId degree
1  439008          Cranston, Bryan  439008     57
2  780051 Gordon-Levitt, Joseph  780051     57
3  854239             Hardy, Tom  854239     57
4  886405        Hemsworth, Chris  886405     57
5 1500789           Neeson, Liam 1500789     57
6  975964         Ivanek, Zeljko  975964     56
```

There are a number of big name actors on this list who appeared in multiple movies in
2012. Why does Bryan Cranston have so many connections? The following quick function
will retrieve the list of movies for a particular actor.

```
getMovies <- function(imdbId, E) {
  E %>%
    filter(src == imdbId | dest == imdbId) %>%
    tally(~ title, data = .)
}
getMovies(439008, E)
```

```
title
       Argo   John Carter Total Recall
         19            19           19
```

Cranston appeared in all three of these movies. Note however, that the distribution of
degrees is not terribly smooth (see Figure 16.5). That is, the number of connections that
each actor has appears to be limited to a few discrete possibilities. Can you think of why
that might be?

The plots created in igraph are flexible, but they don't have some of the nice features
of ggplot2 to which we have become accustomed. For example, we can color the nodes
based on the degree centrality using a ColorBrewer palette, but we would have set those
attributes manually—we can't simply map them to an aesthetic.

Instead, we'll use the ggnetwork package, which provides geom_nodes() and geom_edges()
functions for plotting graphs directly with ggplot2. (Other alternatives include the geomnet
package, which provides a geom_net() function, and GGally, which provides more comprehensive plotting options for both igraph and sna network objects.)

```
library(ggnetwork)
g_df <- ggnetwork(g)
hollywood <- ggplot(g_df, aes(x, y, xend = xend, yend = yend)) +
  geom_edges(aes(size = weight), color = "lightgray") +
  geom_nodes(aes(color = degree), alpha = 0.6) +
```

```
ggplot(data = data.frame(degree = degree(g)), aes(x = degree)) +
  geom_density(size = 2)
```

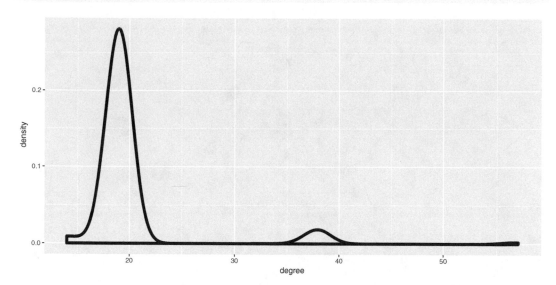

Figure 16.5: Distribution of degrees for actors in the Hollywood network of popular 2012 movies.

```
scale_size_continuous(range = c(0.001, 0.2)) +
theme_blank()
```

The ggnetwork() function transforms our igraph object into a data frame, from which the geom_nodes() and geom_edges() functions can map variables to aesthetics. In this case, since there are so many edges, we use the scale_size_continuous() function to make the edges very thin.

We don't want to show vertex labels for everyone, because that would result in an unreadable mess. However, it would be nice to see the highly central actors. Figure 16.6 shows our completed plot. The thickness of the edges is scaled relatively to the weight measure that we computed earlier.

The ggnetwork() function transforms our igraph object into a data frame, from which the geom_nodes() and geom_edges() functions can map variables to aesthetics. In this case, since there are so many edges, we use the scale_size_continuous() function to make the edges very thin.

```
hollywood +
  geom_nodetext(aes(label = gsub(", ", ",\n", actor_name)),
                data = subset(g_df, degree > 40))
```

16.2.3 Building a Kristen Stewart oracle

Degree centrality does not take into account the weights on the edges. If we want to emphasize the pathways through leading actors and actresses, we could consider *betweenness centrality*.

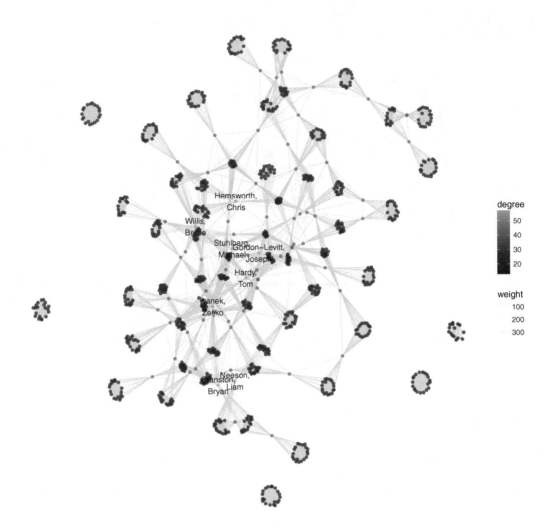

Figure 16.6: The Hollywood network for popular 2012 movies, in `ggplot2`

```
g <- g %>%
  set_vertex_attr("btw", value = igraph::betweenness(g, normalized = TRUE))
get.data.frame(g, what = "vertices") %>%
  arrange(desc(btw)) %>%
  head()
```

```
       name            actor_name imdbId degree   btw
1 3443577       Stewart, Kristen 3443577    38 0.242
2  780051 Gordon-Levitt, Joseph  780051    57 0.221
3  117460       Bale, Christian  117460    19 0.204
4  854239            Hardy, Tom  854239    57 0.198
5 2924441         Kendrick, Anna 2924441   38 0.189
6 1153729         LaBeouf, Shia 1153729    19 0.178
```

```
getMovies(3443577, E)
```

```
title
              Snow White and the Huntsman
                           19
The Twilight Saga: Breaking Dawn - Part 2
                           19
```

Notice that Kristen Stewart has the highest betweenness centrality, while Joseph Gordon–Levitt and Tom Hardy (and others) have the highest degree centrality. Moreover, Christian Bale has the third highest betweenness centrality despite appearing in only one movie. This is because he played the lead in *The Dark Knight Rises*, the movie responsible for the most edges. Thus, most shortest paths through *The Dark Knight Rises* pass through Christian Bale.

If Kristen Stewart (`imdbId 3443577`) is very central to this network, then perhaps instead of a Bacon number, we could consider a Stewart number. Charlize Theron's Stewart number is obviously 1, since they appeared in *Snow White and the Huntsman* together:

```
ks <- V(g)[actor_name == "Stewart, Kristen"]
ct <- V(g)[actor_name == "Theron, Charlize"]
p <- shortest_paths(g, from = ks, to = ct, weights = NA, output = "epath")
edge_attr(g, "title", index = p$epath[[1]])
```

```
[1] "Snow White and the Huntsman"
```

On the other hand, her distance from Joseph Gordon–Levitt is 5. The interpretation here is that Joseph Gordon–Levitt was in *The Dark Knight Rises* with Tom Hardy, who was in *Lawless* with Guy Pearce, who was in *Prometheus* with Charlize Theron, who was in *Snow White and the Huntsman* with Kristen Stewart.

```
jgl <- V(g)[actor_name == "Gordon-Levitt, Joseph"]
p <- shortest_paths(g, from = jgl, to = ks, weights = NA, output = "both")
vertex_attr(g, "actor_name", index = p$vpath[[1]])
```

```
[1] "Gordon-Levitt, Joseph" "Hardy, Tom"          "Pearce, Guy"
[4] "Theron, Charlize"      "Stewart, Kristen"
```

```
edge_attr(g, "title", index = p$epath[[1]])
```

```
[1] "The Dark Knight Rises"   "Lawless"
[3] "Prometheus"              "Snow White and the Huntsman"
```

Note, however, that these shortest paths are not unique. In fact, there are 9 shortest paths between Kristen Stewart and Joseph Gordon–Levitt, each having a length of 5.

```
length(all_shortest_paths(g, from = ks, to = jgl, weights = NA)$res)
```

```
[1] 9
```

As we saw in Figure 16.6, our Hollywood network is not connected, and thus its diameter is infinite. However, the diameter of the largest connected component can be computed. This number (in this case, 10) indicates how many hops separate the two most distant actors in the network.

```
diameter(g, weights = NA)
```

```
[1] 10
```

```
eccentricity(g, vids = ks)
```

```
3443577
      6
```

On the other hand, we note that Kristen Stewart's eccentricity is 6. This means that there is no actor in the connected part of the network who is more than 6 hops away from Kristen Stewart.

16.3   PageRank

For many readers, it may be difficult (or impossible) to remember what search engines on the Web were like *before* Google. Search engines such as Altavista, Web Crawler, Lycos, Excite, and Yahoo! vied for supremacy, but none returned results that were of comparable use to the ones we get today. Frequently, finding what you wanted required sifting through pages of slow-to-load links.

Consider the search problem. A user types in a *search query* consisting of one or more words or terms. Then the search engine produces an ordered list of Web pages ranked by their relevance to that search query. How would you instruct a computer to determine the relevance of a Web page to a query?

This problem is not trivial. Most pre-Google search engines worked by categorizing the words on every Web page, and then determining—based on the search query—which pages were most relevant to that query.

One problem with this approach is that it relies on each Web designer to have the words on its page accurately reflect the content. Naturally, advertisers could easily manipulate search engines by loading their pages with popular search terms, written in the same color as the background (making them invisible to the user), regardless of whether those words were related to the actual content of the page. Thus, naïve search engines might rank these pages more highly, even though they were not relevant to the user.

Google conquered search by thinking about the problem in a fundamentally different way and taking advantage of the network structure of the World Wide Web. The web is a directed graph, in which each webpage (URL) is a node, and edges reflect links from one webpage to another. In 1998, Sergey Brin and Larry Page—while computer science graduate students at Stanford—developed a centrality measure called PageRank that forms the basis of Google's search algorithms [156]. The algorithm led to search results that were so much better than those of its competitors that Google quickly swallowed the entire search market, and is now one of the world's largest companies. The key insight was that one could use the directed links on the Web as a means of "voting" in a way that was much more difficult to exploit. That is, advertisers could only control links on their pages, but not links to their pages from other sites.

Eigenvector centrality   Computing PageRank is a rather simple exercise in linear algebra. It is an example of a *Markov process*. Suppose there are n webpages on the Web. Let $\mathbf{v}_0 = 1/n$ be a vector that gives the initial probability that a randomly chosen Web surfer will be on any given page. In the absence of any information about this user, there is an equal probability that they might be on any page.

But for each of these n webpages, we also know to which pages it links. These are outgoing directed edges in the Web graph. We assume that a random surfer will follow each link with equal probability, so if there are m_i outgoing links on the ith webpage, then the probability that the random surfer goes from page i to page j is $p_{ij} = 1/m_i$. Note that if the ith page doesn't link to the jth page, then $p_{ij} = 0$. In this manner we can form the $n \times n$ *transition matrix* $\mathbf{P}$, wherein each entry describes the probability of moving from page i to page j.

The product $\mathbf{P}\mathbf{v}_0 = \mathbf{v}_1$ is a vector where v_{1i} indicates the probability of being at the ith webpage, after picking a webpage uniformly at random to start, and then clicking on one link chosen at random (with equal probability). The product $\mathbf{P}\mathbf{v}_1 = \mathbf{P}^2\mathbf{v}_0$ gives us the probabilities after two clicks, etc. It can be shown mathematically that if we continue to iterate this process, then we will arrive at a *stationary distribution* $\mathbf{v}^*$ that reflects the long-term probability of being on any given page. Each entry in that vector then represents the popularity of the corresponding webpage—$\mathbf{v}^*$ is the PageRank of each webpage.[1] Because $\mathbf{v}^*$ is an eigenvector of the transition matrix (since $\mathbf{P}\mathbf{v}^* = \mathbf{v}^*$), this measure of centrality is known as *eigenvector centrality*. It was in fact developed earlier, but Page and Brin were the first to apply the idea to the World Wide Web for the purpose of search.

The success of PageRank has led to its being applied in a wide variety of contexts— virtually any problem in which a ranking measure on a network setting is feasible. In addition to the college team sports example below, applications of PageRank include: scholarly citations (eigenfactor.org), doctoral programs, protein networks, and lexical semantics.

Another metaphor that may be helpful in understanding PageRank is that of movable mass. That is, suppose that there is a certain amount of mass in a network. The initial vector $\mathbf{v}_0$ models a uniform distribution of that mass over the vertices. That is, $1/n$ of the total mass is located on each vertex. The transition matrix $\mathbf{P}$ models that mass flowing through the network according to the weights on each edge. After a while, the mass will "settle" on the vertices, but in a non-uniform distribution. The node that has accumulated the most mass has the largest PageRank.

16.4 Extended example: 1996 men's college basketball

Every March, the attention of many sports fans and college students is captured by the NCAA basketball tournament, which pits 68 of the best teams against each other in a winner-take-all, single-elimination tournament. (A *tournament* is a special type of directed graph.) However, each team in the tournament is seeded based on their performance during the regular season. These seeds are important, since getting a higher seed can mean an easier path through the tournament. Moreover, a tournament berth itself can mean millions of dollars in revenue to a school's basketball program. Finally, predicting the outcome of the tournament has become something of a sport unto itself.

Kaggle has held a machine learning (see Chapters 8 and 9) competition each spring to solicit these predictions. We will use their data to build a PageRank metric for team strength for the 1995–1996 regular season (the best season in the history of the University of Massachusetts). To do this, we will build a directed graph whereby each team is a node, and each game creates a directed edge from the losing team to the winning team, which can be weighted based on the margin of victory. The PageRank in such a network is a measure of each team's strength.

First, we need to download the game-by-game results, and a lookup table that translates the team IDs into school names. Note that Kaggle requires a sign-in, so the code below may not work for you without your using your Web browser to authenticate.

---

[1] As we will see below, this is not *exactly* true, but it is the basic idea.

```
prefix <- "https://www.kaggle.com/c/march-machine-learning-mania-2015"
url_teams <- paste(prefix, "download/teams.csv", sep = "/")
url_games <- paste(prefix,
                   "download/regular_season_compact_results.csv", sep = "/")
download.file(url_teams, destfile = "data/teams.csv")
download.file(url_games, destfile = "data/games.csv")
```

Next, we will load this data and `filter()` to select just the 1996 season.

```
library(mdsr)
teams <- readr::read_csv("data/teams.csv")
games <- readr::read_csv("data/games.csv") %>%
  filter(season == 1996)
dim(games)
```

```
[1] 4122    8
```

Since the basketball schedule is very unbalanced (each team does not play the same number of games against each other team), margin of victory seems like an important factor in determining how much better one team is than another. We will use the ratio of the winning team's score to the losing team's score as an edge weight.

```
E <- games %>%
  mutate(score_ratio = wscore/lscore) %>%
  select(lteam, wteam, score_ratio)
V <- teams %>%
  filter(team_id %in% unique(c(E$lteam, E$wteam)))
library(igraph)
g <- graph_from_data_frame(E, directed = TRUE, vertices = V)
summary(g)
```

```
IGRAPH DN-- 305 4122 --
+ attr: name (v/c), team_name (v/c), score_ratio (e/n)
```

Our graph for this season contains 305 teams, who played a total of 4122 games. The `igraph` package contains a `page_rank()` function that will compute PageRank for us. In the results below, we can see that by this measure, George Washington was the highest ranked team, followed by UMass and Georgetown. In reality, the 7th-ranked team, Kentucky, won the tournament by beating Syracuse, the 16th-ranked team. All four semifinalists (Kentucky, Syracuse, UMass, and Mississippi State) ranked in the top 16 according to PageRank, and all 8 quarterfinalists (also including Wake Forest, Kansas, Georgetown, and Cincinnati) were in the top 20.

```
g <- set_vertex_attr(g, "pagerank", value = page_rank(g)$vector)
as_data_frame(g, what = "vertices") %>%
  arrange(desc(pagerank)) %>%
  head(20)
```

```
    name      team_name pagerank
1   1203   G Washington  0.02186
2   1269  Massachusetts  0.02050
```

```
3  1207     Georgetown  0.01642
4  1234           Iowa  0.01434
5  1163    Connecticut  0.01408
6  1437      Villanova  0.01309
7  1246       Kentucky  0.01274
8  1345         Purdue  0.01146
9  1280  Mississippi St  0.01137
10 1210   Georgia Tech  0.01058
11 1112        Arizona  0.01026
12 1448    Wake Forest  0.01008
13 1242         Kansas  0.00992
14 1336        Penn St  0.00975
15 1185     E Michigan  0.00971
16 1393       Syracuse  0.00956
17 1266      Marquette  0.00944
18 1314 North Carolina  0.00942
19 1153     Cincinnati  0.00940
20 1396         Temple  0.00860
```

Note that these rankings are very different than simply assessing each team's record and winning percentage, since it implicitly considers *who beat whom*, and by how much. Using won–loss record alone, UMass was the best team, with a 31–1 record, while Kentucky was 4th at 28–2.

```
wins <- E %>%
  group_by(wteam) %>%
  summarise(N = n())
losses <- E %>%
  group_by(lteam) %>%
  summarise(N = n())
wins %>%
  full_join(losses, by = c("wteam" = "lteam")) %>%
  left_join(teams, by = c("wteam" = "team_id")) %>%
  rename(wins = N.x, losses = N.y) %>%
  mutate(win_pct = wins / (wins + losses)) %>%
  arrange(desc(win_pct)) %>%
  head(20)
```

```
# A tibble: 20  5
   wteam  wins  losses        team_name win_pct
   <int> <int>  <int>            <chr>   <dbl>
1   1269    31       1    Massachusetts   0.969
2   1403    28       1       Texas Tech   0.966
3   1163    30       2      Connecticut   0.938
4   1246    28       2         Kentucky   0.933
5   1180    25       3           Drexel   0.893
6   1453    24       3     WI Green Bay   0.889
7   1158    22       3   Col Charleston   0.880
8   1307    26       4       New Mexico   0.867
9   1153    25       4       Cincinnati   0.862
10  1242    25       4           Kansas   0.862
```

```
11   1172   22     4          Davidson   0.846
12   1345   25     5            Purdue   0.833
13   1448   23     5       Wake Forest   0.821
14   1185   22     5        E Michigan   0.815
15   1439   22     5     Virginia Tech   0.815
16   1437   25     6         Villanova   0.806
17   1112   24     6           Arizona   0.800
18   1428   23     6              Utah   0.793
19   1265   22     6            Marist   0.786
20   1114   21     6  Ark Little Rock   0.778
```

This particular graph has some interesting features. First, UMass beat Kentucky in their first game of the season.

```
E %>%
  filter(wteam == 1269 & lteam == 1246)

# A tibble: 1  3
  lteam wteam score_ratio
  <int> <int>       <dbl>
1  1246  1269        1.12
```

This helps to explain why UMass has a higher PageRank than Kentucky, since the only edge between them points to UMass. Sadly, Kentucky beat UMass in the semifinal round of the tournament—but that game is not present in this regular season data set.

Secondly, George Washington finished the regular season 21–7, yet they had the highest PageRank in the country. How could this have happened? In this case, George Washington was the only team to beat UMass in the regular season. Even though the two teams split their season series, this allows much of the mass that flows to UMass to flow to George Washington.

```
E %>%
  filter(lteam %in% c(1203, 1269) & wteam %in% c(1203, 1269))

# A tibble: 2  3
  lteam wteam score_ratio
  <int> <int>       <dbl>
1  1269  1203        1.13
2  1203  1269        1.14
```

The national network is large and complex, and therefore we will focus on the Atlantic 10 conference to illustrate how PageRank is actually computed. The A-10 consisted of 12 teams in 1996.

```
A_10 <- c("Massachusetts", "Temple", "G Washington", "Rhode Island",
          "St Bonaventure", "St Joseph's PA", "Virginia Tech", "Xavier",
          "Dayton", "Duquesne", "La Salle", "Fordham")
```

We can form an *induced subgraph* of our national network that consists solely of vertices and edges among the A-10 teams. We will also compute PageRank on this network.

```
a10 <- V(g)[ team_name %in% A_10 ]
a <- induced_subgraph(g, vids = a10)
a <- set_vertex_attr(a, "pagerank", value = page_rank(a)$vector)
summary(a)

IGRAPH DN-- 12 107 --
+ attr: name (v/c), team_name (v/c), pagerank (v/n), score_ratio
| (e/n)
```

We visualize this network in Figure 16.7, where the size of the vertices are proportional to each team's PageRank, and the transparency of the edges is based on the ratio of the scores in that game. We note that George Washington and UMass are the largest nodes, and that all but one of the edges connected to UMass point towards it.

```
library(ggnetwork)
a_df <- ggnetwork(a)
ggplot(a_df, aes(x, y, xend = xend, yend = yend)) +
  geom_edges(aes(alpha = score_ratio), color = "lightgray",
             arrow = arrow(length = unit(0.2, "cm")),
             curvature = 0.2) +
  geom_nodes(aes(size = pagerank, color = pagerank), alpha = 0.6) +
  geom_nodetext(aes(label = team_name)) +
  scale_alpha_continuous(range = c(0.4, 1)) +
  scale_size_continuous(range = c(1, 10)) +
  guides(color = guide_legend("PageRank"), size=guide_legend("PageRank")) +
  theme_blank()
```

Now, let's compute PageRank for this network using nothing but matrix multiplication. First, we need to get the transition matrix for the graph. This is the same thing as the *adjacency matrix*, with the entries weighted by the score ratios.

```
P <- t(as_adjacency_matrix(a, sparse = FALSE, attr = "score_ratio"))
```

However, entries in $\mathbf{P}$ need to be probabilities, and thus they need to be normalized so that each column sums to 1. We can achieve this using the scale() function.

```
P <- scale(P, center = FALSE, scale = colSums(P))
round(P, 2)
```

	1173	1182	1200	1203	1247	1269	1348	1382	1386	1396	1439	1462
1173	0.00	0.09	0.00	0.00	0.09	0	0.14	0.11	0.00	0.00	0.00	0.16
1182	0.10	0.00	0.10	0.00	0.10	0	0.00	0.00	0.00	0.00	0.00	0.00
1200	0.11	0.00	0.00	0.00	0.09	0	0.00	0.00	0.00	0.00	0.00	0.00
1203	0.12	0.12	0.11	0.00	0.09	1	0.14	0.11	0.17	0.33	0.27	0.16
1247	0.00	0.09	0.00	0.25	0.00	0	0.00	0.12	0.00	0.00	0.00	0.00
1269	0.13	0.09	0.14	0.26	0.11	0	0.14	0.12	0.16	0.41	0.25	0.15
1348	0.00	0.10	0.13	0.00	0.10	0	0.00	0.13	0.16	0.26	0.21	0.18
1382	0.11	0.08	0.10	0.00	0.00	0	0.14	0.00	0.00	0.00	0.00	0.00
1386	0.11	0.09	0.09	0.24	0.09	0	0.14	0.10	0.00	0.00	0.00	0.00
1396	0.13	0.15	0.12	0.00	0.12	0	0.15	0.10	0.16	0.00	0.27	0.19
1439	0.09	0.10	0.12	0.25	0.11	0	0.14	0.11	0.17	0.00	0.00	0.15

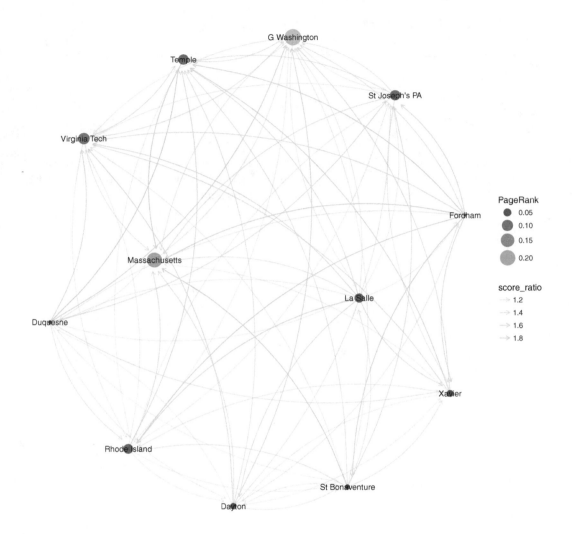

Figure 16.7: Atlantic 10 Conference network, NCAA men's basketball, 1995–1996.

```
1462 0.10 0.09 0.10 0.00 0.10    0 0.00 0.12 0.18 0.00 0.00 0.00
attr(,"scaled:scale")
 1173  1182  1200  1203  1247  1269  1348  1382  1386  1396  1439  1462
10.75 12.19 11.83  4.39 11.76  1.13  7.62 10.47  6.57  4.11  5.11  6.89
```

One shortcoming of this construction is that our graph has multiple edges between pairs of vertices, since teams in the same conference usually play each other twice. Unfortunately, the igraph function as_adjacency_matrix() doesn't handle this well:

> If the graph has multiple edges, the edge attribute of an arbitrarily chosen edge (for the multiple edges) is included.

Thus, even though UMass beat Temple twice, only one of those edges (apparently chosen arbitrarily) will show up in the adjacency matrix. Note also that in the transition matrix shown above, the column labeled 1269 contains a one and eleven zeros. This indicates that the probability of UMass (1269) transitioning to George Washington (1203) is 1—since UMass's only loss was to George Washington. This is not accurate, because the model doesn't handle multiple edges in a sufficiently sophisticated way. It is apparent from the matrix that George Washington is nearly equally likely to move to La Salle, UMass, St. Joseph's, and Virginia Tech—their four losses in the Atlantic 10.

Next, we'll define the initial vector with uniform probabilities—each team has an initial value of 1/12.

```
v0 <- rep(1, vcount(a)) / vcount(a)
v0
```

```
 [1] 0.0833 0.0833 0.0833 0.0833 0.0833 0.0833 0.0833 0.0833 0.0833 0.0833
[11] 0.0833 0.0833
```

To compute PageRank, we iteratively multiply the initial vector $\mathbf{v}_0$ by the transition matrix $\mathbf{P}$. We'll do 20 multiplications with a loop:

```
v <- v0
for (i in 1:20) {
  v <- P %*% v
}
as.vector(v)
```

```
 [1] 0.02538 0.01049 0.00935 0.28601 0.07348 0.18247 0.07712 0.01518
 [9] 0.09192 0.08046 0.11820 0.02995
```

Thus, we find that the fourth vertex—George Washington—has the highest PageRank. Compare these with the values returned by the built-in page_rank() function from igraph:

```
page_rank(a)$vector
```

```
  1173    1182    1200    1203    1247    1269    1348    1382    1386    1396
0.0346  0.0204  0.0193  0.2467  0.0679  0.1854  0.0769  0.0259  0.0870  0.0894
  1439    1462
0.1077  0.0390
```

Why are they different? One limitation of PageRank as we've defined it is that there could be *sinks*, or *spider traps*, in a network. These are individual nodes, or even a collection of nodes, out of which there are no outgoing edges. (UMass is nearly—but not quite—a spider trap in this network.) In this event, if random surfers find themselves in a spider trap, there is no way out, and all of the probability will end up in those vertices. Thus, in practice, PageRank is modified by adding a *random restart*. This means that every so often, the random surfer simply picks up and starts over again. The parameter that controls this in page_rank() is called damping, and it has a default value of 0.85. If we set the damping argument to 1, corresponding to the matrix multiplication we did above, we get a little closer.

```
page_rank(a, damping = 1)$vector
```

```
   1173     1182     1200     1203     1247     1269     1348     1382     1386
0.02290  0.00778  0.00729  0.28605  0.07297  0.20357  0.07243  0.01166  0.09073
   1396     1439     1462
0.08384  0.11395  0.02683
```

Alternatively, we can do the random walk again, but allow for random restarts:

```
w <- v0
d <- 0.85
for (i in 1:20) {
  w <- d * P %*% w + (1 - d) * v0
}
as.vector(w)
```

```
 [1] 0.0381 0.0231 0.0213 0.2468 0.0690 0.1653 0.0825 0.0291 0.0873 0.0859
[11] 0.1102 0.0414
```

```
page_rank(a, damping = 0.85)$vector
```

```
  1173    1182    1200    1203    1247    1269    1348    1382    1386    1396
0.0346 0.0204 0.0193 0.2467 0.0679 0.1854 0.0769 0.0259 0.0870 0.0894
  1439    1462
0.1077 0.0390
```

Again, the results are not exactly the same due to the approximation of values in the adjacency matrix $\mathbf{P}$ mentioned earlier, but they are quite close.

16.5   Further resources

For more sophisticated graph visualization software, see Gephi. In addition to igraph, the sna and network R packages are popular for working with graph objects.

Albert-László Barabási's book *Linked* is a popular introduction to network science [14]. For a broader undergraduate textbook, see [65].

16.6   Exercises

Exercise 16.1

In the CROSS JOIN query in the movies example, how could we have modified the SQL query to include the actor's and actresses' names in the original query? Why would this have been less efficient from a computational and data storage point of view?

Exercise 16.2

Expand the Hollywood network by going further back in time. If you go back to 2000, which actor/actress has the highest degree centrality? Betweenness centrality? Eigenvector centrality?

Exercise 16.3

For a while, Edward Snowden was trapped in a Moscow airport. Suppose that you were trapped not in *one* airport, but in *all* airports. If you were forced to randomly fly around the United States, where would you be most likely to end up?

Exercise 16.4

What information do you need to compute the PageRank of the U.S. airport network? Write an SQL query to retrieve this information for 2012.

Exercise 16.5

Use the data you pulled from SQL in the previous exercise and build the network as a *weighted* `igraph` object, where the weights are proportional to the frequency of flights between each pair of airports.

Exercise 16.6

Compute the PageRank of each airport in your network from the previous exercise. What are the top 10 "most central" airports? Where does Oakland International Airport (`OAK`) rank?

Exercise 16.7

Update the vertex attributes of your network from the previous exercise with the geographic coordinates of each airport (available in the `airports` table).

Exercise 16.8

Use `ggnetwork` to draw the airport network from the previous exercise. Make the thickness or transparency of each edge proportional to its weight.

Exercise 16.9

Overlay your airport network from the previous exercise on a U.S. map (see Chapter 14).

Exercise 16.10

Project the map and the airport network from the previous exercise using the Lambert Conformal Conic projection (see Chapter 14).

Exercise 16.11

Crop the map you created in the previous exercise to zoom in on your local airport.

Chapter 17

Epilogue: Towards "big data"

The terms *data science* and *big data* are often used interchangeably, but this is not correct. Technically, "big data" is a part of data science: the part that deals with data that are so large that they cannot be handled by an ordinary computer. This book provides what we hope is a broad—yet principled—introduction to data science, but it does not specifically prepare the reader to work with big data. Rather, we see the concepts developed in this book as "precursors" to big data [107, 109]. In this epilogue, we explore notions of big data and point the reader towards technologies that scale for truly big data.

17.1 Notions of big data

Big Data is an exceptionally hot topic, but it is not so well-defined. Wikipedia states:

> Big data is a term for data sets that are so large or complex that traditional data processing applications are inadequate ... Relational database management systems and desktop statistics and visualization packages often have difficulty handling big data. The work instead requires "massively parallel software running on tens, hundreds, or even thousands of servers." What is considered "big data" varies depending on the capabilities of the users and their tools, and expanding capabilities make big data a moving target. "For some organizations, facing hundreds of gigabytes of data for the first time may trigger a need to reconsider data management options. For others, it may take tens or hundreds of terabytes before data size becomes a significant consideration" (retrieved March 2016).

Big data is often characterized by the three V's: volume, velocity, and variety [130]. Under this definition, the qualities that make big data different are its *size*, how *quickly* it grows as it is collected, and how many different *formats* it may come in. In big data, the size of tables may be too large to fit on an ordinary computer, the data and queries on it may be coming in too quickly to process, or the data may be distributed across many different systems. Randall Pruim puts in more concisely: "Big data is when your workflow breaks."

Both relative and absolute definitions of big data are meaningful. The absolute definition may be easier to understand: We simply specify a data size and agree that any data that are at least that large are "big"—otherwise they are not. The problem with this definition is that it is a moving target. It might mean *petabytes* (1,000 terabytes) today, but *exabytes* (1,000 petabytes) a few years from now. Regardless of the precise definition, it is increasingly clear that while many organizations like Google, Facebook, and Amazon are working with truly big data, most individuals—even data scientists like you and us—are not.

For us, the relative definition becomes more meaningful. A big data problem occurs when the workflow that you have been using to solve problems becomes infeasible due to the expansion in the size of your data. It is useful in this context to think about *orders of magnitude* of data. The evolution of baseball data illustrates how "big data problems" have arisen as the volume and variety of the data has increased over time.

- Individual game data: Henry Chadwick started collecting boxscores (a tabular summary of each game) in the early 1900s. These data (dozens or even hundreds of rows) can be stored on hand-written pieces of paper, or in a single spreadsheet. Each row might represent one *game*. Thus, a perfectly good workflow for working with data of this size is to store them on paper. A more sophisticated workflow would be to store them in a spreadsheet application.

- Seasonal data: By the 1970s, decades of baseball history were recorded in a seasonal format. Here, the data are aggregated at the *player-team-season* level. An example of this kind of data is the Lahman database we explored in Chapter 4, which has nearly 100,000 rows in the Batting table. Note that in this seasonal format, we know how many home runs each player hit for each team, but we don't know anything about *when* they were hit (e.g., in what month or what inning). Excel is limited in the number of rows that a single spreadsheet can contain. The original limit of $2^{14} = 16,384$ rows was bumped up to $2^{16} = 65,536$ rows in 2003, and the current limit is $2^{20} \approx 1$ million rows. Up until 2003, simply opening the Batting table in Excel would have been impossible. This is a big data problem, because your Excel workflow has broken due to the size of your data. On the other hand, opening the Batting table in R requires far less memory, since R does not try to display all of the data.

- Play-by-play data: By the 1990s, Retrosheet began collecting even more granular play-by-play data. Each row contains information about one *play*. This means that we know exactly when each player hit each home run—what date, what inning, off of which pitcher, which other runners were on base, and even which other players were in the field. As of this writing nearly 100 seasons occupying more than 10 million rows are available. This creates a big data problem for R—you would have a hard time loading these data into R on a typical personal computer. However, SQL provides a scalable solution for data of this magnitude, even on a laptop. Still, you will experience significantly better performance if these data are stored in an SQL cluster with lots of memory.

- Camera-tracking data: The Statcast data set contains (x, y, z)-coordinates for all fielders, baserunners, and the ball every $1/15^{th}$ of a second. Thus, each row is a moment in time. These data indicate not just the outcome of each play, but exactly where each of the players on the field and the ball were as the play evolved. While we still don't know exactly how large these data will be, estimates include several gigabytes per game, which would translate into many terabytes per season. Thus, some sort of distributed server system would be required just to store these data. These data are "big" in the relative sense for any individual, but they are still orders of magnitude away from being "big" in the absolute sense.

What does absolutely big data look like? For an individual user, you might consider the 13.5 terabyte data set of 110 billion events released in 2015 by Yahoo! for use in machine learning research. The grand-daddy of data may be the Large Hadron Collider in Europe, which is generating 25 petabytes of data per year [46]. However, only 0.001% of all of the data that is begin generated by the supercollider is being saved, because to collect it all would mean capturing nearly 500 exabytes *per day*. This is clearly big data.

17.2 Tools for bigger data

By now, you have a working knowledge of both R and SQL. These are battle-tested, valuable tools for working with small and medium data. Both have large user bases, ample deployment, and continue to be very actively developed. Some of that development seeks to make R and SQL more useful for truly large data. While we don't have the space to cover these extensions in detail, in this section we outline some of the most important concepts for working with big data, and highlight some of the tools you are likely to see on this frontier of your working knowledge.

17.2.1 Data and memory structures for big data

An alternative to `dplyr`, `data.table` is a popular R package for fast SQL-style operations on very large data tables (many gigabytes of memory). It is not clear that `data.table` is faster or more efficient than `dplyr`, and it uses a different—but not necessarily better—syntax. At the moment, `dplyr` seems to have the advantage of better access to relational database backends. Moreover, `dplyr` can use `data.table` itself as a backend. We have chosen to highlight `dplyr` in this book primarily because it fits so well syntactically with a number of other R packages we use herein (i.e., the *tidyverse*).

For some problems—more common in machine learning—the number of explanatory variables p can be large (not necessarily relative to the number of observations n). In such cases, the algorithm to compute a least-squares regression model model may eat up quite a bit of memory. The `biglm` package seeks to improve on this by providing a memory-efficient `biglm()` function that can be used in place of `lm()`. In particular, `biglm` can fit generalized linear models with data frames that are larger than memory. It accomplishes this by splitting the computations into more manageable chunks—updating the results iteratively as each chunk is processed. In this manner, you can write a drop-in replacement for your existing code that will scale to data sets larger than the memory on your computer.

```
library(mdsr)
library(biglm)
n <- 20000
p <- 500
d <- as.data.frame(matrix(rnorm(n * (p + 1)), ncol = (p + 1)))
expl_vars <- paste(paste0("V", 2:(p+1)), collapse = " + ")
my_formula <- as.formula(paste("V1 ~ ", expl_vars))
system.time(lm(my_formula, data = d))

   user  system elapsed
  4.457   0.144   4.610

system.time(biglm(my_formula, data = d))

   user  system elapsed
  3.445   0.138   3.588
```

Here we see that the computation completed more quickly (and can be updated to incorporate more observations, unlike `lm()`). The `biglm` package is also useful in settings where there are many observations but not so many predictors. A related package is `bigmemory`. This package extends R's capabilities to map memory to disk, allowing you to work with larger matrices.

17.2.2   Compilation

R, SQL, and Python are *interpreted* programming languages. This means that the code that you write in these languages gets translated into machine language on-the-fly as you execute it. The process is not altogether different than when you hear someone speaking in Russian on the news, and then you hear a halting English translation with a one- or two-second delay. Most of the time, the translation happens so fast that you don't even notice.

Imagine that instead of translating the Russian speaker's words on-the-fly, the translator took dictation, wrote down a thoughtful translation, and then re-recorded the segment in English. You would be able to process the English-speaking segment faster—because you are fluent in English. At the same time, the translation would probably be better, since more time and care went into it, and you would likely pick up subtle nuances that were lost in the on-the-fly translation. Moreover, once the English segment is recorded, it can be watched at any time without incurring the cost of translation again.

This alternative paradigm involves a one-time translation of the code called *compilation*. R code is not compiled (it is interpreted), but C++ code is. The result of compilation is a binary program that can be executed by the CPU directly. This is why, for example, you can't write a desktop application in R, and executables written in C++ will be much faster than scripts written in R or Python. (To continue this line of reasoning, binaries written in assembly language can be faster than those written in C++ , and binaries written in machine language can be faster than those written in assembly.)

If C++ is so much faster than R, then why write code in R? Here again, it is a trade-off. The code written in C++ may be faster, but when your programming time is taken into account you can often accomplish your task much faster by writing in R. This is because R provides extensive libraries that are designed to reduce the amount of code that you have to write. R is also interactive, so that you can keep a session alive and continue to write new code as you run the old code. This is fundamentally different than C++ development, where you have to re-compile every time you change a single line of code. The convenience of R programming comes at the expense of speed.

However, there is a compromise. `Rcpp` allows you to move certain pieces of your R code to C++ . The basic idea is that `Rcpp` provides C++ data structures that correspond to R data structures (e.g., a `data.frame` data structure written in C++). It is thus possible to write functions in R that get compiled into faster C++ code, with minimal additional effort on the part of the R programmer. The `dplyr` package makes extensive use of this functionality to improve performance.

17.2.3   Parallel and distributed computing

Embarrassingly parallel computing

How do you increase a program's capacity to work with larger data? The most obvious way is to add more memory (i.e., RAM) to your computer. This enables the program to read more data at once, enabling greater functionality with any additional programming. But what if the bottleneck is not the memory, but the processor? A processor can only do one thing at a time. So if you have a computation that takes t units of time, and you have to do that computation for many different data sets, then you can expect that it will take many more units of time to complete.

For example, suppose we generate 20 sets of one million (x, y) random pairs and want to fit a regression model to each set.

```
n <- 1e5
k <- 20
d <- data.frame(y = rnorm(n*k), x = rnorm(n*k), set = rep(1:k, each = n))

fit_lm <- function(data, set_id) {
  data %>%
    filter(set == set_id) %>%
    lm(y ~ x, data = .)
}
```

However long it takes to do it for the first set, it will take about 20 times as long to do it for all 20 sets. This is as expected, since the computation procedure was to fit the regression model for the first set, then fit it for the second set, and so on.

```
system.time(fit_lm(d, 1))

   user  system elapsed
  0.081   0.013   0.094

system.time(lapply(1:k, fit_lm, data = d))

   user  system elapsed
  1.664   0.193   1.857
```

However, in this particular case, the data in each of the twenty sets has nothing to do with the data in any of the other sets. This is an example of an *embarrassingly parallel* problem. These data are ripe candidates for a *parallelized* computation. If we had twenty processors, we could fit one regression model on each CPU—all at the same time—and get our final result in about the same time as it takes to fit the model to *one* set of data. This would be a tremendous improvement in speed.

Unfortunately, we don't have twenty CPUs. Nevertheless, most modern computers have multiple cores. (In this case, Nick's computer has four cores.)

```
library(parallel)
my_cores <- detectCores()
my_cores
```

```
[1] 4
```

The `parallel` package provides functionality for parallel computation in R. Specifically, it provides a function `mclapply()` that works just like `lapply()` (see Chapter 5), except that it spreads the computations over multiple cores. The theoretical speed-up is a function of `my_cores`, but in practice this may be less for a variety of reasons (most notably, the overhead associated with combining the parallel results).

The `mc.cores` argument to `mclapply()` controls the number of cores being used for parallel computation. Note that we have set this to one fewer core than exist, to reserve some resources for the operating system.

```
system.time(mclapply(1:k, fit_lm, data = d, mc.cores = my_cores - 1))

  user  system elapsed
  3.42    3.94    9.39
```

The multicore and snow packages also provide support for parallelism in R.

GPU computing and CUDA

Another fruitful avenue to speed up computations is through use of a graphical processing unit (GPU). These devices feature a highly parallel structure that can lead to significant performance gains. CUDA is a parallel computing platform and application programming interface created by NVIDIA (one of the largest manufacturers of GPUs). Access to GPU computing is provided in R through the gputools package. The OpenCL package provides bindings for R to the open-source, general-purpose OpenCL programming language for GPU computing.

```
library(gputools)
```

MapReduce

MapReduce is a programming paradigm for parallel computing. To solve a task using a MapReduce framework, two functions must be written:

1. Map(key_0, value_0): The Map() function reads in the original data (which is stored in key-value pairs), and splits it up into smaller subtasks. It returns a list of key-value pairs $(key_1, value_1)$, where the keys and values are not necessarily of the same type as the original ones.

2. Reduce(key_1, list(value_1)): The MapReduce implementation has a method for aggregating the key-value pairs returned by the Map() function by their keys (i.e., key_1). Thus, you only have to write the Reduce() function, which takes as input a particular key_1, and a list of all the value_1's that correspond to key_1. The Reduce() function then performs some operation on that list, and returns a list of values.

MapReduce is efficient and effective because the Map() step can be highly parallelized. Moreover, MapReduce is also fault tolerant, because if any individual Map() job fails, the controller can simply start another one. The Reduce() step often provides functionality similar to a GROUP BY operation in SQL.

Example   The canonical MapReduce example is to tabulate the frequency of each word in a large number of text documents (i.e., a *corpus* (see Chapter 15)). In what follows we show an implementation written in Python by Bill Howe [112]. Note that at the beginning, this bit of code calls an external MapReduce library that actually implements MapReduce. The user only needs to write the two functions shown in this block of code—not the MapReduce library itself.

```
import MapReduce
import sys
```

```
mr = MapReduce.MapReduce()

def mapper(record):
    key = record[0]
    value = record[1]
    words = value.split()
    for w in words:
      mr.emit_intermediate(w, 1)

def reducer(key, list_of_values):
    total = 0
    for v in list_of_values:
      total += v
    mr.emit((key, total))

if __name__ == '__main__':
  inputdata = open(sys.argv[1])
  mr.execute(inputdata, mapper, reducer)
```

We will use this MapReduce program to compile a word count for the issues raised on GitHub for the ggplot2 package. These are stored in a JSON file (see Chapter 5) as a single JSON array. Since we want to illustrate how MapReduce can parallelize over many files, we will convert this single array into a JSON object for each issue. This will mimic the typical use case. The jsonlite package provides functionality for coverting between JSON objects and native R data structures.

```
library(jsonlite)
gg_issues <- fromJSON("https://api.github.com/repos/hadley/ggplot2/issues")
gg_issues %>%
  select(url, body) %>%
  apply(MARGIN = 1, FUN = toJSON) %>%
  write(file = "code/map-reduce/issues.json")
```

For example, the first issue is displayed below. Note that it consists of two comma-separated character strings within brackets. We can think of this as having the format: [key, value].

```
readLines("code/map-reduce/issues.json")[1] %>%
  stringr::str_wrap(width = 70) %>%
  cat()
```

```
["https://api.github.com/repos/hadley/ggplot2/issues/1734","1. In
the help text of `geom_segment` and `geom_curve` the `arrow` argument
is described. On the other hand, `arrow` is not mentioned in `?
geom_spoke`. I think this argument should be described there as well.
\r\n\r\n2. In the help pages where `arrow` is described, make it clear
that it is `grid::arrow` which is used, i.e. replace \"as created by
arrow()\" with \"as created by `grid::arrow()`\" (note: also add code
formatting). This makes it easier to find more help on the function.\r
\n\r\n3. Possibly add an example on the direction of the arrows, i.e.
the `ends` argument (see e.g. [this Q&A](http://stackoverflow.com/
```

```
questions/39173710/change-direction-of-arrows-in-geom-spoke)). This is
perhaps less important _if_ you refer to `grid::arrow` for description
of further arguments. \r\n\r\n "]
```

In the Python code written above (which is stored in the file wordcount.py), the
mapper() function takes a record argument (i.e., one line of the issues.json file), and
examines its first two elements—the key becomes the first argument (in this case, the URL
of the GitHub issue) and the value becomes the second argument (the text of the issue).
After splitting the value on each space, the mapper() function emits a (*key, value*) pair
for each word. Thus, the first issue shown above would generate the pairs: (In, 1), (the,
1), (help, 1), etc.

The MapReduce library provides a mechanism for efficiently collecting all of the resulting
pairs based on the key, which in this case corresponds to a single word. The reducer()
function simply adds up all of the values associated with each key. In this case, these values
are all 1s, so the resulting pair is a word and the number of times it appears (e.g., (the,
158), etc.).

We can run this Python script from within R and bring the results into R for further analysis. We see that the most common words in this corpus are short articles and prepositions.

```
library(mdsr)
cmd <- "python code/map-reduce/wordcount.py code/map-reduce/issues.json"
res <- system(cmd, intern = TRUE)
freq_df <- res %>%
  lapply(jsonlite::fromJSON) %>%
  lapply(FUN = function(x) { data.frame(word = x[1],
                                        count = as.numeric(x[2]))}) %>%
  bind_rows()
glimpse(freq_df)

Observations: 954
Variables: 2
$ word  <chr> "all", "code", "Unknown", "saves", "results", "existing"...
$ count <dbl> 2, 3, 1, 1, 1, 1, 1, 1, 6, 1, 1, 1, 2, 1, 1, 2, 1, 1, 1,...

freq_df %>%
  filter(grepl(pattern = "[a-z]", word)) %>%
  arrange(desc(count)) %>%
  head(10)

    word count
1    the   100
2     to    63
3     is    38
4      a    38
5     of    34
6     be    25
7     it    23
8    and    22
9     in    22
10   for    22
```

MapReduce has become quite popular and offers some advantages over SQL for some problems. When MapReduce first became popular, and Google used it to redo their webpage ranking system (see Chapter 16), there was great excitement about a coming "paradigm shift" in parallel and distributed computing. Nevertheless, advocates of SQL continue to challenge the notion that it has been superseded by MapReduce [188].

Hadoop

As noted previously, MapReduce requires a software implementation. One popular such implementation is Hadoop MapReduce, which is one of the core components of Apache Hadoop. Hadoop is a larger software ecosystem for storing and processing large data that includes a distributed file system, Pig, Hive, Spark, and other popular open-source software tools. While we won't be able to go into great detail about these items, we will illustrate how to interface with Spark, which has become one of the more notable tools for working with big data.

Spark

One nice feature of Apache Spark—especially for our purposes—is that while it requires a distributed file system, it can implement a pseudo-distributed file system on a single machine. This makes it possible for you to experiment with Spark on your local machine even if you don't have access to a cluster. For obvious reasons, you won't actually see the performance boost that parallelism can bring, but you can try it out and debug your code. Furthermore, the `sparklyr` package makes it painless to install a local Spark cluster from within R, as well as connect to a local or remote cluster. At present, the `sparklyr` package must be downloaded from GitHub.

```
devtools::install_github("rstudio/sparklyr")
```

Once the `sparklyr` package is installed, we can use it to install a local Spark cluster.

```
library(sparklyr)
spark_install() # only once!
```

Next, we make a connection to our local Spark instance from within R. Of course, if we were connecting to a remote Spark cluster, we could modify the `master` argument to reflect that.

```
sc <- spark_connect(master = "local")
class(sc)
```

```
[1] "spark_connection"      "spark_shell_connection"
[3] "DBIConnection"
```

Note that `sc` has class `DBIConnection`—this means that it can do many of the things that other `dplyr` connections can do. For example, the `src_tbls()` function works just like it did on the MySQL connection objects we saw in Chapter 12.

```
src_tbls(sc)
```

```
character(0)
```

In this case, there are no tables present in this Spark cluster, but we can add them using the `copy_to()` command. Here, we will load the babynames table from the babynames package.

```
babynames_tbl <- sc %>% copy_to(babynames::babynames, "babynames")
src_tbls(sc)

[1] "babynames"

class(babynames_tbl)

[1] "tbl_spark" "tbl_sql"   "tbl_lazy"  "tbl"
```

The babynames_tbl object is a tbl_spark, but also a tbl_sql. Again, this is analogous to what we saw in Chapter 12, where a tbl_mysql was also a tbl_sql.

```
babynames_tbl %>%
  filter(name == "Benjamin") %>%
  group_by(year) %>%
  summarize(N = n(), total_births = sum(n)) %>%
  arrange(desc(total_births)) %>%
  head()

Source:    query [?? x 3]
Database: spark connection master=local[4] app=sparklyr local=TRUE

    year      N total_births
   <dbl>  <dbl>        <dbl>
1   1989      2        15783
2   1988      2        15272
3   1987      2        14949
4   2000      2        14862
5   1990      2        14658
6   1981      2        14419
```

As we will see below with Google BigQuery, even though Spark is a parallelized technology designed to supersede SQL, it is still useful to know SQL in order to use Spark. Moreover, unlike BigQuery, sparklyr allows you to work with a Spark cluster using the familiar dplyr interface.

As you might suspect, because babynames_tbl is a tbl_sql, it implements SQL methods common in DBI. Thus, we can also write SQL queries against our Spark cluster.

```
library(DBI)
dbGetQuery(sc, "SELECT year, sum(1) as N, sum(n) as total_births
                FROM babynames WHERE name == 'Benjamin'
                GROUP BY year
                ORDER BY total_births desc
                LIMIT 6")

   year N total_births
1 1989 2        15783
2 1988 2        15272
```

```
3 1987 2      14949
4 2000 2      14862
5 1990 2      14658
6 1981 2      14419
```

Finally, because Spark includes not only a database infrastructure, but also a machine learning library, `sparklyr` allows you fit many of the models we outlined in Chapter 8 and 9 within Spark. This means that you can rely on Spark's big data capabilities without having to bring all of your data into R's memory.

As a motivating example, we fit a multiple regression model for the amount of rainfall at the Smith College MacLeish field station as a function of the temperature, pressure, and relative humidity.

```
library(macleish)
weather_tbl <- copy_to(sc, whately_2015)
weather_tbl %>%
  ml_linear_regression(rainfall ~ temperature + pressure + rel_humidity) %>%
  summary()

Call:
ml_linear_regression(., rainfall ~ temperature + pressure + rel_humidity)

Deviance Residuals::
     Min        1Q     Median        3Q        Max
-0.041290 -0.021761 -0.011632 -0.000576 15.968356

Coefficients:
             Estimate Std. Error t value Pr(>|t|)
(Intercept)  7.18e-01   1.15e-01    6.26 3.8e-10 ***
temperature  4.09e-04   7.77e-05    5.26 1.4e-07 ***
pressure     7.54e-04   1.16e-04   -6.51 7.6e-11 ***
rel_humidity 4.38e-04   3.85e-05   11.38 < 2e-16 ***
---
Signif. codes:  0 '***' 0.001 '**' 0.01 '*' 0.05 '.' 0.1 ' ' 1

R-Squared: 0.004824
Root Mean Squared Error: 0.1982
```

The most recent versions of RStudio include integrated support for management of Spark clusters.

Pro Tip: Use a cloud-based computing service, such as Amazon Web Services or Digital Ocean, for a low-cost alternative to building your own server farm.

17.2.4   Alternatives to SQL

Relational database management systems can be spread across multiple computers into what is called a *cluster*. In fact, it is widely acknowledged that one of the things that allowed Google to grow so fast was its use of the open-source (zero cost) MySQL RDBMS running as a cluster across many identical low-cost servers. That is, rather than investing large amounts of money in big machines, they built a massive MySQL cluster over many small, cheap machines. Both MySQL and PostgreSQL provide functionality for extending a single installation to a cluster.

BigQuery

BigQuery is a Web service offered by Google. Internally, the BigQuery service is supported by Dremel, the open-source version of which is Apache Drill. The `bigrquery` package for R provides access to BigQuery from within R.

To use the BigQuery service, you need to sign up for an account with Google, but you won't be charged unless you exceed the free limit of 10,000 requests per day. If you want to use your own data, you have to upload it to Google Cloud Storage, but Google provides several data sets that you can use for free. Here we illustrate how to query the `shakespeare` data set—which is a list of all of the words that appear in Shakespeare's plays—to find the most common words. Note that BigQuery understands a recognizable dialect of SQL—what makes BigQuery special is that it is built on top of Google's massive computing architecture.

```
library(bigrquery)
project_id <- "my-google-id"

sql <- "
SELECT word, count(distinct corpus) as numPlays
, sum(word_count) as N
FROM [publicdata:samples.shakespeare]
GROUP BY word
ORDER BY N desc
LIMIT 10
"
query_exec(sql, project = project_id)
```

```
4.9 megabytes processed
    word numPlays     N
1    the       42 25568
2      I       42 21028
3    and       42 19649
4     to       42 17361
5     of       42 16438
6      a       42 13409
7    you       42 12527
8     my       42 11291
9     in       42 10589
10    is       42  8735
```

NoSQL

NoSQL refers not to a specific technology, but rather to a class of database architectures that are *not* based on the notion—so central to SQL (and `data.frames` in R)—that a table consists of a rectangular array of rows and columns. Rather than being built around tables, NoSQL databases may be built around columns, key-value pairs, documents, or graphs.

Nevertheless NoSQL databases may (or may not) include an SQL-like query language for retrieving data.

One particularly successful NoSQL database is MongoDB, which is based on a document structure. In particular, MongoDB is often used to store JSON objects (see Chapter 5), which are not necessarily tabular.

17.3   Alternatives to R

Python is a widely used general-purpose, high-level programming language. You will find adherents for both R and Python, and while there are ongoing debates about which is "better," there is no consensus. It is probably true that—for obvious reasons—computer scientists tend to favor Python, while statisticians tend to favor R. We prefer the latter but will not make any claims about its being "better" than Python. A well-rounded data scientist should be competent in both environments.

Python is a modular environment (like R) and includes many libraries for working with data. The most R-like is `Pandas`, but other popular auxiliary libraries include `SciPy` for scientific computation, `NumPy` for large arrays, `matplotlib` for graphics, and `scikit-learn` for machine learning.

Other popular programming languages among data scientists include Scala and Julia. Scala supports a *functional programming* paradigm that has been emphasized by Hadley Wickham [220] and other R users. Julia has a smaller user base but has nonetheless many strong adherents.

17.4   Closing thoughts

Advances in computing power and the Internet have changed the field of statistics in ways that only the greatest visionaries could have imagined. In the 20th century, the science of extracting meaning from data focused on developing inferential techniques that required sophisticated mathematics to squeeze the most information out of small data. In the 21st century, the science of extracting meaning from data has focused on developing powerful computational tools that enable the processing of ever larger and more complex data. While the essential analytical language of the last century—mathematics—is still of great importance, the analytical language of this century is undoubtedly programming. The ability to write code is a necessary but not sufficient condition for becoming a data scientist.

We have focused on programming in R, a well-worn interpreted language designed by statisticians for computing with data. We believe that as an open-source language with a broad following, R has significant staying power. Yet we recognize that all technological tools eventually become obsolete. Nevertheless, by absorbing the lessons in this book, you will have transformed yourself into a competent, ethical, and versatile data scientist—one who possesses the essential capacities for working with a variety of data programmatically. You can build and interpret models, query databases both local and remote, make informative and interactive maps, and wrangle and visualize data in various forms. Internalizing these abilities will allow them to permeate your work in whatever field interests you, for as long as you continue to use data to inform.

17.5   Further resources

Tools for working with big data analytics are developing more quickly than any of the other topics in this book. A special issue of the *The American Statistician* addressed the training of students in statistics and data science [109]. The issue included articles on teaching statistics at "Google-Scale" [47] and on the teaching of data science more generally [20, 95]. In late 2016 the board of directors of the American Statistical Association endorsed the *Curriculum Guidelines for Undergraduate Programs in Data Science* written by the Park City Math Institute (PCMI) Undergraduate Faculty Group [158]. These guidelines recommended fusing statistical thinking into the teaching of techniques to solve big data problems.

A comprehensive survey of R packages for parallel computation and high-performance computing is available through the CRAN task view on that subject. The *Parallel R* book by McCallum and Weston is another resource [141].

More information about Google BigQuery can be found at their website: `https://cloud.google.com/bigquery`. A tutorial for SparkR is available on Apache's website: `https://spark.apache.org/docs/1.6.0/sparkr.html`

Part IV

Appendices

Appendix A

Packages used in this book

A.1 The `mdsr` package

The `mdsr` package contains all of the small data sets used in this book that are not available in other packages. To install it, use `install.packages()` to get the latest release. (See Section B.5.1 for more comprehensive information about R package maintainence.)

```
# this command only needs to be run once
install.packages("mdsr")
```

The list of data sets provided can be retrieved using the `data()` function.

```
library(mdsr)
data(package = "mdsr")
```

While the `mdsr` package does not contain many functions, the `src_scidb()` function provides a shorthand for connecting to the public SQL server at Smith College. We use this function extensively in Chapter 12 and in our classes and research projects.

The other virtue of `mdsr` is that it loads a series of other commonly used and useful packages. Specifically, loading `mdsr` will load `mosaic`, which in turn loads `dplyr`, and `ggplot2`. Thus, a single call to `library()` will generally set up an R session to do most of the things we do in this book and in our work.

A.2 The `etl` package suite

As we discuss in both Chapters 1 and 17, this book is not explicitly about "big data"—it is about mastering data science techniques for small and medium data with an eye towards big data. To that end, we need medium-sized data sets to work with. We have introduced several such data sets in this book, namely `airlines`, `fec`, and `imdb`.

The packages that bring these medium data sets to R belong to a suite of packages that leverage the `etl` framework, which in turn is heavily indebted to `dplyr`. Since medium data are too big to store in memory, but not so big that they can't fit on a single hard drive, a common and appropriate storage solution is SQL (see Chapters 12 and 13, and Appendix F). The process of bringing raw data into SQL is often known as Extract-Transform-Load, or ETL for short. The `etl` package for R facilitates such operations by establishing S3 generic functions that form a clear ETL pipeline.

```
data_source %>%
  etl_extract() %>%
  etl_transform() %>%
  etl_load()
```

The airlines package, which was originally forked from the nycflights13 package, gives R users the ability to download the full 30 years (and counting) of flight data from the United States Bureau of Transportation Statistics and bring it seamlessly into SQL without actually having to write any SQL code. Similarly, the imdb package provides the same functionality for mirroring the Internet Movie Database. The macleish package also uses the etl framework for hourly-updated weather data from the MacLeish field station. As the ecosystem of etl packages grows, more sources of medium data will be available to readers of this book, and R users in general.

A.3   Other packages

The packages we use most commonly include dplyr, mosaic, ggplot2, tidyr, broom, and lubridate. The full list of packages used in this book appears below.

Package	Title
airlines [21]	Data About Flights
alr3 [211]	Data to accompany Applied Linear Regression 3rd edition
ape [157]	Analyses of Phylogenetics and Evolution
aRxiv [172]	Interface to the arXiv API
assertthat [215]	Easy pre and post assertions.
atus [91]	American Time Use Survey 2014 Data
babynames [221]	US Baby Names 1880-2014
base [165]	The R Base Package
benford.analysis [53]	Benford Analysis for Data Validation and Forensic Analytics
biglm [137]	bounded memory linear and generalized linear models
bigrquery [222]	An Interface to Google's 'BigQuery' 'API'
broom [176]	Convert Statistical Analysis Objects into Tidy Data Frames
class [205]	Functions for Classification
DBI [170]	R Database Interface
devtools [232]	Tools to Make Developing R Packages Easier
dplyr [234]	A Grammar of Data Manipulation
DT [240]	A Wrapper of the JavaScript Library 'DataTables'
dygraphs [204]	Interface to 'Dygraphs' Interactive Time Series Charting Library
e1071 [143]	Misc Functions of the Department of Statistics, Probability Theory Group (Formerly: E1071), TU Wien
etl [22]	Extract-Transform-Load Framework for Medium Data
faraway [74]	Functions and Datasets for Books by Julian Faraway
fec [19]	Data about Federal Elections
foreign [164]	Read Data Stored by Minitab, S, SAS, SPSS, Stata, Systat, Weka, dBase, ...
fueleconomy [216]	EPA fuel economy data
GGally [183]	Extension to 'ggplot2'
ggExtra [10]	Add Marginal Histograms to 'ggplot2', and More 'ggplot2' Enhancements
ggmap [122]	Spatial Visualization with ggplot2

ggnetwork [40]	Geometries to Plot Networks with 'ggplot2'
ggplot2 [212]	Create Elegant Data Visualisations Using the Grammar of Graphics
ggthemes [9]	Extra Themes, Scales and Geoms for 'ggplot2'
ggvis [50]	Interactive Grammar of Graphics
googlesheets [42]	Manage Google Spreadsheets from R
graphics [166]	The R Graphics Package
Hmisc [96]	Harrell Miscellaneous
htmlwidgets [201]	HTML Widgets for R
httr [223]	Tools for Working with URLs and HTTP
igraph [61]	Network Analysis and Visualization
jsonlite [154]	A Robust, High Performance JSON Parser and Generator for R
knitr [241]	A General-Purpose Package for Dynamic Report Generation in R
Lahman [80]	Sean 'Lahman' Baseball Database
lars [97]	Least Angle Regression, Lasso and Forward Stagewise
lazyeval [224]	Lazy (Non-Standard) Evaluation
leaflet [51]	Create Interactive Web Maps with the JavaScript 'Leaflet' Library
lubridate [94]	Make Dealing with Dates a Little Easier
macleish [24]	Retrieve Data from MacLeish Field Station
magrittr [11]	A Forward-Pipe Operator for R
maps [29]	Draw Geographical Maps
maptools [32]	Tools for Reading and Handling Spatial Objects
mclust [79]	Gaussian Mixture Modelling for Model-Based Clustering, Classification, and Density Estimation
mdsr [18]	Complement to 'Modern Data Science with R'
methods [167]	Formal Methods and Classes
mosaic [162]	Project MOSAIC Statistics and Mathematics Teaching Utilities
mosaicData [163]	Project MOSAIC Data Sets
nasaweather [217]	Collection of datasets from the ASA 2006 data expo
network [43]	Classes for Relational Data
NeuralNetTools [28]	Visualization and Analysis Tools for Neural Networks
NHANES [161]	Data from the US National Health and Nutrition Examination Study
nnet [206]	Feed-Forward Neural Networks and Multinomial Log-Linear Models
nycflights13 [225]	Flights that Departed NYC in 2013
packrat [200]	A Dependency Management System for Projects and their R Package Dependencies
parallel [168]	Support for Parallel computation in R
partykit [111]	A Toolkit for Recursive Partytioning
plotKML [99]	Visualization of Spatial and Spatio-Temporal Objects in Google Earth
plotly [185]	Create Interactive Web Graphics via 'plotly.js'
randomForest [132]	Breiman and Cutler's Random Forests for Classification and Regression
RColorBrewer [145]	ColorBrewer Palettes
Rcpp [66]	Seamless R and C++ Integration
RCurl [190]	General Network (HTTP/FTP/...) Client Interface for R
readr [236]	Read Tabular Data
readxl [226]	Read Excel Files
rgdal [31]	Bindings for the Geospatial Data Abstraction Library

rgeos [33]	Interface to Geometry Engine - Open Source (GEOS)
RgoogleMaps [134]	Overlays on Static Maps
rmarkdown [4]	Dynamic Documents for R
RMySQL [155]	Database Interface and 'MySQL' Driver for R
ROCR [187]	Visualizing the Performance of Scoring Classifiers
rpart [191]	Recursive Partitioning and Regression Trees
RSQLite [120]	'SQLite' Interface for R
rvest [227]	Easily Harvest (Scrape) Web Pages
scales [228]	Scale Functions for Visualization
shiny [49]	Web Application Framework for R
sna [44]	Tools for Social Network Analysis
sp [34]	Classes and Methods for Spatial Data
sparklyr [138]	R Interface to Apache Spark
streamgraph [181]	streamgraph is an htmlwidget for building streamgraph visualizations
stringr [229]	Simple, Consistent Wrappers for Common String Operations
testthat [214]	Unit Testing for R
tibble [235]	Simple Data Frames
tidyr [230]	Easily Tidy Data with 'spread()' and 'gather()' Functions
tidytext [186]	Text Mining using 'dplyr', 'ggplot2', and Other Tidy Tools
tidyverse [231]	Easily Install and Load 'Tidyverse' Packages
tm [75]	Text Mining Package
twitteR [88]	R Based Twitter Client
UScensus2010 [6]	US Census 2010 Suite of R Packages
UScensus2010tract [7]	US Census 2010 Tract Level Shapefiles and Additional Demographic Data
usdanutrients [219]	USDA nutrient data (release SR26)
webshot [48]	Take Screenshots of Web Pages
wordcloud [76]	Word Clouds
xkcd [192]	Plotting ggplot2 Graphics in an XKCD Style
xtable [62]	Export Tables to LaTeX or HTML
Zelig [52]	Everyone's Statistical Software

Table A.1: List of packages used in this book. Most packages are available on CRAN. Packages available from GitHub include: `airlines`, `fec`, `imdb`, `sparklyr`, and `streamgraph`.

A.4 Further resources

More information on the `mdsr` package and the `etl` packages can be found at `http://www.github.com/beanumber`.

Appendix B

Introduction to **R** and **RStudio**

This chapter provides a (brief) introduction to R and RStudio. R is a free, open-source software environment for statistical computing and graphics [116, 169]. RStudio is an open-source integrated development environment (IDE) for R that adds many features and productivity tools for R. This chapter includes a short history, installation information, a sample session, background on fundamental structures and actions, information about help and documentation, and other important topics.

R is a general purpose package that includes support for a wide variety of modern statistical and graphical methods (many of which have been contributed by users). It is available for Linux, Mac OS X, and Windows. The R Foundation for Statistical Computing holds and administers the copyright of the R software and documentation. R is available under the terms of the Free Software Foundation's GNU General Public License in source code form.

RStudio facilitates use of R by integrating R help and documentation, providing a workspace browser and data viewer, and supporting syntax highlighting, code completion, and smart indentation. It integrates reproducible analysis with knitr and R Markdown (see Appendix D), supports the creation of slide presentations, and includes a debugging environment. It facilitates the creation of dynamic Web applications using Shiny (see Section 11.3). It also provides support for multiple projects as well as an interface to source code control systems such as GitHub. It has become the default interface for many R users, and is our recommended environment for analysis.

RStudio is available as a client (standalone) for Windows, Mac OS X, and Linux. There is also a server version. Commercial products and support are available in addition to the open-source offerings (see http://www.rstudio.com/ide for details).

The first versions of R were written by Ross Ihaka and Robert Gentleman at the University of Auckland, New Zealand, while current development is coordinated by the R Development Core Team, a group of international volunteers.

R is similar to the S language, a flexible and extensible statistical environment originally developed in the 1980s at AT&T Bell Labs (now Alcatel–Lucent). Insightful Corporation has continued the development of S in their commercial software package S-PLUS^{TM}.

B.1 Installation

New users are encouraged to download and install R from the Comprehensive R Archive Network (CRAN, http://www.r-project.org) and install RStudio from http://www.rstudio.com/download. The sample session in the appendix of the *Introduction to R* document, also

421

available from CRAN (see B.2), is highly recommended reading.

The home page for the R project, located at `http://r-project.org`, is the best starting place for information about the software. It includes links to CRAN, which features precompiled binaries as well as source code for R, add-on packages, documentation (including manuals, frequently asked questions, and the R newsletter) as well as general background information. Mirrored CRAN sites with identical copies of these files exist all around the world. Updates to R and packages are regularly posted on CRAN.

B.1.1   Installation under Windows

Versions of R for Windows XP and later—including 64-bit versions—are available at CRAN. The distribution includes `Rgui.exe`, which launches a self-contained windowing system that includes a command-line interface, `Rterm.exe` for a command-line interface only, `Rscript.exe` for batch processing only, and `R.exe`, which is suitable for batch or command-line use. More information on Windows-specific issues can be found in the CRAN *R for Windows FAQ*.

B.1.2   Installation under Mac OS X

A version of R for Mac OS X 10.6 and higher is available at CRAN. This is distributed as a disk image containing the installer. In addition to the graphical interface version, a command line version (particularly useful for batch operations) can be run as the command R. More information on Macintosh-specific issues can be found in the CRAN *R for Mac OS X FAQ*.

B.1.3   Installation under Linux

R is available for most Linux distributions through your distribution's repositories. For example, R is provided on Debian-based distributions like Ubuntu by the `r-base` package. Many additional packages, such as `r-cran-rpart`, are provided at the maintainer's discretion. Installation on Ubuntu is as simple as:

```
sudo apt-get update
sudo apt-get install r-base r-base-dev
```

CRAN provides distribution-specific packages for the Debian, Red Hat, SuSE, and Ubuntu distributions at `https://cran.r-project.org/bin/linux`.

B.1.4   **RStudio**

RStudio for Mac OS X, Windows, or Linux can be downloaded from `http://www.rstudio.com/ide`. RStudio requires R to be installed on the local machine. A server version (accessible from Web browsers) is also available for download. Documentation of the advanced features in the system is available on the RStudio website.

B.2   Running **RStudio** and sample session

Once installation is complete, the recommended next step for a new user would be to start RStudio and run a sample session (see Figure B.1).

The ">" character is the command prompt, and commands are executed once the user presses the RETURN or ENTER key. R can be used as a calculator (as seen from the first

```
R version 3.3.2 (2016-10-31) -- "Sincere Pumpkin Patch"
Copyright (C) 2016 The R Foundation for Statistical Computing
Platform: x86_64-apple-darwin13.4.0 (64-bit)

R is free software and comes with ABSOLUTELY NO WARRANTY.
You are welcome to redistribute it under certain conditions.
Type 'license()' or 'licence()' for distribution details.

  Natural language support but running in an English locale

R is a collaborative project with many contributors.
Type 'contributors()' for more information and
'citation()' on how to cite R or R packages in publications.

Type 'demo()' for some demos, 'help()' for on-line help, or
'help.start()' for an HTML browser interface to help.
Type 'q()' to quit R.

> 3 + 6
[1] 9
> 2 * 3
[1] 6
> x <- c(4, 5, 3, 2)
> x
[1] 4 5 3 2
> y <- seq(1, 4)
> y
[1] 1 2 3 4
> mean(x)
[1] 3.5
> sd(y)
[1] 1.290994
> ds <- read.csv("http://nhorton.people.amherst.edu/r2/datasets/help.csv")
> mean(ds$age)
[1] 35.65342
> # mean(age)     # this will generate an error
> with(ds, mean(age))
[1] 35.65342
> ds$age[1:15]
 [1] 37 37 26 39 32 47 49 28 50 39 34 58 58 60 36
> q()
```

Figure B.1: Sample session in R.

two commands on lines 1 and 3). New variables can be created (e.g., x and y) using the assignment operator <-. If a command generates output, then it is printed on the screen, preceded by a number indicating place in the vector (this is particularly useful if output is longer than one line, e.g., as it is for ds$age[1:25]). Saved data (here assigned the name ds) is read into R on line 15, then summary statistics are calculated (e.g., using mean()) and individual observations are displayed. The $ operator allows access to objects within a data frame. Alternatively, the with() function can be used to access objects within a data set.

As shown in the example below, it is important to remember that R is case-sensitive. A comprehensive sample session in R can be found in Appendix A of *An Introduction to R* [207].

```
x <- 1:3
X <- seq(2, 4)
x

[1] 1 2 3

X

[1] 2 3 4
```

B.3 Learning R

B.3.1 Getting help

R features extensive online documentation, though it can sometimes be challenging to comprehend. Each command has an associated help file that describes usage, lists arguments, provides details of actions, gives references, lists other related functions, and includes examples of its use. The help system is invoked using either the ? or help() commands.

```
?function
help(function)
```

where function is the name of the function of interest (Alternatively, the Help tab in RStudio can be used to access the help system.)

For example, the help file for the mean() function is accessed by the command help(mean). The output from this command is provided in Figure B.2. It describes the mean() function as a generic function for the (trimmed) arithmetic mean, with arguments x (an R object), trim (the fraction of observations to trim, having a default value of 0—setting trim equal to 0.5 is equivalent to calculating the median), and na.rm (should missing values be present, the default behavior na.rm equals FALSE, which leaves missing values as they are).

Some commands (e.g., if) are reserved, so ?if will not generate the desired documentation. Running ?"if" will work (see also ?Reserved and ?Control). Other reserved words include else, repeat, while, function, for, in, next, break, TRUE, FALSE, NULL, Inf, NaN, and NA.

The RSiteSearch() function will search for key words or phrases in many places (including the search engine at http://search.r-project.org). The RSeek.org site can also be helpful in finding more information and examples. Examples of many functions are available using the example() function.

```
example(mean)
```

Other useful resources are help.start(), which provides a set of online manuals, and help.search(), which can be used to look up entries by description. The apropos() command returns any functions in the current search list that match a given pattern (which facilitates searching for a function based on what it does, as opposed to its name).

Other resources for help available from CRAN include the R-help mailing list. The StackOverflow site for R provides a series of questions and answers for common questions that are tagged as being related to R. New users are also encouraged to read the R FAQ (frequently asked questions) list. RStudio provides a curated guide to resources for learning R and its extensions.

```
mean                    package:base                R Documentation
```

Arithmetic Mean

Description:
     Generic function for the (trimmed) arithmetic mean.

Usage:
     mean(x, ...)

     ## Default S3 method:
     mean(x, trim = 0, na.rm = FALSE, ...)

Arguments:
          x: An R object. Currently there are methods for numeric/logical
             vectors and date, date-time and time interval objects.
             Complex vectors are allowed for 'trim = 0', only.

       trim: the fraction (0 to 0.5) of observations to be trimmed from
             each end of 'x' before the mean is computed. Values of trim
             outside that range are taken as the nearest endpoint.

      na.rm: a logical value indicating whether 'NA' values should be
             stripped before the computation proceeds.

        ...: further arguments passed to or from other methods.

Value:
     If 'trim' is zero (the default), the arithmetic mean of the values
     in 'x' is computed, as a numeric or complex vector of length one.
     If 'x' is not logical (coerced to numeric), numeric (including
     integer) or complex, 'NA_real_' is returned, with a warning.

     If 'trim' is non-zero, a symmetrically trimmed mean is computed
     with a fraction of 'trim' observations deleted from each end
     before the mean is computed.

References:
     Becker, R. A., Chambers, J. M. and Wilks, A. R. (1988) _The New S
     Language_. Wadsworth & Brooks/Cole.

See Also:
     'weighted.mean', 'mean.POSIXct', 'colMeans' for row and column
     means.

Examples:
     x <- c(0:10, 50)
     xm <- mean(x)
     c(xm, mean(x, trim = 0.10))
```

Figure B.2: Documentation on the mean() function.
```

B.3.2   swirl

The swirl system is a collection of interactive courses to teach R programming and data science within the R console. It requires the installation of the swirl package, then use of the install_from_swirl() function to download courses. Table B.1 displays some of the courses that were available as of 2016. A sample session is displayed below. After some preliminary introductions, the user is instructed to enter a series of commands and explore in the console. The swirl system detects whether the correct commands have been input.

COURSE	DESCRIPTION
R Programming (beginner)	The basics of programming in R
R Programming Alt (beginner)	Same as the original, but modified for in-class use
Data Analysis (beginner)	Basic ideas in statistics and data visualization
Mathematical Biostatistics Boot Camp (beginner)	One- and two-sample t-tests, power, and sample size
Open Intro (beginner)	A very basic introduction to statistics, data analysis, and data visualization
Regression Models (intermediate)	The basics of regression modeling in R
Getting and Cleaning Data (intermediate)	dplyr, tidyr, lubridate, oh my!

Table B.1: Some of the interactive courses available within swirl.

```
> library(swirl)

| Type swirl() when you are ready to begin.

> install_from_swirl("Getting and Cleaning Data")

| Course installed successfully!

> swirl()

| Welcome to swirl!

| Please sign in. If you've been here before, use the same name as you did
| then. If you are new, call yourself something unique.

What shall I call you? Nick

| Please choose a course, or type 0 to exit swirl.

1: Getting and Cleaning Data
2: R Programming
3: Regression Models
4: Take me to the swirl course repository!

Selection: 1

| Please choose a lesson, or type 0 to return to course menu.
```

```
1: Manipulating Data with dplyr
2: Grouping and Chaining with dplyr
3: Tidying Data with tidyr
4: Dates and Times with lubridate

Selection: 1

| Attempting to load lesson dependencies...

| Package dplyr loaded correctly!

| In this lesson, you'll learn how to manipulate data using dplyr. dplyr is
| a fast and powerful R package written by Hadley Wickham and Romain
| Francois that provides a consistent and concise grammer for manipulating
| tabular data.

...
```

B.4 Fundamental structures and objects

Here we provide a brief introduction to R data structures.

B.4.1 Objects and vectors

Almost everything in R is an object, which may be initially confusing to a new user. An object is simply something stored in R's memory. Common objects include vectors, matrices, arrays, factors, data frames (akin to data sets in other systems), lists, and functions.

The basic variable structure is a vector. Vectors (and other objects) are created using the <- or = assignment operators (which assign the evaluated expression on the right-hand side of the operator to the object name on the left-hand side).

```
x <- c(5, 7, 9, 13, -4, 8) # preferred
x =  c(5, 7, 9, 13, -4, 8)  # equivalent
```

The above code creates a vector of length 6 using the c() function to concatenate scalars. The = operator is used in other contexts for the specification of arguments to functions. Other assignment operators exist, as well as the assign() function (see help("<-") for more information). The exists function conveys whether an object exists in the workspace, and the rm command removes it. In RStudio, the "Environment" tab shows the names (and values) of all objects that exist in the current workspace.

Since vector operations are so fundamental in R, it is important to be able to access (or index) elements within these vectors. Many different ways of indexing vectors are available. Here, we introduce several of these using the x as created above. The command x[2] returns the second element of x (the scalar 7), and x[c(2, 4)] returns the vector $(7, 13)$. The expressions x[c(TRUE, TRUE, TRUE, TRUE, TRUE, FALSE)], x[1:5] and x[-6] all return a vector consisting of the first five elements in x (the last specifies all elements except the 6th). Knowledge and basic comfort with these approaches to vector indexing are important to effective use of R, as they can help with computational efficiency.

```
x[2]
```

```
[1] 7
```

```
x[c(2, 4)]
```

```
[1]   7 13
```

```
x[c(TRUE, TRUE, TRUE, TRUE, TRUE, FALSE)]
```

```
[1]   5   7   9 13 -4
```

```
x[1:5]
```

```
[1]   5   7   9 13 -4
```

```
x[-6]
```

```
[1]   5   7   9 13 -4
```

Vectors are *recycled* if needed; for example, when comparing each of the elements of a vector to a scalar.

```
x > 8
```

```
[1] FALSE FALSE  TRUE  TRUE FALSE FALSE
```

The above expression demonstrates the use of comparison operators (see ?Comparison). Only the third and fourth elements of x are greater than 8. The function returns a logical value of either TRUE or FALSE (see ?Logic).

A count of elements meeting the condition can be generated using the sum() function.

Other comparison operators include == (equal), >= (greater than or equal), <= (less than or equal and != (not equal). Care needs to be taken in the comparison using == if noninteger values are present (see all.equal()).

```
sum(x > 8)
```

```
[1] 2
```

B.4.2   Operators

There are many operators defined in R to carry out a variety of tasks. Many of these were demonstrated in the sample session (assignment, arithmetic) and previous examples (comparison). Arithmetic operations include +, -, *, /, ˆ (exponentiation), %% (modulus), and %/% (integer division). More information about operators can be found using the help system (e.g., ?"+"). Background information on other operators and precedence rules can be found using help(Syntax).

R supports Boolean operations (OR, AND, NOT, and XOR) using the |, ||, &, ! operators and the xor() function. The | is an "or" operator that operates on each element of a vector, while the || is another "or" operator that stops evaluation the first time that the result is true (see ?Logic).

B.4.3   Lists

Lists in R are very general objects that can contain other objects of arbitrary types. List members can be named, or referenced using numeric indices (using the [[operator).

```
newlist <- list(first = "hello", second = 42, Bob = TRUE)
is.list(newlist)

[1] TRUE

newlist

$first
[1] "hello"

$second
[1] 42

$Bob
[1] TRUE

newlist[[2]]

[1] 42

newlist$Bob

[1] TRUE
```

The unlist() function flattens (makes a vector out of) the elements in a list (see also relist()). Note that unlisted objects are coerced to a common type (in this case character).

```
unlisted <- unlist(newlist)
unlisted

  first  second     Bob
"hello"    "42"  "TRUE"
```

B.4.4   Matrices

Matrices are like two-dimensional vectors. Thus, they are rectangular objects where all entries have the same type. We can create a 2×3 matrix, display it, and test for its type.

```
A <- matrix(x, 2, 3)
A

     [,1] [,2] [,3]
[1,]    5    9   -4
[2,]    7   13    8

is.matrix(A)    # is A a matrix?
```

```
[1] TRUE
```

```
is.vector(A)
```

```
[1] FALSE
```

```
is.matrix(x)
```

```
[1] FALSE
```

Note that comments are supported within R (any input given after a # character is ignored).

Indexing for matrices is done in a similar fashion as for vectors, albeit with a second dimension (denoted by a comma).

```
A[2, 3]
```

```
[1] 8
```

```
A[, 1]
```

```
[1] 5 7
```

```
A[1, ]
```

```
[1]  5  9 -4
```

B.4.5 Dataframes

Data sets are often stored in a `data.frame`, which is a special type of `list` that is more general than a `matrix`. This rectangular object, similar to a data table in other systems, can be thought of as a two-dimensional array with columns of vectors of the same length, but of possibly different types (as opposed to a matrix, which consists of vectors of the *same* type; or a list, whose elements needn't be of the same length). The function `read_csv()` in the `readr` package returns a `data.frame` object.

A simple `data.frame` can be created using the `data.frame()` command. Variables can be accessed using the $ operator, as shown below (see also `help(Extract)`). In addition, operations can be performed by column (e.g., calculation of sample statistics). We can check to see if an object is a `data.frame` with `is.data.frame()`.

```
y <- rep(11, length(x))
y
```

```
[1] 11 11 11 11 11 11
```

```
ds <- data.frame(x, y)
ds
```

```
    x  y
1   5 11
2   7 11
3   9 11
4  13 11
```

```
5 -4 11
6  8 11
```

```
ds$x[3]
```

```
[1] 9
```

```
is.data.frame(ds)
```

```
[1] TRUE
```

Note that the use of data.frame() differs from the use of cbind(), which yields a matrix object (unless it is given data frames as inputs).

```
newmat <- cbind(x, y)
newmat
```

```
       x  y
[1,]   5 11
[2,]   7 11
[3,]   9 11
[4,] 13 11
[5,]  -4 11
[6,]   8 11
```

```
is.data.frame(newmat)
```

```
[1] FALSE
```

```
is.matrix(newmat)
```

```
[1] TRUE
```

Data frames are created from matrices using as.data.frame(), while matrices are constructed from data frames using as.matrix().

Although we strongly discourage its use, data frames can be attached to the workspace using the attach() command. The Google R Style guide provides similar advice [90]. Name conflicts are a common problem with attach() (see conflicts(), which reports on objects that exist with the same name in two or more places on the search path).

The search() function lists attached packages and objects. To avoid cluttering the name-space, the command detach() should be used once a data frame or package is no longer needed.

A number of R functions include a data argument to specify a data frame as a local environment. For others, the with() and within() commands can be used to simplify reference to an object within a data frame without attaching.

B.4.6   Attributes and classes

Many objects have a set of associated attributes (such as names of variables, dimensions, or classes) that can be displayed or sometimes changed. For example, we can find the dimension of the matrix defined earlier.

```
attributes(A)
```

```
$dim
[1] 2 3
```

Other types of objects within R include `lists` (ordered objects that are not necessarily rectangular), regression models (objects of class `lm`), and formulae (e.g., y ∼ x1 + x2).

R supports *object-oriented programming* (see `help(UseMethod)`). As a result, objects in R have an associated *class* attribute, which changes the default behavior for some operations on that object. Many functions (called *generics*) have special capabilities when applied to objects of a particular class. For example, when `summary()` is applied to an `lm` object, the `summary.lm()` function is called. Conversely, `summary.aov()` is called when an `aov` object is given as argument. These class-specific implementations of generic functions are called *methods*. The `class()` function returns the classes to which an object belongs, while the `methods()` function displays all of the classes supported by a generic function.

```
head(methods(summary))
```

```
[1] "summary,ANY-method"            "summary,DBIObject-method"
[3] "summary,diagonalMatrix-method" "summary,MySQLConnection-method"
[5] "summary,MySQLDriver-method"    "summary,MySQLResult-method"
```

Objects in R can belong to multiple classes, although those classes need not be nested. As noted above, generic functions are *dispatched* according the class attribute of each object. Thus, in the example below we create the `tbl` object, which belongs to multiple classes. When the `print()` function is called on `tbl`, R looks for a method called `print.tbl_df()`. If no such method is found, R looks for a method called `print.tbl()`. If no such method is found, R looks for a method called `print.data.frame()`. This process continues until a suitable method is found. If there is none, then `print.default()` is called.

```
tbl <- as.tbl(ds)
class(tbl)
```

```
[1] "tbl_df"    "tbl"       "data.frame"
```

```
print(tbl)
```

```
# A tibble: 6   2
      x     y
  <dbl> <dbl>
1     5    11
2     7    11
3     9    11
4    13    11
5    -4    11
6     8    11
```

```
print.data.frame(tbl)
```

```
  x  y
1 5 11
2 7 11
```

```
3  9 11
4 13 11
5 -4 11
6  8 11

print.default(tbl)

$x
[1]  5  7  9 13 -4  8

$y
[1] 11 11 11 11 11 11

attr(,"class")
[1] "tbl_df"      "tbl"         "data.frame"
```

There are a number of functions that assist with learning about an object in R. The `attributes()` command displays the attributes associated with an object. The `typeof()` function provides information about the underlying data structure of objects (e.g., logical, integer, double, complex, character, and list). The `str()` function displays the structure of an object, and the `mode()` function displays its storage mode. For data frames, the `glimpse()` function provides a useful summary of each variable.

A few quick notes on specific types of objects are worth relating here:

- A vector is a one-dimensional array of items of the same data type. There are six basic data types that a vector can contain: `logical`, `character`, `integer`, `double`, `complex`, and `raw`. Vectors have a `length()` but not a `dim()`. Vectors can have—but needn't have—`names()`.

- A `factor` is a special type of vector for categorical data. A factor has `level()`s. We change the reference level of a factor with `relevel()`. Factors are stored internally as integers that correspond to the id's of the factor levels.

Pro Tip: Factors can be problematic and their use is discouraged since they can complicate some aspects of data wrangling. A number of R developers have encouraged the use of the `stringsAsFactors = FALSE` option.

- A `matrix` is a two-dimensional array of items of the same data type. A matrix has a `length()` that is equal to `nrow()` times `ncol()`, or the product of `dim()`.

- A `data.frame` is a `list` of vectors of the same length. This is like a matrix, except that columns can be of different data types. Data frames always have `names()` and often have `row.names()`.

Pro Tip: Do not confuse a `factor` with a `character` vector.

Note that data sets typically have class `data.frame`, but are of type `list`. This is because, as noted above, R stores data frames as special types of lists—a list of several vectors having the same length, but possibly having different types.

```
class(mtcars)
```

```
[1] "data.frame"
```

```
typeof(mtcars)
```

```
[1] "list"
```

Pro Tip: If you ever get confused when working with data frames and matrices, remember that a data.frame is a list, whereas a matrix is more like a vector.

B.4.7 Options

The options() function in R can be used to change various default behaviors. For example, the digits argument controls the number of digits to display in output. The current options are returned when options() is called, to allow them to be restored. The command help(options) lists all of the settable options.

B.4.8 Functions

Fundamental actions within R are carried out by calling *functions* (either built-in or user defined—see Appendix C for guidance on the latter). Multiple *arguments* may be given, separated by commas. The function carries out operations using the provided arguments and returns values (an object such as a vector or list) that are displayed (by default) or which can be saved by assignment to an object.

As an example, the quantile() function takes a numeric vector and returns the minimum, 25th percentile, median, 75th percentile, and maximum of the values in that vector. However, if an optional vector of quantiles is given, those quantiles are calculated instead.

```
vals <- rnorm(1000) # generate 1000 standard normals
quantile(vals)
```

```
    0%     25%     50%     75%    100%
-3.206  -0.665   0.023   0.757   2.891
```

```
quantile(vals, c(.025, .975))
```

```
 2.5% 97.5%
-2.11  1.97
```

```
# Return values can be saved for later use.
res <- quantile(vals, c(.025, .975))
res[1]
```

```
 2.5%
-2.11
```

Arguments (usually named) are available for many functions. The documentation specifies the default action if named arguments are not specified. For the quantile() function, there is a type argument that allows specification of one of nine algorithms for calculating quantiles.

```
res <- quantile(vals, probs = c(.025, .975), type = 3)
res

 2.5% 97.5%
-2.13  1.97
```

Some functions allow a variable number of arguments. An example is the paste() function. The calling sequence is described in the documentation as follows.

```
paste(..., sep = " ", collapse = NULL)
```

To override the default behavior of a space being added between elements output by paste(), the user can specify a different value for sep.

B.5 Add-ons: Packages

B.5.1 Introduction to packages

Additional functionality in R is added through packages, which consist of functions, data sets, examples, vignettes, and help files that can be downloaded from CRAN. The function install.packages() can be used to download and install packages. Alternatively, RStudio provides an easy-to-use Packages tab to install and load packages.

In many cases, add-on packages (see Appendix A) need to be installed prior to running the examples in this book. Packages that are not on CRAN can be installed using the install_github() function in the devtools package.

```
install.packages("mdsr")     # CRAN version
devtools::install_github("beanumber/mdsr")     # dev version
```

The library() function will load an installed package. For example, to install and load Frank Harrell's Hmisc() package, two commands are needed:

```
install.packages("Hmisc")
library(Hmisc)
```

If a package is not installed, running the library() command will yield an error. Here we try to load the Zelig package (which has not been installed):

```
> library(Zelig)
Error in library(Zelig) : there is no package called 'Zelig'
```

To rectify the problem, we install the package from CRAN.

```
> install.packages("Zelig")
trying URL 'https://cran.rstudio.com/macosx/contrib/3.3/Zelig_5.0-12.tgz'
Content type 'application/x-gzip' length 1398050 bytes (1.3 MB)
==================================================
downloaded 1.3 Mb
```

```
library(Zelig)
```

Packages can be installed from other repositories (e.g., OmegaHat or Bioconductor) by specifying the repository using the `repos` argument, or in the case of GitHub, using the `install_github()` function from the `devtools` package.

The `require()` function will test whether a package is available—this will load the library if it is installed, and generate a warning message if it is not (as opposed to `library()`, which will return an error).

B.5.2   CRAN task views

The *Task Views* on CRAN (`http://cran.r-project.org/web/views`) are a very useful resource for finding packages. These are curated listings of relevant packages within a particular application area (such as multivariate statistics, psychometrics, or survival analysis). Table B.2 displays the task views available as of 2016.

B.5.3   Session information

The `sessionInfo()` function provides version information about R as well as details of loaded packages.

```
> sessionInfo()
R version 3.3.2 (2016-10-31)
Platform: x86_64-apple-darwin13.4.0 (64-bit)
Running under: macOS Sierra 10.12.1
locale:
[1] en_US.UTF-8/en_US.UTF-8/en_US.UTF-8/C/en_US.UTF-8/en_US.UTF-8
attached base packages:
[1] methods    stats     graphics  grDevices utils     datasets  base
```

The `R.Version()` function provides access to components of the version and platform status.

```
str(R.Version())

List of 14
 $ platform      : chr "x86_64-apple-darwin13.4.0"
 $ arch          : chr "x86_64"
 $ os            : chr "darwin13.4.0"
 $ system        : chr "x86_64, darwin13.4.0"
 $ status        : chr ""
 $ major         : chr "3"
 $ minor         : chr "3.2"
 $ year          : chr "2016"
 $ month         : chr "10"
 $ day           : chr "31"
 $ svn rev       : chr "71607"
 $ language      : chr "R"
 $ version.string: chr "R version 3.3.2 (2016-10-31)"
 $ nickname      : chr "Sincere Pumpkin Patch"
```

Task View	Subject
Bayesian	Bayesian Inference
ChemPhys	Chemometrics and Computational Physics
ClinicalTrials	Clinical Trial Design, Monitoring, and Analysis
Cluster	Cluster Analysis and Finite Mixture Models
DifferentialEquations	Differential Equations
Distributions	Probability Distributions
Econometrics	Econometrics
Environmetrics	Analysis of Ecological and Environmental Data
ExperimentalDesign	Design of Experiments (DoE) and Analysis of Experimental Data
ExtremeValue	Extreme Value Analysis
Finance	Empirical Finance
Genetics	Statistical Genetics
gR	gRaphical Models in R
Graphics	Graphic Displays and Dynamic Graphics and Graphic Devices and Visualization
HighPerformanceComputing	High-Performance and Parallel Computing with R
MachineLearning	Machine Learning and Statistical Learning
MedicalImaging	Medical Image Analysis
MetaAnalysis	Meta-Analysis
Multivariate	Multivariate Statistics
NaturalLanguageProcessing	Natural Language Processing
NumericalMathematics	Numerical Mathematics
OfficialStatistics	Official Statistics and Survey Methodology
Optimization	Optimization and Mathematical Programming
Pharmacokinetics	Analysis of Pharmacokinetic Data
Phylogenetics	Phylogenetics, Especially Comparative Methods
Psychometrics	Psychometric Models and Methods
ReproducibleResearch	Reproducible Research
Robust	Robust Statistical Methods
SocialSciences	Statistics for the Social Sciences
Spatial	Analysis of Spatial Data
SpatioTemporal	Handling and Analyzing Spatio-Temporal Data
Survival	Survival Analysis
TimeSeries	Time Series Analysis
WebTechnologies	Web Technologies and Services

Table B.2: A complete list of CRAN task views.

Sometimes it is desirable to remove a package (B.5.1) from the workspace. For example, a package might define a function with the same name as an existing function. Packages can be detached using the syntax detach(package:PKGNAME), where PKGNAME is the name of the package. Objects with the same name that appear in multiple places in the environment can be accessed using the location::objectname syntax. As an example, to access the mean() function from the base package, the user would specify base::mean() instead of mean().

The names of all variables within a given data set (or more generally for sub-objects within an object) are provided by the names() command. The names of all objects defined within an R session can be generated using the `objects()` and `ls()` commands, which return a vector of character strings. RStudio includes an `Environment` tab that lists all the objects in the current environment.

The `print()` and `summary()` functions return the object or summaries of that object, respectively. Running `print(object)` at the command line is equivalent to just entering the name of the object, i.e. `object`.

B.5.4   Packages and name conflicts

Different package authors may choose the same name for functions that exist within base R (or within other packages). This will cause the other function or object to be *masked*. This can sometimes lead to confusion, when the expected version of a function is not the one that is called. The `find()` function can be used to determine where in the environment (workspace) a given object can be found.

```
find("mean")
```

```
[1] "package:mosaic" "package:Matrix" "package:base"
```

As an example where this might be useful, there are functions in the `base` and `Hmisc` packages called `units()`. The find command would display both (in the order in which they would be accessed).

```
library(Hmisc)
find("units")
```

```
[1] "package:Hmisc" "package:base"
```

When the `Hmisc` package is loaded, the `units()` function from the `base` package is masked and would not be used by default. To specify that the version of the function from the base package should be used, prefix the function with the package name followed by two colons: `base::units()`. The `conflicts()` function reports on objects that exist with the same name in two or more places on the search path.

B.5.5   Maintaining packages

The `update.packages()` function should be run periodically to ensure that packages are up-to-date (see also `packageVersion()`). The `packrat` package provides a comprehensive dependency system for R. This functionality can be extremely helpful to support reproducible analysis (see Appendix D), as the exact set of packages used for an analysis can be identified and accessed in a project. Support for `packrat` is built into RStudio.

As of December 2016, there were nearly 10,000 packages available from CRAN. This represents a tremendous investment of time and code by many developers [78]. While each of these has met a minimal standard for inclusion, it is important to keep in mind that packages in R are created by individuals or small groups, and not endorsed by the R core group. As a result, they do not necessarily undergo the same level of testing and quality assurance that the core R system does.

B.5.6   Installed libraries and packages

Running the command `library(help="PKGNAME")` will display information about an installed package. Alternatively, the `Packages` tab in RStudio can be used to list, install, and update packages. Entries in the book that utilize packages include a line specifying how to access that library (e.g., `library(mosaic)`). More information about packages used in this book can be found in Appendix A.

B.6 Further resources

Hadley Wickham's *Advanced R* book [220] (`http://adv-r.had.co.nz`) is probably the best source for learning more about how R works. Extensive resources and documentation can be found at the Comprehensive R Archive Network (CRAN).

B.7 Exercises

Exercise B.1

A user has typed the following commands into the RStudio console.

```
obj1 <- 2:10
obj2 <- c(2, 5)
obj3 <- c(TRUE, FALSE)
obj4 <- 42
```

What values are returned by the following commands?

```
obj1 * 10
obj1[2:4]
obj1[-3]
obj1 + obj2
obj1 * obj3
obj1 + obj4
obj2 + obj3
sum(obj2)
sum(obj3)
```

Exercise B.2

A user has typed the following commands into the RStudio console.

```
a <- c(10, 15)
b <- c(TRUE, FALSE)
c <- c("happy", "sad")
```

What do each of the following commands return? Describe the class of the object as well as its value.

```
data.frame(a, b, c)
cbind(a, b)
rbind(a, b)
cbind(a, b, c)
list(a, b, c)[[2]]
```

Exercise B.3

A user has typed the following commands into the RStudio console.

```
mylist <- list(x1="sally", x2=42, x3=FALSE, x4=1:5)
```

What values do each of the following commands return?

```
is.list(mylist)
names(mylist)
length(mylist)
mylist[[2]]
mylist[["x1"]]
mylist$x2
length(mylist[["x4"]])
class(mylist)
typeof(mylist)
class(mylist[[4]])
typeof(mylist[[3]])
```

Exercise B.4

The following code undertakes some data analysis using the HELP (Health Evaluation and Linkage to Primary Care) trial.

```
library(mosaic)
ds <-
   read.csv("http://nhorton.people.amherst.edu/r2/datasets/helpmiss.csv")
summarise(group_by(select(filter(mutate(ds,
  sex = ifelse(female==1, "F", "M")), !is.na(pcs)), age, pcs, sex),
  sex), meanage=mean(age), meanpcs=mean(pcs),n=n())
```

Describe in words what computations are being done. Using the "pipe" notation, translate this code into a more readable version.

Exercise B.5

The following concepts should have some meaning to you: package, function, command, argument, assignment, object, object name, data frame, named argument, quoted character string. Construct an example of R commands that make use of at least four of these. Label which part of your example R command corresponds to each.

Exercise B.6

Which of these kinds of names should be wrapped with quotation marks when used in R?

1. function name

2. file name

3. the name of an argument in a named argument

4. object name

Exercise B.7

What's wrong with this statement?

```
help(NHANES, package <- "NHANES")
```

Exercise B.8

Consult the documentation for CPS85 in the mosaicData package to determine the meaning of CPS.

Exercise B.9

For each of the following assignment statements, describe the error (or note why it does not generate an error).

```
result1 <- sqrt 10
result2 <-- "Hello to you!"
3result <- "Hello to you"
result4 <- "Hello to you
result5 <- date()
```

Appendix C

Algorithmic thinking

C.1 Introduction

Algorithmic thinking can be defined as a set of abilities that are related to constructing and understanding algorithms [81]:

1. the ability to analyze a given problem

2. the ability to precisely specify a problem

3. the ability to find the basic actions that are adequate to solve a problem

4. the ability to construct a correct algorithm to a given problem using basic actions

5. the ability to think about all possible special and normal cases of a problem

6. the ability to improve the efficiency of an algorithm

These important capacities are a necessary but not sufficient component of "computational thinking" and data science.

It is critical that data scientists have the skills to break problems down and code solutions in a flexible and powerful computing environment using *functions*. In this book we focus on the use of R for this task (although other environments such as Python have many adherents and virtues). In this appendix, we presume a basic background in R to the level of Appendix B.

C.2 Simple example

We begin with an example that creates a simple function to complete a statistical task (calculate a confidence interval for an estimate). In R, a new function is defined by the syntax shown below, using the keyword `function`. This creates a new function called `new_function()` in the workspace that takes two arguments (`argument1` and `argument2`). The *body* is made up of a series of commands (or expressions), typically separated by line breaks and enclosed in curly braces.

```
new_function <- function(argument1, argument2) {
  R expression
  another R expression
}
```

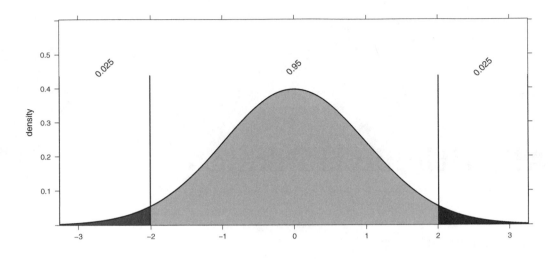

Figure C.1: Illustration of the location of the critical value for a 95% confidence interval for a mean. The critical value of 2.01 corresponds to the location in the t-distribution with 50 degrees of freedom, for which 2.5% of the distribution lies above it.

Here, we create a function to calculate the estimated confidence interval (CI) for a mean, using the formula $\bar{X} \pm t^* s/\sqrt{n}$, where t^* is the appropriate t-value for that particular confidence level. As an example, for a 95% interval with 50 degrees of freedom (equivalent to $n = 51$ observations) the appropriate value of t^* can be calculated using the cdist() function from the mosaic package. This computes the quantiles of the t-distribution between which 95% of the distribution lies. A graphical illustration is shown in Figure C.1.

```
cdist("t", 0.95, df = 50)

[1] -2.01  2.01
```

```
xqt(c(0.025, 0.975), df = 50)
```

```
[1] -2.01  2.01
```

We see that the value is slightly larger than 2. Note that since by construction our confidence interval will be centered around the mean, we want the critical value that corresponds to having 95% of the distribution in the middle.

We will write a function to compute a t-based confidence interval for a mean from scratch. We'll call this function ci_calc(), and it will take a numeric vector x as its first argument, and an optional second argument alpha, which will have a default value of 0.95.

```
# calculate a t confidence interval for a mean
ci_calc <- function(x, alpha = 0.95) {
    samp_size <- length(x)
    t_star <- qt(1 - ((1 - alpha)/2), df = samp_size - 1)
    my_mean <- mean(x)
```

```
    my_sd <- sd(x)
    se <- my_sd/sqrt(samp_size)
    me <- t_star * se
    return(list(ci_vals = c(my_mean - me, my_mean + me),
              alpha = alpha))
}
```

Here the appropriate quantile of the t-distribution is calculated using the `qt()` function, and the appropriate confidence interval is calculated and returned as a list. In this example, we explicitly `return()` a `list` of values. If no return statement is provided, the result of the last expression evaluation is returned by default.

The function has been stored in the object `ci_calc()`. Once created, it can be used like any other function. For example, the expression below will print the CI and confidence level for the object `x1` (a set of 100 random normal variables with mean 0 and standard deviation 1).

```
x1 <- rnorm(100, mean = 0, sd = 1)
ci_calc(x1)

$ci_vals
[1] -0.0867  0.2933

$alpha
[1] 0.95
```

The order of arguments in R matters, since if arguments are not named when a function is called, they are assumed to correspond to the order of the arguments as the function is defined. To see that order, check the documentation, use the `args()` function, or look at the code of the function itself.

```
?ci_calc # won't work because we haven't written any documentation
args(ci_calc)
ci_calc
```

Pro Tip: Consider creating an R package for commonly used functions that you develop so that they can be more easily documented, tested, and reused.

Since we provided only one unnamed argument (`x1`), R passed the value `x1` to the argument `x` of `ci_calc()`. Since we did not specify a value for the `alpha` argument, the default value of `0.95` was used.

User-defined functions nest just as pre-existing functions do. The expression below will return the CI and report that the confidence limit is `0.9` for 100 normal random variates.

```
ci_calc(rnorm(100), 0.9)
```

To change the confidence level, we need only change the `alpha` option by specifying it as a *named argument*.

```
ci_calc(x1, alpha = 0.90)

$ci_vals
[1] -0.0557  0.2623

$alpha
[1] 0.9
```

The output is equivalent to running the command `ci_calc(x1, 0.90)` with two un-named arguments, where the arguments are matched in order. Less intuitive but equivalent would be the following call.

```
ci_calc(alpha = 0.90, x = x1)

$ci_vals
[1] -0.0557  0.2623

$alpha
[1] 0.9
```

The key take-home message is that the order of arguments is not important *if all of the arguments are named.*

Using the pipe operator introduced in Chapter 4 can avoid nesting.

```
rnorm(100, mean = 0, sd = 1) %>%
  ci_calc(alpha = 0.9)

$ci_vals
[1] -0.0175  0.2741

$alpha
[1] 0.9
```

Pro Tip: The `testthat` package can help to improve your functions by writing testing routines to check that the function does what you expect it to.

C.3   Extended example: Law of large numbers

The *Law of large numbers* concerns the convergence of the arithmetic average of a sample to the expected value of a random variable, as the sample size increases. This is an important result in statistics, described in Section 7.2. The convergence (or lack thereof, for certain distributions) can easily be visualized.

We define a function to calculate the running average for a given vector, allowing for variates from many distributions to be generated.

```
runave <- function(n, gendist, ...) {
  x <- gendist(n, ...)
  avex <- numeric(n)
  for (k in 1:n) {
```

```
      avex[k] <- mean(x[1:k])
    }
    return(data.frame(x, avex, n = 1:length(avex)))
}
```

The `runave()` function takes at a minimum two arguments: a sample size `n` and function (see B.4.8) denoted by `gendist` that is used to generate samples from a distribution.

Pro Tip: Note that there are more efficient ways to write this function using vector operations (see for example the `cumsum()` function).

Other options for the function can be specified, using the . . . (dots) syntax. This syntax allows additional options to be provided to functions that might be called downstream. For example, the dots are used to specify the degrees of freedom for the samples generated for the t-distribution in the next code block.

Recall that because the expectation of a Cauchy random variable is undefined [178], the sample average does not converge to the center (see related discussion in Section 7.2). The variance of a Cauchy random variable is also infinite (does not exist). Such a distribution arises when ratios are calculated. Conversely, a t-distribution with more than 1 degree of freedom (a distribution with less of a heavy tail) does converge to the center. For comparison, the two distributions are displayed in Figure C.2.

```
plotDist("t", params = list(df = 4), xlim = c(-5, 5), lty = 2, lwd = 3)
plotDist("cauchy", xlim = c(-10, 10), lwd = 3, add = TRUE)
```

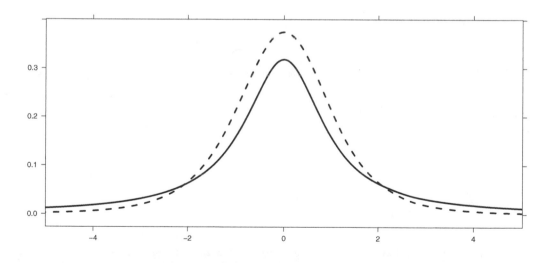

Figure C.2: Cauchy distribution (solid line) and t-distribution with 4 degrees of freedom (dashed line).

To make sure we can replicate our results for this simulation, we first set a fixed seed (see Section 10.7). Next, we generate some data, using our new `runave()` function.

```
nvals <- 1000
set.seed(1984)
```

```
sims <- bind_rows(
  runave(nvals, rt, 4),
  runave(nvals, rcauchy)) %>%
  mutate(dist = rep(c("t4", "cauchy"), each = nvals))
```

In this example, the value 4 is provided to the `rt()` function using the ... mechanism. This is used to specify the `df` argument to `rt()`. The results are plotted in Figure C.3. While the running average of the t-distribution converges to the true mean of zero, the running average of the Cauchy distribution does not.

```
ggplot(data = sims, aes(x = n, y = avex, color = dist)) +
  geom_hline(yintercept = 0, color = "black", linetype = 2) +
  geom_line() + geom_point() +
  ylab("mean") + xlab("sample size") + xlim(c(0, 600))
```

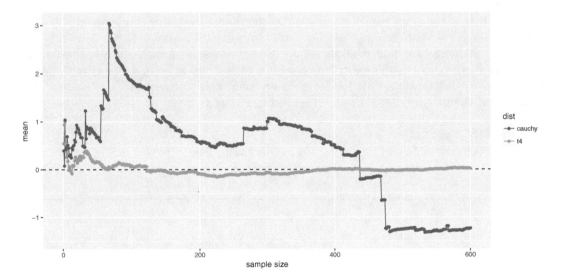

Figure C.3: Running average for t-distribution with four degrees of freedom and a Cauchy random variable (equivalent to a t-distribution with one degree of freedom). Note that while the former converges, the latter does not.

C.4   Non-standard evaluation

When evaluating expressions, R searches for objects in an *environment*. The most general environment is the *global environment*, the contents of which are displayed in the environment tab in RStudio or through the `ls()` command. When you try to access an object that cannot be found in the global environment, you get an error.

We will use a subset of the NHANES data frame from the NHANES package to illustrate a few of these subtleties. This data frame has a variety of data types.

```
library(NHANES)
NHANESsubset <- NHANES %>%
  select(ID, SurveyYr, Gender, Age, AgeMonths, Race1, Poverty)
NHANESsubset
```

```
# A tibble: 10,000  7
      ID SurveyYr Gender   Age AgeMonths  Race1 Poverty
   <int>   <fctr> <fctr> <int>     <int> <fctr>   <dbl>
1  51624  2009_10   male    34       409  White    1.36
2  51624  2009_10   male    34       409  White    1.36
3  51624  2009_10   male    34       409  White    1.36
4  51625  2009_10   male     4        49  Other    1.07
5  51630  2009_10 female    49       596  White    1.91
6  51638  2009_10   male     9       115  White    1.84
7  51646  2009_10   male     8       101  White    2.33
8  51647  2009_10 female    45       541  White    5.00
9  51647  2009_10 female    45       541  White    5.00
10 51647  2009_10 female    45       541  White    5.00
# ... with 9,990 more rows
```

Consider the differences between trying to access the ID variable each of the three ways shown below. In the first case, we are simply creating a character vector of length one that contains the single string ID. The second command causes R to search the global environment for an object called ID—which does not exist. In the third command, we correctly access the ID variable within the NHANESsubset data frame, which *is* accessible in the global environment. These are different examples of how R uses *scoping* to identify objects.

```
"ID"    # string variable

[1] "ID"

ID      # generates an error

Error in eval(expr, envir, enclos):  object 'ID' not found

NHANESsubset$ID %>% summary()      # access within a data frame

   Min. 1st Qu.  Median    Mean 3rd Qu.    Max.
  51600   56900   62200   61900   67000   71900
```

How might this be relevant? Notice that several of the variables in NHANESsubset are factors. We might want to convert each of them to type character. Typically, we would do this using the mutate() command that we introduced in Chapter 4.

```
NHANESsubset %>% mutate(SurveyYr = as.character(SurveyYr)) %>%
  select(ID, SurveyYr) %>%
  glimpse()

Observations: 10,000
Variables: 2
$ ID       <int> 51624, 51624, 51624, 51625, 51630, 51638, 51646, 5164...
$ SurveyYr <chr> "2009_10", "2009_10", "2009_10", "2009_10", "2009_10"...
```

Note however, that in this construction we have to know the name of the variable we wish to convert (i.e., SurveyYr) and list it explicitly. This is unfortunate if the goal is to automate our data wrangling.

If we tried instead to set the name of the column (i.e., `SurveyYr`) to a variable (i.e., varname) and use that variable to change the names, it would not work as intended. In this case, rather than changing the data type of `SurveyYr`, we have created a new variable called varname that is a character vector of the values `SurveyYr`.

```
varname <- "SurveyYr"
mutate(NHANESsubset, varname = as.character(varname)) %>%
  select(ID, SurveyYr, varname) %>%
  glimpse()

Observations: 10,000
Variables: 3
$ ID       <int> 51624, 51624, 51624, 51625, 51630, 51638, 51646, 5164...
$ SurveyYr <fctr> 2009_10, 2009_10, 2009_10, 2009_10, 2009_10, 2009_10...
$ varname  <chr> "SurveyYr", "SurveyYr", "SurveyYr", "SurveyYr", "Surv...
```

This behavior is a consequence of a feature of the R language called *non-standard evaluation* (NSE). This approach provides a principled way to work with expressions in functions, and is used extensively in the `dplyr` package. The `dplyr` functions `mutate()` (and `select()`) use non-standard evaluation—this is why R is able to locate `SurveyYr`, even though there is no object called `SurveyYr` in the global environment. In this case, `mutate()` knows to look for `SurveyYr` within the `NHANESsubset` data frame.

Each `dplyr` verb has a counterpart that does not use NSE—these functions all have the same name but end with an underscore. For example, consider the following ways to summarize the `ID` variable in `NHANESsubset`. In the first case, we use `select()` in the familiar way, using NSE. In the second example, we use the `select_()` function to access the `ID` variable without using NSE. Here, the tilde means that the second argument to `select_()` is a formula.

```
select(NHANESsubset, ID) %>% summary()

      ID
 Min.    :51624
 1st Qu.:56904
 Median :62160
 Mean    :61945
 3rd Qu.:67039
 Max.    :71915

select_(NHANESsubset, ~ID) %>% summary()

      ID
 Min.    :51624
 1st Qu.:56904
 Median :62160
 Mean    :61945
 3rd Qu.:67039
 Max.    :71915
```

We compare the actions of four variants of these functions below. In the call to `mutate()`, two variables are created using NSE. The first is a naïve attempt to pass the variable name (`SurveyYr`) as a character that results in the assignment of a vector consisting entirely of

SurveyYr. The second definition gives the desired result, but does not involve the `varname` argument, and thus won't work for any other variable. The call to `mutate_()` uses standard evaluation. Notice that unlike in the first expression, the correct values are returned, but the new variable (var_se_wrong_type) is still a factor. Only after the second expression do we obtain the desired result.

```
factor_to_char <- function(data, varname) {
    data %>%
    mutate(var_nse_wrong_values = varname,
           var_nse_hard_coded = as.character(SurveyYr)) %>%
    mutate_(var_se_wrong_type = varname,
            var_se_correct = ~as.character(var_se_wrong_type))
}
factor_to_char(NHANESsubset, "SurveyYr") %>%
  select(SurveyYr, contains("var")) %>%
  glimpse()
```

```
Observations: 10,000
Variables: 5
$ SurveyYr             <fctr> 2009_10, 2009_10, 2009_10, 2009_10, 2009...
$ var_nse_wrong_values <chr> "SurveyYr", "SurveyYr", "SurveyYr", "Surv...
$ var_nse_hard_coded   <chr> "2009_10", "2009_10", "2009_10", "2009_10...
$ var_se_wrong_type    <fctr> 2009_10, 2009_10, 2009_10, 2009_10, 2009...
$ var_se_correct       <chr> "2009_10", "2009_10", "2009_10", "2009_10...
```

Only the last approach correctly modifies the SurveyYr variable. The process also requires two steps. How might we create a function to find all of the factors in NHANESsubset and modify them in place? To do this, we want to employ algorithmic thinking by breaking the problem down into small pieces. First, we will use `sapply()` to identify the variables in NHANESsubset that are factors.

```
is_factor <- sapply(NHANESsubset, class) == "factor"
sum(is_factor)
```

```
[1] 3
```

We find that three of the seven variables are factors. We will then store the names of these variables.

```
var_names <- names(NHANESsubset[is_factor])
```

Finally, we can use the `mutate_at()` function to apply an arbitrary function (in this case, `as.character()`) to each of those variables. Note that there are no factors in the resulting data frame.

```
NHANESsubset %>%
  mutate_at(.funs = as.character, .cols = var_names)
```

```
# A tibble: 10,000  7
      ID SurveyYr Gender   Age AgeMonths Race1 Poverty
   <int>    <chr>  <chr> <int>     <int> <chr>   <dbl>
```

```
1  51624  2009_10    male    34      409 White    1.36
2  51624  2009_10    male    34      409 White    1.36
3  51624  2009_10    male    34      409 White    1.36
4  51625  2009_10    male     4       49 Other    1.07
5  51630  2009_10  female    49      596 White    1.91
6  51638  2009_10    male     9      115 White    1.84
7  51646  2009_10    male     8      101 White    2.33
8  51647  2009_10  female    45      541 White    5.00
9  51647  2009_10  female    45      541 White    5.00
10 51647  2009_10  female    45      541 White    5.00
# ... with 9,990 more rows
```

For a different example, consider the task of squaring every integer-valued column. We could accomplish this in a manner similar to the above using a user-defined function.

```
convert_types <- function(data, type, convert_fun) {
  col_idx <- sapply(data, class) == type
  mutate_at(data, .funs = convert_fun, .cols = names(data[col_idx]))
}
convert_types(NHANESsubset, type = "integer", convert_fun = function(x) x^2)
```

```
# A tibble: 10,000  7
           ID SurveyYr Gender    Age AgeMonths  Race1 Poverty
        <dbl>    <fctr> <fctr> <dbl>     <dbl> <fctr>   <dbl>
1  2.67e+09  2009_10    male  1156    167281  White    1.36
2  2.67e+09  2009_10    male  1156    167281  White    1.36
3  2.67e+09  2009_10    male  1156    167281  White    1.36
4  2.67e+09  2009_10    male    16      2401  Other    1.07
5  2.67e+09  2009_10  female  2401    355216  White    1.91
6  2.67e+09  2009_10    male    81     13225  White    1.84
7  2.67e+09  2009_10    male    64     10201  White    2.33
8  2.67e+09  2009_10  female  2025    292681  White    5.00
9  2.67e+09  2009_10  female  2025    292681  White    5.00
10 2.67e+09  2009_10  female  2025    292681  White    5.00
# ... with 9,990 more rows
```

C.5   Debugging and defensive coding

R and RStudio include extensive support for debugging functions and code. Calling the browser() function in the body of a function will cause execution to stop and set up an R interpreter. Once at the browser prompt, the analyst can enter either commands (such as c to continue execution, f to finish execution of the current function, n to evaluate the next statement (without stepping into function calls), s to evaluate the next statement (stepping into function calls), Q to exit the browser, or help to print this list. Other commands entered at the browser are interpreted as R expressions to be evaluated (the function ls() lists available objects). Calls to the browser can be set using the debug() or debugonce() functions (and turned off using the undebug() function). RStudio includes a debugging mode that is displayed when debug() is called.

Adopting *defensive coding* techniques is always recommended: They will tend to identify problems early and minimize errors. The try() function can be used to evaluate an expression while allowing for error recovery. The stop() function can be used to stop evaluation of the current expression and execute an error action (typically displaying an error message). More flexible testing is available in the assertthat package.

Let's revisit the ci_calc() function we defined to calculate a confidence interval. How might we make this more robust? We can begin by confirming that the calling arguments are sensible.

```
library(assertthat)
# calculate a t confidence interval for a mean
ci_calc <- function(x, alpha = 0.95) {
  if (length(x) < 2) {
    stop("Need to provide a vector of length at least 2.\n")
  }
  if (alpha < 0 | alpha > 1) {
    stop("alpha must be between 0 and 1.\n")
  }
  assert_that(is.numeric(x))
  samp_size <- length(x)
  t_star <- qt(1 - ((1 - alpha)/2), df = samp_size - 1)
  my_mean <- mean(x)
  my_sd <- sd(x)
  se <- my_sd / sqrt(samp_size)
  me <- t_star * se
  return(list(ci_vals = c(my_mean - me, my_mean + me),
              alpha = alpha))
}
ci_calc(1)    # will generate error
```

```
Error in ci_calc(1):  Need to provide a vector of length at least 2.
```

```
ci_calc(1:3, alpha = -1)    # will generate error
```

```
Error in ci_calc(1:3, alpha = -1):  alpha must be between 0 and 1.
```

```
ci_calc(c("hello", "goodbye"))    # will generate error
```

```
Error:  x is not a numeric or integer vector
```

C.6 Further resources

More examples of functions can be found in Chapter 10. The American Statistical Association's *Guidelines for Undergraduate Programs in Statistics* [8] stress the importance of algorithmic thinking (see also [149]). Texts by Rizzo [175] and Wickham [220] provide useful reviews of statistical computing. A variety of online resources are available to describe how to create R packages and to deploy them on GitHub (see for example http://kbroman.org/pkg_primer). The testthat package is helpful in structuring more extensive unit tests for functions. The dplyr package documentation includes a vignette detailing its use of the lazyeval package for performing non-standard evaluation.

C.7   Exercises

Exercise C.1

Write another function called `grab_name()` that, when given a name and a year as an argument, returns the rows from the `babynames` data frame in the `babynames` package that match that name for that year (and returns an error if that name and year combination does not match any rows). Run the function once with the arguments `Ezekiel` and `1883` and once with `Ezekiel` and `1983`.

Exercise C.2

Write a function called `count_name()` that, when given a name as an argument, returns the total number of births by year from the `babynames` data frame in the `babynames` package that match that name. The function should return one row per year that matches (and generate an error message if there are no matches). Run the function once with the argument `Ezekiel` and once with `Ezze`.

Exercise C.3

Write a function called `count_na()` that, when given a vector as an argument, will count the number of NAs in that vector. Count the number of missing values in the `SEXRISK` variable in the `HELPfull` data frame in the `mosaicData` package.

Exercise C.4

Apply `count_na()` to the columns of the `Teams` data frame from the `Lahman` package. How many of the columns have missing data?

Exercise C.5

Write a function called `cum_min()` that, when given a vector as an argument, returns the cumulative minimum of that vector. Compare the result of your function to the built-in `cummin()` function for the vector `c(4, 7, 9, -2, 12)`.

Exercise C.6

Write a function called `prop_cancel()` that takes as arguments a month number and destination airport and returns the proportion of flights missing arrival delay for each day to that destination. Apply this function to the `nycflights13` package for February and Atlanta airport (`ATL`) and again with an invalid month number.

Exercise C.7

Write a function called `map_negative()` that takes as arguments a data frame and the name of a variable and returns that data frame with the negative values of the variable replaced by zeroes. Apply this function the `cyl` variable in the `mtcars` data set.

Exercise C.8

Benford's law concerns the frequency distribution of leading digits from numerical data. Write a function that takes a vector of numbers and returns the empirical distribution of the first digit. Apply this function to data from the `corporate.payment` data set in the `benford.analysis` package.

Appendix D

Reproducible analysis and workflow

The notion that scientific findings can be confirmed repeatedly through *replication* is fundamental to the centuries-old paradigm of science. The underlying logic is that if you have identified a truth about the world, that truth should persist upon further investigation by other observers. In the physical sciences, there are two challenges in replicating a study: replicating the experiment itself, and *reproducing* the subsequent data analysis that led to the conclusion. More concisely, replicability means that different people get the same results with *different* data. Reproducibility means that the same person (or different people) get the same results with the *same* data.

It is easy to imagine why replicating a physical experiment might be difficult, and not being physical scientists ourselves, we won't tackle those issues here. On the other hand, the latter challenge of reproducing the data analysis is most certainly our domain. It seems like a much lower hurdle to clear—isn't this just a matter of following a few steps? Upon review, for a variety of reasons many scientists are in fact tripping over even this low hurdle.

To further explicate the distinction between *replicability* and *reproducibility*, recall that scientists are legendary keepers of lab notebooks. These notebooks are intended to contain all of the information needed to carry out the study again (i.e., replicate): reagents and other supplies, equipment, experimental material, etc. Modern software tools enable scientists to carry this same ethos to data analysis: Everything needed to repeat the analysis (i.e., reproduce) should be recorded in one place.

Even better, modern software tools allow the analysis to be repeated at the push of a button. This provides a proof that the analysis being documented is in fact exactly the same as the analysis that was performed. Moreover, this capability is a boon to those generating the analysis. It enables them to draft and redraft the analysis until they get it exactly right. Even better, when the analysis is written appropriately, it's straightforward to apply the analysis to new data. Spreadsheet software, despite its popularity, is not suitable for this. Spreadsheet software references specific rows and columns of data, and so the analysis commands themselves need to be updated to conform to new data.

The "replication crisis" is a very real problem for modern science. More than ten years ago, John Ioannidis argued that "most published research findings are false." [118] More recently, the journal *Nature* ran a series of editorials bemoaning the lack of replicability in published research [67]. It now appears that even among peer-reviewed, published scientific articles, many of the findings—which are supported by experimental and statistical evidence—do not hold up under the scrutiny of replication. That is, when other researchers try to do the same study, they don't reliably reach the same conclusions.

Some of the issues leading to irreproducibility are hard to understand, let alone solve. Much of the blame involves multiplicity and the "garden of forking paths" introduced in Chapter 7. While we touch upon issues related to null hypothesis testing in Chapter 7, the focus of this chapter is on modern workflows for *reproducible data analysis*, since the ability to regenerate a set of results at a later point in time is a necessary but not sufficient condition for reproducible results.

Reproducible workflows consist of three components: a fully scriptable statistical programming environment (such as R or Python), reproducible analysis (first described as literate programming), and version control (commonly implemented using GitHub).

D.1   Scriptable statistical computing

In order for data analysis to be reproducible, all of the steps taken in the analysis have to be recorded in a linear fashion. Scriptable applications like Python, R, SAS, and Stata do this by default. Even when graphical user interfaces to these programs are used, they add the automatically generated code to the history so that it too can be recorded. Thus, the full series of commands that make up the data analysis can be recorded, reviewed, and transmitted. Contrast this with the behavior of spreadsheet applications like Microsoft Excel and Google Sheets, where it is not always possible to fully retrace one's steps.

D.2   Reproducible analysis with R Markdown

The concept of *literate programming* was introduced by Knuth decades ago [127]. His advice was:

> "Instead of imagining that our main task is to instruct a computer what to do, let us concentrate rather on explaining to human beings what we want a computer to do."

Central to this prescription is the idea that the relevant documentation for the code—which is understandable not just to the programmer, but to other human beings as well—occurs alongside the code itself. In data analysis, this is manifest as the need to have three kinds of things in one document: the code, the results of that code, and the written analysis. We belong to a growing group of people who find the `rmarkdown` [5] and `knitr` packages [239] to be an environment that is ideally suited to support a reproducible analysis workflow [23].

The `rmarkdown` and `knitr` packages use a *source file* and *output file* paradigm. This approach is common in programming, but is fundamentally different than a "what-you-see-is-what-you-get" editor like Microsoft Word or Google Drive. Code is typed into the source document, which is then rendered into an output format that is readable by anyone. The principles of literate programming stipulates that the source file should *also* be readable by anyone.

We favor the simple document markup language R Markdown [4] for most applications. An R Markdown source file can be rendered (by `knitr`, leveraging `pandoc`) into PDF, HTML, and Microsoft Word formats. The resulting document will contain the R code, the results of that code, and the analyst's written analysis.

Markdown is well-integrated with RStudio, and both LaTeX and Markdown source files can be rendered via a single-click mechanism. More details can be found in [239] and [82] as well as the CRAN reproducible research task view [128] (see also `http://yihui.name/knitr`).

As an example of how these systems work, we demonstrate a document written in the Markdown format using data from the `SwimRecords` data frame. Within RStudio, a new template R Markdown file can be generated by selecting `R Markdown` from the `New File` option on the `File` menu. This generates the dialog box displayed in Figure D.1. The default output format is HTML, but other options (PDF or Microsoft Word) are available.

Pro Tip: The R Markdown templates included with the `mosaic` package are useful to set up more appropriate defaults for graph and font size. These can be accessed using the "From Template" option when opening a new R Markdown file.

New R Markdown

| Document |
| Presentation |
| Shiny |
| From Template |

Title: Sample R Markdown example

Author: Nick Horton

Default Output Format:

◉ **HTML**
Recommended format for authoring (you can switch to PDF or Word output anytime).

○ **PDF**
PDF output requires TeX (MiKTeX on Windows, MacTeX 2013+ on OS X, TeX Live 2013+ on Linux).

○ **Word**
Previewing Word documents requires an installation of MS Word (or Libre/Open Office on Linux).

OK   Cancel

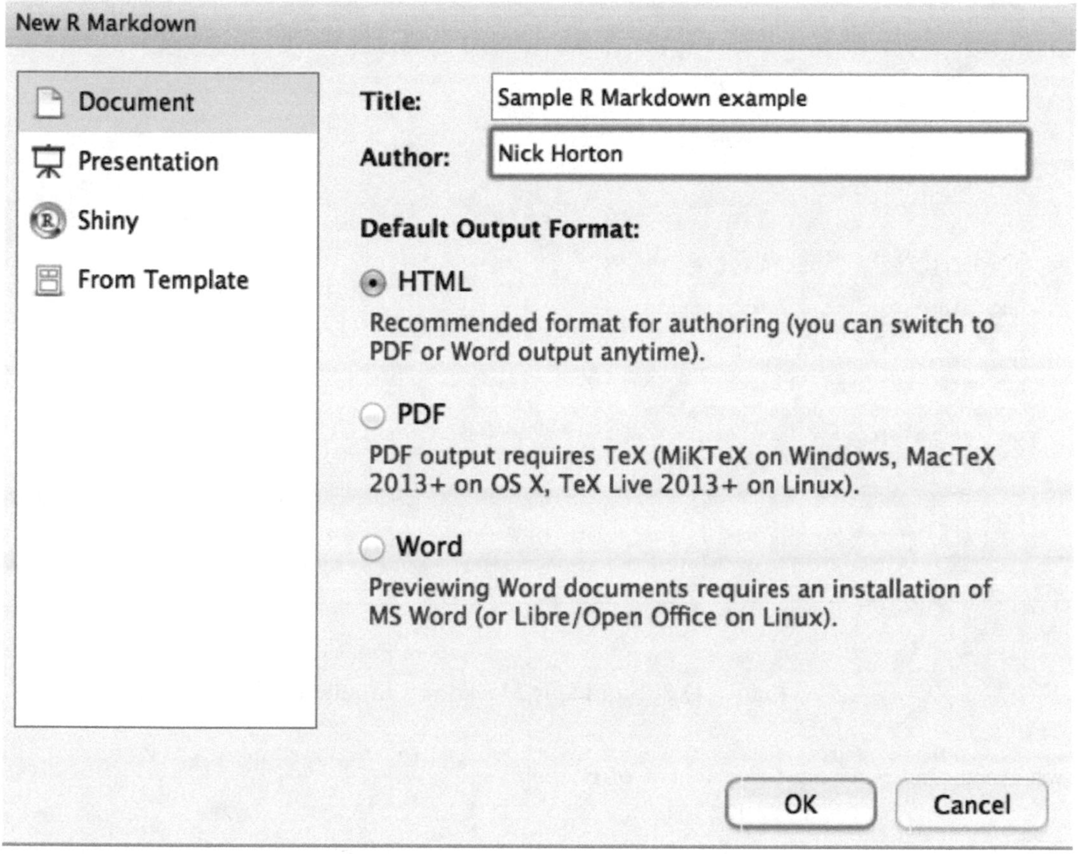

Figure D.1: Generating a new R Markdown file in RStudio

Figure D.2 displays a modified version of the default R Markdown input file. The file is given a title (`Sample R Markdown example`) with output format set by default to HTML. Simple markup (such as bolding) is added through use of the `**` characters before and after the word `Help`. Blocks of code are begun using the ' ' '{r} command and closed with a ' ' ' command (three back quotes).

The formatted output can be generated and displayed by clicking the `Knit HTML` button in RStudio, or by using the commands in the following code block, which can also be used

```
---
title: "Sample R Markdown example"
author: "Sample User"
date: "November 8, 2016"
output: html_document
---

```{r setup, include=FALSE}
knitr::opts_chunk$set(echo = TRUE)
library(mdsr)
```

R Markdown

This is an R Markdown document. Markdown is a simple formatting syntax for
authoring HTML, PDF, and MS Word documents. For more details on using R
Markdown see http://rmarkdown.rstudio.com.

When you click the **Knit** button a document will be generated that
includes both content as well as the output of any embedded R code chunks
within the document. You can embed an R code chunk like this:

```{r display}
glimpse(SwimRecords)
```

Including Plots

You can also embed plots, for example:

```{r scatplot, echo=FALSE}
ggplot(data = SwimRecords, aes(x = year, y = time)) +
  geom_point() + aes(colour = sex)  +
  stat_smooth(method = loess, se = FALSE) + theme(legend.position = "right") +
  labs(title = "100m Swimming Records over time")
```

There are n=`r nrow(SwimRecords)` rows in the Swim records dataset.

Note that the `echo = FALSE` parameter was added to the code chunk to
prevent printing of the R code that generated the plot.
```

Figure D.2: Sample R Markdown input file.

when running R without the benefit of RStudio.

```
library(rmarkdown)
render("filename.Rmd")    # creates filename.html
browseURL("filename.html")
```

The render() function extracts the R commands from a specially formatted R Markdown
input file (filename.Rmd), evaluates them, and integrates the resulting output, including
text and graphics, into an output file (filename.html). A screenshot of the results of
performing these steps on the .Rmd file displayed in Figure D.2 is displayed in Figure D.3.
render() uses the value of the output: option to determine what format to generate. If
the .Rmd file specified output: word_document, then a Microsoft Word document would
be created.

Alternatively, a PDF or Microsoft Word document can be generated in RStudio by
selecting New from the R Markdown menu, then clicking on the PDF or Word options.

RStudio also supports the creation of R Presentations using a variant of the R Markdown syntax. Instructions and an example can be found by opening a new `R presentations` document in RStudio.

D.3 Projects and version control

A useful feature of RStudio is projects. A project provides a separate workspace. Selecting a project also reorients your RStudio environment to a specified directory, in the process reorienting the Files tab, the working directory, etc. Once you start working on multiple projects, being able to switch back and forth becomes very helpful.

Given that data science has been called a "team sport," the ability to track changes to files and discuss issues in a collaborative manner is an important prerequisite to reproducible analysis. Projects can be tied to a *version control system*, such as Subversion or Git. These systems help you and your collaborators keep track of changes to files, so that you can go back in time to review changes to previous pieces of code, compare versions, and retrieve older versions as needed.

Pro Tip: While critical for collaboration with others, source code version control systems are also useful for individual projects because they document changes and maintain version histories. In such a setting, the collaboration is with your future self!

GitHub is a cloud-based implementation of Git that is tightly integrated into RStudio. It works efficiently, without cluttering your workspace with duplicate copies of old files or compressed archives. RStudio users can collaborate on projects hosted on GitHub without having to use the command line. This has proven to be an effective way of ensuring a consistent, reproducible workflow, even for beginners. This book was written collaboratively through a private repository on GitHub, just as the `mdsr` package is maintained in a public repository.

D.4 Further resources

Project TIER is an organization at Haverford College that has developed a protocol [12] for reproducible research. Their efforts originated in the social sciences using Stata, but have since expanded to include R.

R Markdown is under active development. For the latest features see the R Markdown authoring guide at `http://rmarkdown.rstudio.com`. The RStudio cheat sheet serves as a useful reference.

GitHub can be challenging to learn but is now the default in many (most?) data science research settings. Jenny Bryan's resources on "Happy Git and GitHub for the useR" (`http://happygitwithr.com`) are particularly relevant for new data scientists beginning to use GitHub.

Another challenge for reproducible analyses concerns versions of R and other R packages. The `packrat` package helps ensure that projects can maintain a particular version of R and set of packages. This functionality is tied in closely with RStudio.

Sample R Markdown example

Sample User

November 8, 2016

R Markdown

This is an R Markdown document. Markdown is a simple formatting syntax for authoring HTML, PDF, and MS Word documents. For more details on using R Markdown see http://rmarkdown.rstudio.com (http://rmarkdown.rstudio.com).

When you click the **Knit** button a document will be generated that includes both content as well as the output of any embedded R code chunks within the document. You can embed an R code chunk like this:

```
glimpse(SwimRecords)
```

```
## Observations: 62
## Variables: 3
## $ year <int> 1905, 1908, 1910, 1912, 1918, 1920, 1922, 1924, 1934, 193...
## $ time <dbl> 65.80, 65.60, 62.80, 61.60, 61.40, 60.40, 58.60, 57.40, 5...
## $ sex  <fctr> M, M, M, M, M, M, M, M, M, M, M, M, M, M, M, M, M, M, M,...
```

Including Plots

You can also embed plots, for example:

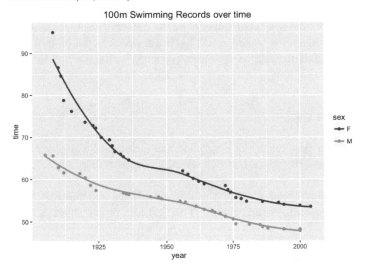

There are n=62 rows in the Swim records dataset.

Note that the `echo = FALSE` parameter was added to the code chunk to prevent printing of the R code that generated the plot.

Figure D.3: Formatted output from R Markdown example.

D.5 Exercises

Exercise D.1

The following exercises provide practice with R Markdown files and introduce useful features. Consider an R Markdown file that includes the following code chunks. What will be output when this file is rendered?

````
```{r}
x <- 1:5
```
````

````
```{r}
x <- x + 1
```
````

````
```{r}
x
```
````

Exercise D.2

Consider an R Markdown file that includes the following code chunks. What will be output when this file is rendered?

````
```{r echo=FALSE}
x <- 1:5
```
````

````
```{r echo=FALSE}
x <- x + 1
```
````

````
```{r echo=FALSE}
x
```
````

Exercise D.3

Consider an R Markdown file that includes the following code chunks. What will be output when the file is rendered?

````
```{r echo=FALSE}
x <- 1:5
```
````

````
```{r echo=FALSE, eval=FALSE}
x <- x + 1
```
````

````
```{r echo=FALSE}
x
```
````

Exercise D.4

Consider an R Markdown file that includes the following code chunks. What will be output when the file is rendered?

```{r echo=FALSE}
x <- 1:5
```

```{r echo=FALSE}
x <- x + 1
```

```{r include=FALSE}
x
```

Exercise D.5

Describe in words what the following excerpt from an R Markdown file will display when rendered.

```{r echo=FALSE}
n <- 679
```

The data set has n=`r n` observations.

Exercise D.6

Describe in words what the following excerpt from an R Markdown file will display when rendered.

$\hat{y} = \hat{\beta}_0 + \hat{\beta}_1 \cdot x + \epsilon$

Exercise D.7

Describe the implications of changing `warning=TRUE` to `warning=FALSE` in the following code chunk.

```{r warning=TRUE}
sqrt(-1)
```

Exercise D.8

Why does the `mosaic` package plain R Markdown template include the code chunk option `message=FALSE` when the `mosaic` package is loaded?

Exercise D.9

Describe how the `fig.width` and `fig.height` chunk options can be used to control the size of graphical figures.

Exercise D.10

Explain what the following code chunks will display and why this might be useful for technical reports from a data science project.

````
```{r chunk1, eval=TRUE, include=FALSE}
x <- 15
x
```
````

````
```{r chunk2}
x <- x + 3
x
```
````

````
```{r chunk1, eval=FALSE, include=TRUE}
```
````

Exercise D.11

The `xtable` package allows the analyst to display nicely formatted tables and results when outputting to pdf files. Use the following code chunks as an example to create a similar display using your own data.

````
```{r results="asis"}
library(xtable)
library(mdsr)
options(xtable.comment = FALSE)
mod <- lm(cesd ~ mcs + sex, data = HELPrct)
xtable(mod)
```
````

Exercise D.12

Insert a chunk in your .Rmd document so that it renders even when there are errors. Some errors are easier to diagnose if you can execute specific R statements during rendering and leave more evidence behind for forensic examination.

Put this chunk near the top of your R Markdown document if you want to soldier on through errors, i.e., turn `foo.Rmd` into `foo.md` and/or `foo.html` no matter what.

````
```{r setup, include = FALSE, cache = FALSE}
knitr::opts_chunk$set(error = TRUE)
```
````

This is also helpful if you are writing a tutorial and want to demo code that throws an error.

It's also possible to set things so that errors are tolerated in a specific chunk.

````
```{r alwaysrun, error = TRUE}
## code goes here
```
````

Use either of these strategies to generate an R Markdown file that includes an error but compiles nonetheless. (Kudos to Jenny Bryan for describing this approach.)

Appendix E

Regression modeling

Regression analysis is a powerful and flexible framework that allows an analyst to model an outcome (the *response variable*) as a function of one or more *explanatory variables* (or predictors). Regression forms the basis of many important statistical models described in Chapters 7 and 8. This appendix provides a brief review of linear and logistic *regression models*, beginning with a single predictor, then extending to multiple predictors.

E.1 Simple linear regression

Linear regression can help us understand how values of a quantitative (numerical) outcome (or response) are associated with values of a quantitative explanatory (or predictor) variable. This technique is often applied in two ways: to generate predicted values or to make inferences regarding associations in the dataset.

In some disciplines the outcome is called the dependent variable and the predictor the independent variable. We avoid such usage since the words dependent and independent have many meanings in statistics.

A simple linear regression model for an outcome y as a function of a predictor x takes the form:

$$y_i = \beta_0 + \beta_1 x_i + \epsilon_i, \text{ for } i = 1, \ldots, n,$$

where n represents the number of observations (rows) in the data set. For this model, β_0 is the population parameter corresponding to the *intercept* (i.e., the predicted value when $x = 0$) and β_1 is the true (population) *slope* coefficient (i.e., the predicted increase in y for a unit increase in x). The ϵ_i's are the *errors* (these are assumed to be random noise with mean 0).

We almost never know the true values of the population parameters β_0 and β_1, but we estimate them using data from our sample. The lm() function finds the "best" coefficients $\hat{\beta}_0$ and $\hat{\beta}_1$ where the the *fitted values* (or expected values) are given by $\hat{y}_i = \hat{\beta}_0 + \hat{\beta}_1 x_i$. What is left over is captured by the *residuals* ($\hat{\epsilon}_i = y_i - \hat{y}_i$). The model almost never fits perfectly—if it did there would be no need for a model.

The best fitting regression line is usually determined by a *least squares* criteria that minimizes the sum of the squared residuals. The least squares regression line (defined by the values of $\hat{\beta}_0$ and $\hat{\beta}_1$) is unique.

E.1.1   Motivating example: Modeling usage of a rail trail

The Pioneer Valley Planning Commission (PVPC) collected data north of Chestnut Street
in Florence, Massachusetts for a ninety day period. Data collectors set up a laser sensor
that recorded when a rail-trail user passed the data collection station.

```
glimpse(RailTrail)

Observations: 90
Variables: 10
$ hightemp   <int> 83, 73, 74, 95, 44, 69, 66, 66, 80, 79, 78, 65, 41,...
$ lowtemp    <int> 50, 49, 52, 61, 52, 54, 39, 38, 55, 45, 55, 48, 49,...
$ avgtemp    <dbl> 66.5, 61.0, 63.0, 78.0, 48.0, 61.5, 52.5, 52.0, 67....
$ spring     <int> 0, 0, 1, 0, 1, 1, 1, 1, 0, 0, 0, 1, 1, 0, 0, 1, 0, ...
$ summer     <int> 1, 1, 0, 1, 0, 0, 0, 0, 1, 1, 1, 0, 0, 0, 0, 0, 1, ...
$ fall       <int> 0, 0, 0, 0, 0, 0, 0, 0, 0, 0, 0, 0, 0, 1, 1, 0, 0, ...
$ cloudcover <dbl> 7.6, 6.3, 7.5, 2.6, 10.0, 6.6, 2.4, 0.0, 3.8, 4.1, ...
$ precip     <dbl> 0.00, 0.29, 0.32, 0.00, 0.14, 0.02, 0.00, 0.00, 0.0...
$ volume     <int> 501, 419, 397, 385, 200, 375, 417, 629, 533, 547, 4...
$ weekday    <fctr> 1, 1, 1, 0, 1, 1, 1, 0, 0, 1, 1, 1, 1, 1, 1, 1, 0,...
```

The PVPC wants to understand the relationship between daily ridership (i.e., the number of riders and walkers who use the bike path on any given day) and a collection of
explanatory variables, including the temperature, rainfall, cloud cover, and day of the week.

In a simple linear regression model, there is a single quantitative explanatory variable.
It seems reasonable that the high temperature for the day (hightemp, measured in degrees
Fahrenheit) might be related to ridership, so we will explore that first. Figure E.1 shows a
scatterplot between ridership (volume) and high temperature (hightemp), with the simple
linear regression line overlaid. The fitted coefficients are shown below by providing a formula
to the lm() function.

```
mod <- lm(volume ~ hightemp, data = RailTrail)
coef(mod)

(Intercept)    hightemp
   -17.079       5.702
```

The first coefficient is $\hat{\beta}_0$, the estimated y-intercept. The interpretation is that if the
high temperature was 0 degrees Fahrenheit, then the estimated ridership would be about -17
riders. This is doubly non-sensical in this context, since it is impossible to have a negative
number of riders and this represents a substantial extrapolation to far colder temperatures
than are present in the data set (recall the *Challenger* discussion from Chapter 2). It turns
out that the monitoring equipment didn't work when it got too cold, so values for those
days are unavailable.

Pro Tip: In this case, it is not appropriate to simply multiply the average number of
users on the observed days by the number of days in a year, since cold days that are likely
to have fewer trail users are excluded due to instrumentation issues. Such missing data can
lead to selection bias.

The second coefficient (the slope) is usually more interesting. This coefficient ($\hat{\beta}_1$) is
interpreted as the predicted increase in trail users for each additional degree in temperature.

```
plotModel(mod, system = "ggplot2")
```

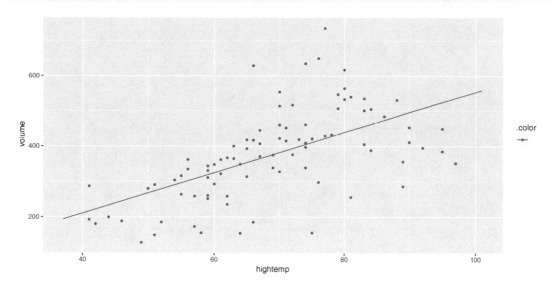

Figure E.1: Scatterplot of number of trail crossings as a function of highest daily temperature (in degrees Fahrenheit).

We expect to see about 5.7 additional riders use the rail trail on a day that is one degree warmer than another day.

E.1.2   Model visualization

Figure E.1 allows us to visualize our model in the data space. How does our model compare to a null model? That is, how do we know that our model is useful?

In Figure E.2, we compare the least squares regression line (right) with the null model that simply returns the average for every input (left). That is, on the left, the average temperature of the day is ignored. The model simply predicts an average ridership every day, regardless of the temperature. However, on the right, the model takes the average ridership into account, and accordingly makes a different prediction for each input value.

Obviously, the regression model works better than the null model (that forces the slope to be zero), since it is more flexible. But how much better?

E.1.3   Measuring the strength of fit

The correlation coefficient, r, is used to quantify the strength of the linear relationship between two variables. We can quantify the proportion of variation in the response variable (y) that is explained by the model in a similar fashion. This quantity is called the *coefficient of determination* and is denoted R^2. It is a common measure of goodness-of-fit for regression models. Like any proportion, R^2 is always between 0 and 1. For simple linear regression

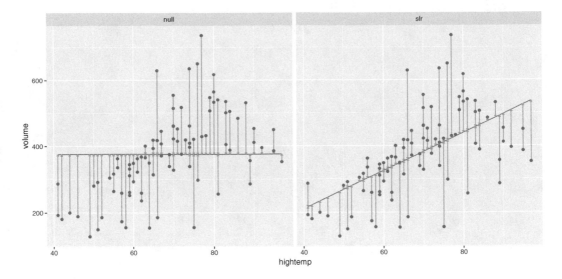

Figure E.2: At left, the model based on the overall average high temperature. At right, the simple linear regression model.

(one explanatory variable), $R^2 = r^2$. The definition of R^2 is given by:

$$R^2 = 1 - \frac{SSE}{SST} = \frac{SSM}{SST}$$
$$= 1 - \frac{\sum_{i=1}^{n}(y_i - \hat{y}_i)^2}{\sum_{i=1}^{n}(y_i - \bar{y})^2}$$
$$= 1 - \frac{SSE}{(n-1)Var(y)},$$

where SSE is the sum of the squared residuals, SSM is the sum of the squares attributed to the model, and SST is the total sum of the squares. Let's calculate these values for the rail trail example.

```
n <- nrow(RailTrail)
SST <- var(~volume, data = RailTrail) * (n - 1)
SSE <- var(residuals(mod)) * (n - 1)
1 - SSE / SST

[1] 0.3394

rsquared(mod)

[1] 0.3394
```

In Figure E.2, the null model on the left has an R^2 of 0, because $\hat{y}_i = \bar{y}$ for all i, and so $SSE = SST$. On the other hand, the R^2 of the regression model on the right is 0.3394. We say that the regression model based on average daily temperature explained about 34% of the variation in daily ridership.

E.1.4 Categorical explanatory variables

Suppose that instead of using temperature as our explanatory variable for ridership on the rail trail, we only considered whether it was a weekday or not a weekday (e.g., weekend or holiday). The indicator variable `weekday` is *binary* (or dichotomous) in that it only takes on the values 0 and 1. (Such variables are sometimes called *indicator* variables or more pejoratively *dummy* variables.) This new linear regression model has the form:

$$\widehat{volume} = \hat{\beta}_0 + \hat{\beta}_1 \cdot weekday,$$

where the fitted coefficients are given below.

```
coef(lm(volume ~ weekday, data = RailTrail))

(Intercept)    weekday1
    430.71      -80.29
```

Note that these coefficients could have been calculated from the means of the two groups (since the regression model has only two possible predicted values). The average ridership on weekdays is 350.4 while the average on non-weekdays is 430.7.

```
mean(volume ~ weekday, data = RailTrail)

    0     1
430.7 350.4

diff(mean(volume ~ weekday, data = RailTrail))

      1
-80.29
```

In the coefficients listed above, the `weekday1` variable corresponds to rows in which the value of the `weekday` variable was 1 (i.e., weekdays). Because this value is negative, our interpretation is that 80 fewer riders are expected on a weekday as opposed to a weekend or holiday.

To improve the readability of the output we can create a new variable with more mnemonic values.

```
RailTrail <- RailTrail %>%
  mutate(day = ifelse(weekday == 1, "weekday", "weekend/holiday"))
```

Pro Tip: Care was needed to recode the `weekday` variable because it was a `factor`. Avoid the use of factors unless they are needed.

```
coef(lm(volume ~ day, data = RailTrail))

      (Intercept) dayweekend/holiday
           350.42              80.29
```

The model coefficients have changed (although they still provide the same interpretation). By default, the `lm()` function will pick the alphabetically lowest value of the categorical predictor as the *reference group* and create indicators for the other levels (in this case `dayweekend/holiday`). As a result the intercept is now the predicted number of trail crossings on a `weekday`. In either formulation, the interpretation of the model remains the same: On a weekday, 80 fewer riders are expected than on a weekend or holiday.

E.2   Multiple regression

Multiple regression is a natural extension of simple linear regression that incorporates multiple explanatory (or predictor) variables. It has the general form:

$$y = \beta_0 + \beta_1 x_1 + \beta_2 x_2 + \cdots + \beta_p x_p + \epsilon, \text{ where } \epsilon \sim N(0, \sigma_\epsilon).$$

The estimated coefficients (i.e., $\hat{\beta}_i$'s) are now interpreted as "conditional on" the other variables—each β_i reflects the *predicted* change in y associated with a one-unit increase in x_i, conditional upon the rest of the x_i's. This type of model can help to disentangle more complex relationships between three or more variables. The value of R^2 from a multiple regression model has the same interpretation as before: the proportion of variability explained by the model.

Pro Tip: Interpreting conditional regression parameters can be challenging. The analyst needs to ensure that comparisons that hold other factors constant do not involve extrapolations beyond the observed data.

E.2.1   Parallel slopes: Multiple regression with a categorical variable

Consider first the case where x_2 is an *indicator* variable that can only be 0 or 1 (e.g., `weekday`). Then,

$$\hat{y} = \hat{\beta}_0 + \hat{\beta}_1 x_1 + \hat{\beta}_2 x_2.$$

In the case where x_1 is quantitative but x_2 is an indicator variable, we have:

$$\text{For weekends,} \quad \hat{y}|_{x_1, x_2 = 0} = \hat{\beta}_0 + \hat{\beta}_1 x_1$$

$$\text{For weekdays,} \quad \hat{y}|_{x_1, x_2 = 1} = \hat{\beta}_0 + \hat{\beta}_1 x_1 + \hat{\beta}_2 \cdot 1$$

$$= \left(\hat{\beta}_0 + \hat{\beta}_2 \right) + \hat{\beta}_1 x_1.$$

This is called a *parallel slopes* model (see Figure E.3), since the predicted values of the model take the geometric shape of two parallel lines with slope $\hat{\beta}_1$: one with y-intercept $\hat{\beta}_0$ for weekends, and another with y-intercept $\hat{\beta}_0 + \hat{\beta}_2$ for weekdays.

```
mod_parallel <- lm(volume ~ hightemp + weekday, data = RailTrail)
coef(mod_parallel)

(Intercept)      hightemp      weekday1
     42.807         5.348       -51.553

rsquared(mod_parallel)

[1] 0.3735
```

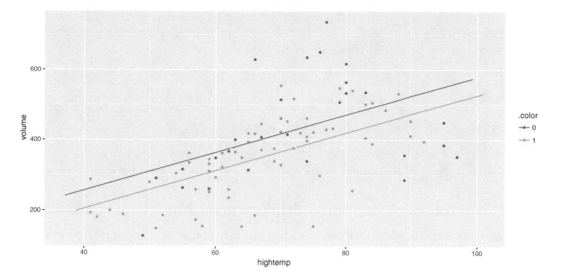

Figure E.3: Visualization of parallel slopes model for the rail trail data.

```
plotModel(mod_parallel, system = "ggplot2")
```

E.2.2   Parallel planes: Multiple regression with a second quantitative variable

If x_2 is a quantitative variable, then we have:

$$\hat{y} = \hat{\beta}_0 + \hat{\beta}_1 x_1 + \hat{\beta}_2 x_2 \,.$$

Notice that our model is no longer a line, rather it is a *plane* that exists in three dimensions.

Now suppose that we want to improve our model for ridership by considering not only the average temperature, but also the amount of precipitation (rain or snow, measured in inches). We can do this in R by simply adding this variable to our regression model.

```
mod_planes <- lm(volume ~ hightemp + precip, data = RailTrail)
coef(mod_planes)
```

```
(Intercept)     hightemp        precip
   -31.520        6.118      -153.261
```

Note that the coefficient on `hightemp` (6.1 riders per degree) has changed from its value in the simple linear regression model (5.7 riders per degree). This is due to the moderating effect of precipitation. Our interpretation is that for each additional degree in temperature, we expect an additional 6.1 riders on the rail trail, after controlling for the amount of precipitation.

Pro Tip: Note that since the median precipitation on days when there was precipitation was only 0.15 inches, a predicted change for an additional inch may be misleading. It may be better to report a predicted difference of 0.15 additional inches or replace the continuous term in the model with a dichotomous indicator of any precipitation.

As you can imagine, the effect of precipitation is strong—some people may be less likely to bike or walk in the rain. Thus, even after controlling for temperature, an inch of rainfall is associated with a drop in ridership of about 153.

```
mod_p_planes <- lm(volume ~ hightemp + precip + weekday, data = RailTrail)
coef(mod_p_planes)

(Intercept)      hightemp        precip       weekday1
     19.319         5.801      -145.609        -43.144
```

If we added all three explanatory variables to the model we would have parallel planes.

E.2.3   Non-parallel slopes: Multiple regression with interaction

Let's return to a model that includes `weekday` and `hightemp` as predictors. What if the parallel slopes model doesn't fit well? Adding an additional term into the model can make it more flexible and allow there to be a different slope on the two different types of days:

$$\hat{y} = \hat{\beta}_0 + \hat{\beta}_1 x_1 + \hat{\beta}_2 x_2 + \hat{\beta}_3 x_1 x_2.$$

We then have:

For weekends, $\hat{y}|_{x_1, x_2=0} = \hat{\beta}_0 + \hat{\beta}_1 x_1$

For weekdays, $\hat{y}|_{x_1, x_2=1} = \hat{\beta}_0 + \hat{\beta}_1 x_1 + \hat{\beta}_2 \cdot 1 + \hat{\beta}_3 \cdot x_1$

$$= \left(\hat{\beta}_0 + \hat{\beta}_2 \right) + \left(\hat{\beta}_1 + \hat{\beta}_3 \right) x_1.$$

This is called an *interaction model* (see Figure E.4). The predicted values of the model take the geometric shape of two non-parallel lines with different slopes.

```
mod_interact <- lm(volume ~ hightemp + weekday + hightemp * weekday,
  data = RailTrail)
coef(mod_interact)

        (Intercept)           hightemp      weekday1 hightemp:weekday1
            135.153              4.075      -186.377             1.906
```

```
rsquared(mod_interact)
```

```
[1] 0.3816
```

```
plotModel(mod_interact, system = "ggplot2")
```

We see that the slope on weekdays is about two riders per degree higher than on weekends and holidays. This may indicate that trail users on weekends and holidays are less concerned about the temperature than on weekdays.

E.2.4   Modelling non-linear relationships

A linear model with a single parameter fits well in many situations but is not appropriate in others. Consider modeling height (in centimeters) as a function of age (in years) using data from a subset of female subjects included in the National Health and Nutrition Examination

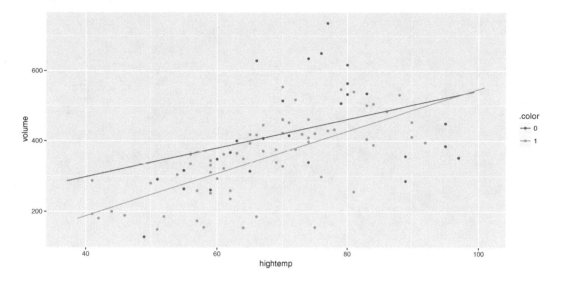

Figure E.4: Visualization of interaction model for the rail trail data.

Study (from the NHANES package) with a linear term. Another approach uses a *smoother* instead of a linear model. Unlike the straight line, the smoother can bend to better fit the points when modeling the functional form of a relationship (see Figure E.5).

```
library(NHANES)
NHANES %>%
  sample(300) %>%
  filter(Gender == "female") %>%
ggplot(aes(x = Age, y = Height)) +
  geom_point() +
  stat_smooth(method = lm, se = 0) +
  stat_smooth(method = loess, se = 0, color = "green") +
  xlab("Age (in years)") + ylab("Height (in cm)")
```

The fit of the linear model (denoted in blue) is poor: A straight line does not account for the dramatic increases in height during puberty to young adulthood or for the gradual decline in height for older subjects. The smoother (in green) does a much better job of describing the functional form.

The improved fit does come with a cost. Compare the results for linear and smoothed models in Figure E.6. Here the functional form of the relationship between high temperature and volume of trail use is closer to linear (with some deviation for warmer temperatures).

```
ggplot(data = RailTrail, aes(x = hightemp, y = volume)) +
  geom_point() +
  stat_smooth(method = lm) + stat_smooth(method = loess, color = "green") +
  ylab("Number of trail crossings") + xlab("High temperature (F)")
```

The width of the confidence bands for the smoother tend to be wider than that for the linear model. This is the cost of the additional flexibility in modeling. The other cost is interpretation: It is more complicated to explain the results from the smoother than to interpret a slope coefficient.

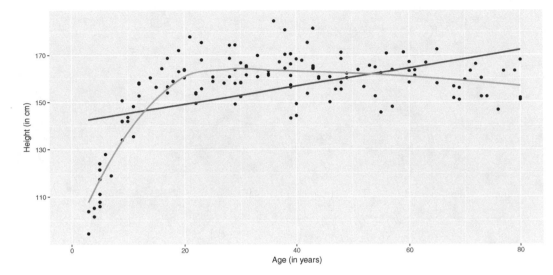

Figure E.5: Scatterplot of height as a function of age with superimposed linear model (blue) and smoother (green).

E.3   Inference for regression

Thus far, we have fit several models and interpreted their estimated coefficients. However, with the exception of the confidence bands in Figure E.6, we have only made statements about the estimated coefficients (i.e., the $\hat{\beta}$'s)—we have made no statements about the true coefficients (i.e., the β's), the values of which of course remain unknown.

However, we can use our understanding of the t-distribution to make *inferences* about the true value of regression coefficients. In particular, we can test a hypothesis about β_1 (most commonly that it is equal to zero) and find a confidence interval (range of plausible values) for it.

```
msummary(mod_p_planes)

            Estimate Std. Error t value Pr(>|t|)
(Intercept)   19.319     60.339    0.32  0.74961
hightemp       5.801      0.799    7.26  1.6e-10 ***
precip      -145.609     38.894   -3.74  0.00033 ***
weekday1     -43.144     22.194   -1.94  0.05517 .

Residual standard error: 95.2 on 86 degrees of freedom
Multiple R-squared:  0.461,Adjusted R-squared:  0.443
F-statistic: 24.6 on 3 and 86 DF,  p-value: 1.44e-11
```

In the output above, the p-value that is associated with the `hightemp` coefficient is displayed as 1.6e-10 (or nearly zero). That is, if the true coefficient (β_1) was in fact zero, then the probability of observing an association on ridership due to average temperature as large or larger than the one we actually observed in the data, after controlling for precipitation and day of the week, is essentially zero. This suggests that the hypothesis that β_1 was in fact zero is dubious based on these data. Perhaps there is a real association between ridership and average temperature.

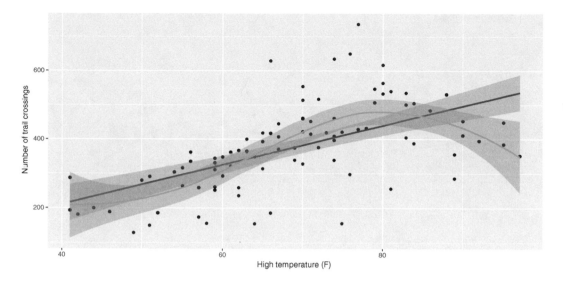

Figure E.6: Scatterplot of volume as a function of high temperature with superimposed linear and smooth models for the rail trail data.

Pro Tip: Very small p-values should be rounded to the nearest 0.0001. We suggest reporting this p-value as $p < 0.0001$.

Another way of thinking about this process is to form a confidence interval around our estimate of the slope coefficient $\hat{\beta}_1$. Here we can say with 95% confidence that the value of the true coefficient β_1 is between 4.21 and 7.39 riders per degree. That this interval does not contain zero confirms the previous hypothesis test.

```
confint(mod_p_planes)
```

	2.5 %	97.5 %
(Intercept)	-100.631	139.2684
hightemp	4.213	7.3881
precip	-222.927	-68.2909
weekday1	-87.265	0.9764

E.4 Assumptions underlying regression

The inferences we made above were predicated upon our assumption that the slope follows a t-distribution. This follows from the assumption that the errors follow a normal distribution (with mean 0 and standard deviation σ_ϵ, for some constant σ_ϵ). Inferences from the model are only valid if the following assumptions hold:

Linearity: The functional form of the relationship between the predictors and the outcome follows a linear combination of regression parameters that are correctly specified (this assumption can be verified by bivariate graphical displays).

Independence: Are the errors uncorrelated? Or do they follow a pattern (perhaps over time or within clusters of subjects)?

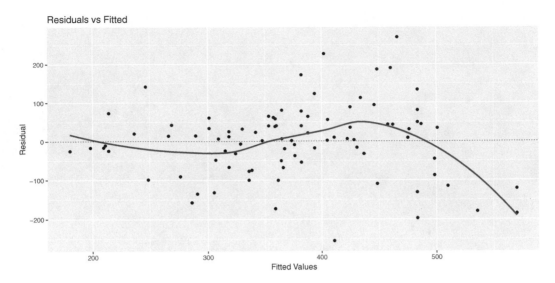

Figure E.7: Assessing linearity using a scatterplot of residuals versus fitted (predicted) values.

Normality of residuals: Do the residuals follow a distribution that is approximately normal? This assumption can be verified using univariate displays.

Equal variance of residuals: Is the variance in the residuals constant across the explanatory variables (*homoscedastic errors*)? Or does the variance in the residuals depend on the value of one or more of the explanatory variables (*heteroscedastic errors*)? This assumption can be verified using residual diagnostics.

These conditions are sometimes called the "LINE" assumptions. All but the independence assumption can be assessed using diagnostic plots.

How might we assess the mod_p_planes model? Figure E.7 displays a scatterplot of residuals versus fitted (predicted) values. As we observed in Figure E.6, the number of crossings does not increase as much for warm temperatures as it does for more moderate ones. We may need to consider a more sophisticated model with a more complex model for temperature.

```
mplot(mod_p_planes, which = 1, system = "ggplot2")
```

Figure E.8 displays the quantile–quantile plot for the residuals from the regression model. The plot deviates from the straight line: This indicates that the residuals have heavier tails than a normal distribution.

```
mplot(mod_p_planes, which = 2, system = "ggplot2")
```

Figure E.9 displays the scale–location plot for the residuals from the model: The results indicate that there is evidence of heteroscedasticity (the variance of the residuals increases as a function of predicted value).

```
mplot(mod_p_planes, which = 3, system = "ggplot2")
```

When performing model diagnostics, it is important to identify any outliers and understand their role in determining the regression coefficients.

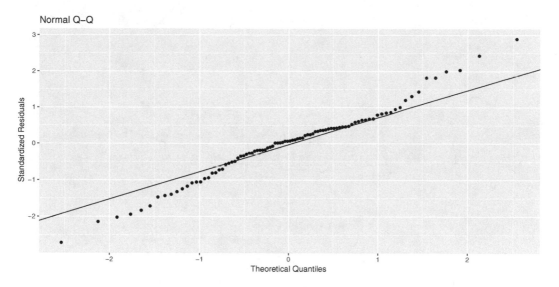

Figure E.8: Assessing normality assumption using a Q–Q plot.

- An *outlier* is an observation that doesn't seem to fit the general pattern of the data.

- An observation with an extreme value of the explanatory variable is a point of high *leverage*.

- A high leverage point that exerts disproportionate influence on the slope of the regression line is an *influential point*.

Figure E.10 displays the values for Cook's distance (a common measure of influential points in a regression model).

```
mplot(mod_p_planes, which = 4, system = "ggplot2")
```

We use the augment() function from the broom package to calculate the value of this statistic and identify the most extreme Cook's distance.

```
library(broom)
augment(mod_p_planes) %>%
  filter(.cooksd > 0.4)

  volume hightemp precip weekday .fitted .se.fit .resid   .hat .sigma
1    388       84   1.49       1   246.5   54.84  141.5 0.3321  93.87
  .cooksd .std.resid
1  0.4116       1.82
```

The outlier corresponds to a day with nearly one and a half inches of rain (the most recorded in the dataset) and a high temperature of 84 degrees.

E.5 Logistic regression

Our previous examples had quantitative (or continuous) outcomes. What happens when we are interested in modeling a dichotomous outcome? For example, we might model the

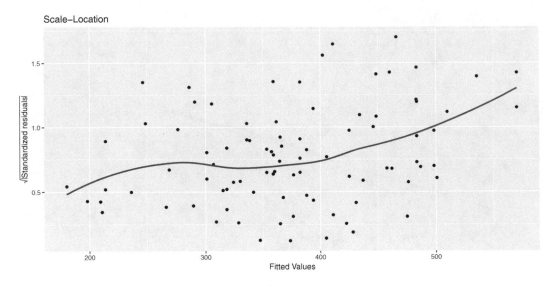

Figure E.9: Assessing equal variance using a scale–location plot.

probability of developing diabetes as a function of age and BMI (we explored this question further in Chapter 8). Figure E.11 displays the scatterplot of diabetes status as a function of age, while Figure E.12 displays the scatterplot of diabetes as a function of BMI (body mass index). Note that each subject can either have diabetes or not, so all of the points are displayed at zero or one on the y-axis.

```
NHANES <- NHANES %>%
  mutate(has_diabetes = as.numeric(Diabetes == "Yes"))
log_plot <- ggplot(data = NHANES, aes(x = Age, y = has_diabetes)) +
  geom_jitter(alpha = 0.1, height = 0.05) +
  geom_smooth(method = "glm", method.args = list(family = "binomial")) +
  ylab("Diabetes status")
```

Which variable is more important: Age or BMI? We can use a logistic regression model to model the probability of diabetes as a function of both predictors.

```
logreg <- glm(has_diabetes ~ BMI + Age, family = "binomial", data = NHANES)
msummary(logreg)
```

```
Coefficients:
            Estimate Std. Error z value Pr(>|z|)
(Intercept) -8.08029    0.24445   -33.1   <2e-16 ***
BMI          0.09433    0.00552    17.1   <2e-16 ***
Age          0.05728    0.00249    23.0   <2e-16 ***

(Dispersion parameter for binomial family taken to be 1)

    Null deviance: 5263.8  on 9628  degrees of freedom
Residual deviance: 4146.0  on 9626  degrees of freedom
  (371 observations deleted due to missingness)
AIC: 4152
```

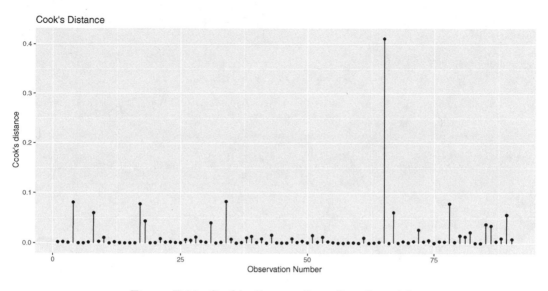

Figure E.10: Cook's distance for rail trail model.

```
Number of Fisher Scoring iterations: 7
```

The answer is that both are important (both are statistically significant predictors). To interpret the findings, we might consider a visual display of predicted probabilities as displayed in Figure E.13 (compare with Figure 8.11).

```
ages <- range(~Age, data = NHANES)
bmis <- range(~BMI, data = NHANES, na.rm = TRUE)
res <- 100
fake_grid <- expand.grid(
  Age = seq(from = ages[1], to = ages[2], length.out = res),
  BMI = seq(from = bmis[1], to = bmis[2], length.out = res)
)
y_hats <- fake_grid %>%
  mutate(y_hat = predict(logreg, newdata = ., type = "response"))
```

```
ggplot(data = NHANES, aes(x = Age, y = BMI)) +
  geom_tile(data = y_hats, aes(fill = y_hat), color = NA) +
  geom_count(aes(color = as.factor(has_diabetes)), alpha = 0.4) +
  scale_fill_gradient(low = "white", high = "dodgerblue") +
  scale_color_manual("Diabetes", values = c("gray", "gold")) +
  scale_size(range = c(0, 2))
```

We see that very few young adults have diabetes, even if they have moderately high BMI scores. As we look at older subjects while holding BMI fixed, the probability of diabetes increases.

```
log_plot + xlab("Age (in years)")
```

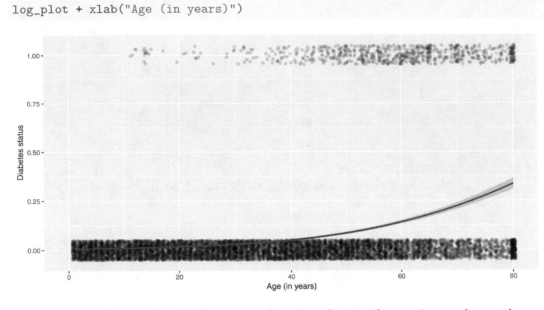

Figure E.11: Scatterplot of diabetes as a function of age with superimposed smoother.

```
log_plot + aes(x = BMI) + xlab("BMI (body mass index)")
```

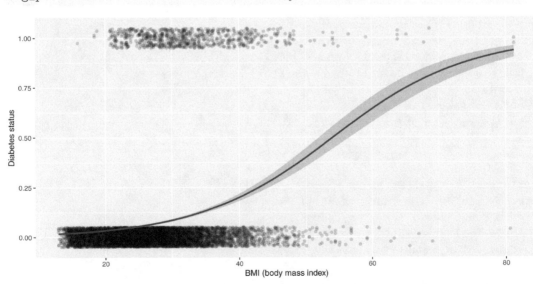

Figure E.12: Scatterplot of diabetes as a function of BMI with superimposed smoother.

Figure E.13: Predicted probabilities for diabetes as a function of BMI and age.

E.6 Further resources

Regression is described in many books. An introduction is found in most introductory statistics textbooks, including *Open Intro Statistics* [63]. For a deeper but still accessible treatment, we suggest [45]. Modern texts by James et al. [121] and Hastie, Tibshirani, and Friedman [98] also cover regression from a modeling and machine learning perspective. Hoaglin [103] details how conditional regression parameters should be interpreted. Cook [59] reviews regression diagnostics. An accessible introduction to smoothing can be found in Ruppert et al. [182].

E.7   Exercises

Exercise E.1

In the HELP (Health Evaluation and Linkage to Primary Care) study, investigators were interested in determining predictors of severe depressive symptoms (measured by the Center for Epidemiologic Studies—Depression scale, cesd) amongst a cohort enrolled at a substance abuse treatment facility. These predictors include substance of abuse (alcohol, cocaine, or heroin), mcs (a measure of mental well-being), gender, and housing status (housed or homeless). Answer the following questions regarding the following multiple regression model.

```
library(mdsr)
fm <- lm(cesd ~ substance + mcs + sex + homeless, data = HELPrct)
msummary(fm)

                  Estimate Std. Error t value Pr(>|t|)
(Intercept)        57.7794     1.4664   39.40   <2e-16 ***
substancecocaine   -3.5406     1.0101   -3.51   0.0005 ***
substanceheroin    -1.6818     1.0731   -1.57   0.1178
mcs                -0.6407     0.0338  -18.97   <2e-16 ***
sexmale            -3.3239     1.0075   -3.30   0.0010 **
homelesshoused     -0.8327     0.8686   -0.96   0.3383


Residual standard error: 8.97 on 447 degrees of freedom
Multiple R-squared:  0.492,Adjusted R-squared:  0.486
F-statistic: 86.4 on 5 and 447 DF,  p-value: <2e-16

confint(fm)

                  2.5 % 97.5 %
(Intercept)      54.898 60.661
substancecocaine -5.526 -1.555
substanceheroin  -3.791  0.427
mcs              -0.707 -0.574
sexmale          -5.304 -1.344
homelesshoused   -2.540  0.874
```

1. Write out the linear model.

2. Calculate the predicted CESD for a female homeless cocaine-involved subject with an MCS score of 20.

3. Interpret the 95% confidence interval for the substancecocaine coefficient.

4. Make a conclusion and summarize the results of a test of the homeless parameter.

5. Report and interpret the R^2 (coefficient of determination) for this model.

6. What do we conclude about the distribution of the residuals?

7. What do we conclude about the relationship between the fitted values and the residuals?

8. What do we conclude about the relationship between the MCS score and the residuals?

9. What other things can we learn from the residual diagnostics?

10. Which observations should we flag for further study?

Exercise E.2

Investigators in the HELP (Health Evaluation and Linkage to Primary Care) study were interested in modeling predictors of being homeless (one or more nights spent on the street or in a shelter in the past six months vs. housed) using baseline data from the clinical trial. Fit and interpret a parsimonious model that would help the investigators identify predictors of homelessness.

Exercise E.3

The Gestation data set contains birth weight, date, and gestational period collected as part of the Child Health and Development Studies. Information about the baby's parents— age, education, height, weight, and whether the mother smoked is also recorded.

```
library(mdsr)
glimpse(Gestation)
```

```
Observations: 1,236
Variables: 23
$ id       <int> 15, 20, 58, 61, 72, 100, 102, 129, 142, 148, 164, 17...
$ pluralty <int> 5, 5, 5, 5, 5, 5, 5, 5, 5, 5, 5, 5, 5, 5, 5, 5, 5, 5...
$ outcome  <int> 1, 1, 1, 1, 1, 1, 1, 1, 1, 1, 1, 1, 1, 1, 1, 1, 1, 1...
$ date     <int> 1411, 1499, 1576, 1504, 1425, 1673, 1449, 1562, 1408...
$ gestation <int> 284, 282, 279, NA, 282, 286, 244, 245, 289, 299, 351...
$ sex      <int> 1, 1, 1, 1, 1, 1, 1, 1, 1, 1, 1, 1, 1, 1, 1, 1, 1, 1...
$ wt       <int> 120, 113, 128, 123, 108, 136, 138, 132, 120, 143, 14...
$ parity   <int> 1, 2, 1, 2, 1, 4, 4, 2, 3, 3, 2, 4, 3, 5, 3, 4, 3, 3...
$ race     <int> 8, 0, 0, 0, 0, 0, 7, 7, 0, 0, 0, 0, 0, 8, 7, 7, 4, 3...
$ age      <int> 27, 33, 28, 36, 23, 25, 33, 23, 25, 30, 27, 32, 23, ...
$ ed       <int> 5, 5, 2, 5, 5, 2, 2, 1, 4, 5, 5, 2, 1, 5, 2, 2, 7, 2...
$ ht       <int> 62, 64, 64, 69, 67, 62, 62, 65, 62, 66, 68, 64, 63, ...
$ wt.1     <int> 100, 135, 115, 190, 125, 93, 178, 140, 125, 136, 120...
$ drace    <fctr> 8, 0, 5, 3, 0, 3, 7, 7, 3, 0, 5, 0, 5, 0, 7, 7, 7, ...
$ dage     <int> 31, 38, 32, 43, 24, 28, 37, 23, 26, 34, 28, 36, 28, ...
$ ded      <int> 5, 5, 1, 4, 5, 2, 4, 4, 1, 5, 4, 1, 2, 5, 0, 0, 1, 2...
$ dht      <int> 65, 70, NA, 68, NA, 64, NA, 71, 70, NA, NA, 74, NA, ...
$ dwt      <int> 110, 148, NA, 197, NA, 130, NA, 192, 180, NA, NA, 18...
$ marital  <int> 1, 1, 1, 1, 1, 1, 1, 1, 0, 1, 1, 1, 1, 1, 1, 1, 1, 1...
$ inc      <int> 1, 4, 2, 8, 1, 4, NA, 2, 2, 2, NA, 2, 2, 2, 1, 1, 1,...
$ smoke    <int> 0, 0, 1, 3, 1, 2, 0, 0, 0, 1, 3, 1, 1, 1, 0, 0, 1, 1...
$ time     <int> 0, 0, 1, 5, 1, 2, 0, 0, 0, 1, 4, 1, 1, 1, 0, 0, 1, 1...
$ number   <int> 0, 0, 1, 5, 5, 2, 0, 0, 0, 4, 2, 1, 1, 2, 0, 0, 5, 5...
```

1. Fit a linear regression model for birthweight (wt) as a function of the mother's age (age).

2. Find a 95% confidence interval and p-value for the slope coefficient.

3. What do you conclude about the association between a mother's age and her baby's birthweight?

Exercise E.4

The Child Health and Development Studies investigate a range of topics. One study, in particular, considered all pregnancies among women in the Kaiser Foundation Health Plan in the San Francisco East Bay area. The goal is to model the weight of the infants (bwt, in ounces) using variables including length of pregnancy in days (gestation), mother's age in years (age), mother's height in inches (height), whether the child was the first born (parity), mother's pregnancy weight in pounds (weight), and whether the mother was a smoker (smoke). The summary table below shows the results of a regression model for predicting the average birth weight of babies based on all of the variables included in the data set.

```
library(mdsr)
babies <- Gestation %>%
  rename(bwt = wt, height = ht, weight = wt.1) %>%
  mutate(parity = parity == 0, smoke = smoke > 0) %>%
  select(id, bwt, gestation, age, height, weight, parity, smoke)
mod <- lm(bwt ~ gestation + age + height + weight + parity + smoke,
  data = babies)
coef(mod)
```

(Intercept)	gestation	age	height	weight	parityTRUE
-85.4729	0.4567	0.0116	1.1605	0.0540	-3.0726
smokeTRUE					
-5.9976					

Answer the following questions regarding this linear regression model.

1. The coefficient for parity is different than if you fit a linear model predicting weight using only that variable. Why might there be a difference?

2. Calculate the residual for the first observation in the data set.

3. The variance of the residuals is 249.28, and the variance of the birth weights of all babies in the data used to build the model is 335.94. Calculate the R^2 and the adjusted R^2. Note that there are 1,236 observations in the data set, but there was missing data in 62 of those observations, so only 1,174 observations were used to build the regression model.

```
var(~residuals(mod))
```

```
[1] 257
```

```
var(~bwt, data = mod$model)
```

```
[1] 336
```

```
# rsquared(mod)
```

4. This data set contains missing values. What happens to these rows when we fit the model?

Exercise E.5

In 1966 Cyril Burt published a paper called "The genetic determination of differences in intelligence: A study of monozygotic twins reared apart." The data consist of IQ scores for [an assumed random sample of] 27 identical twins, one raised by foster parents, the other by the biological parents.

Here is the regression output for using `Biological` IQ to predict `Foster` IQ:

```
library(mdsr)
library(faraway)
mod <- lm(Foster ~ Biological, data = twins)
coef(mod)

(Intercept)   Biological
      9.208        0.901

rsquared(mod)

[1] 0.778
```

Which of the following is **FALSE**? Justify your answers.

1. Alice and Beth were raised by their biological parents. If Beth's IQ is 10 points higher than Alice's, then we would expect that her foster twin Bernice's IQ is 9 points higher than the IQ of Alice's foster twin Ashley.

2. Roughly 78% of the foster twins' IQs can be accurately predicted by the model.

3. The linear model is $\widehat{Foster} = 9.2 + 0.9 \times Biological$.

4. Foster twins with IQs higher than average are expected to have biological twins with higher than average IQs as well.

Exercise E.6

The atus package includes data from the American Time Use Survey (ATUS). Use the atusresp dataset to model `hourly_wage` as a function of other predictors in the dataset.

Appendix F

Setting up a database server

Setting up a local or remote database server is neither trivial nor difficult. In this chapter we provide instructions as to how to set up a local database server on a computer that you control. While everything that is done in this chapter can be accomplished on any modern operating system, many tools for data science are designed for Unix-like operating systems, and can be a challenge to set up on Windows. This is no exception. In particular, comfort with the command line is a plus and the material presented here will make use of *shell* commands. On Mac OS X and other Unix-like operating systems (e.g., Ubuntu), the command line is acessible using a Terminal application. On Windows, some of these shell commands might work at a DOS prompt, but others will not.[1] Unfortunately, providing Windows-specific setup instructions is outside the scope of this book.

Three open-source SQL database systems are most commonly encountered. These include SQLite, MySQL, and PostgreSQL. While MySQL and PostgreSQL are full-featured relational database systems that employ a strict client-server model, SQLite is a lightweight program that runs only locally and requires no initial configuration. However, while SQLite is certainly the easiest system to set up, it has has far fewer functions, lacks a caching mechanism, and is not likely to perform as well under heavy usage. Please see the official documentation for appropriate uses of SQLite for assistance with choosing the right SQL implementation for your needs.

Both MySQL and PostgreSQL employ a *client-server* architecture. That is, there is a server program running on a computer somewhere, and you can connect to that server from any number of client programs—from either that same machine or over the Internet. Still, even if you are running MySQL or PostgreSQL on your local machine, there are always two parts: the client and the server. This chapter provides instructions for setting up the server on a machine that you control—which for most analysts, is your local machine.

F.1 SQLite

For SQLite, there is nothing to configure, but it must be installed. On Linux systems, sqlite is likely already installed, but the source code, as well as pre-built binaries for Mac OS X and Windows, are available at https://www.sqlite.org/download.html.

---

[1]Note that Cygwin provides a Unix-like shell for Windows.

F.2   MySQL

We will focus on the use of MySQL (with brief mention of PostgreSQL in the next section). The steps necessary to install a PostgreSQL server will follow similar logic, but the syntax will be importantly different.

F.2.1   Installation

If you are running Mac OS X or a Linux-based operating system, then you probably already have a MySQL server installed and running on your machine. You can check to see if this is the case by running the following from your operating system's shell (i.e., the command line, in Mac OS X parlance, using the "Terminal" application).

```
ps aux | grep "mysql"

mysql     17218   4472   1620 ?       Jan26   0:00 /bin/sh /usr/bin/mysqld_safe
mysql     17580 794460 127624 ?       Jan26   1:25 /usr/sbin/mysqld
bbaumer   18977  16672   2880 pts/1 11:05   0:00 bash -c ps aux | grep "mysql"
bbaumer   18979  13692   2204 pts/1 11:05   0:00 grep mysql
```

If you see anything like the first line of this output (i.e., containing `mysqld`), then MySQL is already running. (If you don't see anything like that, then it is not. The last three lines are all related to the `ps` command we just ran.)

If MySQL is not installed, then you can install it by downloading the relevant version of the MySQL Community Server for your operating system at `http://dev.mysql.com/downloads/mysql/`. If you run into trouble, please consult the instructions at `https://dev.mysql.com/doc/refman/5.6/en/installing.html`.

For Mac OS X, there are more specific instructions available. After installation, you will want to install the Preference Pane, open it, check the box, and start the server.

It is also helpful to add the `mysql` binary directory to your `PATH` environment variable, so you can launch `msyql` easily from the shell. To do this, execute the following command in your shell:

```
export PATH=$PATH:/usr/local/mysql/bin
echo $PATH
```

You may have to modify the path to the `mysql bin` directory to suit your local setup.

F.2.2   Access

In most cases, the installation process will result in a server process being launched on your machine, such as the one that we saw above in the output of the `ps` command. Once the server is running, we need to configure it properly for our use. The full instructions for post-installation provide great detail on this process. However, in our case, we will mostly stick with the default configuration, so there are only a few things to check.

The most important thing is to gain access to the server. MySQL maintains a set of user accounts just like your operating system. After installation, there is usually only one account created: `root`. In order to create other accounts, we need to log into MySQL as `root`. *Please read the documentation on Securing the Initial MySQL Accounts for your setup.* From that documentation:

Some accounts have the user name `root`. These are superuser accounts that have all privileges and can do anything. If these root accounts have empty passwords, anyone can connect to the MySQL server as root without a password and be granted all privileges.

If this is your first time accessing MySQL, typing this into your shell might work:

```
mysql -u root
```

If you see an `Access denied` error, it means that the `root` MySQL user has a password, but you did not supply it. You may have created a password during installation. If you did, try:

```
mysql -u root -p
```

and then enter that password (it may well be blank). If you don't know the `root` password, try a few things that might be the password. If you can't figure it out, contact your system administrator or re-install MySQL.

You might—on Windows especially—get an error that says something about "command not found." This means that the program `mysql` is not accessible from your shell. You have two options: 1) you can specify the full path to the MySQL application; or 2) you can append your PATH variable to include the directory where the MySQL application is. The second option is preferred, and is illustrated above.

If you don't know where the application is, you can try to find it using the `find` program provided by your operating system.

```
find / -name "mysql"
```

On Linux or Mac OS X, it is probably in `/usr/bin/` or `/usr/local/mysql/bin` or something similar, and on Windows, it is probably in `\Applications\MySQL Server 5.6\bin` or something similar. Once you find the path to the application and the password, you should be able to log in. You will know when it works if you see a `mysql` prompt instead of your usual one.

```
bbaumer@bbaumer-Precision-Tower-7810:~$ mysql -u root -p
Enter password:
Welcome to the MySQL monitor.  Commands end with ; or \g.
Your MySQL connection id is 47
Server version: 5.5.44-0ubuntu0.14.04.1 (Ubuntu)

Copyright (c) 2000, 2015, Oracle and/or its affiliates. All rights reserved.

Oracle is a registered trademark of Oracle Corporation and/or its
affiliates. Other names may be trademarks of their respective
owners.

Type 'help;' or '\h' for help. Type '\c' to clear the current input
statement.

mysql>
```

Once you are logged into MySQL, try running the following command at the `mysql>` prompt (do not forget the trailing semi-colon):[2]

```
SELECT User, Host, Password  FROM mysql.user;
```

This command will list the users on the MySQL server, their encrypted passwords, and the hosts from which they are allowed to connect. Next, if you want to change the root password, set it to something else (in this example `mypass`).

```
UPDATE mysql.user SET Password = PASSWORD('mypass') WHERE User = 'root';
FLUSH PRIVILEGES;
```

The most responsible thing to do now is to create a new account for yourself. You should probably choose a different password than the one for the `root` user. Do this by running:

```
CREATE USER 'r-user'@'localhost' IDENTIFIED BY 'mypass';
```

It is important to understand that MySQL's concept of users is really a $\{user, host\}$ pair. That is, the user `'bbaumer'@'localhost'` can have a different password and set of privileges than the user `'bbaumer'@'%'`. The former is only allowed to connect to the server from the machine on which the server is running. (For most of you, that is your computer.) The latter can connect from anywhere (`'%'` is a wildcard character). Obviously, the former is more secure. Use the latter only if you want to connect to your MySQL database from elsewhere.

You will also want to make yourself a superuser.

```
GRANT ALL PRIVILEGES ON *.* TO 'r-user'@'localhost' WITH GRANT OPTION;
```

Now, flush the privileges:

```
FLUSH PRIVILEGES;
```

Finally, log out by typing `quit`. You should now be able to log in to MySQL as yourself by typing the following into your shell:

```
mysql -u yourusername -p
```

Using an option file

A relatively safe and convenient method of connecting to MySQL servers (whether local or remote) is by using an option file. This is a simple text file located at `~/.my.cnf` that may contain various connection parameters. Your entire file might look like this:

```
[client]
user=r-user
password="mypass"
```

These options will be read by MySQL automatically anytime you connect from a client program. Thus, instead of having to type:

---

[2]NB: as of version 5.7, the `mysql.user` table include the field `authentication_string` instead of `password`.

```
mysql -u yourusername -p
```

you should be automatically logged on with just `mysql`. Moreover, you can have `dplyr` read your MySQL option file using the `default.file` argument (see Section F.4.3).

F.2.3 Running scripts from the command line

MySQL will run SQL scripts contained in a file via the command line client. If the file `myscript.sql` is a text file containing MySQL commands, you can run it using the following command from your shell:

```
mysql -u yourusername -p dbname < myscript.sql
```

The result of each command in that script will be displayed in the terminal. Please see Section 13.3 for an example of this process in action.

F.3 PostgreSQL

Setting up a PostgreSQL server is logically analogous to the procedure demonstrated above for MySQL. The default user in a PostgreSQL installation is `postgres` and the default password is either `postgres` or blank. Either way, you can log into the PostgreSQL command line client—which is called `psql`—using the `sudo` command in your shell.

```
sudo -u postgres psql
```

This means: "Launch the `psql` program as if I was the user `postgres`." If this is successful, then you can create a new account for yourself from inside PostgreSQL. Here again, the procedure is similar to the procedure demonstrated above for MySQL in section F.2.2.

You can list all of the PostgreSQL users by typing at your `postgres` prompt:

```
\du
```

You can change the password for the `postgres` user:

```
ALTER USER postgres PASSWORD 'some_pass';
```

Create a new account for yourself:

```
CREATE USER yourusername SUPERUSER CREATEDB PASSWORD 'some_pass';
```

Create a new database called `airlines`:

```
CREATE DATABASE airlines;
```

Quit the `psql` client by typing:

```
\q
```

Now that your user account is created, you can log out and back in with the shell command:

```
psql -U yourusername -W
```

If this doesn't work, it is probably because the client authentication is set to `ident` instead of `md5`. Please see the documentation on client authentication for instructions on how to correct this on your installation, or simply continue to use the `sudo` method described above.

F.4   Connecting to SQL

There are many different options for connecting to and retrieving data from an SQL server. In all cases, you will need to specify at least four pieces of information:

host the name of the SQL server. If you are running this server locally, that name is `localhost`

dbname the name of the database on that server to which you want to connect (e.g., `airlines`)

user your username on the SQL server

password your password on the SQL server

F.4.1   The command line client

From the command line, the syntax is:

```
mysql -u username -p -h localhost dbname
```

After entering your password, this will bring you to an interactive MySQL session, where you can bounce queries directly off of the server and see the results in your terminal. This is often useful for debugging, because you can see the error messages directly, and you have the full suite of MySQL directives at your disposal. On the other hand, it is a fairly cumbersome route to database development, since you are limited to text-editing capabilities of the command line.

Command-line access to PostgreSQL is provided via the `psql` program described above.

F.4.2   GUIs

The MySQL Workbench is a graphical user interface (GUI) that can be useful for configuration and development. This software is available on Windows, Linux, and Mac OS X (see `https://www.mysql.com/products/workbench`). The analogous tool for PostgreSQL is pgAdmin, and it is similarly cross-platform. `sqlitebrowser` is another cross-platform GUI for SQLite databases.

These programs provide full-featured access to the underlying database system, with many helpful and easy-to-learn drop-down menus. We recommend developing queries and databases in these programs, especially when learning SQL.

F.4.3   R and RStudio

The downside to the previous approaches is that you don't actually capture the data returned by your queries, so you can't do anything with them. Using the GUIs, you can of course save the results of any query to a CSV. But a more elegant solution is to pull the data directly into R. This functionality is provided by the RMySQL, RPostgreSQL, and RSQLite packages. The DBI package provides a common interface to all three of the SQL back-ends listed above, and the dplyr package provides a slicker interface to DBI. A schematic of these dependencies is displayed in Figure F.1. We recommend using either the dplyr or the DBI interfaces whenever possible, since they are implementation agnostic.

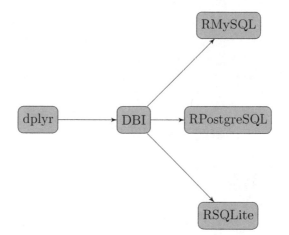

Figure F.1: Schematic of SQL-related R packages and their dependencies.

For most purposes (e.g., SELECT queries) there may be significant performance advantages to using the dplyr interface. However, the functionality of this construction is limited to SELECT queries. Thus, other SQL directives (e.g., EXPLAIN, INSERT, UPDATE, etc.) will not work in the dplyr construction. This functionality must be accessed using DBI.

In what follows, we illustrate how to connect to a MySQL backend using dplyr and DBI. However, the instructions for connecting to a PostgreSQL and SQLite are perfectly analogous. First, you will need to load the relevant package.

```
library(RMySQL)
```

Using dplyr

To set up a connection to a MySQL database using dplyr, we must specify the four parameters outlined above, and save the resulting object using the src_mysql() function.

```
library(dplyr)
db <- src_mysql(dbname = "airlines", host = "localhost",
                user = "r-user", password = "mypass")
```

If you have a MySQL option file already set up (see Section F.2.2), then you can alternatively connect using the default.file argument. This enables you to connect without having to type your password, or save it in plaintext in your R scripts.

```
db <- src_mysql(dbname = "airlines", host = "localhost",
                default.file = "~/.my.cnf",
                user = NULL, password = NULL)
```

Next, we can retrieve data using the tbl function and the sql() command.

```
res <- tbl(db, sql("SELECT faa, name FROM airports"))
res

Source:    query [?? x 2]
Database: mysql 5.5.47-0ubuntu0.14.04.1 [r-user@localhost:/airlines]

       faa                        name
     <chr>                      <chr>
1     04G            Lansdowne Airport
2     06A  Moton Field Municipal Airport
3     06C          Schaumburg Regional
4     06N              Randall Airport
5     09J         Jekyll Island Airport
6     0A9 Elizabethton Municipal Airport
7     0G6         Williams County Airport
8     0G7  Finger Lakes Regional Airport
9     0P2   Shoestring Aviation Airfield
10    0S9          Jefferson County Intl
# ... with more rows
```

Note that the resulting object has class `tbl_sql`.

```
class(res)

[1] "tbl_mysql" "tbl_sql"    "tbl_lazy"  "tbl"
```

Note also that the derived table is described as having an unknown (??) number of rows. This is because dplyr is smart (and lazy) about evaluation. It hasn't actually pulled all of the data into R. To force it to do so, use `collect()`.

```
collect(res)

# A tibble: 1,458   2
     faa                        name
     <chr>                      <chr>
1     04G            Lansdowne Airport
2     06A  Moton Field Municipal Airport
3     06C          Schaumburg Regional
4     06N              Randall Airport
5     09J         Jekyll Island Airport
6     0A9 Elizabethton Municipal Airport
7     0G6         Williams County Airport
8     0G7  Finger Lakes Regional Airport
9     0P2   Shoestring Aviation Airfield
10    0S9          Jefferson County Intl
# ... with 1,448 more rows
```

Using DBI

For a closer connection to the SQL server, we use DBI. A connection object can be created using the `dbConnect()` function, which works similarly to the dplyr connection we created above.

```
library(DBI)
con <- dbConnect(MySQL(), dbname = "airlines", host = "localhost",
                user = "r-user", password = "mypass")
```

Next, we use the dbGetQuery() function to send an SQL command to the server and retrieve the results.

```
res <- dbGetQuery(con, "SELECT faa, name FROM airports")
head(res, 10)
```

```
     faa                          name
1    04G              Lansdowne Airport
2    06A  Moton Field Municipal Airport
3    06C            Schaumburg Regional
4    06N                Randall Airport
5    09J           Jekyll Island Airport
6    0A9 Elizabethton Municipal Airport
7    0G6         Williams County Airport
8    0G7  Finger Lakes Regional Airport
9    0P2    Shoestring Aviation Airfield
10   0S9           Jefferson County Intl
```

Note that this time, the results are stored as a data.frame.

```
class(res)
```

```
[1] "data.frame"
```

Unlike the tbl() function from dplyr, dbGetQuery() can execute arbitrary SQL commands, not just SELECT statements. So we can also run EXPLAIN, DESCRIBE, and SHOW commands.

```
dbGetQuery(con, "EXPLAIN SELECT faa, name FROM airports")
```

```
  id select_type    table type possible_keys  key key_len  ref rows Extra
1  1      SIMPLE airports  ALL          <NA> <NA>    <NA> <NA> 1458
```

```
dbGetQuery(con, "DESCRIBE airports")
```

```
     Field          Type Null Key Default Extra
1      faa    varchar(3)   NO PRI
2     name  varchar(255)  YES         <NA>
3      lat decimal(10,7)  YES         <NA>
4      lon decimal(10,7)  YES         <NA>
5      alt       int(11)  YES         <NA>
6       tz  smallint(4)  YES         <NA>
7      dst       char(1)  YES         <NA>
8     city  varchar(255)  YES         <NA>
9  country  varchar(255)  YES         <NA>
```

```
dbGetQuery(con, "SHOW DATABASES")
```

```
          Database
1 information_schema
2          airlines
3              imdb
4            lahman
5              math
6         retrosheet
7              yelp
```

Connection objects

Note that the db object that we created with dplyr is of class src_mysql.

```
db
```

```
src:  mysql 5.5.47-0ubuntu0.14.04.1 [r-user@localhost:/airlines]
tbls: airports, carriers, flights, planes, summary, weather
```

```
class(db)
```

```
[1] "src_mysql" "src_sql"   "src"
```

However, the con connection object we created with DBI is of class MySQL Connection.

```
con
```

```
<MySQLConnection:0,1>
```

```
class(con)
```

```
[1] "MySQLConnection"
attr(,"package")
[1] "RMySQL"
```

Although they were created with all of the same information, they are not the same. However, the db object contains an object functionally equivalent to con. Namely, db$con.

```
class(db$con)
```

```
[1] "MySQLConnection"
attr(,"package")
[1] "RMySQL"
```

Thus, once you have a created a connection to your database through dplyr, you can use all of the DBI functions without having to create a new connection.

```
dbGetQuery(db$con, "SHOW TABLES")
```

```
  Tables_in_airlines
1           airports
2           carriers
```

```
3                flights
4                 planes
5                summary
6                weather
```

F.4.4   Load into SQLite database

A process similar to the one we exhibit in Section 13.3 can be used to create a SQLite database, although in this case it is not even necessary to specify the table schema in advance. Launch `sqlite3` from the command line using the shell command:

```
sqlite3
```

Create a new database called `babynames` in the current directory using the `.open` command:

```
.open babynamesdata.sqlite3
```

Next, set the `.mode` to `csv`, import the two tables, and exit.

```
.mode csv
.import babynames.csv babynames
.import births.csv births
.exit
```

This should result in an SQLite database file called `babynamesdata.sqlite3` existing in the current directory that contains two tables. We can connect to this database and query it using `dplyr`.

```
db <- src_sqlite(path = "babynamesdata.sqlite3")
babynames <- tbl(db, "babynames")
babynames %>% filter(name == "Benjamin")
```

```
Source:   query [?? x 5]
Database: sqlite 3.8.6 [babynamesdata.sqlite3]
```

	year	sex	name	n	prop
	<chr>	<chr>	<chr>	<chr>	<chr>
1	1976	F	Benjamin	53	3.37186805943904e-05
2	1976	M	Benjamin	10680	0.0065391571834601
3	1977	F	Benjamin	63	3.83028784917178e-05
4	1977	M	Benjamin	12112	0.00708409319279004
5	1978	F	Benjamin	73	4.44137806835342e-05
6	1978	M	Benjamin	11411	0.00667764880752091
7	1979	F	Benjamin	79	4.58511127310548e-05
8	1979	M	Benjamin	12516	0.00698620342042644
9	1980	F	Benjamin	80	4.49415983928884e-05
10	1980	M	Benjamin	13630	0.00734980487697031

```
# ... with more rows
```

Bibliography

[1] D. Alberani. Imdbpy, 2014. http://imdbpy.sourceforge.net.

[2] J. Albert. *Teaching statistics using baseball.* Mathematical Association of America: Washington, DC, 2003.

[3] R. Albert and A.-L. Barabási. Statistical mechanics of complex networks. *Reviews of modern physics*, 74(1):47, 2002.

[4] J. Allaire, J. Cheng, Y. Xie, J. McPherson, W. Chang, J. Allen, H. Wickham, A. Atkins, and R. Hyndman. *rmarkdown: Dynamic Documents for R*, 2016. R package version 1.0.

[5] J. J. Allaire, J. Horner, V. Marti, and N. Porte. *markdown: Markdown rendering for R*, 2014. R package version 0.7.4.

[6] Z. W. Almquist. US Census spatial and demographic data in R: The UScensus2000 suite of packages. *Journal of Statistical Software*, 37(6):1–31, 2010.

[7] Z. W. Almquist. *US Census 2000 tract level shapefiles and additional demographic data*, 2012. R package version 0.03.

[8] American Statistical Association Undergraduate Guidelines Workgroup. *2014 Curriculum Guidelines for Undergraduate Programs in Statistical Science*, 2014. http://www.amstat.org/education/curriculumguidelines.cfm.

[9] J. B. Arnold. *ggthemes: Extra Themes, Scales and Geoms for ggplot2*, 2016. R package version 3.0.1.

[10] D. Attali. *ggExtra: Add Marginal Histograms to 'ggplot2', and More 'ggplot2' Enhancements*, 2016. R package version 0.5.

[11] S. B. Bache and H. Wickham. *magrittr: a forward-pipe operator for R*, 2014. R package version 1.0.1.

[12] R. Ball and N. Medeiros. Teaching integrity in empirical research: A protocol for documenting data management and analysis. *The Journal of Economic Education*, 43(2):182–189, 2012.

[13] A.-L. Barabási and R. Albert. Emergence of scaling in random networks. *Science*, 286(5439):509–512, 1999.

[14] A.-L. Barabási and J. Frangos. *Linked: the new science of networks.* Basic Books, 2014.

[15] P. Basu, B. S. Baumer, A. Bar-Noy, C.-K. Chau, and M. City. Social-communication composite networks. In *Opportunistic Mobile Social Networks*, pages 1–36. CRC Press, 2014.

[16] B. Baumer. In a Moneyball world, a number of teams remain slow to buy into sabermetrics. In R. Webb, editor, *The Great Analytics Rankings*. ESPN.com, 2 2015. `http://espn.go.com/espn/feature/story/_/id/12331388/the-great-analytics-rankings#!mlb`.

[17] B. Baumer, P. Basu, and A. Bar-Noy. Modeling and analysis of composite network embeddings. In A. Helmy, B. Landfeldt, and L. Bononi, editors, *MSWiM*, pages 341–350. ACM, 2011.

[18] B. Baumer, N. J. Horton, and D. Kaplan. *mdsr: Complement to 'Modern Data Science with R'*, 2016. R package version 0.1.3.

[19] B. S. Baumer. *fec: Data about Federal Elections*. R package version 0.0.0.9005.

[20] B. S. Baumer. A data science course for undergraduates: Thinking with data. *The American Statistician*, 69(4):334–342, 2015.

[21] B. S. Baumer. *airlines: Data About Flights*, 2016. R package version 0.2.2.9010.

[22] B. S. Baumer. *etl: Extract-Transform-Load Framework for Medium Data*, 2016. R package version 0.3.3.

[23] B. S. Baumer, M. Çetinkaya Rundel, A. Bray, L. Loi, and N. J. Horton. R markdown: Integrating a reproducible analysis tool into introductory statistics. *Technology Innovations in Statistics Education*, 8(1), 2014.

[24] B. S. Baumer, R. Goueth, W. Li, and W. Zhang. *Weather and spatial data from the MacLeish field station*, 2016. R package version 0.9.

[25] B. S. Baumer, G. Rabanca, A. Bar-Noy, and P. Basu. Star search: Effective subgroups in collaborative social networks. In J. Pei, F. Silvestri, and J. Tang, editors, *Proceedings of the 2015 IEEE/ACM International Conference on Advances in Social Networks Analysis and Mining, ASONAM 2015, Paris, France, August 25–28, 2015*, pages 729–736. ACM, 2015.

[26] B. S. Baumer, Y. Wei, and G. S. Bloom. The smallest non-autograph. *Discussiones Mathematicae Graph Theory*, 36(3):577–602, 2016.

[27] B. S. Baumer and A. Zimbalist. *The Sabermetric Revolution: Assessing the Growth of Analytics in Baseball*. University of Pennsylvania Press: Philadelphia, PA, 2014.

[28] M. Beck. *NeuralNetTools: Visualization and Analysis Tools for Neural Networks*, 2015. R package version 1.4.0.

[29] R. A. Becker, A. R. Wilks, R. Brownrigg, T. P. Minka, and A. Deckmyn. *maps: Draw Geographical Maps*, 2016. R package version 3.1.1.

[30] D. R. Bild, Y. Liu, R. P. Dick, Z. M. Mao, and D. S. Wallach. Aggregate characterization of user behavior in Twitter and analysis of the retweet graph. *ACM Transactions on Internet Technology (TOIT)*, 15(1):4, 2015.

[31] R. Bivand, T. Keitt, and B. Rowlingson. *rgdal: Bindings for the Geospatial Data Abstraction Library*, 2016. R package version 1.1-10.

[32] R. Bivand and N. Lewin-Koh. *maptools: Tools for Reading and Handling Spatial Objects*, 2016. R package version 0.8-39.

[33] R. Bivand and C. Rundel. *rgeos: Interface to Geometry Engine - Open Source (GEOS)*, 2016. R package version 0.3-19.

[34] R. S. Bivand, E. Pebesma, and V. Gómez-Rubio. *Applied Spatial Data Analysis with R (second edition)*. Springer Verlag: New York, NY, 2013.

[35] P. Bogdanov, B. S. Baumer, P. Basu, A. Bar-Noy, and A. K. Singh. As strong as the weakest link: Mining diverse cliques in weighted graphs. In H. Blockeel, K. Kersting, S. Nijssen, and F. Zelezný, editors, *Machine Learning and Knowledge Discovery in Databases - European Conference, ECML PKDD 2013, Prague, Czech Republic, September 23–27, 2013, Proceedings, Part I*, volume 8188 of *Lecture Notes in Computer Science*, pages 525–540. Springer Verlag: New York, NY, 2013.

[36] C. Bombardier, L. Laine, A. Reicin, D. Shapiro, R. Burgos-Vargas, B. Davis, R. Day, M. B. Ferraz, C. J. Hawkey, M. C. Hochberg, T. K. Kvien, and T. J. Schnitzer. Comparison of upper gastrointestinal toxicity of rofecoxib and naproxen in patients with rheumatoid arthritis. *New England Journal of Medicine*, 343:1520–1528, 2000.

[37] L. Breiman. Statistical modeling: The two cultures. *Statistical Science*, 16(3):199–215, 2001. http://www.jstor.org/stable/2676681.

[38] C. A. Brewer. Color use guidelines for mapping and visualization. *Visualization in modern cartography*, 2:123–148, 1994.

[39] C. A. Brewer. Color use guidelines for data representation. In *Proceedings of the Section on Statistical Graphics, American Statistical Association*, pages 55–60, 1999.

[40] F. Briatte. *ggnetwork: Geometries to Plot Networks with 'ggplot2'*, 2016. R package version 0.5.1.

[41] L. C. Bridgeford. Q&A: Statistical proof of discrimination isn't static, June 20, 2014. http://www.bna.com/qa-statistical-proof-b17179891425.

[42] J. Bryan and J. Zhao. *googlesheets: Manage Google Spreadsheets from R*, 2016. R package version 0.2.1.

[43] C. T. Butts. network: a package for managing relational data in R. *Journal of Statistical Software*, 24(2), 2008.

[44] C. T. Butts. *sna: Tools for Social Network Analysis*, 2016. R package version 2.4.

[45] A. R. Cannon, G. W. Cobb, B. A. Hartlaub, J. M. Legler, R. H. Lock, T. L. Moore, A. J. Rossman, and J. Witmer. *STAT2: Building Models for a World of Data*. W. H. Freeman and Company: New York, NY, 2013.

[46] CERN. Lhc guide: A collection of facts and figures about the large hadron collider (lhc) in the form of questions and answers, 2008. http://cds.cern.ch/record/1092437/files/CERN-Brochure-2008-001-Eng.pdf?version=1.

[47] N. Chamandy, O. Muraldharan, and S. Wager. Teaching statistics at 'Google-Scale'. *The American Statistician*, 69(4):283–291, 2015.

[48] W. Chang. *webshot: Take Screenshots of Web Pages*, 2016. R package version 0.3.2.

[49] W. Chang, J. Cheng, J. J. Allaire, Y. Xie, and J. McPherson. *shiny: Web Application Framework for R*, 2016. R package version 0.13.2.

[50] W. Chang and H. Wickham. *ggvis: Interactive grammar of graphics*, 2016. R package version 0.4.3.

[51] J. Cheng and Y. Xie. *leaflet: Create Interactive Web Maps with the JavaScript 'Leaflet' Library*, 2016. R package version 1.0.1.

[52] C. Choirat, J. Honaker, K. Imai, G. King, and O. Lau. *Zelig: Everyone's Statistical Software*, 2016. R package version 5.0-12.

[53] C. Cinelli. *benford.analysis: Benford Analysis for Data Validation and Forensic Analytics*, 2016. R package version 0.1.4.

[54] W. S. Cleveland. Data science: an action plan for expanding the technical areas of the field of statistics. *International statistical review*, 69(1):21–26, 2001. http://www.jstor.org/stable/1403527.

[55] W. S. Cleveland and R. McGill. Graphical perception: Theory, experimentation, and application to the development of graphical methods. *Journal of the American Statistical Association*, 79(387):531–554, 1984.

[56] G. W. Cobb. The introductory statistics course: A Ptolemaic curriculum? *Technology Innovations in Statistics Education (TISE)*, 1(1), 2007. http://escholarship.org/uc/item/6hb3k0nz.

[57] G. W. Cobb. Mere renovation is too little too late: We need to rethink our undergraduate curriculum from the ground up. *The American Statistician*, 69(4):266–282, 2015.

[58] Committee on Professional Ethics. *Ethical Guidelines for Statistical Practice*, August 1999. http://www.amstat.org/about/ethicalguidelines.cfm.

[59] R. D. Cook. *Residuals and Influence in Regression*. Chapman & Hall, London, 1982.

[60] N. Cressie. *Statistics for spatial data*. John Wiley & Sons: Hoboken, NJ, 1993.

[61] G. Csardi and T. Nepusz. The igraph software package for complex network research. *InterJournal*, Complex Systems:1695, 2006.

[62] D. B. Dahl. *xtable: Export Tables to LaTeX or HTML*, 2016. R package version 1.8-2.

[63] D. M. Diez, C. D. Barr, and M. Çetinkaya Rundel. *OpenIntro Statistics (third edition)*. OpenIntro.org, 2015.

[64] D. Donoho. 50 years of data science. Technical report, Stanford University, 9 2015. http://courses.csail.mit.edu/18.337/2015/docs/50YearsDataScience.pdf.

[65] D. Easley and J. Kleinberg. *Networks, crowds, and markets: Reasoning about a highly connected world*. Cambridge University Press: Cambridge, UK, 2010.

[66] D. Eddelbuettel and R. François. Rcpp: Seamless R and C++ integration. *Journal of Statistical Software*, 40(8):1–18, 2011.

[67] Editorial. Announcement: Reducing our irreproducibility. *Nature*, 496, April 2013.

[68] B. Efron and T. Hastie. *Computer Age Statistical Inference: Algorithms, Evidence, and Data Science*. Cambridge University Press: Cambridge, UK, 2016.

[69] B. Efron and R. J. Tibshirani. *An Introduction to the Bootstrap*. Chapman & Hall, London, 1993.

[70] J. H. Ellenberg. Ethical guidelines for statistical practice: a historical perspective. *The American Statistician*, 37(1):1–4, 1983.

[71] R. L. Engstrom and J. K. Wildgen. Pruning thorns from the thicket: an empirical test of the existence of racial gerrymandering. *Legislative Studies Quarterly*, pages 465–479, 1977.

[72] P. Erdős and A. Rényi. On random graphs. *Publicationes Mathematicae Debrecen*, 6:290–297, 1959.

[73] L. Euler. Leonhard Euler and the Königsberg bridges. *Scientific American*, 189(1):66–70, 1953.

[74] J. Faraway. *faraway: Functions and Datasets for Books by Julian Faraway*, 2016. R package version 1.0.7.

[75] I. Feinere, K. Hornik, and D. Meyer. Text mining infrastructure in R. *Journal of Statistical Software*, 25(5):1–54, 2008.

[76] I. Fellows. *wordcloud: Word Clouds*, 2014. R package version 2.5.

[77] W. Finzer. The data science education dilemma. *Technology Innovations in Statistics Education*, 7(2), 2013. http://escholarship.org/uc/item/7gv0q9dc.pdf.

[78] J. Fox. Aspects of the social organization and trajectory of the R Project. *The R Journal*, 1(2):5–13, December 2009.

[79] C. Fraley, A. E. Raftery, T. B. Murphy, and L. Scrucca. *mclust: Normal Mixture Modeling for Model-Based Clustering, Classification, and Density Estimation*, 2016. R package version 5.2.

[80] M. Friendly. *Lahman: Sean Lahman's Baseball Database*, 2015. R package version 4.0-1.

[81] G. Futschek. *Algorithmic thinking: the key for understanding computer science, in R. Mittermeir (Ed.) Informatics education–the bridge between using and understanding computers, (Vol. 4226)*. Springer Verlag: New York, NY, 2006.

[82] C. Gandrud. *Reproducible Research with R and RStudio*. CRC Press: Boca Raton, FL, 2014.

[83] M. R. Garey and D. S. Johnson. *Computers and Intractability: A Guide to the Theory of NP-completeness*. W. H. Freeman and Company: New York, NY, 1979.

[84] A. Gelman. Ethics and statistics: Open data and open methods. *Chance*, 24(4):51–53, 2011.

[85] A. Gelman. Ethics and statistics: Ethics and the statistical use of prior information. *Chance*, 25(4):52–54, 2012.

[86] A. Gelman and E. Loken. Ethics and statistics: Statisticians: When we teach, we don't practice what we preach. *Chance*, 25(1):47–48, 2012.

[87] A. Gelman, C. Pasarica, and R. Dodhia. Let's practice what we preach: turning tables into graphs. *The American Statistician*, 56(2):121–130, 2002.

[88] J. Gentry. *twitteR: R based Twitter client*, 2015. R package version 1.1.9.

[89] P. L. Good and J. W. Hardin. *Common Errors in Statistics (and How to Avoid Them, fourth edition)*. John Wiley & Sons: Hoboken, NJ, 2012.

[90] Google. R style guide. Technical report, Google, 2016. https://google.github.io/styleguide/Rguide.xml.

[91] P. Gramieri, A. Titelbaum, and X. Wang. *atus: American Time Use Survey 2014 Data*, 2015. R package version 0.1.

[92] A. S. Graphodatsky, V. A. Trifonov, and R. Stanyon. The genome diversity and karyotype evolution of mammals. *Molecular cytogenetics*, 4(1):1, 2011.

[93] J. L. Green and E. E. Blankenship. Fostering conceptual understanding in mathematical statistics. *The American Statistician*, 69(4):315–325, 2015.

[94] G. Grolemund and H. Wickham. Dates and times made easy with lubridate. *Journal of Statistical Software*, 40(3):1–25, 2011.

[95] J. Hardin, R. Hoerl, N. J. Horton, D. Nolan, B. S. Baumer, O. Hall-Holt, P. Murrell, R. Peng, P. Roback, D. Temple Lang, and M. D. Ward. Data science in statistics curricula: Preparing students to 'think with data'. *The American Statistician*, 69(4):343–353, 2015.

[96] F. E. Harrell. *Hmisc: Harrell Miscellaneous*, 2016. R package version 3.17-4.

[97] T. Hastie and B. Efron. *lars: Least Angle Regression, Lasso and Forward Stagewise*, 2013. R package version 1.2.

[98] T. Hastie, R. Tibshirani, and J. J. H. Friedman. *The elements of statistical learning*. Springer Verlag: New York, NY, 2nd edition, 2009. http://www-stat.stanford.edu/~tibs/ElemStatLearn/.

[99] T. Hengl, P. Roudier, D. Beaudette, and E. Pebesma. plotKML: Scientific visualization of spatio-temporal data. *Journal of Statistical Software*, 63(5):1–25, 2015.

[100] T. Herndon, M. Ash, and R. Pollin. Does high public debt consistently stifle economic growth? a critique of Reinhart and Rogoff. *Cambridge journal of economics*, 38(2):257–279, 2014.

[101] T. Hesterberg. What teachers should know about the bootstrap: Resampling in the undergraduate statistics curriculum. *The American Statistician*, 69(4):371–386, 2015.

[102] T. C. Hesterberg, D. S. Moore, S. Monaghan, A. Clipson, and R. Epstein. *Bootstrap Methods and Permutation Tests*. W. H. Freeman and Company: New York, NY, 2005.

[103] D. C. Hoaglin. Regressions are commonly misinterpreted. *Stata Journal*, 16(1):5–22, 2016.

[104] J. K. Hodge, E. Marshall, and G. Patterson. Gerrymandering and convexity. *The College Mathematics Journal*, 41(4):312–324, 2010.

[105] N. J. Horton. I hear, I forget. I do, I understand: A modified Moore-method mathematical statistics course. *The American Statistician*, 67(3):219–228, 2013.

[106] N. J. Horton. Challenges and opportunities for statistics and statistical education: looking back, looking forward. *The American Statistician*, 69(2):138–145, 2015.

[107] N. J. Horton, B. S. Baumer, and H. Wickham. Setting the stage for data science: integration of data management skills in introductory and second courses in statistics. *Chance*, 28(2), 2015. `http://chance.amstat.org/2015/04/setting-the-stage/`.

[108] N. J. Horton, E. R. Brown, and L. Qian. Use of R as a toolbox for mathematical statistics exploration. *The American Statistician*, 58(4):343–357, 2004.

[109] N. J. Horton and J. S. Hardin. Teaching the next generation of statistics students to "think with data": special issue on statistics and the undergraduate curriculum. *The American Statistician*, 69(4):259–265, 2015.

[110] N. J. Horton and K. P. Kleinman. Much ado about nothing: A comparison of missing data methods and software to fit incomplete data regression models. *The American Statistician*, 61:79–90, 2007.

[111] T. Hothorn and A. Zeileis. *partykit: A Toolkit for Recursive Partytioning*, 2016. R package version 1.1-0.

[112] B. Howe. Data manipulation at scale: Systems and algorithms. Coursera, 2014. public materials at `https://github.com/uwescience/datasci_course_materials`.

[113] L. Hubert and H. Wainer. *A statistical guide for the ethically perplexed*. CRC Press, 2012.

[114] D. Huff. *How to lie with statistics*. W.W. Norton & Company: New York, NY, 1954.

[115] L. Hyafil and R. L. Rivest. Constructing optimal binary decision trees is NP-complete. *Information Processing Letters*, 5(1):15–17, 1976.

[116] R. Ihaka and R. Gentleman. R: A language for data analysis and graphics. *Journal of Computational and Graphical Statistics*, 5(3):299–314, 1996.

[117] IMDB.com. Internet movie database, 2013. `http://www.imdb.com/help/show_article?conditions`.

[118] J. P. Ioannidis. Why most published research findings are false. *Chance*, 18(4):40–47, 2005. `http://journals.plos.org/plosmedicine/article?id=10.1371/journal.pmed.0020124`.

[119] B. James. *The Bill James Historical Baseball Abstract*. Random House: New York, NY, 1986.

[120] D. A. James and S. Falcon. *RSQLite: SQLite interface for R*, 2013. R package version 0.11.4.

[121] G. James, D. Witten, T. Hastie, and R. Tibshirani. *An introduction to statistical learning*. Springer Verlag: New York, NY, 2013. `http://www-bcf.usc.edu/~gareth/ISL/`.

[122] D. Kahle and H. Wickham. ggmap: Spatial visualization with ggplot2. *The R Journal*, 5(1):144–161, 2013.

[123] S. Kern, I. Skoog, A. Börjesson-Hanson, K. Blennow, H. Zetterberg, S. Östling, J. Kern, P. Gudmundsson, T. Marlow, L. Rosengren, et al. Higher CSF interleukin-6 and CSF interleukin-8 in current depression in older women. results from a population-based sample. *Brain, behavior, and immunity*, 41:55–58, 2014.

[124] S. Kern, I. Skoog, A. Börjesson-Hanson, S. Östling, J. Kern, P. Gudmundsson, T. Marlow, M. Waern, K. Blennow, H. Zetterberg, et al. Retraction notice to "lower CSF interleukin-6 predicts future depression in a population-based sample of older women followed for 17 years". *Brain, Behavior, and Immunity*, 32:153–158, 2013.

[125] A. Y. Kim. OKCupid data for introductory statistics and data science courses. *Journal of Statistics Education*, 23(2), 2015.

[126] K. E. Kline, D. Kline, B. Hunt, and D. Heymann-Reder. *SQL in a nutshell*. O'Reilly Media: Sebastopol, CA, 2008. 3rd edition.

[127] D. Knuth. Literate programming. *CSLI Lecture Notes, Stanford University*, 27, 1992.

[128] M. Kuhn. Cran task view: Reproducible research, 12 2015. accessed November 22, 2016.

[129] S. Kuiper and J. Sklar. *Practicing statistics: Guided investigations for the second course*. Pearson Education: New York, NY, 2012.

[130] D. Laney. 3D data management: Controlling data volume, velocity and variety. *META Group Research Note*, 6:70, 2001.

[131] M. Lewis. *Moneyball: The Art of Winning an Unfair Game*. W.W. Norton & Company: New York, NY, 2003.

[132] A. Liaw and M. Wiener. Classification and regression by randomforest. *R News*, 2(3):18–22, 2002.

[133] R. J. A. Little and D. B. Rubin. *Statistical Analysis With Missing Data (second edition)*. John Wiley & Sons: Hoboken, NJ, 2002.

[134] M. Loecher and K. Ropkins. RgoogleMaps and loa: Unleashing R graphics power on map tiles. *Journal of Statistical Software*, 63(4):1–18, 2015.

[135] Y. Lu, P. Zhang, Y. Cao, Y. Hu, and L. Guo. On the frequency distribution of retweets. *Procedia Computer Science*, 31:747–753, 2014.

[136] L. Ludwig. Technically speaking, 2012. `http://techspeaking.denison.edu/Technically_Speaking`.

[137] T. Lumley. *biglm: Bounded Memory Linear and Generalized Linear Models*, 2013. R package version 0.9-1.

[138] J. Luraschi, K. Ushey, J. J. Allaire, and The Apache Software Foundation. *sparklyr: R Interface to Apache Spark*. R package version 0.2.32.

[139] J. Mackenzie. Gerrymandering and legislator efficiency. Technical report, University of Delaware, 2009. `https://www.udel.edu/johnmack/research/gerrymandering.pdf`.

[140] M. Marchi and J. Albert. *Analyzing Baseball Data with R*. CRC Press: Boca Raton, FL, 2013.

[141] E. McCallum and S. Weston. *Parallel R.* O'Reilly Media: Sebastopol, CA, 2011.

[142] J. McGregor. Glassdoor ranks the 25 best jobs in America. *The Washington Post*, 1 2016.

[143] D. Meyer, E. Dimitriadou, K. Hornik, A. Weingessel, and F. Leisch. *e1071: Misc Functions of the Department of Statistics, Probability Theory Group (Formerly: E1071), TU Wien*, 2015. R package version 1.6-7.

[144] F. Mosteller. *Fifty Challenging Problems in Probability with Solutions.* Dover Publications: Mineola, NY, 1987.

[145] E. Neuwirth. *RColorBrewer: ColorBrewer Palettes*, 2014. R package version 1.1-2.

[146] R. G. Niemi, B. Grofman, C. Carlucci, and T. Hofeller. Measuring compactness and the role of a compactness standard in a test for partisan and racial gerrymandering. *The Journal of Politics*, 52(4):1155–1181, 1990.

[147] D. Nolan and J. Perrett. Teaching and learning data visualization: Ideas and assignments. *The American Statistician*, 70(3):260–269, 2016.

[148] D. Nolan and T. P. Speed. Teaching statistics theory through applications. *The American Statistician*, 53:370–375, 1999.

[149] D. Nolan and D. Temple Lang. Computing in the statistics curricula. *The American Statistician*, 64(2):97–107, 2010. `http://www.stat.berkeley.edu/users/statcur/Preprints/ComputingCurric3.pdf`.

[150] R. Nuzzo. Scientific method: statistical errors. *Nature*, 506:150–152, 2014.

[151] P. Ohm. Broken promises of privacy: Responding to the surprising failure of anonymization. *UCLA Law Review*, 57:1701, 2010.

[152] R. Olford and W. Cherry. Picturing probability: the poverty of Venn diagrams, the richness of Eikosograms, 9 2003. `http://sas.uwaterloo.ca/~rwoldfor/papers/venn/eikosograms/paper.pdf`.

[153] C. O'Neil. *Weapons of Math Destruction: How Big Data Increases Inequality and Threatens Democracy.* Crown Publishing, New York, NY, 2016.

[154] J. Ooms. The jsonlite package: A practical and consistent mapping between JSON data and R objects. *arXiv:1403.2805 [stat.CO]*, 2014.

[155] J. Ooms, D. James, S. DebRoy, H. Wickham, and J. Horner. *RMySQL: Database Interface and MySQL Driver for R*, 2016. R package version 0.10.9.

[156] L. Page, S. Brin, R. Motwani, and T. Winograd. The PageRank citation ranking: bringing order to the web. Technical report, Stanford University InfoLab, 1999. `http://ilpubs.stanford.edu:8090/422`.

[157] E. Paradis, J. Claude, and K. Strimmer. APE: analyses of phylogenetics and evolution in R language. *Bioinformatics*, 20:289–290, 2004.

[158] Park City Math Institute (PCMI) Undergraduate Faculty Group: R. De Veaux (chair), M. Agarwal, M. Averett, B. Baumer, A. Bray, T. Bressoud, L. Bryant, L. Cheng, A. Francis, R. Gould, A. Y. Kim, M. Kretchmar, Q. Lu, A. Moskol, D. Nolan, R. Pelayo, S. Raleigh, R. J. Sethi, M. Sondjaja, N. Tiruviluamala, P. Uhlig,

T. Washington, C. Wesley, D. White, and P. Ye. Curriculum guidelines for under-graduate programs in data science. *Annual Review of Statistics and Its Applications*, 4, 2016. in press.

[159] S. Pierson. Jordan urges both computational and inferential thinking in data science. *Amstat News*, 2016. http://magazine.amstat.org/blog/2016/03/01/jordan16.

[160] F. Provost and T. Fawcett. Data science and its relationship to big data and data-driven decision making. *Big Data*, 1(1):51–59, 2013.

[161] R. Pruim. *NHANES: Data from the US National Health and Nutrition Examination Study*, 2015. R package version 2.1.0.

[162] R. Pruim, D. Kaplan, and N. J. Horton. *mosaic: Project MOSAIC (mosaic-web.org) Statistics and Mathematics Teaching Utilities*, 2016. R package version 0.14.9000.

[163] R. Pruim, D. Kaplan, and N. J. Horton. *mosaicData: Project MOSAIC Data Sets*, 2016. R package version 0.14.9000.

[164] R Core Team. *foreign: Read Data Stored by Minitab, S, SAS, SPSS, Stata, Systat, Weka, dBase, ...*, 2014. R package version 0.8-61.

[165] R Core Team. *base package: A Language and Environment for Statistical Computing*. R Foundation for Statistical Computing, Vienna, Austria, 2016.

[166] R Core Team. *graphics package: A Language and Environment for Statistical Computing*. R Foundation for Statistical Computing, Vienna, Austria, 2016.

[167] R Core Team. *methods package: A Language and Environment for Statistical Computing*. R Foundation for Statistical Computing, Vienna, Austria, 2016.

[168] R Core Team. *parallel package: Parallel computing in R*. R Foundation for Statistical Computing, Vienna, Austria, 2016.

[169] R Core Team. *R: A Language and Environment for Statistical Computing*. R Foundation for Statistical Computing, Vienna, Austria, 2016. ISBN 3-900051-07-0.

[170] R Special Interest Group on Databases (R-SIG-DB), H. Wickham, and K. Müller. *DBI: R Database Interface*, 2016. R package version 0.5.

[171] T. E. Raghunathan. What do we do with missing data? some options for analysis of incomplete data. *Annual Review of Public Health*, 25:99–117, 2004.

[172] K. Ram and K. Broman. *aRxiv: Interface to the arXiv API*, 2015. R package version 0.5.10.

[173] K. Ram and H. Wickham. *wesanderson: A Wes Anderson Palette Generator*, 2015. R package version 0.3.2.

[174] J. A. Rice. *Mathematical Statistics and Data Analysis (third edition)*. Cengage Learning: Boston, MA, 2006.

[175] M. L. Rizzo. *Statistical Computing with R*. CRC Press: Boca Raton, FL, 2007.

[176] D. Robinson. *broom: Convert Statistical Analysis Objects into Tidy Data Frames*, 2016. R package version 0.4.1.

[177] K. Rogoff and C. Reinhart. Growth in a time of debt. *American Economic Review*, 100(2):573–8, 2010.

[178] J. P. Romano and A. F. Siegel. *Counterexamples in Probability and Statistics*. Cengage Learning: Boston, MA, 1986.

[179] K. Roose. Meet the 28-year-old grad student who just shook the global austerity movement, April 2013.

[180] O. Rosling, A. R. Rönnlund, and H. Rosling. Gapminder, 2005. `http://gapminder.org`.

[181] B. Rudis. *streamgraph: htmlwidget for building streamgraph visualizations*, 2015. R package version 0.8.1.

[182] D. Ruppert, M. P. Wand, and R. J. Carroll. *Semiparametric regression*. Cambridge University Press, Cambridge, UK, 2003.

[183] B. Schloerke, J. Crowley, D. Cook, H. Hofmann, H. Wickham, F. Briatte, and M. Marbach. *GGally: Extension to ggplot2*, 2014. R package version 0.4.8.

[184] A. Schwarz. *The Numbers Game: Baseball's Lifelong Fascination With Statistics*. St. Martin's Press: New York, NY, 2005.

[185] C. Sievert, C. Parmer, T. Hocking, S. Chamberlain, K. Ram, M. Corvellec, and P. Despouy. *plotly: Create Interactive Web Graphics via 'plotly.js'*, 2016. R package version 3.6.0.

[186] J. Silge and D. Robinson. tidytext: Text mining and analysis using tidy data principles in r. *JOSS*, 1(3), 2016.

[187] T. Sing, O. Sander, N. Beerenwinkel, and T. Lengauer. ROCR: visualizing classifier performance in R. *Bioinformatics*, 21(20):7881, 2005.

[188] M. Stonebraker, D. Abadi, D. J. DeWitt, S. Madden, E. Paulson, A. Pavlo, and A. Rasin. MapReduce and parallel DBMSs: friends or foes? *Communications of the ACM*, 53(1):64–71, 2010.

[189] P.-N. Tan, M. Steinbach, and V. Kumar. *Introduction to Data Mining*. Pearson Education: New York, NY, 1st edition, 2006. `http://www-users.cs.umn.edu/~kumar/dmbook/index.php`.

[190] D. Temple Lang. *RCurl: General network (HTTP/FTP/...) client interface for R*, 2014. R package version 1.95-4.3.

[191] T. Therneau, B. Atkinson, and B. Ripley. *rpart: Recursive Partitioning*, 2015. R package version 4.1-10.

[192] E. Torres-Manzanera. *xkcd: Plotting ggplot2 Graphics in an XKCD Style*, 2016. R package version 0.0.5.

[193] J. Travers and S. Milgram. An experimental study of the small world problem. *Sociometry*, pages 425–443, 1969.

[194] E. R. Tufte. *Envisioning Information*. Graphics Press: Cheshire, CT, 1990.

[195] E. R. Tufte. *Visual Explanations: Images and Quantities, Evidence and Narrative*. Graphics Press: Cheshire, CT, 1997.

[196] E. R. Tufte. *Visual Display of Quantitative Information (second edition)*. Graphics Press: Cheshire, CT, 2001.

[197] E. R. Tufte. *The Cognitive Style of PowerPoint*. Graphics Press: Cheshire, CT, 2003.

[198] E. R. Tufte. *Beautiful Evidence*. Graphics Press: Cheshire, CT, 2006.

[199] J. W. Tukey. Data-based graphics: visual display in the decades to come. *Statistical Science*, 5(3):327–339, 1990.

[200] K. Ushey, J. McPherson, J. Cheng, and J. J. Allaire. *packrat: A dependency management system for projects and their R package dependencies*, 2014. R package version 0.4.1-1.

[201] R. Vaidyanathan, Y. Xie, J. J. Allaire, J. Cheng, and K. Russell. *htmlwidgets: HTML Widgets for R*, 2016. R package version 0.7.

[202] G. van Belle. *Statistical Rules of Thumb (second edition)*. John Wiley & Sons: Hoboken, NJ, 2008.

[203] M. van der Laan. Statistics as a science, not an art: the way to survive in data science. *Amstat News*, 2015. `http://magazine.amstat.org/blog/2015/02/01/statscience_feb2015`.

[204] D. Vanderkam, J. Allaire, J. Owen, D. Gromer, P. Shevtsov, and B. Thieurmel. *dygraphs: Interface to 'Dygraphs' Interactive Time Series Charting Library*, 2016. R package version 1.1.1-1.

[205] W. N. Venables and B. D. Ripley. *class package: Modern Applied Statistics with S*. Springer Verlag: New York, NY, fourth edition, 2002.

[206] W. N. Venables and B. D. Ripley. *nnet package: Modern Applied Statistics with S*. Springer Verlag: New York, NY, fourth edition, 2002.

[207] W. N. Venables, D. M. Smith, and the R Core Team. An introduction to R: Notes on R: A programming environment for data analysis and graphics, version 3.3.2. *http://cran.r-project.org/doc/manuals/R-intro.pdf*, 2016.

[208] V. Wang. The OBP/SLG ratio: What does history say? *By the Numbers*, 16(3):3, 2006.

[209] R. L. Wasserstein and N. A. Lazar. The ASA's statement on p–values: Context, process, and purpose. *The American Statistician*, 70(2):129–133, 2016.

[210] D. J. Watts and S. H. Strogatz. Collective dynamics of 'small-world' networks. *Nature*, 393(6684):440–442, 1998.

[211] S. Weisberg. *Applied Linear Regression*. John Wiley & Sons: Hoboken, NJ, third edition, 2005.

[212] H. Wickham. *ggplot2: Elegant Graphics for Data Analysis*. Springer Verlag: New York, NY, 2009.

[213] H. Wickham. Asa 2009 data expo. *Journal of Computational and Graphical Statistics*, 20(2):281–283, 2011.

[214] H. Wickham. testthat: Get started with testing. *The R Journal*, 3:5–10, 2011.

[215] H. Wickham. *assertthat: Easy pre and post assertions*, 2013. R package version 0.1.

[216] H. Wickham. *fueleconomy: EPA fuel economy data*, 2014. R package version 0.1.

[217] H. Wickham. *nasaweather: Collection of datasets from the ASA 2006 data expo*, 2014. R package version 0.1.

[218] H. Wickham. Tidy data. *The Journal of Statistical Software*, 59(10), 2014. `http://vita.had.co.nz/papers/tidy-data.html`.

[219] H. Wickham. *usdanutrients: USDA nutrient data (release SR26)*, 2014. R package version 0.1.

[220] H. Wickham. *Advanced R*. CRC Press: Boca Raton, FL, 2015.

[221] H. Wickham. *babynames: US Baby Names 1880-2014*, 2016. R package version 0.2.1.

[222] H. Wickham. *bigrquery: An Interface to Google's 'BigQuery' 'API'*, 2016. R package version 0.3.0.

[223] H. Wickham. *httr: Tools for working with URLs and HTTP*, 2016. R package version 1.2.1.

[224] H. Wickham. *lazyeval: Lazy (Non-Standard) Evaluation*, 2016. R package version 0.2.0.

[225] H. Wickham. *nycflights13: Flights that Departed NYC in 2013*, 2016. R package version 0.2.0.

[226] H. Wickham. *readxl: Read Excel Files*, 2016. R package version 0.1.1.

[227] H. Wickham. *rvest: Easily Harvest (Scrape) Web Pages*, 2016. R package version 0.3.2.

[228] H. Wickham. *scales: Scale Functions for Visualization*, 2016. R package version 0.4.0.

[229] H. Wickham. *stringr: Simple, Consistent Wrappers for Common String Operations*, 2016. R package version 1.1.0.

[230] H. Wickham. *tidyr: Easily Tidy Data with 'spread()' and 'gather()' Functions*, 2016. R package version 0.6.0.

[231] H. Wickham. *tidyverse: Easily Install and Load 'Tidyverse' Packages*, 2016. R package version 1.0.0.

[232] H. Wickham and W. Chang. *devtools: Tools to make developing R packages easier*, 2016. R package version 1.12.0.9000.

[233] H. Wickham, D. Cook, and H. Hofmann. Visualizing statistical models: Removing the blindfold. *Statistical Analysis and Data Mining: The ASA Data Science Journal*, 8(4):203–225, 2015.

[234] H. Wickham and R. Francois. *dplyr: a grammar of data manipulation*, 2016. R package version 0.5.0.9000.

[235] H. Wickham, R. Francois, and K. Mller. *tibble: Simple Data Frames*, 2016. R package version 1.2.

[236] H. Wickham, J. Hester, and R. Francois. *readr: Read Tabular Data*, 2016. R package version 1.0.0.

[237] Wikipedia.   Hippocratic   oath,   2016.   `https://en.wikipedia.org/wiki/Hippocratic_Oath`.

[238] L. Wilkinson, D. Wills, D. Rope, A. Norton, and R. Dubbs. *The Grammar of Graphics (second edition)*. Springer Verlag: New York, NY, 2005.

[239] Y. Xie. *Dynamic Documents with R and knitr*. CRC Press: Boca Raton, FL, 2014.

[240] Y. Xie. *DT: A Wrapper of the JavaScript Library 'DataTables'*, 2016. R package version 0.2.

[241] Y. Xie. *knitr: A General-Purpose Package for Dynamic Report Generation in R*, 2016. R package version 1.14.

[242] N. Yau. *Visualize this: the Flowing Data guide to design, visualization, and statistics*. John Wiley & Sons: Hoboken, NJ, 2011.

[243] N. Yau. *Data points: visualization that means something*. John Wiley & Sons: Hoboken, NJ, 2013.

[244] A. M. Zaslavsky and N. J. Horton. Balancing disclosure risk against the loss of nonpublication. *Journal of Official Statistics*, 14(4):411–419, 1998.

Indices

Separate indices are provided for subject (concept or task) and R command. References to the examples are denoted in *italics*.

Subject index

R index